Jacobians of
Matrix Transformations
and
Functions of
Matrix Argument

Jacobians of Matrix Transformations and Functions of Matrix Argument

A. M. Mathai

Department of Mathematics & Statistics,
McGill University

World Scientific

Singapore • New Jersey • London • Hong Kong

Published by

World Scientific Publishing Co. Pte. Ltd.
5 Toh Tuck Link, Singapore 596224
USA office: 27 Warren Street, Suite 401-402, Hackensack, NJ 07601
UK office: 57 Shelton Street, Covent Garden, London WC2H 9HE

Library of Congress Cataloging-in-Publication Data
Mathai, A. M.
    Jacobians of matrix transformations and functions of matrix argument / A. M. Mathai.
      p. cm.
    Includes bibliographical references and indexes.
    ISBN-13 978-981-02-3095-1 -- ISBN-10 9810230958
    1. Jacobians. 2. Matrices. 3. Transformations (Mathematics) I. Title.
    QA191.M36 1997
    512.9'434--dc21   97-3779
                                                    CIP

British Library Cataloguing-in-Publication Data
A catalogue record for this book is available from the British Library.

Preface

A large number of statisticians, physicists, engineers, applied mathematicians, econometricians and others shy away from dealing with integrals involving real-valued scalar functions of matrix argument because the integrations often require the evaluations of Jacobians of matrix transformations. These could be linear matrix transformations or nonlinear ones which may contain special types of matrices such as diagonal, upper or lower triangular, symmetric, skew symmetric, orthogonal, hermitian, skew hermitian, unitary, semiunitary and so on. Computation of the Jacobians in such matrix transformations is usually quite difficult. The aim of this book is to develop some techniques systematically so that anyone with a little bit of mathematical maturity and exposure to multivariable calculus can easily follow through the steps and understand the various methods by which the Jacobians in complicated matrix transformations are evaluated. Then they should be able to make use of these results in their own fields of endeavor. The material in this book is developed slowly with lots of worked examples, aimed at self-study. Some exercises are also given at the end of each section. The author hopes that the material in this book will form a great reference source for statisticians, engineers, physicists, econometricians, applied mathematicians and people working in many other areas. The material in this book can be used for a one semester graduate level course on Jacobians and functions of matrix argument.

Numerical analysts, computer scientists and people working on nonlinear least-squares problems often need to look into the existence, properties and inversions of Jacobian matrices. Such existence conditions, the types of transformations giving rise to a given Jacobian, conditions for a certain transformation to be unique, particular types of mappings such as Samuelson map, and so on, will not be discussed in this book. Topics such as Jacobians on algebraic structures and dynamical systems, the Jacobian conjecture, maps with special types of Jacobian matrices such as sparse matrices, parallel maps on hypersurfaces, Jacobians in differential equations and algebraic geometry, the decomposition of a Jacobian matrix for applications to specific problems in the study of energy levels in physics, game theory, time series analysis and so on, will not be treated in this book. Some sample references to such topics are given at the end of Section 1.2 of Chapter 1. This book will concentrate on the topic of the evaluation of Jacobians in some specific linear as well as nonlinear matrix transformations, in the real and complex cases, which are widely applied in the statistical, physical, engineering, biological and social sciences.

The evaluation of an integral of a real valued scalar function of a real or complex matrix, or of several matrices, will involve matrix transformations and the associated Jacobians at each step. Hence to compile a full bibliography of functions of matrix argument and papers where specific Jacobians appear, is not a viable proposition. Sample references to various topics are listed in the bibliography so that the reader can get exposure to the large number of topics where Jacobians appear in various contexts. Since the materials in this book on Jacobians as well as on functions of matrix argument are developed from first principles, not many references are actually used in working out the details. Hence sample references are listed at the end of each section so that the reader can get a foothold to other related topics. No particular criteria are used in selecting these sample references and hence they cannot be interpreted as the most important or the most typical ones.

Chapters 1 and 2 deal with matrix transformations and the associated Jacobians in the real case. After giving some preliminaries on matrix derivatives, Chapter 1 concentrates on linear and some smooth nonlinear transformations. Outlines of the proofs of the various results are also given in all the chapters. Chapter 2 covers more complicated nonlinear transformations, in the real case, involving diagonal, triangular, skew symmetric, orthogonal and semiorthogonal matrices. Singular value decomposition, and transformations containing many matrices and submatrices are also discussed in this chapter.

Chapters 3 and 4 deal with matrix transformations in the complex case. Complex cases, which are analogous to the results in the real case discussed in Chapters 1 and 2, are dealt with in Chapters 3 and 4. Transformations involving hermitian, skew hermitian, unitary and semiunitary matrices are also considered. Thus the materials in Chapters 3 and 4 are parallel to those dealt with in Chapters 1 and 2.

Chapter 5 deals with real-valued scalar functions of matrix argument in the real case. After introducing elementary functions such as exponential, gamma and beta functions, hypergeometric functions are discussed. Generalized hypergeometric functions in the category of Meijer's G- and Fox's H-functions are also introduced. Various approaches available in the literature for defining hypergeometric functions of matrix argument, such as the ones through the generalized Laplace transform, zonal polynomials and the M-transform, are illustrated. Some results on fractional integrals and fractional derivatives are also given. Scalar functions of many matrix arguments in the categories of Appell's functions, Humbert's functions, Kampé de Fériet's functions and Lauricella's functions are introduced and many of their properties are also established in this chapter. Some applications in statistics and physics are also pointed out.

The material in Chapter 6 is parallel to that in Chapter 5 but deals with real-valued scalar functions of one or more matrices in the complex case. Thus the complex analogues of most of the materials in Chapter 5 can be found in Chapter 6.

The material in this book has been developed over the past several years. From time to time several people read various parts of the book and gave their comments. The author would like to express his sincere thanks to Takesi Hayakawa, Serge Provost and Giorgio Pederzoli for reading through the first draft of the whole book and making many valuable suggestions. The author acknowledges with thanks the time spent by William J. Anderson, Neville Sancho and my post-doctoral fellow Shagufta A. Sultan who read through various chapters and provided some valuable comments. The help given by my graduate students Avner Bar-Hen, Moses Njoroge, David Chu, Melvin Munsaka, Grace Mwawasi, Kalyanee Viraswami, and Alfred A. Nnadozie Jr, on various occasions, is much appreciated. The author assumes responsibility for the remaining mistakes and imperfections, if any.

The author would like to acknowledge with thanks the financial assistance received from the Natural Sciences and the Engineering Research Council of Canada.

Montreal   A.M.M

4 October 1996

Contents

CHAPTER 1

Jacobians of matrix transformations

1.0   Introduction

When dealing with scalar functions of matrix arguments it is often neces-
sary to transform a matrix to another matrix and calculate the Jacobian
associated with such a matrix transformation. (The concept of Jacobian
is due to the German mathematician C.G.J. Jacobi (1804-1851)). For this
purpose one needs vector and matrix derivatives. Some such derivatives
will be developed in Section 1.1. These will include derivatives of a scalar
or vector or matrix with respect to a scalar and scalar functions of a vector
or matrix with respect to a vector or a matrix. Sections 1.2 and 1.3 will deal
with the Jacobians associated with the various types of linear and nonlinear
matrix transformations. Outlines of the proofs for some key results will be
given and others will be stated without proofs. Unless specified otherwise,
all matrices appearing in this chapter are real in the sense that the elements
are not complex quantities.

   The following standard notations will be used. Scalars will be denoted
by lower-case letters, vectors and matrices by capital letters. As far as
possible variable matrices will be denoted by $X, Y, ...$ and constant matrices
by $A, B, ...$ Let $A = (a_{ij})$ be a $p \times p$ matrix then $\operatorname{tr}(A) = \operatorname{tr} A = $ trace of
$A = a_{11} + a_{22} + ... + a_{pp}$, $|A| = \det(A) = $ determinant of A, and a prime
over a vector or a matrix will denote its transpose. Let $X = (x_{ij})$ be a $p \times q$
matrix of functionally independent real variables x_{ij}'s. Then dX stands for
the product of the $p \times q$ differential elements (see also (1.2.8)–(1.2.10))

$$dX = dx_{11}dx_{12}...dx_{1q}dx_{21}...dx_{2q}...dx_{pq}$$

and when X is a real square symmetric matrix, that is, $p = q$, $X = X'$,
then dX is the product of the $p(p+1)/2$ differential elements, that is,

$$dX = dx_{11}dx_{21}dx_{22}...dx_{p1}...dx_{pp}.$$

Integration over X will be denoted by $\int_X$. Positive definite and positive
semidefinite matrices will be denoted by $X > 0$ and $X \geq 0$ respectively.
$0 < A < X < B \Rightarrow A > 0, X > 0, B > 0, X - A > 0, B - X >$
0. $\int_{A<X<B} f(X)dX = \int_A^B f(X)dX$ where $f(X)$ is a scalar function of the

1

matrix X. The Kronecker product of two matrices X and Y will be denoted
by $X \otimes Y$. The Jacobian of the transformation of Y written as a function
of X will be denoted by $J(Y : X)$. An identity matrix will be denoted by
I and a null matrix as well as zero by 0. Whenever there is no possibility
of confusion $\text{tr}(AB)$ will be written without the brackets as $\text{tr}AB$. Other
notations will be explained as they appear, see also the glossary of symbols
given at the end of this book.

1.1   Vector and matrix derivatives

In this section we will consider the following: (i) The derivative of a matrix
with respect to a scalar, which includes the derivative of a scalar or vector
or matrix with respect to a scalar; (ii) The derivative of a scalar function
of a vector with respect to a vector; (iii) The derivative of a scalar function
of a matrix with respect to a matrix.

1.1.1   The derivative of a matrix with respect to a scalar

Let $X = (x_{ij})$ where x_{ij}'s are functions of a real scalar variable x. Then the
derivative of X with respect to x is the corresponding matrix of derivatives
of the elements, denoted by

$$\frac{\partial X}{\partial x} = \left(\frac{\partial x_{ij}}{\partial x} \right).$$

Example 1.1

$$X = \begin{bmatrix} x^2 & 5x \\ e^x & \sin x \end{bmatrix} \Rightarrow \frac{\partial X}{\partial x} = \begin{bmatrix} 2x & 5 \\ e^x & \cos x \end{bmatrix}.$$

1.1.2   The derivative of a scalar function of a vector with respect to this vector

Consider a scalar function of a $p \times 1$ real vector X. For example (i) $y_1 = a_1 x_1 + ... + a_p x_p = a'X = X'a$; ($ii$) $y_2 = x_1^2 + ... + x_p^2 = X'X$ are two scalar
functions of X where $X' = (x_1, ..., x_p)$, $a' = (a_1, ..., a_p)$ and $a_1, ..., a_p$ are
constants and a prime denotes a transpose. We will consider derivatives of
such scalar functions with respect to the vector X first, and then in the

next section such derivatives of scalar functions of a matrix X with respect to the matrix X. The partial derivative vector operator $\frac{\partial}{\partial X}$ is defined as the column vector of partial derivative operators with respect the elements x_j's of X. That is

$$\left(\frac{\partial}{\partial X}\right)' = \left(\frac{\partial}{\partial x_1}, \frac{\partial}{\partial x_2}, ..., \frac{\partial}{\partial x_p}\right).$$

The following are some simple examples but these are needed later on and hence these will be stated as a theorem.

Theorem 1.1 *Let X be a $p \times 1$ vector of functionally independent real variables $x_1, ..., x_p$ and $a' = (a_1, ..., a_p)$ a vector of constants. Then*

$$y_1 = a'X = X'a = a_1x_1 + ... + a_px_p \Rightarrow \frac{\partial y_1}{\partial X} = a, \qquad (1.1.1)$$

$$y_2 = X'X = x_1^2 + ... + x_p^2 \Rightarrow \frac{\partial y_2}{\partial X} = 2X, \qquad (1.1.2)$$

and

$$y_3 = X'AX = \sum_{i=1}^{p}\sum_{j=1}^{p} a_{ij}x_ix_j, A = A', \Rightarrow \frac{\partial y_3}{\partial X} = 2AX \qquad (1.1.3)$$

where $A = (a_{ij})$ is a symmetric matrix of constants.

Proof The first and second results are obvious since the partial derivatives are $\frac{\partial y_1}{\partial x_j} = a_j$, $j = 1, ..., p$, and $\frac{\partial y_2}{\partial x_j} = 2x_j$. For the third result note that

$$\frac{\partial y_3}{\partial x_1} = 2a_{11}x_1 + (a_{12} + a_{21})x_2 + ... + (a_{1p} + a_{p1})x_p$$

$$= 2a_{11}x_1 + 2a_{12}x_2 + ... + 2a_{1p}x_p$$

$$= 2a'_{(1)}X$$

where $a'_{(i)}$ denotes the i-th row of A since, due to the assumption of symmetry, $a_{ij} = a_{ji}$ for all i and j. Thus the partial derivative with respect to x_i gives $2a'_{(i)}X$, $i = 1, ..., p$ and hence the result.

If A is not assumed to be symmetric then

$$\frac{\partial y_3}{\partial X} = (A + A')X. \qquad (1.1.4)$$

If the transpose of the operator $\frac{\partial}{\partial X}$ is denoted by $(\frac{\partial}{\partial X})'$ then we have the following interesting results.

$$y_3 = X'AX \Rightarrow \frac{\partial y_3}{\partial X} = (A + A')X$$

$$\Rightarrow \left(\frac{\partial y_3}{\partial X}\right)' = X'(A' + A) \tag{1.1.5}$$

$$\Rightarrow \left(\frac{\partial}{\partial X}\right)\left(\frac{\partial}{\partial X}\right)' y_3 = A' + A$$

$$\Rightarrow \frac{\partial^2 y_3}{\partial X \partial X'} = A' + A$$

where the operator $\frac{\partial}{\partial X}$ is applied on each element of the row vector $(\frac{\partial}{\partial X})'$ to get the second order derivatives and the matrix of second order partial derivatives is given by

$$\frac{\partial^2}{\partial X \partial X'} = \begin{bmatrix} \frac{\partial^2}{\partial x_1^2} & \cdots & \frac{\partial^2}{\partial x_1 \partial x_p} \\ \vdots & \ddots & \vdots \\ \frac{\partial^2}{\partial x_p \partial x_1} & \cdots & \frac{\partial^2}{\partial x_p^2} \end{bmatrix}.$$

Example 1.2   Evaluate the following:

$$(i) \; \frac{\partial u_1}{\partial X}, \; (ii) \; \frac{\partial u_2}{\partial X}, \; (iii) \; \frac{\partial^2 u_2}{\partial X \partial X'}, \; (iv) \; \frac{\partial u_3}{\partial X},$$

where

$$u_1 = 2x_1 - x_2 + 5x_3,$$
$$u_2 = x_1^2 + 3x_2^2 - x_3^2 + 4x_1x_2 - 2x_1x_3,$$

and

$$u_3 = b'AX, \; b' = (1, 0, -1), \; A = \begin{bmatrix} 1 & 0 & 0 \\ 0 & 1 & 1 \\ -1 & 2 & 0 \end{bmatrix}.$$

Solution   Writing $u_1 = a'X$ one has $a' = (2, -1, 5)$ and hence from (1.1.1) the answer for (i) is a. The matrix appearing in a quadratic form can be taken to be symmetric without any loss of generality since $X'AX = X'(\frac{A+A'}{2})X$. Thus from (1.1.3) the answer for (ii) is

$$2AX = 2 \begin{bmatrix} 1 & 2 & -1 \\ 2 & 3 & 0 \\ -1 & 0 & -1 \end{bmatrix} X = 2 \begin{bmatrix} x_1 + 2x_2 - x_3 \\ 2x_1 + 3x_2 \\ -x_1 - x_3 \end{bmatrix}.$$

For (iii) note that

$$\frac{\partial u_2}{\partial X'} = 2(x_1 + 2x_2 - x_3, 2x_1 + 3x_2, -x_1 - x_3).$$

Hence

$$\frac{\partial^2 u_2}{\partial X \partial X'} = 2\left(\frac{\partial}{\partial X}(x_1 + 2x_2 - x_3), \frac{\partial}{\partial X}(2x_1 + 3x_2), \frac{\partial}{\partial X}(-x_1 - x_3)\right)$$

$$= 2 \begin{bmatrix} 1 & 2 & -1 \\ 2 & 3 & 0 \\ -1 & 0 & -1 \end{bmatrix} = \begin{bmatrix} 2 & 4 & -2 \\ 4 & 6 & 0 \\ -2 & 0 & -2 \end{bmatrix}.$$

Writing $u_3 = b'AX = X'A'b$ and from (1.1.1) the answer for (iv) is

$$A'b = \begin{bmatrix} 1 & 0 & -1 \\ 0 & 1 & 2 \\ 0 & 1 & 0 \end{bmatrix} \begin{bmatrix} 1 \\ 0 \\ -1 \end{bmatrix} = \begin{bmatrix} 2 \\ -2 \\ 0 \end{bmatrix}.$$

Example 1.3 Check for the maxima and minima of the function $f(X) = -x_1^2 - 2x_2^2 + 4x_2 - x_3^2 + 5$.

Solution Take the first order partial derivatives and equate each to zero to get the turning points or the critical points and check for the definiteness of the matrix of second order derivatives at the turning points to see maxima or minima.

$$\frac{\partial f}{\partial X} = \begin{bmatrix} -2x_1 \\ -4x_2 + 4 \\ -2x_3 \end{bmatrix}; \ \frac{\partial f}{\partial X} = 0 \Rightarrow x_1 = 0, \ x_2 = 1, \ x_3 = 0.$$

Thus the only turning point is $X' = (0, 1, 0)$. If the second order partial derivative matrix is negative definite (positive definite) at this point then the point corresponds to a maximum (minimum). But

$$\frac{\partial^2 f}{\partial X \partial X'} = \begin{bmatrix} -2 & 0 & 0 \\ 0 & -4 & 0 \\ 0 & 0 & -2 \end{bmatrix} < 0.$$

Hence there is a maximum at this point.

1.1.3   The derivative of a scalar function of a matrix with respect to the matrix

If $f(X)$ is a scalar function of the real matrix $X = (x_{ij})$ then the derivative of $f(X)$ with respect to X is defined as

$$\frac{\partial f}{\partial X} = \left(\frac{\partial f}{\partial x_{ij}}\right). \tag{1.1.6}$$

For example $\text{tr}(X), |X|, \text{tr}(AX)$ are all scalar functions of X. If for example $f(X) = \text{tr}(X) = x_{11} + ... + x_{pp}$, where $X = (x_{ij})$ is a $p \times p$ matrix, means $\frac{\partial f}{\partial x_{ij}} = 1$ if $i = j$ and zero if $i \neq j$. Then $\frac{\partial(\text{tr}(X))}{\partial X} = I$ where I is the identity matrix. This as well as several other results are needed for our discussion of Jacobians of matrix transformations. Hence these will be given as theorems. The ones which can be proved by straight multiplication and differentiation will be given without proofs. For others outlines of the proofs will be given.

Theorem 1.2   *Let $X = (x_{ij})$ be a matrix of functionally independent real variables and $A = (a_{ij})$ be a matrix of constants. Then*

(i) $f(X) = \text{tr}(X), \quad X : p \times p \Rightarrow \dfrac{\partial f}{\partial X} = I,$

(ii) $f(X) = \text{tr}(A'X), \quad X : p \times q, \ A : p \times q$

$$\Rightarrow \frac{\partial f}{\partial X} = A,$$

(iii) $f(X) = a'Xb, \quad X : p \times p, \ a : p \times 1, \ b : p \times 1$

$$\Rightarrow \frac{\partial f}{\partial X} = ab'$$

where a and b are vectors of constants,

(iv) $f(X) = \text{tr}(XA) \Rightarrow \dfrac{\partial f}{\partial X} = A'$

if X is not symmetric where X and A are $p \times p$ and

(v) $\dfrac{\partial f}{\partial X} = A + A' - \text{diag}(A)$

if X is symmetric where diag(A) *means a diagonal matrix formed with the diagonal elements of A.*

Proofs   For proving (iii) note that one can look at this as a bilinear form in the vectors a and b. Then

$$a'Xb = \sum_{i=1}^{p}\sum_{j=1}^{p} a_i b_j x_{ij}$$

and

$$\frac{\partial(a'Xb)}{\partial x_{ij}} = a_i b_j.$$

Thus for example when $i = 1$ the elements are $a_1 b_1, ..., a_1 b_p$. The matrix configuration is then

$$\begin{bmatrix} a_1 b_1 & \cdots & a_1 b_p \\ a_2 b_1 & \cdots & a_2 b_p \\ \vdots & \ddots & \vdots \\ a_p b_1 & \cdots & a_p b_p \end{bmatrix} = \begin{bmatrix} a_1 \\ a_2 \\ \vdots \\ a_p \end{bmatrix} \begin{bmatrix} b_1 \cdots & b_p \end{bmatrix} = ab'.$$

For proving (ii) premultiply X by A' and take the trace to get

$$\begin{aligned} \text{tr}(A'X) = {} & a_{11}x_{11} + a_{21}x_{21} + ... + a_{p1}x_{p1} \\ & + a_{12}x_{12} + a_{22}x_{22} + ... + a_{p2}x_{p2} \\ & + ... \\ & + a_{1p}x_{1p} + a_{2p}x_{2p} + ... + a_{pp}x_{pp}. \end{aligned}$$

Then if x_{ij}'s are all distinct then

$$\frac{\partial(\text{tr}(A'X))}{\partial x_{ij}} = a_{ij}.$$

Hence the result.

     For proving (iv) note from above that if all x_{ij}'s are distinct then

$$\frac{\partial(\text{tr}(XA))}{\partial x_{ij}} = a_{ji} \Rightarrow \frac{\partial(\text{tr}(XA))}{\partial X} = A'.$$

But if $x_{ij} = x_{ji}$ for all i and j then

$$\frac{\partial(\text{tr}(XA))}{\partial x_{ij}} = a_{ji} + a_{ij},\ i \neq j$$

$$= a_{ii},\ i = j.$$

Thus the matrix configuration of the partial derivatives is $A + A'$ except that the diagonal elements are taken only once each. Hence it is $A + A' - \text{diag}(A)$.

From the properties of trace note that for any two matrices where the following products are defined

$$\text{tr}(BX) = \text{tr}(XB) = \text{tr}(BX)'$$
$$= \text{tr}(X'B') = \text{tr}(B'X').$$

Example 1.4   Let

$$A = \begin{bmatrix} 1 & 2 \\ 2 & 5 \end{bmatrix}, \quad X = \begin{bmatrix} x_{11} & x_{12} \\ x_{21} & x_{22} \end{bmatrix}, \quad Y = \begin{bmatrix} y_{11} & y_{12} \\ y_{12} & y_{22} \end{bmatrix},$$

$$a = \begin{bmatrix} 1 \\ 0 \end{bmatrix}, \quad b = \begin{bmatrix} 1 \\ 1 \end{bmatrix}.$$

Evaluate the following:

$$(i) \ \frac{\partial}{\partial X} \text{tr}(AX); \quad (ii) \ \frac{\partial}{\partial X}(a'Xb); \quad (iii) \ \frac{\partial}{\partial Y} \text{tr}(AY).$$

Solution   Note that

$$\frac{\partial}{\partial X} = \begin{bmatrix} \frac{\partial}{\partial x_{11}} & \frac{\partial}{\partial x_{12}} \\ \frac{\partial}{\partial x_{21}} & \frac{\partial}{\partial x_{22}} \end{bmatrix},$$

$$\text{tr}(AX) = x_{11} + 2x_{21} + 2x_{12} + 5x_{22}$$
$$= \text{tr}(XA) = \text{tr}(X'A') = \text{tr}(A'X'),$$

$$\frac{\partial}{\partial x_{11}} \text{tr}(AX) = 1, \quad \frac{\partial}{\partial x_{12}} \text{tr}(AX) = 2,$$

$$\frac{\partial}{\partial x_{21}} \text{tr}(AX) = 2, \quad \frac{\partial}{\partial x_{22}} \text{tr}(AX) = 5.$$

Thus

(i)
$$\frac{\partial}{\partial X} \text{tr}(AX) = \begin{bmatrix} 1 & 2 \\ 2 & 5 \end{bmatrix}.$$

Observe that

$$a'Xb = (1,0)X \begin{pmatrix} 1 \\ 1 \end{pmatrix} = x_{11} + x_{12}.$$

Hence

$$\frac{\partial}{\partial x_{11}}(a'Xb) = 1, \quad \frac{\partial}{\partial x_{12}}(a'Xb) = 1,$$

$$\frac{\partial}{\partial x_{21}}(a'Xb) = 0, \quad \frac{\partial}{\partial x_{22}}(a'Xb) = 0.$$

That is,

(ii)
$$\frac{\partial}{\partial X}(a'Xb) = \begin{bmatrix} 1 & 1 \\ 0 & 0 \end{bmatrix} = \begin{bmatrix} 1 \\ 0 \end{bmatrix}[1,1] = ab'.$$

Note that
$$\text{tr}(AY) = y_{11} + 4y_{12} + 5y_{22}.$$

Hence

(iii)
$$\frac{\partial}{\partial Y}\text{tr}(AY) = \begin{bmatrix} 1 & 4 \\ 4 & 5 \end{bmatrix}$$
$$= 2\begin{bmatrix} 1 & 2 \\ 2 & 5 \end{bmatrix} - \begin{bmatrix} 1 & 0 \\ 0 & 5 \end{bmatrix}.$$

Theorem 1.3   *When X is a $p \times p$ nonsingular matrix of functionally independent real variables then*

$$\frac{\partial |X|}{\partial X} = |X|\left(X^{-1}\right)', \text{ for a general } X,$$
$$= |X|\left(2X^{-1} - \text{diag}(X^{-1})\right), \text{ for } X = X'$$

where $\text{diag}(X^{-1})$ means the diagonal matrix formed with the diagonal elements of X^{-1}.

Proof   Consider the expansion of a determinant in terms of the elements and their cofactors of any row or column. Let the cofactor of x_{ij} be denoted by $|X_{ij}|$. Then for example

$$|X| = x_{i1}|X_{i1}| + \ldots + x_{ip}|X_{ip}|$$

for fixed i, and the partial derivative of $|X|$ with respect to x_{ik} will be $|X_{ik}|$ if X is not symmetric; if X is symmetric then we get $|X_{ik}|$ for $i = k$ and $2|X_{ik}|$ for $i \neq k$ because another term comes from the derivative with respect to x_{ki}. This may be seen from the representation

$$|X| = \sum \ldots \sum \pm x_{1i_1} x_{2i_2} \ldots x_{pi_p}$$

where $i_1, \ldots, i_p$ are distinct permutations of the natural numbers $1, 2, \ldots, p$. Thus the matrix of partial derivatives gives the matrix of cofactors when X is not symmetric. The transpose of this cofactor matrix divided by $|X|$ is the inverse X^{-1}, hence the result for a general X. When $X = X'$ the matrix of partial derivatives is two times the cofactor matrix with one set of diagonal elements removed, hence the result.

Example 1.5   Let

$$X = \begin{bmatrix} u & v \\ w & z \end{bmatrix} \text{ with } uz - vw \neq 0,$$

$$Y = \begin{bmatrix} u & v \\ v & z \end{bmatrix} \text{ with } uz - v^2 \neq 0.$$

Evaluate

$$(i) \ \frac{\partial}{\partial X}|X|; \ (ii) \ \frac{\partial}{\partial Y}|Y|; \ (iii) \ \frac{\partial}{\partial T}|T|; \ (iv) \ \frac{\partial^*}{\partial T}|T|$$

where

$$\frac{\partial^*}{\partial T} = \begin{bmatrix} \frac{\partial}{\partial u} & \frac{\partial}{\partial v} \\ \frac{\partial}{\partial v} & \frac{\partial}{\partial z} \end{bmatrix}, \ T = \begin{bmatrix} u & \frac{1}{2}v \\ \frac{1}{2}v & z \end{bmatrix} \text{ with } uz - \frac{1}{4}v^2 \neq 0.$$

Solution   Note that $|X| = uz - vw$, $|Y| = uz - v^2$ and $|T| = uz - \frac{1}{4}v^2$. Then

$$\frac{\partial}{\partial X} = \begin{bmatrix} \frac{\partial}{\partial u} & \frac{\partial}{\partial v} \\ \frac{\partial}{\partial w} & \frac{\partial}{\partial z} \end{bmatrix}, \ X^{-1} = \frac{1}{|X|} \begin{bmatrix} z & -v \\ -w & u \end{bmatrix},$$

$$Y^{-1} = \frac{1}{|Y|} \begin{bmatrix} z & -v \\ -v & u \end{bmatrix}, \ T^{-1} = \frac{1}{|T|} \begin{bmatrix} z & -\frac{1}{2}v \\ -\frac{1}{2}v & u \end{bmatrix}.$$

Hence

(i)

$$\frac{\partial}{\partial X}|X| = \begin{bmatrix} z & -w \\ -v & u \end{bmatrix} = |X|(X^{-1})'.$$

But

$$\frac{\partial}{\partial u}|Y| = z, \ \frac{\partial}{\partial z}|Y| = u, \ \frac{\partial}{\partial v}|Y| = -2v.$$

Thus

(ii)

$$\frac{\partial}{\partial Y}|Y| = \begin{bmatrix} z & -2v \\ -2v & u \end{bmatrix}.$$

But this can be written as follows:

$$\begin{bmatrix} z & -2v \\ -2v & u \end{bmatrix} = 2 \begin{bmatrix} z & -v \\ -v & u \end{bmatrix} - \begin{bmatrix} z & 0 \\ 0 & u \end{bmatrix}$$

$$= |Y| \{2Y^{-1} - \text{diag}(Y^{-1})\}$$

where $\text{diag}(Y^{-1})$ means the diagonal matrix with the diagonal elements equal to the diagonal elements of Y^{-1}. According to our notation

$$\frac{\partial}{\partial T} = \begin{bmatrix} \frac{\partial}{\partial u} & \frac{\partial}{\partial(\frac{1}{2}v)} \\ \frac{\partial}{\partial(\frac{1}{2}v)} & \frac{\partial}{\partial z} \end{bmatrix} = \left(\frac{\partial}{\partial t_{ij}}\right).$$

Note that

$$\frac{\partial}{\partial(\frac{1}{2}v)} = 2\frac{\partial}{\partial v}.$$

Thus

$$\frac{\partial}{\partial(\frac{1}{2}v)}|T| = 2\left(-\frac{2}{4}v\right) = -v.$$

Hence from (ii) we have

$$\frac{\partial}{\partial T}|T| = \begin{bmatrix} z & -v \\ -v & u \end{bmatrix} = 2\begin{bmatrix} z & -\frac{1}{2}v \\ -\frac{1}{2}v & u \end{bmatrix} - \begin{bmatrix} z & 0 \\ 0 & u \end{bmatrix}$$

which can be written as

(iii) $$\frac{\partial}{\partial T}|T| = |T|\left\{2T^{-1} - \mathrm{diag}\,(T^{-1})\right\}.$$

But

$$\frac{\partial^*}{\partial u}|T| = z, \quad \frac{\partial^*}{\partial v}|T| = -\frac{1}{2}v \quad \text{and} \quad \frac{\partial^*}{\partial z}|T| = u.$$

Hence

$$\frac{\partial^*}{\partial T}|T| = \begin{bmatrix} z & -\frac{1}{2}v \\ -\frac{1}{2}v & u \end{bmatrix} = |T|T^{-1}.$$

That is,

(iv) $$\frac{\partial^*}{\partial T}|T| = |T|T^{-1},$$

the structure of which corresponds to that in the nonsymmetric situation.

Theorem 1.4   *For a $p \times p$ matrix X of functionally independent real scalar variables*

$$\frac{\partial\big(\mathrm{tr}(X^2)\big)}{\partial X} = 2X'.$$

Theorem 1.5   *For X a $p \times q$ matrix of functionally independent real variables and A a $q \times q$ matrix of constants*

$$\frac{\partial\big(\mathrm{tr}(XAX')\big)}{\partial X} = X(A + A').$$

Proof   Let the i-th row of X be denoted by $x_{(i)}$ and its transpose by $x'_{(i)}$. Then XAX' is equivalent to writing

$$XAX' = \begin{bmatrix} x_{(1)} \\ x_{(2)} \\ \vdots \\ x_{(p)} \end{bmatrix} A[x'_{(1)}, \quad \cdots, \quad x'_{(p)}]$$

$$= \begin{bmatrix} x_{(1)} A x'_{(1)} & \cdots & x_{(1)} A x'_{(p)} \\ \vdots & \ddots & \vdots \\ x_{(p)} A x'_{(1)} & \cdots & x_{(p)} A x'_{(p)} \end{bmatrix}.$$

Hence

$$\mathrm{tr}(X A X') = x_{(1)} A x'_{(1)} + \ldots + x_{(p)} A x'_{(p)}.$$

From equations (1.1.4) and (1.1.5) the derivative of $x_{(i)} A x'_{(i)}$ with respect to the row vector $x_{(i)}$ gives $x_{(i)}(A + A')$. Thus

$$\frac{\partial(\mathrm{tr}(X A X'))}{\partial X} = X(A + A').$$

In the above theorem there is a multiplication of X with AX' and then the trace is taken. Thus it is a function of a function of X. This is a simple case and one can easily see the trace by straight multiplication. But if complicated functions are involved we need some other way of handling such situations. We can see that the chain rule for derivatives can be applied. Let $y = f(X)$ be a scalar function of the real matrix X and $u = g(y)$ be a scalar function of this real scalar variable y. Then

$$\frac{\partial u}{\partial X} = \frac{\partial u}{\partial y} \times \frac{\partial y}{\partial X}. \tag{1.1.7}$$

This is easily seen by observing that the derivative with respect to any element x_{ij} is given by

$$\frac{\partial u}{\partial x_{ij}} = \frac{\partial u}{\partial y} \times \frac{\partial y}{\partial x_{ij}}$$

where the only parts changing with x_{ij}'s are $\frac{\partial u}{\partial x_{ij}}$ and $\frac{\partial y}{\partial x_{ij}}$. By using this chain rule we establish the following theorems. Note that since $\frac{\partial u}{\partial y}$ involves only a scalar function of a scalar variable the results in the following theorems are straightforward and hence these are stated without proofs.

Theorem 1.6   *For a $p \times p$ real nonsingular matrix X*

$$\frac{\partial |X|^k}{\partial X} = k|X|^k (X^{-1})', \text{ for a general } X,$$

$$= k|X|^k (2X^{-1} - \mathrm{diag}(X^{-1})), \text{ for } X = X';$$

$$\frac{\partial(e^{\mathrm{tr}(X^2)})}{\partial X} = 2e^{\mathrm{tr}(X^2)} X';$$

and

$$\frac{\partial(\ln|X|)}{\partial X} = (X^{-1})' \text{ for a general } X,$$
$$= 2X^{-1} - \text{diag}(X^{-1}) \text{ for } X = X'.$$

Example 1.6 Verify Theorems 1.4, 1.5 and 1.6 if

$$X = \begin{bmatrix} x & y \\ u & v \end{bmatrix} \text{ and } A = \begin{bmatrix} 1 & 1 \\ 0 & 2 \end{bmatrix}.$$

Solution

$$X = \begin{bmatrix} x & y \\ u & v \end{bmatrix} \Rightarrow X^2 = \begin{bmatrix} x^2 + yu & xy + yv \\ ux + vu & uy + v^2 \end{bmatrix} \Rightarrow$$
$$\text{tr}\left(X^2\right) = x^2 + v^2 + 2uy.$$

Hence

$$\frac{\partial}{\partial X}\text{tr}\left(X^2\right) = \begin{bmatrix} 2x & 2u \\ 2y & 2v \end{bmatrix}$$
$$= 2\begin{bmatrix} x & u \\ y & v \end{bmatrix} = 2X'.$$

This verifies Theorem 1.4.

$$XAX' = \begin{bmatrix} x & y \\ u & v \end{bmatrix}\begin{bmatrix} 1 & 1 \\ 0 & 2 \end{bmatrix}\begin{bmatrix} x & u \\ y & v \end{bmatrix}$$
$$= \begin{bmatrix} x^2 + xy + 2y^2 & xu + xv + 2yv \\ ux + uy + 2vy & u^2 + uv + 2v^2 \end{bmatrix}.$$

Hence

$$\text{tr}(XAX') = x^2 + 2y^2 + u^2 + 2v^2 + xy + uv.$$
$$\frac{\partial}{\partial X}\text{tr}(XAX') = \begin{bmatrix} 2x + y & 4y + x \\ 2u + v & 4v + u \end{bmatrix}$$
$$= \begin{bmatrix} x & y \\ u & v \end{bmatrix}\begin{bmatrix} 2 & 1 \\ 1 & 4 \end{bmatrix}$$
$$= X\left\{\begin{bmatrix} 1 & 1 \\ 0 & 2 \end{bmatrix} + \begin{bmatrix} 1 & 0 \\ 1 & 2 \end{bmatrix}\right\}$$
$$= X(A + A').$$

This verifies Theorem 1.5. Note that

$$e^{\text{tr}(X^2)} = e^{x^2 + v^2 + 2uy}$$

and

$$\frac{\partial}{\partial X}\left(e^{\text{tr}(X^2)}\right) = e^{x^2+v^2+2uy}\begin{bmatrix} 2x & 2u \\ 2y & 2v \end{bmatrix}$$

$$= 2e^{\text{tr}(X^2)}\begin{bmatrix} x & u \\ y & v \end{bmatrix}$$

$$= 2e^{\text{tr}(X^2)}X'.$$

1.1.4   The derivative of a matrix with respect to an element of the same matrix

In computing the Jacobians of matrix transformations often we need the derivative of a matrix with respect to an element of the same matrix to see the pattern of the matrix of partial derivatives. Let X be a $p \times q$ real matrix of functionally independent elements x_{ij}'s. Let one of the x_{ij}'s be denoted by θ. Then the derivative of the matrix X with respect to θ will yield a $p \times q$ matrix of zeros except a unity at the position of θ. Let this matrix be denoted by Δ. If the position of the element θ is to be indicated then we will denote Δ by Δ_{ij} when θ is the i-th row j-th column element. That is,

$$\frac{\partial X}{\partial \theta} = \Delta$$

$$= \Delta_{ij}, \text{ when } \theta = x_{ij}. \tag{1.1.8}$$

The following results follow from the definition itself and hence these will be stated without proofs.

Theorem 1.7   *Let A and B be matrices of constants, X a matrix of functionally independent real variables, and θ an element of X. The orders of the matrices are set at each stage so that the products in each statement makes sense. Then*

$$Y = XA \Rightarrow \frac{\partial Y}{\partial \theta} = \Delta A,$$

$$Y = AX \Rightarrow \frac{\partial Y}{\partial \theta} = A\Delta,$$

$$Y = AX' \Rightarrow \frac{\partial Y}{\partial \theta} = A\Delta',$$

$$Y = X'A \Rightarrow \frac{\partial Y}{\partial \theta} = \Delta'A,$$

$$Y = AXB \Rightarrow \frac{\partial Y}{\partial \theta} = A\Delta B, \tag{1.1.9}$$

and

$$Y = AX'B \Rightarrow \frac{\partial Y}{\partial \theta} = A\Delta'B.$$

Example 1.7   If $X = (x_{ij})$ is 3×2, $A = (a_{ij})$ is 2×2 and $\theta = x_{12}$ then

$$Y = XA \Rightarrow \frac{\partial Y}{\partial \theta} = \Delta A$$

$$= \begin{bmatrix} 0 & 1 \\ 0 & 0 \\ 0 & 0 \end{bmatrix} \begin{bmatrix} a_{11} & a_{12} \\ a_{21} & a_{22} \end{bmatrix} = \begin{bmatrix} a_{21} & a_{22} \\ 0 & 0 \\ 0 & 0 \end{bmatrix}.$$

In the following results X is a square matrix when X^2 and $X'X$ appear. In the other results X could also be rectangular. These results can be easily established by writing, for example, $X^2 = XX$ and applying the rule for differentiation with respect to scalars, keeping in mind the orders of the matrix multiplications. Hence these will be stated without proofs.

Theorem 1.8

$$Y = X^2 \Rightarrow \frac{\partial Y}{\partial \theta} = \Delta X + X\Delta,$$

$$Y = X'X \Rightarrow \frac{\partial Y}{\partial \theta} = \Delta'X + X'\Delta,$$

$$Y = XX' \Rightarrow \frac{\partial Y}{\partial \theta} = \Delta X' + X\Delta',$$

and

$$Y = X'X' \Rightarrow \frac{\partial Y}{\partial \theta} = \Delta'X' + X'\Delta'.$$

Example 1.8   Evaluate $\frac{\partial Y}{\partial \theta}$ if $Y = XX'$, $X = \begin{bmatrix} x & y \\ u & v \end{bmatrix}$ and $\theta = x$.

Solution

$$Y = XX' = \begin{bmatrix} x & y \\ u & v \end{bmatrix} \begin{bmatrix} x & u \\ y & v \end{bmatrix}$$

$$= \begin{bmatrix} x^2 + y^2 & xu + yv \\ xu + yv & u^2 + v^2 \end{bmatrix}.$$

$$\frac{\partial Y}{\partial \theta} = \frac{\partial}{\partial x}(XX') = \begin{bmatrix} 2x & u \\ u & 0 \end{bmatrix}.$$

$$\frac{\partial X}{\partial \theta} = \begin{bmatrix} 1 & 0 \\ 0 & 0 \end{bmatrix} = \Delta.$$

Then

$$\Delta X' + X\Delta' = \begin{bmatrix} 1 & 0 \\ 0 & 0 \end{bmatrix} \begin{bmatrix} x & u \\ y & v \end{bmatrix} + \begin{bmatrix} x & y \\ u & v \end{bmatrix} \begin{bmatrix} 1 & 0 \\ 0 & 0 \end{bmatrix}$$

$$= \begin{bmatrix} x & u \\ 0 & 0 \end{bmatrix} + \begin{bmatrix} x & 0 \\ u & 0 \end{bmatrix} = \begin{bmatrix} 2x & u \\ u & 0 \end{bmatrix}.$$

This verifies the result.

Theorem 1.9   *Let X be a matrix of functionally independent real variables and A and B matrices of constants so that all the following products make sense. Then*

$$Y = XX'X$$
$$\Rightarrow \frac{\partial Y}{\partial \theta} = \Delta X'X + X\Delta'X + XX'\Delta,$$

and

$$Y = AXX'B$$
$$\Rightarrow \frac{\partial Y}{\partial \theta} = A\Delta X'B + AX\Delta'B.$$

Note that in the above results if the matrices are permuted among themselves the corresponding results hold good as long as the products are defined.

The next result can be proved by using $XX^{-1} = I$ and then proceeding as before.

Theorem 1.10   *For a nonsingular matrix X of distinct elements,*

$$Y = X^{-1} \Rightarrow \frac{\partial Y}{\partial \theta} = -X^{-1}\Delta X^{-1}. \qquad (1.1.10)$$

Example 1.9   Let $X = \begin{bmatrix} x & y \\ u & v \end{bmatrix}$ with $xv - uy \neq 0$ and $\theta = v$. Verify the result in (1.1.10).

Solution

$$\Delta = \frac{\partial}{\partial v}X = \begin{bmatrix} 0 & 0 \\ 0 & 1 \end{bmatrix}.$$

$$X^{-1} = \frac{1}{(xv - uy)} \begin{bmatrix} v & -y \\ -u & x \end{bmatrix}.$$

Then

$$-X^{-1}\Delta X^{-1} = -\frac{1}{(xv-uy)^2}\begin{bmatrix} v & -y \\ -u & x \end{bmatrix}\begin{bmatrix} 0 & 0 \\ 0 & 1 \end{bmatrix}\begin{bmatrix} v & -y \\ -u & x \end{bmatrix}$$

$$= \frac{1}{(xv-uy)^2}\begin{bmatrix} -uy & xy \\ ux & -x^2 \end{bmatrix}.$$

$$\frac{\partial}{\partial v}X^{-1} = \frac{\partial}{\partial v}\left\{\frac{1}{(xv-uy)}\begin{bmatrix} v & -y \\ -u & x \end{bmatrix}\right\}.$$

Note that

$$\frac{\partial}{\partial v}\frac{v}{(xv-uy)} = -\frac{uy}{(xv-uy)^2}, \quad \frac{\partial}{\partial v}\frac{(-u)}{(xv-uy)} = \frac{ux}{(xv-uy)^2},$$

$$\frac{\partial}{\partial v}\frac{(-y)}{(xv-uy)} = \frac{xy}{(xv-uy)^2}$$

and

$$\frac{\partial}{\partial v}\frac{x}{(xv-uy)} = -\frac{x^2}{(xv-uy)^2}.$$

Thus the result is verified.

By using the ideas in Theorems 1.7–1.10 one can establish the following result.

Theorem 1.11 *Let A, B, C, D be matrices of constants and X a matrix of functionally independent real variables so that all the following products are defined and BXC is nonsingular. Then*

$$Y = A(BXC)^{-1}D \Rightarrow$$
$$\frac{\partial Y}{\partial\theta} = -A(BXC)^{-1}B\Delta C(BXC)^{-1}D.$$

Proof From (1.1.9)

$$\frac{\partial Y}{\partial\theta} = A\left[\frac{\partial}{\partial\theta}(BXC)^{-1}\right]D.$$

From (1.1.10)

$$\frac{\partial}{\partial\theta}(BXC)^{-1} = -(BXC)^{-1}\left[\frac{\partial}{\partial\theta}(BXC)\right](BXC)^{-1}.$$

Again from (1.1.9)

$$\frac{\partial}{\partial\theta}(BXC) = B\Delta C$$

and hence the result.

Example 1.10   Verify Theorem 1.11 for

$$X = \begin{bmatrix} x & y \\ u & v \end{bmatrix}, \text{ with } xv - uy \neq 0, \ A = \begin{bmatrix} 1 & 1 \\ 0 & 0 \end{bmatrix},$$

$$B = \begin{bmatrix} 1 & 0 \\ 0 & 1 \end{bmatrix}, \ C = \begin{bmatrix} 1 & 0 \\ 0 & 2 \end{bmatrix}, \ D = \begin{bmatrix} 1 & 0 \\ 1 & 0 \end{bmatrix} \text{ and } \theta = x.$$

Solution

$$\begin{aligned} BXC &= \begin{bmatrix} 1 & 0 \\ 0 & 1 \end{bmatrix} \begin{bmatrix} x & y \\ u & v \end{bmatrix} \begin{bmatrix} 1 & 0 \\ 0 & 2 \end{bmatrix} \\ &= \begin{bmatrix} x & 2y \\ u & 2v \end{bmatrix}. \end{aligned}$$

$$(BXC)^{-1} = \frac{1}{2(xv - uy)} \begin{bmatrix} 2v & -2y \\ -u & x \end{bmatrix}.$$

$$\begin{aligned} A(BXC)^{-1}D &= \frac{1}{2(xv - uy)} \begin{bmatrix} 1 & 1 \\ 0 & 0 \end{bmatrix} \begin{bmatrix} 2v & -2y \\ -u & x \end{bmatrix} \begin{bmatrix} 1 & 0 \\ 1 & 0 \end{bmatrix} \\ &= \frac{1}{2(xv - uy)} \begin{bmatrix} 2v - 2y - u + x & 0 \\ 0 & 0 \end{bmatrix}. \end{aligned}$$

$$\frac{\partial}{\partial x}[A(BXC)^{-1}D] = \begin{bmatrix} \alpha & 0 \\ 0 & 0 \end{bmatrix}$$

where

$$\alpha = \frac{uv - uy + 2vy - 2v^2}{2(xv - uy)^2}.$$

$$\Delta = \frac{\partial}{\partial x}X = \begin{bmatrix} 1 & 0 \\ 0 & 0 \end{bmatrix}.$$

But

$$B\Delta C = \begin{bmatrix} 1 & 0 \\ 0 & 1 \end{bmatrix} \begin{bmatrix} 1 & 0 \\ 0 & 0 \end{bmatrix} \begin{bmatrix} 1 & 0 \\ 0 & 2 \end{bmatrix} = \begin{bmatrix} 1 & 0 \\ 0 & 0 \end{bmatrix}.$$

$$(BXC)^{-1}(B\Delta C)(BXC)^{-1} = \frac{1}{2(xv - uy)^2} \begin{bmatrix} 2v^2 & -2yv \\ -uv & uy \end{bmatrix}.$$

Then

$$\begin{aligned} \frac{\partial}{\partial x}Y &= -A\left[(BXC)^{-1}(B\Delta C)(BXC)^{-1}\right]D \\ &= \frac{(-1)}{2(xv - uy)^2} \begin{bmatrix} 1 & 1 \\ 0 & 0 \end{bmatrix} \begin{bmatrix} 2v^2 & -2yv \\ -uv & uy \end{bmatrix} \begin{bmatrix} 1 & 0 \\ 1 & 0 \end{bmatrix} \\ &= \begin{bmatrix} \alpha & 0 \\ 0 & 0 \end{bmatrix} \end{aligned}$$

and the result is verified.

Matrix differential operators are widely used in many applied areas especially in econometrics and statistics, and in the evaluation of the Jacobians of matrix transformations. There are many papers and books on this topic. Some of them are the following: Dwyer (1967), Dwyer and Macphail (1948), Henderson and Searle (1979), Jinadasa and Tracy (1986), MacRae (1974), McCulloch (1982), Magnus and Neudecker (1988), Nel (1980), Neudecker (1969), Neudecker and Wansbeek (1983), Polasek (1985), Pollock (1987), Rao (1973), Rogers (1980), Srivastava and Khatri (1979), Tracy and Dwyer (1969), Tracy and Jinadasa (1988), Tracy and Singh (1971), Tracy and Sultan (1993) and Wong (1980).

Exercises

1.1.1 Let X be a matrix of functionally independent real variables, C and D be matrices of constants. When $XCXD$ is defined show that

$$\frac{\partial}{\partial X}\text{tr}(XCXD) = U', \text{ for a general } X$$

$$= U + U' - \text{diag}(U), \text{ if } X \text{ is symmetric}$$

where $U = DXC + CXD$ and $\text{diag}(U)$ is a diagonal matrix with the same diagonal as in U.

1.1.2 Solve for maxima or minima

(i) $f(X) = X'X, \quad X' = (x_1, x_2)$;

(ii) $f(X) = X'X$ subject to $X'AX = 3$, $A = \begin{bmatrix} 1 & \frac{1}{2} \\ \frac{1}{2} & 1 \end{bmatrix}$.

1.1.3 For the linear model $Y = X\beta + e$ where X is an $n \times m$ matrix of known constants, β is an $m \times 1$ vector of unknown parameters, e is the error vector, $n \geq m$ and the rank of X is m, show that $e'e$ is minimized at $\beta = (X'X)^{-1}X'Y$.

1.1.4 (Maximum likelihood estimation) Let

$$L = \prod_{j=1}^{n} \left\{ \frac{e^{-\frac{1}{2}(x_j' V^{-1} x_j)}}{(2\pi)^{\frac{p}{2}} |V|^{\frac{1}{2}}} \right\}$$

where V is a $p \times p$ symmetric positive definite matrix of unknown parameters and X_j, $j = 1, ..., n$ are $p \times 1$ vectors of known observations, $n > p$. Evaluate the maximum of L and the corresponding point V.

1.1.5   Let X be a $p \times p$ real matrix with distinct eigenvalues $\lambda_1, \lambda_2, ..., \lambda_p$. Show that the differential of $\text{tr}\left(X^k\right)$ is given by

$$d\left(\text{tr}\left(X^k\right)\right) = k\left(\sum_{j=1}^{p}\lambda_j^{k-1}d\lambda_j\right).$$

1.1.6   (Discriminant analysis) Let α be a $p \times 1$ vector of unknown real scalars, μ_1 and μ_2 be $p \times 1$ vectors of known constants and V be a $p \times p$ nonsingular real symmetric matrix of known elements. Evaluate α for which $(\alpha'\mu_1 - \alpha'\mu_2)^2$ is a maximum subject to $\alpha'V\alpha = 1$.

1.1.7   (Principal component analysis) Let β be a $p \times 1$ vector of unknown real scalars and V be a $p \times p$ real symmetric positive definite matrix of known elements. Evaluate β so that $\beta'V\beta$ is maximized subject to $\beta'\beta = 1$.

1.1.8   (Canonical correlation analysis) Let V be a $p \times p$ real symmetric positive definite matrix of known elements. Let it be partitioned as follows:

$$V = \begin{pmatrix} V_{11} & V_{12} \\ V_{21} & V_{22} \end{pmatrix}, \quad \begin{matrix} V_{11} : r \times r, \\ V_{22} : (p-r) \times (p-r), \end{matrix} \quad \begin{matrix} V_{12} : r \times (p-r) \\ V_{21}' = V_{12} \end{matrix}.$$

Let α be $r \times 1$ and γ be $(p-r) \times 1$ vectors of real scalars. Find α and γ such that $\alpha'V_{12}\gamma$ is maximized subject to the conditions $\alpha'V_{11}\alpha = 1$ and $\gamma'V_{22}\gamma = 1$.

1.1.9   Consider the following matrices of differentials:

$$(dX) = (dx_{ij}), \quad (dY) = (dy_{ij}), \quad (dZ) = (dz_{ij}).$$

Show that

(i) $Y = X^{-1}$, $|X| \neq 0 \Rightarrow (dY) = -X^{-1}(dX)X^{-1}$;

(ii) $Z = TT' \Rightarrow T^{-1}(dZ)T'^{-1} = U + U'$, $U = T^{-1}(dT)$

where T is a lower triangular matrix with positive diagonal elements and $Z = Z' > 0$.

1.1.10   For an orthogonal matrix H show that $H(dH')$ is skew symmetric where (dH') denotes the matrix of differentials in H'.

1.2   Jacobians of linear matrix transformations

In this section we consider the Jacobians associated with transforming one
matrix to another matrix through a linear transformation. Nonlinear transformations and transformations involving many matrices will be considered
in the coming sections. All the matrices appearing in this section and in
the remaining sections are assumed to be with real elements (not complex
numbers or complex variables) unless specified otherwise.

1.2.1   Jacobians of vector transformations

It is assumed that the reader is familiar with the calculation of Jacobians
when a vector of scalar variables is transformed to a vector of scalar variables. The result is stated here for the sake of completeness. Let the vector
of scalar variables $X' = (x_1, ..., x_p)$ be transformed to $Y' = (y_1, ..., y_p)$ by a
one-to-one transformation. Let the matrix of partial derivatives be denoted
by

$$\frac{\partial Y}{\partial X} = \left(\frac{\partial y_i}{\partial x_j}\right).$$

Then the determinant of the matrix $\left(\frac{\partial y_i}{\partial x_j}\right)$ is known as the Jacobian of the
transformation X going to Y or Y as a function of X and it is written as

$$J(Y : X) = \left|\left(\frac{\partial y_i}{\partial x_j}\right)\right| \text{ or } dY = J\, dX, \ J \neq 0$$

and

$$J(Y : X) = \frac{1}{J(X : Y)}$$

or

$$1 = J(Y : X)J(X : Y).$$

Note that when transforming X to Y the variables can be taken in any
order because a permutation brings only a change of sign in the determinant
and the magnitude remains the same, that is, $|J|$ remains the same where
$|J|$ denotes the absolute value of J. When evaluating integrals involving
functions of matrix arguments one often needs only the absolute value of J.
Hence in all the theorems in this chapter the notation $dY = J\, dX$ means
that the relation is written ignoring the sign. The simplest case of a linear
transformation is a scalar multiplication of X. This will be given in the
next theorem.

Theorem 1.12   *Let X and Y be $p \times 1$ vectors of functionally independent real variables and A be a $p \times p$ nonsingular matrix of constants. Then, ignoring the sign,*

$$Y = AX \Rightarrow dY = |A|dX,$$
$$= a^p dX \text{ for } A = aI \qquad (1.2.1)$$

where a is a scalar and I is the identity matrix.

Proof   The result follows from the definition itself. Note that when $Y = AX$, $A = (a_{ij})$ one has

$$y_i = a_{i1}x_1 + \ldots + a_{ip}x_p, \quad i = 1, \ldots, p$$

where x_j's and y_j's denote the elements of the vectors X and Y respectively. Thus the partial derivative of y_i with respect to x_j is a_{ij}, and then the determinant of the Jacobian matrix is $|A|$. When $A = aI$, $|A| = a^p$ and hence the result. Note that $A = aI$ is nothing but the scalar multiplication of X by a.

Before we proceed with more complicated cases let us examine the effect of permuting the differentials among themselves. Consider a simple case $p = 2$. Then from Theorem 1.12 we have

$$dy_1 dy_2 = \begin{vmatrix} a_{11} & a_{12} \\ a_{21} & a_{22} \end{vmatrix} dx_1 dx_2. \qquad (1.2.2)$$

If y_2 is taken as the first equation and y_1 as the second equation then we end up with

$$dy_2 dy_1 = \begin{vmatrix} a_{21} & a_{22} \\ a_{11} & a_{12} \end{vmatrix} dx_1 dx_2$$
$$= -\begin{vmatrix} a_{11} & a_{12} \\ a_{21} & a_{22} \end{vmatrix} dx_1 dx_2.$$

In other words

$$dy_1 dy_2 = -dy_2 dy_1. \qquad (1.2.3)$$

This should then imply that

$$dy_1(dy_1) = -(dy_1)dy_1 \Rightarrow dy_1 dy_1 = 0 = (dy_1)^2$$
$$dy_2 dy_2 = 0 = (dy_2)^2. \qquad (1.2.4)$$

Under the skew symmetric multiplication given by (1.2.3) can a direct substitution for dy_1 and dy_2 yield the equation (1.2.2)? Let us check the defining equations.

$$dy_1 = \frac{\partial y_1}{\partial x_1} dx_1 + \frac{\partial y_1}{\partial x_2} dx_2$$

and

$$dy_2 = \frac{\partial y_2}{\partial x_1} dx_1 + \frac{\partial y_2}{\partial x_2} dx_2.$$

Thus

$$dy_1 dy_2 = \left(\frac{\partial y_1}{\partial x_1} dx_1 + \frac{\partial y_1}{\partial x_2} dx_2 \right) \left(\frac{\partial y_2}{\partial x_1} dx_1 + \frac{\partial y_2}{\partial x_2} dx_2 \right). \quad (1.2.5)$$

Now multiply the right side keeping the order in which the differentials enter into the products. That is,

$$dy_1 dy_2 = \left(\frac{\partial y_1}{\partial x_1} \right) \left(\frac{\partial y_2}{\partial x_1} \right) dx_1 dx_1 + \left(\frac{\partial y_1}{\partial x_1} \right) \left(\frac{\partial y_2}{\partial x_2} \right) dx_1 dx_2$$

$$+ \left(\frac{\partial y_1}{\partial x_2} \right) \left(\frac{\partial y_2}{\partial x_1} \right) dx_2 dx_1 + \left(\frac{\partial y_1}{\partial x_2} \right) \left(\frac{\partial y_2}{\partial x_2} \right) dx_2 dx_2. \quad (1.2.6)$$

Now use the results in (1.2.3) and (1.2.4) to get

$$dy_1 dy_2 = \left(\frac{\partial y_1}{\partial x_1} \right) \left(\frac{\partial y_2}{\partial x_2} \right) dx_1 dx_2 - \left(\frac{\partial y_1}{\partial x_2} \right) \left(\frac{\partial y_2}{\partial x_1} \right) dx_1 dx_2$$

$$= \begin{vmatrix} \frac{\partial y_1}{\partial x_1} & \frac{\partial y_1}{\partial x_2} \\ \frac{\partial y_2}{\partial x_1} & \frac{\partial y_2}{\partial x_2} \end{vmatrix} dx_1 dx_2.$$

This is our equation (1.2.2). Thus our product is a skew symmetric product which is often known as the *wedge product or exterior product* and it is written as $\wedge_{j=1}^{p} dx_j$ indicating that

$$dx_i \wedge dx_j = -dx_j \wedge dx_i, \; i \neq j$$
$$= 0, \; i = j. \quad (1.2.7)$$

For convenience we will denote this wedge product by dX in our discussions. This can create a little confusion when we will use the same notation ($d\mathbf{X}$) with X typed in bold character to denote the matrix of differentials later on. But from the context itself the notation will be clear and hence there won't be any confusion. Thus our notation will be the following: Let $X = (x_{ij})$ be a $p \times q$ matrix of functionally independent real variables. Then

$$dX = \wedge_{i=1}^{p} \wedge_{j=1}^{q} dx_{ij} \quad (1.2.8)$$

when dX appears with integrals or Jacobians of transformations;

$$dX = \Pi_{i=1}^{p} \Pi_{j=1}^{q} dx_{ij} \qquad (1.2.9)$$

when dX appears with integrals involving density functions where the functions are nonnegative and the absolute value of the Jacobian is automatically taken. Note that

$$(dX) = (dx_{ij}) \qquad (1.2.10)$$

that is , the matrix of differentials.

Example 1.11   If the $p \times 1$ real random vector X has a p-variate real nonsingular normal density $N_p(\mu, V)$ (see Definition A4 in Appendix A) and if A is a $p \times p$ nonsingular matrix of constants then show that $Y = AX + C$ has a nonsingular normal density with the parameters $A\mu + C$ and AVA' where C is a $p \times 1$ vector of constants.

Solution   Let $g(Y)$ be the density of Y. Since $|A| \neq 0$ the transformation $Y = AX + C \Rightarrow X = A^{-1}(Y - C)$ and thus it is a one-to-one transformation and hence $g(Y) = f(X)|J(X:Y)|$. But from Theorem 1.12, $|J(Y:X)| = |A|_+$ where $|A|_+$ denotes the absolute value of $|A|$ and

$$(X - \mu)'V^{-1}(X - \mu) = (Y - \mu^*)'(AVA')^{-1}(Y - \mu^*)$$

where

$$\mu^* = A\mu + C.$$

Hence

$$g(Y) = \frac{1}{|A|_+} f(X)$$
$$= \frac{e^{-\frac{1}{2}(Y - \mu^*)'(AVA')^{-1}(Y - \mu^*)}}{(2\pi)^{\frac{p}{2}} |AVA'|^{\frac{1}{2}}} \Rightarrow$$
$$Y \sim N_p(\mu^*, AVA').$$

This establishes the result.

1.2.2   Linear matrix transformations

The simplest linear matrix transformation is a generalization of Theorem 1.12 to $p \times q$ matrices.

Theorem 1.13   *Let X and Y be $p \times q$ matrices of functionally independent real variables and A be a $p \times p$ nonsingular matrix of constants. Then,*

ignoring the sign,
$$Y = AX \Rightarrow dY = |A|^q dX. \tag{1.2.11}$$

Proof   Let the columns of Y and X be denoted by Y_i and X_i, $i = 1, ..., q$ respectively and treat the problem as a transformation of the column vectors whose transposes are $(Y_1', ..., Y_q')$ and $(X_1', ..., X_q')$ respectively. The matrix of partial derivatives will then be a block diagonal matrix with q diagonal blocks each equal to A and the determinant will then be $|A|^q$. Hence the result.

We can also treat the problem as the problem of transforming the q column vectors of X to the q column vectors of Y. That is, $Y = AX$ means

$$\begin{bmatrix} y_{11} & \cdots & y_{1q} \\ y_{21} & \cdots & y_{2q} \\ \vdots & \ddots & \vdots \\ y_{p1} & \cdots & y_{pq} \end{bmatrix} = \begin{bmatrix} a_{11} & \cdots & a_{1p} \\ a_{21} & \cdots & a_{2p} \\ \vdots & \ddots & \vdots \\ a_{p1} & \cdots & a_{pp} \end{bmatrix} \begin{bmatrix} x_{11} & \cdots & x_{1q} \\ x_{21} & \cdots & x_{2q} \\ \vdots & \ddots & \vdots \\ x_{p1} & \cdots & x_{pq} \end{bmatrix}.$$

Thus

$$[Y_1 \quad Y_2 \quad \cdots \quad Y_q] = A[X_1 \quad X_2 \quad \cdots \quad X_q]$$
$$= [AX_1 \quad AX_2 \quad \cdots \quad AX_q] \tag{1.2.12}$$

and hence from Theorem 1.12

$$\frac{\partial Y_i}{\partial X_j} = A, \ i = j,$$
$$= 0, \ i \neq j \tag{1.2.13}$$

where 0 denotes a null matrix. We may also note that when $A = aI$ where a is a scalar and I the identity matrix we have Y a scalar multiple of X. Also if the transformation is $Y = AX + B$ where B is a matrix of constants the Jacobian remains the same. Hence we have the following resuts.

$$Y = AX + B \Rightarrow dY = |A|^q dX,$$
$$Y = aX + B \Rightarrow dY = a^{pq} dX,$$

and

$$Y = X + B \Rightarrow dY = dX. \tag{1.2.14}$$

The following result can be established in exactly the same way by considering the transformation as a transformation of a long column vector formed by the various columns of X to that of Y or take the transposes and apply Theorem 1.13.

Theorem 1.14   *Let X and Y be $p \times q$ matrices of functionally independent real variables and B a $q \times q$ nonsingular matrix of constants. Then, ignoring the sign,*

$$Y = XB \Rightarrow dY = |B|^p dX. \tag{1.2.15}$$

Example 1.12   Let the $p \times q$ real matrix X have a matrix-variate real nonsingular normal density as given in Definition A5 of Appendix A with the parameters M, W and V. Let $Y_1 = AX + C$, $Y_2 = XB + C$ and $Y = AXB + C$ where C is a $p \times q$ matrix of constants, and A and B are $p \times p$ and $q \times q$ nonsingular matrices of constants respectively. Then show that Y_1, Y_2 and Y have matrix-variate real nonsingular normal densities.

Solution   Recall the density $f(X)$ of X. That is,

$$f(X) = (2\pi)^{-\frac{pq}{2}} |V|^{-\frac{q}{2}} |W|^{-\frac{p}{2}}$$
$$\times \exp\left\{-\frac{1}{2}\text{tr}\left[V^{-1}(X - M)W^{-1}(X - M)'\right]\right\}.$$
$$Y = AXB + C \Rightarrow$$
$$X = A^{-1}YB^{-1} - A^{-1}CB^{-1}$$

and

$$X - M = A^{-1}[Y - C - AMB]B^{-1}.$$

The exponent reduces to $-\frac{1}{2}\text{tr}\left[V^{*-1}(Y - M^*)W^{*-1}(Y - M^*)'\right]$ where

$$V^* = AVA', \quad W^* = BWB', \text{ and } M^* = AMB + C.$$

Let $g(Y)$ be the density of Y. Since the transformation is one-to-one we have $g(Y) = f(X)|J(X : Y)|$. But by combining Theorems 1.13 and 1.14 or from Theorem 1.18 one has

$$dY = |A|_+^q |B|_+^p dX$$
$$= |AA'|^{\frac{q}{2}} |BB'|^{\frac{p}{2}} dX$$

where $|\cdot|_+$ indicates that the absolute value is taken. That is,

$$g(Y) = (2\pi)^{-\frac{pq}{2}} |V^*|^{-\frac{q}{2}} |W^*|^{-\frac{p}{2}}$$
$$\times \exp\left\{-\frac{1}{2}\text{tr}\left[V^{*-1}(Y - M^*)W^{*-1}(Y - M^*)'\right]\right\}.$$

Put $B = I_q$ to get the density of Y_1 and $A = I_p$ to get the density of Y_2.

Theorem 1.15   *Let X, A be lower triangular $p \times p$ matrices where $A = (a_{ij})$ is a constant matrix with $a_{jj} > 0$, $j = 1, ..., p$ and X is a matrix of functionally independent real variables. Then*

$$Y = X + X' \Rightarrow dY = 2^p dX,$$

$$Y = XA \Rightarrow dY = \left\{ \prod_{j=1}^{p} a_{jj}^{p-j+1} \right\} dX,$$

$$Y = AX \Rightarrow dY = \left\{ \prod_{j=1}^{p} a_{jj}^{j} \right\} dX,$$

and

$$Y = aX \Rightarrow dY = a^{p(p+1)/2} dX \qquad (1.2.16)$$

where a is a scalar quantity.

Proof $Y = X + X' \Rightarrow$

$$\begin{bmatrix} x_{11} & \cdots & 0 \\ x_{21} & \cdots & 0 \\ \vdots & \ddots & \vdots \\ x_{p1} & \cdots & x_{pp} \end{bmatrix} + \begin{bmatrix} x_{11} & \cdots & x_{p1} \\ 0 & \cdots & x_{p2} \\ \vdots & \ddots & \vdots \\ 0 & \cdots & x_{pp} \end{bmatrix} = \begin{bmatrix} 2x_{11} & \cdots & x_{p1} \\ x_{21} & \cdots & x_{p2} \\ \vdots & \ddots & \vdots \\ x_{p1} & \cdots & 2x_{pp} \end{bmatrix}.$$

When taking the partial derivatives the p diagonal elements give 2 each and others unities and hence $dY = 2^p dX$.

$Y = XA$

$$= \begin{bmatrix} x_{11} & 0 & \cdots & 0 \\ x_{21} & x_{22} & \cdots & 0 \\ \vdots & \vdots & \ddots & \vdots \\ x_{p1} & x_{p2} & \cdots & x_{pp} \end{bmatrix} \begin{bmatrix} a_{11} & 0 & \cdots & 0 \\ a_{21} & a_{22} & \cdots & 0 \\ \vdots & \vdots & \ddots & \vdots \\ a_{p1} & a_{p2} & \cdots & a_{pp} \end{bmatrix}$$

$$= \begin{bmatrix} x_{11}a_{11} & 0 & \cdots & 0 \\ x_{21}a_{11} + x_{22}a_{21} & x_{22}a_{22} & \cdots & 0 \\ \vdots & \vdots & \ddots & \vdots \\ \sum_{j=1}^{p} x_{pj}a_{j1} & \sum_{j=2}^{p} x_{pj}a_{j2} & \cdots & x_{pp}a_{pp} \end{bmatrix}. \qquad (1.2.17)$$

The matrices of the configurations of the partial derivatives, by taking the elements in the orders $y_{11}, y_{21}, y_{22}, ..., y_{pp}$ and $x_{11}, x_{21}, x_{22}, ..., x_{pp}$, are the following:

$$\frac{\partial y_{11}}{\partial x_{11}} = a_{11}, \quad \frac{\partial(y_{21}, y_{22})}{\partial(x_{21}, x_{22})} = \begin{bmatrix} a_{11} & a_{21} \\ 0 & a_{22} \end{bmatrix}'$$

$$\frac{\partial(y_{31}, y_{32}, y_{33})}{\partial(x_{31}, x_{32}, x_{33})} = \begin{bmatrix} a_{11} & a_{21} & a_{31} \\ 0 & a_{22} & a_{32} \\ 0 & 0 & a_{33} \end{bmatrix}'$$

and so on. Thus the whole configuration is a triangular matrix with a_{11} appearing p times and a_{22} appearing $p-1$ times and so on in the diagonal. Hence the result.

One can also use the ideas in (1.2.3) and (1.2.4) to establish this result. Note that

$$y_{11} = a_{11}x_{11} \qquad\qquad dy_{11} = a_{11}dx_{11}$$
$$y_{21} = a_{11}x_{21} + a_{21}x_{22} \qquad dy_{21} = a_{11}dx_{21} + a_{21}dx_{22}$$
$$y_{22} = a_{22}x_{22} \qquad\qquad dy_{22} = a_{22}dx_{22}.$$

Now multiply the right sides, neglecting the higher powers of dx_{11} and dx_{22} by using (1.2.4), to get

$$\begin{aligned} dy_{11}dy_{21}dy_{22} &= (a_{11}dx_{11})(a_{11}dx_{21} + a_{21}dx_{22})(a_{22}dx_{22}) \\ &= (a_{11}dx_{11})(a_{11}a_{22}dx_{21}dx_{22}) \\ &= a_{11}^2 a_{22}dx_{11}dx_{21}dx_{22}. \end{aligned}$$

The result follows by induction. In a similar manner one can prove the case for $Y = AX$.

In Theorem 1.15 we considered the cases of lower triangular matrices. What happens if all are upper triangular matrices? Note that when X and A are lower triangular and if $Y = XA$ then $Y' = A'X'$ is upper triangular. Since $dY = dY'$ the results in (1.2.16) will be reversed. This will be stated as a theorem.

Theorem 1.16   Let X, B be upper triangular $p \times p$ matrices where $B = (b_{ij})$ is a constant matrix with $b_{jj} > 0$, $j = 1, ..., p$ and X is a matrix of functionally independent real variables. Then

$$Y = X + X' \Rightarrow dY = 2^p dY,$$

$$Y = XB \Rightarrow dY = \left\{ \prod_{j=1}^{p} b_{jj}^{j} \right\} dX,$$

$$Y = BX \Rightarrow dY = \left\{ \prod_{j=1}^{p} b_{jj}^{p+1-j} \right\} dX,$$

and

$$Y = bX \Rightarrow dY = b^{p(p+1)/2} dX \qquad (1.2.18)$$

where b is a scalar quantity.

Let us see what happens in the transformation $Y = X + X'$ where X and Y are $p \times p$ matrices. When $X = X'$, $Y = 2X$ and this case is covered in (1.2.14). If $X \neq X'$ then Y has become symmetric with $p(p+1)/2$ functionally independent variables whereas in X there are p^2 variables and hence this is not a one-to-one transformation.

Theorem 1.17   Let X, A, B be $p \times p$ lower triangular matrices where A and B are nonsingular matrices of constants with positive diagonal elements and X is a matrix of $p(p+1)/2$ functionally independent real variables. Then

$$Y = AXB \Rightarrow dY = \left\{ \prod_{j=1}^{p} a_{jj}^{j} b_{jj}^{p+1-j} \right\} dX,$$

and

$$Z = A'X'B' \Rightarrow dZ = \left\{ \prod_{j=1}^{p} b_{jj}^{j} a_{jj}^{p+1-j} \right\} dX. \qquad (1.2.19)$$

The proof for dY follows by applying Theorem 1.15 in two steps and that for dZ follows by applying Theorem 1.16. Note that the upper triangular case is covered in Z. Observe that when the diagonal elements of A and B are unities then $dY = dX$ and $dZ = dX' = dX$.

Example 1.13   Verify Theorems 1.15, 1.16 and 1.17 for

$$X = \begin{bmatrix} x_{11} & 0 \\ x_{21} & x_{22} \end{bmatrix}, \quad Y = \begin{bmatrix} y_{11} & y_{12} \\ 0 & y_{22} \end{bmatrix},$$

$$A = \begin{bmatrix} 2 & 0 \\ 1 & 1 \end{bmatrix}, \quad B = \begin{bmatrix} 1 & 1 \\ 0 & 2 \end{bmatrix}, \quad C = \begin{bmatrix} 2 & 0 \\ 1 & 3 \end{bmatrix}.$$

Solution   Let

$$U = \begin{bmatrix} u_{11} & 0 \\ u_{21} & u_{22} \end{bmatrix} = XA = \begin{bmatrix} x_{11} & 0 \\ x_{21} & x_{22} \end{bmatrix} \begin{bmatrix} 2 & 0 \\ 1 & 1 \end{bmatrix} \Rightarrow$$

$$u_{11} = 2x_{11}, \ u_{21} = 2x_{21} + x_{22}, \ u_{22} = x_{22} \Rightarrow$$

$$du_{11} = 2dx_{11}, \ du_{21} = 2dx_{21} + dx_{22}, \ du_{22} = dx_{22}.$$

Hence

$$du_{11}du_{21}du_{22} = (2dx_{11})(2dx_{21} + dx_{22})(dx_{22})$$
$$= 4dx_{11}dx_{21}dx_{22}$$

neglecting higher powers of dx_{22}. But

$$\prod_{j=1}^{p} a_{jj}^{p-j+1} = a_{11}^2 a_{22} = 4.$$

$$V = \begin{bmatrix} v_{11} & 0 \\ v_{21} & v_{22} \end{bmatrix} = AX = \begin{bmatrix} 2 & 0 \\ 1 & 1 \end{bmatrix} \begin{bmatrix} x_{11} & 0 \\ x_{21} & x_{22} \end{bmatrix} \Rightarrow$$

$$v_{11} = 2x_{11}, \ v_{21} = x_{11} + x_{21}, \ v_{22} = x_{22} \Rightarrow$$

$$dv_{11} = 2dx_{11}, \ dv_{21} = dx_{11} + dx_{21}, \ dv_{22} = dx_{22}.$$

Hence

$$dv_{11}dv_{21}dv_{22} = (2dx_{11})(dx_{11} + dx_{21})(dx_{22})$$
$$= 2dx_{11}dx_{21}dx_{22}.$$

But

$$\prod_{j=1}^{p} a_{jj}^{j} = a_{11} a_{22}^2 = 2.$$

These verify Theorem 1.15. Use Y and B and proceed as above to verify Theorem 1.16. Let

$$W = \begin{bmatrix} w_{11} & 0 \\ w_{21} & w_{22} \end{bmatrix} = AXC \Rightarrow$$

$$w_{11} = 4x_{11}, \ w_{21} = 2x_{11} + 2x_{21} + x_{22}, \ w_{22} = 3x_{22}.$$

Then

$$dw_{11}dw_{21}dw_{22} = (4dx_{11})(2dx_{11} + 2dx_{21} + dx_{22})(3dx_{22})$$
$$= 24dx_{11}dx_{21}dx_{22}$$

neglecting higher powers of the differentials. But

$$\prod_{j=1}^{p} a_{jj}^{j} c_{jj}^{p+1-j} = a_{11} c_{11}^{2} a_{22}^{2} c_{22}$$

$$= 24.$$

Let

$$Z = \begin{bmatrix} z_{11} & z_{12} \\ 0 & z_{22} \end{bmatrix} = A'X'C'$$

$$= \begin{bmatrix} 2 & 1 \\ 0 & 1 \end{bmatrix} \begin{bmatrix} x_{11} & x_{21} \\ 0 & x_{22} \end{bmatrix} \begin{bmatrix} 2 & 1 \\ 0 & 3 \end{bmatrix}.$$

Then

$$z_{11} = 4x_{11}, \ z_{12} = 2x_{11} + 6x_{21} + 3x_{22}, \ z_{22} = 3x_{22}.$$

Then from the previous steps the results follow. These verify Theorem 1.17.

1.2.3 Function of a function, linear case

Here we consider transformations of the type $Y \to Z$, that is, Y as a function of Z, and then $Z \to X$ where both the transformations are linear and we are interested in the relationship between dX and dY. Thus we have a linear transformation of a linear transformation.

Theorem 1.18 *Let X and Y be $p \times q$ matrices of functionally independent real variables and A a $p \times p$ and B a $q \times q$ nonsingular matrices of constants. Then, ignoring the sign,*

$$Y = AXB \Rightarrow dY = |A|^{q}|B|^{p}dX. \tag{1.2.20}$$

Proof Consider the transformation $Z = AX$. Then by Theorem 1.13 we have $dZ = |A|^{q}dX$, ignoring the sign. Now consider $Y = ZB \Rightarrow dY = |B|^{p}dZ$ from Theorem 1.14, ignoring the sign. Now substituting for dZ one has the result. As a corollary we have for $q = p$ and $B = A$ or $B = A'$

$$Y = AXA' \Rightarrow dY = |A|^{2p}dX, \tag{1.2.21}$$

ignoring the sign. This two step operation and the resulting Jacobian can be written as follows:

$$J(Y : X) = J(Y : Z)J(Z : X). \tag{1.2.22}$$

By using this idea and combining the results in Theorem 1.15 we have the following:

Theorem 1.19   *Let X be a lower triangular matrix of functionally independent real variables and $A = (a_{ij})$ be a lower triangular matrix of constants with $a_{jj} > 0$, $j = 1, ..., p$. Then*

$$Y = XA + A'X' \Rightarrow dY = 2^p \left\{ \prod_{j=1}^{p} a_{jj}^{p-j+1} \right\} dX,$$

and

$$Y = AX + X'A' \Rightarrow dY = 2^p \left\{ \prod_{j=1}^{p} a_{jj}^{j} \right\} dX. \qquad (1.2.23)$$

When A and X are lower triangular then A' and X' are upper triangular. Hence Theorem 1.19 covers the case of two upper triangular matrices also. In equation (1.2.21) we have a result for arbitrary X and Y. What happens if X and Y are symmetric matrices of functionally independent variables? This is the most commonly appearing transformation when dealing with functions of real symmetric positive definite matrices. Hence this very important transformation is given in the next theorem.

Theorem 1.20   *Let X and Y be $p \times p$ symmetric matrices of functionally independent real variables and A a $p \times p$ nonsingular matrix of constants. Then, ignoring the sign,*

$$Y = AXA' \Rightarrow dY = |A|^{p+1}dX,$$

and

$$Y = aX \Rightarrow dY = a^{p(p+1)/2}dX \qquad (1.2.24)$$

where a is a scalar quantity.

Proof   Since both X and Y are symmetric matrices and A is nonsingular we can split A and A' as products of elementary matrices and write in the form

$$Y = ...E_2E_1XE_1'E_2'....$$

where $E_j, j = 1, 2...$ are elemetary matrices. Write $Y = AXA'$ as a sequence of transformations of the type

$$Y_1 = E_1 X E_1', \ Y_2 = E_2 Y_1 E_2', ...$$
$$\Rightarrow dY_1 = J(Y_1 : X)dX, \ dY_2 = J(Y_2 : Y_1)dY_1, ...$$

Now successive substitutions give the final result as long as the Jacobians of the type $J(Y_k : Y_{k-1})$ are computed. Note that the elementary matrices are formed by multiplying any row (or column) of an identity matrix with a scalar, adding a row (column) to another row (column) and combinations of these operations. Hence we need to consider only these two basic elementary matrices. Let us consider a 3×3 case and compute the Jacobians. Let E_1 be the elementary matrix obtained by multiplying the first row by α and E_2 by adding the first row to the second row of an identity matrix. That is,

$$E_1 = \begin{bmatrix} \alpha & 0 & 0 \\ 0 & 1 & 0 \\ 0 & 0 & 1 \end{bmatrix}, \ E_2 = \begin{bmatrix} 1 & 0 & 0 \\ 1 & 1 & 0 \\ 0 & 0 & 1 \end{bmatrix}$$

and

$$E_1 X E_1' = \begin{bmatrix} \alpha^2 x_{11} & \alpha x_{12} & \alpha x_{13} \\ \alpha x_{21} & x_{22} & x_{23} \\ \alpha x_{31} & x_{32} & x_{33} \end{bmatrix} ;$$

$$E_2 U E_2' = \begin{bmatrix} u_{11} & u_{11} + u_{12} & u_{13} \\ u_{11} + u_{21} & u_{11} + u_{21} + u_{12} + u_{22} & u_{13} + u_{23} \\ u_{31} & u_{31} + u_{32} & u_{33} \end{bmatrix} ,$$

where $Y_1 = E_1 X E_1'$ and $Y_2 = E_2 Y_1 E_2'$ and the elements of Y_1 are denoted by u_{ij}'s for convenience. The matrix of partial derivatives in the transformation Y_1 written as a function of X is then

$$\frac{\partial Y_1}{\partial X} = \begin{bmatrix} \alpha^2 & 0 & 0 & 0 & 0 & 0 \\ 0 & \alpha & 0 & 0 & 0 & 0 \\ 0 & 0 & 1 & 0 & 0 & 0 \\ 0 & 0 & 0 & \alpha & 0 & 0 \\ 0 & 0 & 0 & 0 & 1 & 0 \\ 0 & 0 & 0 & 0 & 0 & 1 \end{bmatrix} .$$

This is obtained by taking the x_{ij}'s in the order $x_{11}, x_{21}, x_{22}, x_{31}, x_{32}, x_{33}$ and the u_{ij}'s also in the same order. Thus the Jacobian is given by

$$J(Y_1 : X) = \alpha^4 = \alpha^{3+1} = |E_1|^{3+1}.$$

For a $p \times p$ matrix it will be α^{p+1}. Let the elements of Y_2 be denoted by v_{ij}'s. Then again taking the variables in the order as in the case of Y_1 written as a function of X the matrix of partial derivatives in this transformation is the following:

$$\frac{\partial Y_2}{\partial Y_1} = \begin{bmatrix} 1 & 0 & 0 & 0 & 0 & 0 \\ 1 & 1 & 0 & 0 & 0 & 0 \\ 1 & 2 & 1 & 0 & 0 & 0 \\ 0 & 0 & 0 & 1 & 0 & 0 \\ 0 & 0 & 0 & 1 & 1 & 0 \\ 0 & 0 & 0 & 0 & 0 & 1 \end{bmatrix}.$$

The determinant of this matrix is $1 = 1^{3+1} = |E_2|^{3+1}$. In general such a transformation gives the Jacobian, in absolute value, as $1 = 1^{p+1}$. Thus the Jacobian is given by

$$J(Y : X) = |...E_2 E_1|^{p+1} = |A|^{p+1}.$$

Example 1.14   Let X be a $p \times p$ real symmetric positive definite matrix having a matrix-variate gamma distribution with parameters $(\alpha, B = B' > 0)$ (see Definition A7 in Appendix A). Show that

$$|B|^{-\alpha} = \frac{1}{\Gamma_p(\alpha)} \int_{X>0} |X|^{\alpha - \frac{p+1}{2}} e^{-\mathrm{tr}(BX)} dX, \quad \mathrm{Re}(\alpha) > \frac{p-1}{2}.$$

Solution   Since B is symmetric positive definite there exists a nonsingular matrix C such that $B = CC'$. Note that

$$\mathrm{tr}(BX) = \mathrm{tr}(CC'X) = \mathrm{tr}(C'XC).$$

Let

$$U = C'XC \Rightarrow dU = |C|^{p+1} dX$$

from Theorem 1.20 and

$$|X| = |C'^{-1} U C^{-1}| = |CC'|^{-1}|U| = |B|^{-1}|U|.$$

The integral on the right reduces to the following:

$$\int_{X>0} |X|^{\alpha - \frac{p+1}{2}} e^{-\mathrm{tr}(BX)} dX = |B|^{-\alpha} \int_{U>0} |U|^{\alpha - \frac{p+1}{2}} e^{-\mathrm{tr}(U)} dU.$$

But

$$\int_{U>0} |U|^{\alpha - \frac{p+1}{2}} e^{-\mathrm{tr}(U)} dU = \Gamma_p(\alpha)$$

for $\text{Re}(\alpha) > \frac{p-1}{2}$ (see Appendix A). Hence the result.

Example 1.15   Let X_1 and X_2 be two $p \times p$ real matrix-variate gamma random variables with the parameters (α_1, I) and (α_2, I) respectively, and independently distributed. Show that

$$V = (X_1 + X_2)^{-\frac{1}{2}} X_1 (X_1 + X_2)^{-\frac{1}{2}}$$

has a real matrix-variate beta distribution (see Definition A9 in Appendix A for the definition of a real matrix-variate beta) where $(X_1 + X_2)^{-\frac{1}{2}}$ denotes the symmetric square root of $(X_1 + X_2)^{-1}$.

Solution   Let $g(X_1, X_2)$ denote the joint density of X_1 and X_2. Since X_1 and X_2 are independently distributed the joint density is the product of the marginal densities. That is,

$$g(X_1, X_2) = \frac{|X_1|^{\alpha_1 - \frac{p+1}{2}} |X_2|^{\alpha_2 - \frac{p+1}{2}} e^{-\text{tr}(X_1 + X_2)}}{\Gamma_p(\alpha_1)\Gamma_p(\alpha_2)}$$

for $\text{Re}(\alpha_1) > \frac{p-1}{2}$, $\text{Re}(\alpha_2) > \frac{p-1}{2}$ where $\text{Re}(\cdot)$ denotes the real part of $(\cdot)$. Let $U = X_1 + X_2$ for fixed X_2. Then $dU = dX_1$ and $X_2 = U - X_1$. Denoting the joint density of U and X_1 by $g_1(U, X_1)$ we have

$$g_1(U, X_1) dU dX_1 = \frac{|X_1|^{\alpha_1 - \frac{p+1}{2}} |U - X_1|^{\alpha_2 - \frac{p+1}{2}} e^{-\text{tr}(U)}}{\Gamma_p(\alpha_1)\Gamma_p(\alpha_2)} dU dX_1$$

for $X_1 > 0$, $U > X_1$. Note that

$$|U - X_1| = |U| \, |I - U^{-\frac{1}{2}} X_1 U^{-\frac{1}{2}}|$$

where $U^{\frac{1}{2}}$ is the symmetric square root of U. Put $V = U^{-\frac{1}{2}} X_1 U^{-\frac{1}{2}}$ for fixed U then $dV = |U|^{-\frac{p+1}{2}} dX_1$ by Theorem 1.20. Let $g_2(U, V)$ denote the joint density of U and V. Then

$$g_2(U, V) dU dV = \frac{|U|^{\alpha_1 + \alpha_2 - \frac{p+1}{2}} e^{-\text{tr}(U)}}{\Gamma_p(\alpha_1)\Gamma_p(\alpha_2)}$$
$$\times |I - V|^{\alpha_2 - \frac{p+1}{2}} |V|^{\alpha_1 - \frac{p+1}{2}} dU dV, \ U > 0, \ 0 < V < I.$$

The density of V, denoted by $g_V(V)$, is available by integrating out U in $g_2(U, V)$. But

$$\int_{U>0} |U|^{\alpha_1 + \alpha_2 - \frac{p+1}{2}} e^{-\text{tr}(U)} dU = \Gamma_p(\alpha_1 + \alpha_2)$$

from the definition of the real matrix-variate gamma and hence

$$g_V(V) = \frac{\Gamma_p(\alpha_1 + \alpha_2)}{\Gamma_p(\alpha_1)\Gamma_p(\alpha_2)}|V|^{\alpha_1 - \frac{p+1}{2}}|I - V|^{\alpha_2 - \frac{p+1}{2}}, \ 0 < V < I$$

and $g_V(V) = 0$ elsewhere. This establishes the result.

The proof in the following theorem is similar to that in Theorem 1.20 and hence it will be stated without proof.

Theorem 1.21   *Let X and Y be $p \times p$ skew symmetric matrices of functionally independent real variables and A be a $p \times p$ nonsingular matrix of constants. Then, ignoring the sign,*

$$Y = AXA' \Rightarrow dY = |A|^{p-1}dX,$$

and

$$Y = aX \Rightarrow dY = a^{p(p-1)/2}dX \qquad (1.2.25)$$

where a is a scalar quantity.

Note that when X is skew symmetric the diagonal elements are zeros and hence there are only $p(p-1)/2$ functionally independent variables in X.

Example 1.16   Verify Theorem 1.21 for

$$X = \begin{bmatrix} 0 & x_{12} \\ -x_{12} & 0 \end{bmatrix}, \quad A = \begin{bmatrix} 1 & 2 \\ 1 & 5 \end{bmatrix}.$$

Solution

$$Y = AXA' = \begin{bmatrix} 1 & 2 \\ 1 & 5 \end{bmatrix}\begin{bmatrix} 0 & x_{12} \\ -x_{12} & 0 \end{bmatrix}\begin{bmatrix} 1 & 1 \\ 2 & 5 \end{bmatrix}$$

$$= \begin{bmatrix} 0 & 3x_{12} \\ -3x_{12} & 0 \end{bmatrix}.$$

Then

$$dY = dy_{12} = 3dx_{12}.$$

But

$$|A|^{p-1} = |A| = \begin{vmatrix} 1 & 2 \\ 1 & 5 \end{vmatrix} = 3.$$

This verifies Theorem 1.21.

Theorem 1.22 *Let X and A be $p \times p$ lower triangular matrices where A is a nonsingular constant matrix with positive diagonal elements and X is a matrix of functionally independent real variables. Then*

$$Y = A'X + X'A \Rightarrow dY = 2^p \left\{ \prod_{j=1}^{p} a_{jj}^{j} \right\} dX, \qquad (1.2.26)$$

and

$$Y = AX' + XA' \Rightarrow dY = 2^p \left\{ \prod_{j=1}^{p} a_{jj}^{p-j+1} \right\} dX. \qquad (1.2.27)$$

Proof Consider $Y = X'A + A'X$. Premultiply by $(A')^{-1}$ and postmultiply by A^{-1} to get the following:

$$Y = A'X + X'A$$
$$\Rightarrow (A')^{-1}YA^{-1} = (A')^{-1}X' + XA^{-1}.$$

Let

$$U = XA^{-1} + (A')^{-1}X'$$
$$\Rightarrow dU = 2^p \left\{ \prod_{j=1}^{p} a_{jj}^{-(p-j+1)} \right\} dX$$

by Theorem 1.19 and

$$U = (A')^{-1}YA^{-1} \Rightarrow dU = |A|^{-(p+1)}dY$$

by Theorem 1.20. Now writing dY in terms of dX one has

$$dY = \left\{ \prod_{j=1}^{p} a_{jj}^{-(p-j+1)} \right\} 2^p \left\{ \prod_{j=1}^{p} a_{jj}^{p+1} \right\} dX$$

$$= 2^p \left\{ \prod_{j=1}^{p} a_{jj}^{j} \right\} dX$$

since $|A| = a_{11}...a_{pp}$ because A is lower triangular. Thus the first result follows. The second is proved in a similar way.

Remark 1.1 Note that the transformations in Theorems 1.1– 1.22 are nonsingular linear transformations and hence one-to-one. In other words, Y written as a function of X also gives X as a unique function of Y.

The basic transformations discussed in this section are widely applied in statistical distribution problems and related areas. Books on multivariate statistical analysis, and related papers, provide many applications of this type, see for example, Anderson (1971), Deemer and Olkin (1951), Muirhead (1982), and Srivastava and Khatri (1979).

In this section we have illustrated how to evaluate the Jacobians of linear matrix transformations. Such Jacobians will be evaluated for various types of linear and nonlinear transformations in Chapters 1,2,3 and 4. Properties of Jacobians, the existence of Jacobians, the types of possible transformations with a given Jacobian, conditions for a certain transformation to be unique, particular types of mappings, structure theorems and so on will not be discussed in this book. Some sample references on these topics are the following and more could be found therein. For Jacobians on Samuelson map see Campbell (1993, 1994), for invertibility of the Jacobian matrix on algebraic structures and dynamical systems see Angermüller (1983), Fleiss (1993), Morris and Wang (1981) and Wang (1980), for maps with a sparse Jacobian matrix see Coleman and More (1983) and Ypma (1987), for parallel maps on hypersurfaces see Amur (1986), for Jacobians in differential equations and algebraic geometry see Meisters (1982), for the Jacobian conjecture see Lang and Mandall (1993) and Xavier (1993), for the study of energy levels by diagonalizing a Jacobian matrix see Dean (1956), for specific Jacobians in econometrics see Benhabib and Nishimura (1979) and Nishimura (1981), for those in game theory see Parthasarathy and Ravindran (1986), and those in time series see Anderson and Takemura (1986). Extensively large number of such topics can be listed where Jacobians enter into the picture directly or indirectly. In order to confine our discussion to a general area we will concentrate on the topic of evaluating the Jacobians for some specific transformations which are readily applicable to statistical sciences, physical and engineering sciences, social sciences, and biological and medical sciences.

Exercises

1.2.1 If the $p \times 1$ real random vector X has a unit normal distribution, that is, $X \sim N_p(0, I)$ show that the density is invariant under orthogonal transformations. That is, show that $Y = AX \sim N_p(0, I)$, $AA' = I$ (see Appendix A for the definition of a normal density).

1.2.2 Let X be a $p \times 1$ vector of real variables and A a $p \times p$ symmetric positive definite matrix of constants. Then show that

$$\int_X e^{-X'AX} dX = \pi^{\frac{p}{2}} |A|^{-\frac{1}{2}}.$$

1.2.3 If the $p \times p$ real symmetric positive definite matrix X, $0 < X < I$, has a type-1 beta distribution show that $Y = (I - X)^{-\frac{1}{2}} X (I - X)^{-\frac{1}{2}}$ has a type-2 beta distribution where $(I - X)^{-\frac{1}{2}}$ is the symmetric square root of $(I - X)^{-1}$ (see the definition of a beta density from Appendix A).

1.2.4 Verify Theorems 1.15, 1.16, 1.17 and 1.19 from first principles for

$$X = \begin{bmatrix} x_{11} & 0 & 0 \\ x_{21} & x_{22} & 0 \\ x_{31} & x_{32} & x_{33} \end{bmatrix}, \quad Y = \begin{bmatrix} y_{11} & y_{12} & y_{13} \\ 0 & y_{22} & y_{23} \\ 0 & 0 & y_{33} \end{bmatrix}$$

$$A = \begin{bmatrix} 2 & 0 & 0 \\ 1 & 1 & 0 \\ 1 & 1 & 3 \end{bmatrix}, \quad B = \begin{bmatrix} 2 & 1 & 0 \\ 0 & 1 & 1 \\ 0 & 0 & 2 \end{bmatrix}.$$

1.2.5 Verify Theorem 1.21 from first principles for

$$X = \begin{bmatrix} 0 & x_{12} & x_{13} \\ -x_{12} & 0 & x_{23} \\ -x_{13} & -x_{23} & 0 \end{bmatrix}, \quad A = \begin{bmatrix} 1 & 0 & 1 \\ 2 & 1 & 0 \\ 1 & 2 & 1 \end{bmatrix}.$$

1.2.6 Let X and Y be $p \times q$ matrices of functionally independent real variables. Let A be a $p \times p$ and B a $q \times q$ nonsingular matrices of constants. Consider the transformation

$$Y = AXB = AU, \quad U = XB.$$

The Jacobians are then

$$J(Y : X) = J(Y : U)J(U : X).$$

Note that $J(Y : U)$ is a function of A only, denoted by $f(A)$ and $J(U : X)$ is a function of B. Thus one has

$$f(AB) = f(A)f(B).$$

By solving this functional equation for f one can evaluate the Jacobians. Solve the following functional equations. See also Olkin and Sampson (1972).

(i)   Show that the functional equation

$$f(D_a)f(D_b) = f(D_a D_b) \Rightarrow$$
$$f(D_a) = \prod_{j=1}^{p} |a_j|^{c_j}$$

where D_a is a real $p \times p$ diagonal matrix with the i-th diagonal element a_i, $i = 1, ..., p$, f is a measurable, positive and nonconstant scalar function and c_j's are arbitrary constants.

(ii)   Show that the functional equation

$$f(U_1)f(U_2) = f(U_1 U_2) \Rightarrow$$
$$f(U) = \exp\left\{ \sum_{j=1}^{p-1} c_j u_{j+1,j} \right\}$$

where U_1, U_2 are $p \times p$ arbitrary real lower triangular matrices with unit diagonals, f is a measurable nonzero scalar function, $U = (u_{ij})$ and c_j's are arbitrary constants.

(iii)   Show that the functional equation

$$f(V_1)f(V_2) = f(V_1 V_2) \Rightarrow$$
$$f(V) = \prod_{j=1}^{p} |v_{jj}|^{c_j}$$

where V_1, V_2 are arbitrary $p \times p$ real nonsingular lower triangular matrices, $V = (v_{ij})$, f is a measurable, positive and nonconstant scalar function and c_j's are arbitrary constants.

(iv) Show that the functional equation

$$f(A)f(B) = f(AB) \Rightarrow$$
$$f(A) = |A|^c$$

where A, B are arbitrary $p \times p$ real nonsingular matrices, f is a measurable, positive and nonconstant scalar function and c is an arbitrary constant.

(v) Show that the functional equation

$$f(A) + f(B) = f(A + B) \text{ and } f(AB) = f(BA) \Rightarrow$$
$$f(A) = c \operatorname{tr}(A)$$

where A and B are arbitrary $p \times p$ real matrices, f is a measurable scalar function and c is an arbitrary constant.

1.2.7 By using Exercise 1.2.6((i),(iv)) prove the following results: Theorems 1.13, 1.14, 1.18, 1.20 and 1.21.

1.2.8 By using Exercise 1.2.6 ((ii),(iii)) prove the following results: Theorems 1.15, 1.16, 1.17, 1.19 and 1.22. (Some functional equations involving matrices may also be seen from Olkin (1959) and Rasch (1948)).

1.2.9 Show that

(i)
$$\int_{0<X<I} |X|^{\alpha - \frac{p+1}{2}} |I - X|^{\beta - \frac{p+1}{2}} dX = B_p(\alpha, \beta),$$

and

(ii)
$$\int_{X>0} |X|^{\alpha - \frac{p+1}{2}} |I + X|^{-(\alpha + \beta)} dX = B_p(\alpha, \beta)$$

where $X = X' > 0$, $\operatorname{Re}(\alpha) > \frac{p-1}{2}$, $\operatorname{Re}(\beta) > \frac{p-1}{2}$,

$$B_p(\alpha, \beta) = \frac{\Gamma_p(\alpha)\Gamma_p(\beta)}{\Gamma_p(\alpha + \beta)}$$

and

$$\Gamma_p(\alpha) = \int_{X>0} |X|^{\alpha - \frac{p+1}{2}} e^{-\text{tr}(X)} dX, \ X = X' > 0, \ \text{Re}(\alpha) > \frac{p-1}{2}.$$

1.2.10   If for $X = X' > 0$, $\text{Re}(\alpha) > \frac{p-1}{2}$, $\text{Re}(\beta) > \frac{p-1}{2}$,

$$f(X) = \frac{1}{B_p(\alpha, \beta)} |X|^{\alpha - \frac{p+1}{2}} |I + X|^{-(\alpha+\beta)}$$

find the domain of h for which $\int_{X>0} |X|^h f(X) dX$ exists.

1.3   Nonlinear transformations: one matrix case

Consider a transformation of the type $Y = X^{-1}$ where X is a $p \times p$ matrix and X^{-1} denotes its inverse. This transformation is not linear. Nonlinear transformations can sometimes be handled as linear transformations of differentials and then the results of Section 1.2 can be conveniently applied. We will consider transformations of this type which will be called smooth transformations. First we will deal with some nonlinear vector transformations which are frequently used in the literature. Then we will deal with smooth transformations and other transformations involving function of a function.

1.3.1   Nonlinear vector transformations

Some nonlinear transformations which are frequently used are transformations to general polar coordinates and transformations to elementary symmetric functions. Some of these will be discussed in this section.

Theorem 1.23   *Let $x_1, ..., x_p$ be real scalar variables which are functionally independent and consider the symmetric functions $y_j = x_{(j)}, j = 1, ..., p$ where the notation $x_{(j)}$ stands for the sum of all products taken j at a time. That is, $y_1 = x_1 + ... + x_p, y_2 = x_1 x_2 + x_1 x_3 + ... + x_{p-1} x_p, ..., y_p = x_1 x_2 ... x_p$. Then for $x_j > 0, j = 1, ..., p$*

$$dy_1 ... dy_p = \left\{ \prod_{i=1}^{p-1} \prod_{j=i+1}^{p} |x_i - x_j| \right\} dx_1 ... dx_p.$$

Proof   For $p = 2$ the result is obvious. Take $p = 3$ and the variables in the natural order. Then the matrix of partial derivatives, denoted by $\frac{\partial Y}{\partial X}$ will be the following.

$$\frac{\partial Y}{\partial X} = \begin{bmatrix} 1 & 1 & 1 \\ x_2 + x_3 & x_1 + x_3 & x_1 + x_2 \\ x_2 x_3 & x_1 x_3 & x_1 x_2 \end{bmatrix}.$$

For evaluating the determinant add (-1) times the first column to the second and third columns, then take out $(x_1 - x_2)$ from the second column and $(x_1 - x_3)$ from the third column. Repeat the process once again with the 2×2 submatrix to get the result. The general case follows by induction.

Example 1.17   Suppose that the joint density of two real scalar random variables x_1 and x_2 is given by

$$f(x_1, x_2) = 2(x_2 - x_1)e^{-(x_1+x_2)}$$

for $0 < x_1 < x_2 < \infty$ and $f(x_1, x_2) = 0$ elsewhere. Evaluate the density of $u = x_1 x_2$.

Solution   Change x_1, x_2 to the elementary symmetric functions $u = x_1 x_2$ and $v = x_1 + x_2$. Then from Theorem 1.23 we have

$$(x_2 - x_1)dx_1 dx_2 = du dv.$$

Let the joint density of u and v be denoted by $g(u, v)$. Since the transformation is one-to-one we have

$$g(u, v)du dv = f(x_1, x_2)dx_1 dx_2 \Rightarrow$$
$$g(u, v) = 2e^{-v}, \quad v^2 > 4u.$$

This condition is obtained by solving for x_1 and x_2 in terms of u and v. Note that $v^2 - 4u = (x_2 - x_1)^2 > 0$. Thus the marginal density of u, denoted by $g_u(u)$, is given by

$$g_u(u) = 2 \int_{2\sqrt{u}}^{\infty} e^{-v}dv = 2e^{-2\sqrt{u}}$$

for $0 < u < \infty$ and $g_u(u) = 0$ elsewhere.

Theorem 1.24   Let the functionally independent real variables $x_1, ..., x_p$ be transformed to the general polar coordinates $r, \theta_1, ..., \theta_{p-1}$ as follows, where $r > 0$, $-\frac{\pi}{2} < \theta_j \le \frac{\pi}{2}$, $j = 1, ..., p-2$, $-\pi < \theta_{p-1} \le \pi$.

$$x_1 = r \sin\theta_1,$$
$$x_j = r \cos\theta_1 \cos\theta_2 ... \cos\theta_{j-1} \sin\theta_j, \quad j = 2, 3, ..., p-1,$$
$$x_p = r \cos\theta_1 \cos\theta_2 ... \cos\theta_{p-1}.$$

Then, ignoring the sign,

$$dx_1...dx_p = r^{p-1} \left\{ \prod_{j=1}^{p-1} |\cos\theta_j|^{p-j-1} \right\} dr \, d\theta_1...d\theta_{p-1}.$$

Proof   Take $p = 3$. Then

$$x_1 = r \sin\theta_1, \qquad\qquad x_1^2 + x_2^2 + x_3^3 = r^2,$$
$$x_2 = r \cos\theta_1 \sin\theta_2, \Rightarrow \qquad x_3^2 + x_2^2 = r^2 \cos^2\theta_1,$$
$$x_3 = r \cos\theta_1 \cos\theta_2, \qquad\qquad x_3^2 = r^2 \cos^2\theta_1 \cos^2\theta_2.$$

Take the differentials to get

(i) $2x_1 dx_1 + 2x_2 dx_2 + 2x_3 dx_3 = 2r\ dr;$

(ii) $2x_3 dx_3 + 2x_2 dx_2 = 2r\ dr(\cos^2\theta_1)$
$$- 2r^2 \cos\theta_1 \sin\theta_1 d\theta_1;$$

(iii) $2x_3 dx_3 = 2r\ dr(\cos^2\theta_1 \cos^2\theta_2)$
$$- 2r^2 \cos\theta_1 \sin\theta_1 d\theta_1 \cos^2\theta_2$$
$$- 2r^2 \cos^2\theta_1 \cos\theta_2 \sin\theta_2 d\theta_2.$$

Multiply the left sides and right sides of $(i) - (iii)$ separately deleting higher order powers of the differentials by using (1.2.4) to get

$$x_1 x_2 x_3 dx_1\ dx_2\ dx_3 = r^5 \cos^3\theta_1 \sin\theta_1 \cos\theta_2 \sin\theta_2 dr\ d\theta_1\ d\theta_2.$$

Now divide through by $x_1 x_2 x_3$ to get

$$dx_1\ dx_2\ dx_3 = r^2 \cos\theta_1 dr\ d\theta_1\ d\theta_2.$$

Thus the result for $p = 3$ is established. In a similar manner we can prove for the general p.

Theorem 1.25   *Let $r > 0$, $0 < \theta_j \leq \pi$, $j = 1, ..., p - 2$, $0 < \theta_{p-1} \leq 2\pi$. Then the transformation*

$$x_1 = r \sin\theta_1 \sin\theta_2... \sin\theta_{p-2} \sin\theta_{p-1},$$
$$x_2 = r \sin\theta_1 \sin\theta_2... \sin\theta_{p-2} \cos\theta_{p-1},$$
$$x_3 = r \sin\theta_1 \sin\theta_2... \cos\theta_{p-2},$$

$$\vdots$$

$$x_{p-1} = r \sin\theta_1 \cos\theta_2,$$
$$x_p = r \cos\theta_1,$$

$\Rightarrow$

$$dx_1...dx_p = r^{p-1}\left\{\prod_{j=1}^{p-1} |\sin\theta_j|^{p-j-1}\right\} dr\ d\theta_1...d\theta_{p-1}.$$

Example 1.18   Let x_1, x_2, x_3 be three independently distributed real scalar standard normal random variables. Let

$$u_1 = \frac{x_1^2}{x_1^2 + x_2^2 + x_3^2}, \quad u_2 = \frac{x_2^2}{x_1^2 + x_2^2 + x_3^2} \text{ and } u_3 = x_1^2 + x_2^2 + x_3^2.$$

Show that u_3 is independently distributed of u_1 and u_2.

Solution   Let $f(x_1, x_2, x_3)$ be the joint density of x_1, x_2 and x_3. Then from elementary statistical theory

$$f(x_1, x_2, x_3) = \frac{e^{-\frac{1}{2}(x_1^2 + x_2^2 + x_3^2)}}{(2\pi)^{\frac{3}{2}}}, \quad -\infty < x_j < \infty, \ j = 1, 2, 3.$$

Change to general polar coordinates. Let

$$x_1 = r\sin\theta_1\sin\theta_2, \quad x_2 = r\sin\theta_1\cos\theta_2, \quad x_3 = r\cos\theta_1,$$

$0 < \theta_1 \le \pi, 0 < \theta_2 \le 2\pi$. Then from Theorem 1.25 one has

$$dx_1 dx_2 dx_3 = r^2\sin\theta_1 dr d\theta_1 d\theta_2.$$

Let the joint density of r, θ_1 and θ_2 be denoted by $g(r, \theta_1, \theta_2)$. Since the transformation is one-to-one we have

$$f(x_1, x_2, x_3)dx_1 dx_2 dx_3 = g(r, \theta_1, \theta_2)dr d\theta_1 d\theta_2 \Rightarrow$$

$$g(r, \theta_1, \theta_2) = \frac{e^{-\frac{r^2}{2}}r^2\sin\theta_1}{(2\pi)^{\frac{3}{2}}},$$

$$0 < r < \infty, \ 0 < \theta_1 \le \pi, \ 0 < \theta_2 \le 2\pi.$$

Since the right side factorizes into functions of r, θ_1 and θ_2 we note that r, θ_1 and θ_2 are mutually independently distributed. But

$$u_1 = \sin^2\theta_1\sin^2\theta_2, \quad u_2 = \sin^2\theta_1\cos^2\theta_2 \text{ and } u_3 = r^2.$$

Thus u_1 and u_2 are free of r and hence u_3 is independently distributed of (u_1, u_2). Note that we could have established this result by using Theorem 1.24 as well.

Theorem 1.26   *For the functionally independent real variables $x_j >$ 0, $j = 1, 2, ..., p$, consider the transformation*

$$y_1 = x_1 + ... + x_p$$
$$y_2 = x_1^2 + ... + x_p^2$$
$$\vdots$$
$$y_{p-1} = x_1^{p-1} + ... + x_p^{p-1}$$
$$y_p = x_1...x_p.$$

Then, ignoring the sign,

$$dy_1...dy_p = (p-1)! \left\{ \prod_{i=1}^{p-1} \prod_{j=i+1}^{p} |x_i - x_j| \right\} dx_1...dx_p.$$

This can be proved either by taking the differentials and multiplying or by forming the matrix of partial derivatives by taking the x_j's and y_j's in the natural order.

Remark 1.2   It is easy to note that all the transformations used in Theorems 1.23–1.26 are one-to-one under the conditions stated therein.

1.3.2   Smooth nonlinear transformations

Let $Y = f(X)$ be a matrix function of the real matrix variable X. For example, $Y = X^2$, and $Y = X^{-1}$ when X is nonsingular, are two matrix functions of X. Let the matrices of differentials be denoted by (dX) and (dY). Then for certain nonlinear transformations it can be shown that $J(Y : X) = J((dY) : (dX))$. Such transformations will be called *smooth transformations*. In this case, for computing the Jacobian in Y written as a function of X we need to compute only the Jacobian in (dY) written as a function of (dX) taking the differentials as variables and all other quantities, including the original variables, as constants. Such cases will be considered in this section. For the sake of illustration let us redo the problem in Theorem 1.12. Taking the differentials we have the following result.

$$Y = AX \Rightarrow (dY) = A(dX)$$

where Y and X are $p \times 1$ vectors of variables and A is a $p \times p$ nonsingular constant matrix. That is,

$$
\begin{bmatrix} dy_1 \\ \vdots \\ dy_p \end{bmatrix} = \begin{bmatrix} a_{11} & \cdots & a_{1p} \\ \vdots & \ddots & \vdots \\ a_{p1} & \cdots & a_{pp} \end{bmatrix} \begin{bmatrix} dx_1 \\ \vdots \\ dx_p \end{bmatrix}.
\tag{1.3.1}
$$

Take the exterior product to get

$$
dy_1 ... dy_p = \left\{ \prod_{i=1}^{p} (a_{i1} dx_1 + ... + a_{ip} dx_p) \right\}.
\tag{1.3.2}
$$

Simplify the right side by using (1.2.3) and (1.2.4), that is, by taking the higher powers of differentials as zeros and keeping the order in which the differentials enter into the product. We note that in each term of the product on the right side of (1.3.2) one and only one element comes from each row and column of the matrix A and the sign depends on the number of permutations (one interchange of two differentials is one permutation) needed to bring the differentials to the natural order $dx_1, ..., dx_p$. This, by definition, is the determinant of A and thus

$$
dy_1 ... dy_p = |A| dx_1 ... dx_p.
$$

Hence by taking $(d\mathbf{X})$ and $(d\mathbf{Y})$ as new variables the equation $(d\mathbf{Y}) = A(d\mathbf{X})$ gives the same Jacobian as the equation $Y = AX$.

In order to see the results in the more complicated cases we need the concept of a *Kronecker product*.

Definition 1.1   Kronecker product   Let $A = (a_{ij})$ be a $p \times q$ matrix and $B = (b_{ij})$ be a $m \times n$ matrix. Then the *Kronecker product*, denoted by $\otimes$, is a $pm \times qn$ matrix formed as follows.

$$
A \otimes B = \begin{bmatrix} a_{11}B & a_{12}B & \cdots & a_{1q}B \\ a_{21}B & a_{22}B & \cdots & a_{2q}B \\ \vdots & \vdots & \ddots & \vdots \\ a_{p1}B & a_{p2}B & \cdots & a_{pq}B \end{bmatrix}
\tag{1.3.3}
$$

and

$$
B \otimes A = \begin{bmatrix} b_{11}A & b_{12}A & \cdots & b_{1n}A \\ b_{21}A & b_{22}A & \cdots & b_{2n}A \\ \vdots & \vdots & \ddots & \vdots \\ b_{m1}A & b_{m2}A & \cdots & b_{mn}A \end{bmatrix}.
$$

(This concept of Kronecker product is due to the German mathematician L. Kronecker (1823-1891)). Note that in general $A \otimes B \neq B \otimes A$. Many properties are enjoyed by the Kronecker product of two matrices. Some of these may be seen, for example, from Rogers (1980), Magnus and Neudecker (1988) and Olkin and Sampson (1972). The ones that we need for our discussions will be given as lemmas. Before we give these results we need one more concept which will be stated after an illustrative example.

Example 1.19   Write the Kronecker products $A \otimes B$ and $B \otimes A$ for

$$A = \begin{bmatrix} 1 & -2 & -1 \\ 0 & 3 & 2 \end{bmatrix}, \quad B = \begin{bmatrix} 1 & -1 \\ 2 & 0 \end{bmatrix}.$$

Solution

$$A \otimes B = (a_{ij}B) = \begin{bmatrix} (1)B & (-2)B & (-1)B \\ (0)B & (3)B & (2)B \end{bmatrix}$$

$$= \begin{bmatrix} 1 & -1 & -2 & 2 & -1 & 1 \\ 2 & 0 & -4 & 0 & -2 & 0 \\ 0 & 0 & 3 & -3 & 2 & -2 \\ 0 & 0 & 6 & 0 & 4 & 0 \end{bmatrix}.$$

$$B \otimes A = (b_{ij}A) = \begin{bmatrix} (1)A & (-1)A \\ (2)A & (0)A \end{bmatrix}$$

$$= \begin{bmatrix} 1 & -2 & -1 & -1 & 2 & 1 \\ 0 & 3 & 2 & 0 & -3 & -2 \\ 2 & -4 & -2 & 0 & 0 & 0 \\ 0 & 6 & 4 & 0 & 0 & 0 \end{bmatrix}.$$

Definition 1.2   vec(X)   Let $X = (x_{ij})$ be a $p \times q$ matrix. Let the j-th column of X be denoted by $x_{(j)}$. Consider a $pq \times 1$ vector formed by appending $x_{(1)}, x_{(2)}, ..., x_{(q)}$ and forming a long string. This vector will be denoted by vec(X). That is,

$$\text{vec}(X) = \begin{bmatrix} x_{(1)} \\ x_{(2)} \\ \vdots \\ x_{(q)} \end{bmatrix}. \qquad (1.3.4)$$

Example 1.20   Let $A = \begin{bmatrix} 1 & -1 & 2 \\ 0 & 4 & 5 \end{bmatrix}$. Evaluate the transpose of vec(A).

Solution $\text{vec}(A)$ is formed by appending the columns one after the other. Hence the transpose is available by writing the transposes of the column vectors in a row. That is,

$$[\text{vec}(A)]' = (1, 0, -1, 4, 2, 5).$$

Lemma 1.1   *If A is $p \times q$, X is $q \times r$ and B is $r \times s$ then the $ps \times 1$ vector*

$$\text{vec}(AXB) = (B' \otimes A)\text{vec}(X) \tag{1.3.5}$$

where B' is the transpose of B.

Proof   Let the r columns of X be denoted by $x_{(1)}, x_{(2)}, ..., x_{(r)}$. Then

$$\begin{aligned}(AXB) &= A\begin{bmatrix} x_{(1)} & x_{(2)} & \cdots & x_{(r)} \end{bmatrix} B \\ &= \begin{bmatrix} Ax_{(1)} & Ax_{(2)} & \cdots & Ax_{(r)} \end{bmatrix} B. \end{aligned} \tag{1.3.6}$$

Note that $Ax_{(j)}, j = 1, ..., r$ are $p \times 1$ vectors. Let the j-th row of B be denoted by $b^{(j)} = (b_{j1}, b_{j2}, ..., b_{js})$ which is a $1 \times s$ vector for $j = 1, ..., r$. Now

$$AXB = Ax_{(1)}b^{(1)} + Ax_{(2)}b^{(2)} + ... + Ax_{(r)}b^{(r)}.$$

Let us look at the first column of AXB. These are formed by the first elements in $b^{(1)}, ..., b^{(r)}$ and $Ax_{(1)}, ..., Ax_{(r)}$. That is, $b_{11}Ax_{(1)} + b_{21}Ax_{(2)} + ... + b_{r1}Ax_{(r)}$. Denoting the j-th column of AXB by $(AXB)_{(j)}$ we have

$$\begin{aligned}\text{vec}(AXB) &= \begin{bmatrix} (AXB)_{(1)} \\ \vdots \\ (AXB)_{(s)} \end{bmatrix} \\ &= \begin{bmatrix} b_{11}Ax_{(1)} + b_{21}Ax_{(2)} + ... + b_{r1}Ax_{(r)} \\ \vdots \\ b_{1s}Ax_{(1)} + b_{2s}Ax_{(2)} + ... + b_{rs}Ax_{(r)} \end{bmatrix} \\ &= \begin{bmatrix} b_{11}A & b_{21}A & \cdots & b_{r1}A \\ \vdots & \vdots & \ddots & \vdots \\ b_{1s}A & b_{2s}A & \cdots & b_{rs}A \end{bmatrix} \begin{bmatrix} x_{(1)} \\ x_{(2)} \\ \vdots \\ x_{(r)} \end{bmatrix} \\ &= (B' \otimes A)\text{vec}(X). \end{aligned} \tag{1.3.7}$$

Example 1.21   Verify Lemma 1.1 for

$$A = (1, 2), \quad X = \begin{bmatrix} x_{11} & x_{12} & x_{13} \\ x_{21} & x_{22} & x_{23} \end{bmatrix}, \quad B = \begin{bmatrix} 1 & -1 \\ 0 & 2 \\ -1 & 4 \end{bmatrix}.$$

Solution

$$AXB = (1, 2) \begin{bmatrix} x_{11} & x_{12} & x_{13} \\ x_{21} & x_{22} & x_{23} \end{bmatrix} \begin{bmatrix} 1 & -1 \\ 0 & 2 \\ -1 & 4 \end{bmatrix}$$

$$= (x_{11} + 2x_{21}, x_{12} + 2x_{22}, x_{13} + 2x_{23}) \begin{bmatrix} 1 & -1 \\ 0 & 2 \\ -1 & 4 \end{bmatrix}$$

$$= (\alpha, \beta)$$

where

$$\alpha = x_{11} + 2x_{21} - x_{13} - 2x_{23}$$

and

$$\beta = -x_{11} - 2x_{21} + 2x_{12} + 4x_{22} + 4x_{13} + 8x_{23}.$$

$$\text{vec}(AXB) = \begin{bmatrix} \alpha \\ \beta \end{bmatrix}.$$

But

$$B' \otimes A = \begin{bmatrix} 1 & 0 & -1 \\ -1 & 2 & 4 \end{bmatrix} \otimes (1, 2)$$

$$= \begin{bmatrix} 1 & 2 & 0 & 0 & -1 & -2 \\ -1 & -2 & 2 & 4 & 4 & 8 \end{bmatrix}.$$

$$[\text{vec}(X)]' = (x_{11}, x_{21}, x_{12}, x_{22}, x_{13}, x_{23})$$

and

$$[B' \otimes A]\text{vec}(X) = \begin{bmatrix} x_{11} + 2x_{21} - x_{13} - 2x_{23} \\ -x_{11} - 2x_{21} + 2x_{12} + 4x_{22} + 4x_{13} + 8x_{23} \end{bmatrix}$$

$$= \begin{bmatrix} \alpha \\ \beta \end{bmatrix}.$$

This verifies Lemma 1.1.

Now let us try to redo (1.2.11). Let $Y = AX$ where Y and X are $p \times q$ matrices of functionally independent real variables and A is a $p \times p$ nonsingular matrix of constants. Then by taking differentials we get

$$Y = AX \Rightarrow (dY) = A(dX) \Rightarrow$$
$$\text{vec}(dY) = (I \otimes A)\text{vec}(dX) \tag{1.3.8}$$

from Lemma 1.1. Now by using (1.3.1) and (1.3.2) we have

$$dY = |I \otimes A| dX.$$

Note that I in this case is $q \times q$ and

$$I \otimes A = \begin{bmatrix} A & 0 & \cdots & 0 \\ 0 & A & \cdots & 0 \\ \vdots & \vdots & \ddots & \vdots \\ 0 & 0 & \cdots & A \end{bmatrix}$$

$$\Rightarrow |I \otimes A| = |A|^q$$

and hence the result.

If $Y = AXB$ where X and Y are $p \times q$ matrices of real variables, A is a $p \times p$ and B is a $q \times q$ nonsingular matrices of constants then we can arrive at the same result in Theorem 1.18 by using the above procedures and the following lemma.

Lemma 1.2   *If A is a $p \times p$ and B is a $q \times q$ matrices then*

$$|A \otimes B| = |A|^q |B|^p.$$

Proof   By definition

$$A \otimes B = \begin{bmatrix} a_{11}B & \cdots & a_{1p}B \\ \vdots & \ddots & \vdots \\ a_{p1}B & \cdots & a_{pp}B \end{bmatrix}$$

$$= \begin{bmatrix} a_{11}I & \cdots & a_{1p}I \\ \vdots & \ddots & \vdots \\ a_{p1}I & \cdots & a_{pp}I \end{bmatrix} \begin{bmatrix} B & 0 & \cdots & 0 \\ 0 & B & \cdots & 0 \\ \vdots & \vdots & \ddots & \vdots \\ 0 & 0 & \cdots & B \end{bmatrix}$$

where I is $q \times q$ and in the last matrix there are p diagonal blocks consisting of the matrix B each. Note that by adding suitable multiples of rows and columns to other rows and columns, the determinant is unaltered, the first matrix above can be brought to the following form.

$$\begin{bmatrix} a_{11}I & \cdots & a_{1p}I \\ \vdots & \ddots & \vdots \\ a_{p1}I & \cdots & a_{pp}I \end{bmatrix} \rightarrow \begin{bmatrix} A & 0 & \cdots & 0 \\ 0 & A & \cdots & 0 \\ \vdots & \vdots & \ddots & \vdots \\ 0 & 0 & \cdots & A \end{bmatrix}$$

and there are q such diagonal blocks of A. Hence,

$$|A \otimes B| = \begin{vmatrix} A & 0 & \cdots & 0 \\ 0 & A & \cdots & 0 \\ \vdots & \vdots & \ddots & \vdots \\ 0 & 0 & \cdots & A \end{vmatrix} \begin{vmatrix} B & 0 & \cdots & 0 \\ 0 & B & \cdots & 0 \\ \vdots & \vdots & \ddots & \vdots \\ 0 & 0 & \cdots & B \end{vmatrix} = |A|^q |B|^p.$$

Example 1.22   Verify Lemma 1.2 for

$$A = \begin{bmatrix} 1 & 1 \\ 1 & 2 \end{bmatrix} \text{ and } B = \begin{bmatrix} 1 & 1 & 1 \\ 1 & -1 & 1 \\ 1 & 2 & 0 \end{bmatrix}.$$

Solution

$$A \otimes B = \begin{bmatrix} 1 & 1 & 1 & 1 & 1 & 1 \\ 1 & -1 & 1 & 1 & -1 & 1 \\ 1 & 2 & 0 & 1 & 2 & 0 \\ 1 & 1 & 1 & 2 & 2 & 2 \\ 1 & -1 & 1 & 2 & -2 & 2 \\ 1 & 2 & 0 & 2 & 4 & 0 \end{bmatrix}.$$

Evaluating the determinant by adding suitable combinations of rows to other rows one has

$$|A \otimes B| = \begin{vmatrix} 1 & 1 & 1 & 1 & 1 & 1 \\ 0 & -2 & 0 & 0 & -2 & 0 \\ 0 & 1 & -1 & 0 & 1 & -1 \\ 0 & 0 & 0 & 1 & 1 & 1 \\ 0 & -2 & 0 & 1 & -3 & 1 \\ 0 & 1 & -1 & 1 & 3 & -1 \end{vmatrix}$$

$$= -2 \begin{vmatrix} 1 & 1 & 1 & 1 & 1 & 1 \\ 0 & 1 & 0 & 0 & 1 & 0 \\ 0 & 0 & -1 & 0 & 0 & -1 \\ 0 & 0 & 0 & 1 & 1 & 1 \\ 0 & 0 & 0 & 0 & -2 & 0 \\ 0 & 0 & 0 & 0 & 0 & -2 \end{vmatrix} = 8.$$

$$|A| = 1, \ |B| = 2, \Rightarrow |A|^2 |B|^3 = 8.$$

Hence the result is verified.

The following theorems will be established by using these ideas and by treating the differentials as new variables and everything else fixed.

Theorem 1.27   *Let X be a $p \times p$ nonsingular matrix of functionally independent real variables. Then, ignoring the sign,*

$$Y = X^{-1} \Rightarrow dY = |X|^{-2p}dX \text{ for a general } X,$$
$$= |X|^{-(p+1)}dX \quad \text{for } X = X',$$
$$= |X|^{-(p-1)}dX \quad \text{for } X = -X',$$
$$= |X|^{-(p+1)}dX \quad \text{for } X \text{ lower} \qquad (1.3.9)$$

or upper triangular.

Proof   Since $XX^{-1} = I$ where I denotes an identity matrix from Theorem 1.10 one has

$$XX^{-1} = XY = I \Rightarrow (dX)Y + X(dY) = 0$$
$$\Rightarrow (dY) = -X^{-1}(dX)X^{-1}. \qquad (1.3.10)$$

Now treat (dX) and (dY) as variables and X fixed since X is free of (dX) or (dY) and apply (1.2.21) and Theorems 1.20, 1.21, 1.15 and 1.16 to establish the result. Note that when X is skew symmetric $-X^{-1} = (X^{-1})'$.

Remark 1.3   Note that the transformation $Y = X^{-1}$ is one-to-one when X is nonsingular.

Example 1.23   Verify Theorem 1.27 from first principles for

(i) $X_1 = \begin{bmatrix} x & y \\ u & v \end{bmatrix}$,   (ii) $X_2 = \begin{bmatrix} x & y \\ y & z \end{bmatrix}$ and (iii) $X_3 = \begin{bmatrix} 0 & y \\ -y & 0 \end{bmatrix}$.

Solution

$$|X_1| = xv - yu$$

and let

$$U = (u_{ij}) = X_1^{-1} = \frac{1}{xv - yu} \begin{bmatrix} v & -y \\ -u & x \end{bmatrix} \Rightarrow$$

$$u_{11} = \frac{v}{xv - yu}, \quad u_{12} = -\frac{y}{xv - yu},$$

$$u_{21} = -\frac{u}{xv - yu}, \quad u_{22} = \frac{x}{xv - yu} \Rightarrow$$

$$du_{11} = \frac{\partial u_{11}}{\partial x}dx + \frac{\partial u_{11}}{\partial y}dy + \frac{\partial u_{11}}{\partial u}du + \frac{\partial u_{11}}{\partial v}dv$$

$$= \frac{1}{|X_1|^2}\left[-v^2dx + uvdy + vydu - yudv\right],$$

$$du_{21} = \frac{1}{|X_1|^2}\left[uvdx - u^2dy - xvdu + uxdv\right],$$

$$du_{12} = \frac{1}{|X_1|^2}\left[yvdx - xvdy - y^2du + xydv\right],$$

$$du_{22} = \frac{1}{|X_1|^2}\left[-yudx + uxdy + xydu - x^2dv\right].$$

Hence the product, after neglecting higher powers of differentials, is given by

$$du_{11}du_{12}du_{21}du_{22} = \frac{1}{|X_1|^8}(xv - uy)^4 dx_{11}dx_{12}dx_{21}dx_{22}$$

$$= |X_1|^{-4}dx_{11}dx_{12}dx_{21}dx_{22}.$$

This verifies the result for X_1.

$$|X_2| = xz - y^2.$$

Let

$$V = (v_{ij}) = X_2^{-1} = \frac{1}{xz - y^2}\begin{bmatrix} z & -y \\ -y & x \end{bmatrix} \Rightarrow$$

$$dv_{11} = \frac{1}{|X_2|^2}\left[-z^2dx + 2yzdy - y^2dz\right],$$

$$dv_{12} = \frac{1}{|X_2|^2}\left[yzdx - (xz + y^2)dy + yxdz\right],$$

$$dv_{22} = \frac{1}{|X_2|^2}\left[-y^2dx + 2yxdy - x^2dz\right].$$

Hence

$$dv_{11}dv_{12}dv_{22} = \frac{1}{|X_2|^6}\begin{vmatrix} -z^2 & 2yz & -y^2 \\ yz & -(xz + y^2) & yx \\ -y^2 & 2yx & -x^2 \end{vmatrix}dxdydz$$

$$= \frac{(y^2 - xz)^3}{|X_2|^6} = -\frac{1}{|X_2|^3}dxdydz.$$

Thus the result is verified for X_2.

$$|X_3| = y^2.$$

Let

$$W = (w_{ij}) = X_3^{-1} = \frac{1}{y^2} \begin{bmatrix} 0 & -y \\ y & 0 \end{bmatrix}.$$

$$W = \begin{bmatrix} 0 & w_{12} \\ -w_{12} & 0 \end{bmatrix}.$$

$$dw_{12} = -\frac{1}{y^2} dy = -|X_3|^{-1} dy.$$

Thus the result is verified for X_3.

Theorem 1.28   Let X be a $p \times p$ symmetric positive definite matrix of functionally independent real variables and $T = (t_{ij})$ a real lower triangular matrix with $t_{jj} > 0$, $j = 1, ..., p$, and t_{ij}, $i \geq j$ functionally independent. Then

$$X = TT' \Rightarrow dX = 2^p \left\{ \prod_{j=1}^p t_{jj}^{p+1-j} \right\} dT, \qquad (1.3.11)$$

and

$$X = T'T \Rightarrow dX = 2^p \left\{ \prod_{j=1}^p t_{jj}^{j} \right\} dT. \qquad (1.3.12)$$

Proof   By considering the matrix of differentials one has

$$X = TT' \Rightarrow (dX) = (dT)T' + T(dT').$$

Now treat this as a linear transformation in the differentials, that is, (dX) and (dT) as variables and T a constant. Apply the results in equations (1.2.27) and (1.2.26) to complete the proof.

These results can also be proved by evaluating the matrices of partial derivatives directly, treating these as transformations of the $p(p+1)/2$ variables in T to the $p(p+1)/2$ variables in X.

Note that the upper triangular case is available from (1.3.12) and thus a separate theorem is not required to cover this case.

Remark 1.4   The transformations $X = TT'$ and $X = T'T$ are one-to-one when $t_{jj} > 0$, $j = 1, ..., p$ and X is positive definite.

Example 1.24   Let X be a real $p \times p$ symmetric positive definite matrix and $\text{Re}(\alpha) > \frac{p-1}{2}$. Show that

$$\Gamma_p(\alpha) = \int_{X>0} |X|^{\alpha - \frac{p+1}{2}} e^{-\text{tr}(X)} dX$$

$$= \pi^{\frac{p(p-1)}{4}} \Gamma(\alpha) \Gamma\left(\alpha - \frac{1}{2}\right) ... \Gamma\left(\alpha - \frac{p-1}{2}\right).$$

Solution Let T be a real lower triangular matrix with positive diagonal elements. Then the unique representation

$$X = TT' \Rightarrow dX = 2^p \left\{ \prod_{j=1}^{p} t_{jj}^{p+1-j} \right\} dT.$$

Note that

$$\text{tr}(X) = \text{tr}(TT')$$
$$= t_{11}^2 + \left(t_{21}^2 + t_{22}^2\right) + ... + \left(t_{p1}^2 + ... + t_{pp}^2\right)$$

and

$$|X| = |TT'| = |T| \, |T'| = \prod_{j=1}^{p} t_{jj}^2.$$

When $X > 0$ we have $TT' > 0$, but $t_{jj} > 0$, $j = 1, ..., p$ which means that $-\infty < t_{ij} < \infty$, $i > j$, $0 < t_{jj} < \infty$, $j = 1, ..., p$. The integral splits into p integrals on t_{jj}'s and $p(p-1)/2$ integrals on t_{ij}'s, $i > j$. That is,

$$\Gamma_p(\alpha) = \left\{ \prod_{j=1}^{p} 2 \int_0^{\infty} \left(t_{jj}^2\right)^{\alpha - \frac{p+1}{2}} t_{jj}^{p+1-j} e^{-t_{jj}^2} dt_{jj} \right\}$$
$$\times \left\{ \prod_{i>j} \int_{-\infty}^{\infty} e^{-t_{ij}^2} dt_{ij} \right\}.$$

But

$$2 \int_0^{\infty} \left(t_{jj}^2\right)^{\alpha - \frac{j}{2}} e^{-t_{jj}^2} dt_{jj} = \Gamma\left(\alpha - \frac{j-1}{2}\right),$$

for $\text{Re}(\alpha) > \frac{j-1}{2}$, $j = 1, ..., p$ and

$$\int_{-\infty}^{\infty} e^{-t_{ij}^2} dt_{ij} = \sqrt{\pi}.$$

Multiplying them together the result follows. Note that $\text{Re}(\alpha) > \frac{j-1}{2}$, $j = 1, ..., p \Rightarrow \text{Re}(\alpha) > \frac{p-1}{2}$.

Theorem 1.29 Let X be a $p \times p$ symmetric positive definite matrix of functionally independent real variables with $x_{jj} = 1$, $j = 1, ..., p$. Let

$T = (t_{ij})$ be a lower triangular matrix of functionally independent real variables with $t_{jj} > 0$, $j = 1, ..., p$. Then

$$X = TT', \text{ with } \sum_{j=1}^{i} t_{ij}^2 = 1, \ i = 1, ..., p$$

$$\Rightarrow dX = \left\{ \prod_{j=2}^{p} t_{jj}^{p-j} \right\} dT, \qquad (1.3.13)$$

and

$$X = T'T, \text{ with } \sum_{i=j}^{p} t_{ij}^2 = 1, \ j = 1, ..., p$$

$$\Rightarrow dX = \left\{ \prod_{j=1}^{p-1} t_{jj}^{j-1} \right\} dT. \qquad (1.3.14)$$

Proof   Since X is symmetric with $x_{jj} = 1$, $j = 1, ..., p$ there are only $p(p-1)/2$ variables in X. When $X = TT'$ take the x_{ij}'s in the order $x_{21}, ..., x_{p1}, x_{32}, ..., x_{p2}, ..., x_{pp-1}$ and the t_{ij}'s also in the same order and form the matrix of partial derivatives. Then we get a triangular formulation with diagonal elements having t_{11} repeated $p-1$ times, t_{22} repeated $p-2$ times and so on. Note that $t_{11} = 1$. The determinant establishes (1.3.13). The steps are parallel for proving (1.3.14) and in this case $t_{pp} = 1$.

Note that (1.3.14) covers the upper triangular case and hence there is no need for a separate theorem. Also note that when $t_{jj} > 0$, $j = 1, ..., p$ the transformations $X = TT'$ and $X = T'T$ are one-to-one when X is positive definite.

Example 1.25   Let $R = (r_{ij})$ be a $p \times p$ real symmetric positive definite matrix such that $r_{jj} = 1$, $j = 1, ..., p$, $-1 < r_{ij} = r_{ji} < 1$, $i \neq j$. (This is known as the correlation matrix in statistical theory). Then show that

$$f(R) = \frac{[\Gamma(\alpha)]^p}{\Gamma_p(\alpha)} |R|^{\alpha - \frac{p+1}{2}}$$

for $\text{Re}(\alpha) > \frac{p-1}{2}$ is a density function.

Solution   Since R is positive definite $f(R) \geq 0$ for all R. It remains to show that the total integral is unity. Let T be a lower triangular matrix as described in Theorem 1.29 such that $R = TT'$. Then

$$dR = \left\{ \prod_{j=2}^{p-1} t_{jj}^{p-j} \right\} dT$$

and

$$|R| = \prod_{j=2}^{p} t_{jj}^2.$$

Note that $t_{11}^2 = 1$. Now

$$\int_R |R|^{\alpha - \frac{p+1}{2}} dR = \int_T \left\{ \prod_{j=2}^{p} t_{jj}^2 \right\}^{\alpha - \frac{p+1}{2}} \left\{ \prod_{j=2}^{p-1} t_{jj}^{p-j} \right\} dT$$

$$= \int_T \left\{ \prod_{j=2}^{p} (t_{jj}^2)^{\alpha - \frac{j+1}{2}} \right\} dT.$$

Since there are only $p(p-1)/2$ free variables in T we may take them as t_{ij}'s, $i > j$. Then we may write

$$t_{jj}^2 = 1 - t_{j1}^2 - t_{j2}^2 - \ldots - t_{jj-1}^2$$

where $-1 < t_{ij} < 1$, $i > j$. Let the total integral be denoted by A. Then

$$A = \int_R |R|^{\alpha - \frac{p+1}{2}} dR = \prod_{i>j} \Delta_j$$

where

$$\Delta_j = \int_{w_j} \left(1 - t_{j1}^2 - \ldots - t_{jj-1}^2 \right)^{\alpha - \frac{j+1}{2}} dt_{j1} \ldots dt_{jj-1}$$

where $w_j = \{ t_{jk}, -1 < t_{jk} < 1, \ k = 1, \ldots, j-1, \ \sum_{k=1}^{j-1} t_{jk}^2 < 1 \}$

$$= 2^{j-1} \int_{w_j^*} \left(1 - t_{j1}^2 - \ldots - t_{jj-1}^2 \right)^{\alpha - \frac{j+1}{2}} dt_{j1} \ldots dt_{jj-1}$$

where $w_j^* = \{ t_{jk}, \ 0 < t_{jk} < 1, \ k = 1, \ldots, j-1, \ \sum_{k=1}^{j-1} t_{jk}^2 < 1 \}$. For $0 < t_{jk} < 1$ let

$$u_{jk} = t_{jk}^2 \Rightarrow \frac{1}{2} u_{jk}^{-\frac{1}{2}} du_{jk} = dt_{jk}$$

and hence

$$\Delta_j = \int_{w_j} \left\{ \prod_{k=1}^{j-1} u_{jk}^{\frac{1}{2}-1} \right\}$$

$$\times (1 - u_{j1} - \dots - u_{jj-1})^{\alpha - \frac{i-1}{2} - 1} du_{j1} \dots du_{jj-1}$$

$$= \frac{[\Gamma(\frac{1}{2})]^{j-1} \Gamma(\alpha - \frac{i-1}{2})}{\Gamma(\alpha)}.$$

The multiple integral in Δ_j is evaluated by using type-1 Dirichlet integral, see Definition B1 of Appendix B for details. Multiplying together we get

$$A = \frac{[\Gamma(\frac{1}{2})]^{\frac{p(p-1)}{2}} \Gamma(\alpha - \frac{1}{2}) \Gamma(\alpha - 1) \dots \Gamma(\alpha - \frac{p-1}{2})}{[\Gamma(\alpha)]^{p-1}}$$

$$= \frac{\Gamma_p(\alpha)}{[\Gamma(\alpha)]^p}$$

which establishes the result.

1.3.3   Function of a function

In (1.2.21) and (1.2.24) we considered a transformation of the type $Y = AXA'$. What will be the Jacobian if X is replaced by X^{-1} for $|X| \neq 0$? This can be evaluated in a two-step process.

Theorem 1.30   *Let* X, A, B *be* $p \times p$ *nonsingular matrices where* A *and* B *are constant matrices and* X *is a matrix of functionally independent real variables. Then, ignoring the sign,*

$$Y = AX^{-1}B \Rightarrow dY = |AB|^p |X|^{-2p} dX \quad \text{for a general } X,$$
$$= |AX^{-1}|^{p+1} dX \quad \text{for} \quad X = X', \ B = A',$$
$$= |AX^{-1}|^{p-1} dX \quad \text{for} \quad X = -X', \ B = A'.$$

Proof   Let $U = X^{-1}$. From Theorem 1.18 we have

$$Y = AUB \Rightarrow dY = |AB|^p dU \text{ for a general } U,$$
$$= |A|^{p+1} dU \text{ for } A = B', \ U = U',$$
$$= |A|^{p-1} dU \text{ for } A = B', \ U = -U'.$$

From Theorem 1.27 we have

$$U = X^{-1} \Rightarrow dU = |X|^{-2p}dX \quad \text{for a general } X,$$
$$= |X|^{-(p+1)}dX \quad \text{for } X = X',$$
$$= |X|^{-(p-1)}dX \quad \text{for } X = -X'.$$

Now the result follows from these by using the fact

$$J(Y : X) = J(Y : U)J(U : X).$$

Theorem 1.31 *Let X and A be $p \times p$ matrices where A is a nonsingular constant matrix and X is a matrix of functionally independent real variables such that $A + X$ is nonsingular. Then, ignoring the sign,*

$$Y = (A + X)^{-1}(A - X) \text{ or } (A - X)(A + X)^{-1} \Rightarrow$$
$$dY = 2^{p^2}|A|^p|A + X|^{-2p}dX \quad \text{for a general } X,$$
$$= 2^{\frac{p(p+1)}{2}}|I + X|^{-(p+1)}dX \quad \text{for } A = I, \ X = X'.$$

Proof Write

$$Y = (A + X)^{-1}(A - X) = (A + X)^{-1}(2A - (A + X))$$
$$= 2(A + X)^{-1}A - I.$$

Note that the transformation is one-to-one. Put

$$U = (A + X)^{-1} \Rightarrow$$
$$dY = d(2UA - I) = d(2UA) = |2A|^p dU$$
$$= 2^{\frac{p(p+1)}{2}}dU \quad \text{for } A = I, \ X = X'$$

by Theorem 1.14. But

$$dU = |A + X|^{-2p}dX \quad \text{for a general } A + X,$$
$$= |A + X|^{-(p+1)}dX \quad \text{for } A + X = (A + X)'$$

from Theorem 1.27. Hence the result.

Example 1.26 Let the real symmetric positive definite $p \times p$ matrix X have a type-2 beta density given by

$$f(X) = \begin{cases} \frac{1}{B_p(\alpha,\beta)}|X|^{\alpha - \frac{p+1}{2}}|I + X|^{-(\alpha+\beta)}, & X > 0 \\ 0 \text{ elsewhere,} \end{cases}$$

where

$$B_p(\alpha, \beta) = \frac{\Gamma_p(\alpha)\Gamma_p(\beta)}{\Gamma_p(\alpha + \beta)}, \quad \text{Re}(\alpha) > \frac{p-1}{2}, \quad \text{Re}(\beta) > \frac{p-1}{2}.$$

Evaluate the density of $Y = 2(I + X)^{-1} - I$.

Solution   The Jacobian in this transformation is available from Theorem 1.31.

$$Y = 2(I + X)^{-1} - I \Rightarrow I + X = 2(I + Y)^{-1}$$
$$\Rightarrow X = (I + Y)^{-1}(I - Y), \quad -I < Y < I.$$

Substituting these one has

$$f(X)dX = \frac{2^{-p(\alpha+\beta)+\frac{p(p+1)}{2}}}{B_p(\alpha, \beta)} |I - Y|^{\alpha - \frac{p+1}{2}} |I + Y|^{\beta - \frac{p+1}{2}} dY.$$

Hence the density of Y, denoted by $g(Y)$, is given by

$$g(Y) = \begin{cases} \frac{2^{-p(\alpha+\beta)+\frac{p(p+1)}{2}}}{B_p(\alpha, \beta)} |I - Y|^{\alpha - \frac{p+1}{2}} |I + Y|^{\beta - \frac{p+1}{2}}, & -I < Y < I \\ 0, & \text{elsewhere.} \end{cases}$$

What happens if A is a lower triangular matrix?

Theorem 1.32   *Let A and X be real lower triangular matrices where A is a constant matrix and X is composed of functionally independent real variables such that A and $A + X$ are nonsingular. Then, ignoring the sign,*

$$Y = (A + X)^{-1}(A - X) \Rightarrow$$
$$dY = 2^{\frac{p(p+1)}{2}} |A + X|_+^{-(p+1)} \left\{ \prod_{j=1}^{p} |a_{jj}|^{p-j+1} \right\} dX, \qquad (1.3.15)$$

and

$$Y = (A - X)(A + X)^{-1} \Rightarrow$$
$$dY = 2^{\frac{p(p+1)}{2}} |A + X|_+^{-(p+1)} \left\{ \prod_{j=1}^{p} |a_{jj}|^{j} \right\} dX \qquad (1.3.16)$$

where $|\cdot|_+$ indicates that the absolute value is taken.

Proof   Note that the transformation is one-to-one. From Theorem 1.31 we have seen that

$$Y = (A + X)^{-1}(A - X) = 2(A + X)^{-1}A - I.$$

Taking the differentials by using (1.3.10) we have

$$(dY) = -2(A + X)^{-1}(dX)(A + X)^{-1}A.$$

Now treat (dY) and (dX) as variables and all other quantities as constants. This will give the same Jacobian as for the original transformation of X to Y. Let

$$B = -2(A + X)^{-1}, \ C = (A + X)^{-1}A, \ (dU) = B(dX).$$

Then the equation of differentials is now $(dY) = (dU)C, \ C = (c_{ij})$. From Theorem 1.15, ignoring the sign,

$$dY = \left\{ \prod_{j=1}^{p} |c_{jj}|^{p-j+1} \right\} dU = \left\{ \prod_{j=1}^{p} |a_{jj} + x_{jj}|^{-(p-j+1)} |a_{jj}|^{p-j+1} \right\} dU$$

and

$$dU = \left\{ \prod_{j=1}^{p} |b_{jj}|^j \right\} dX = \left\{ \prod_{j=1}^{p} (-2)^j |a_{jj} + x_{jj}|^{-j} \right\} dX.$$

Substitute for dU, simplify and take the absolute value of (-2) to get

$$dY = 2^{\frac{p(p+1)}{2}} |A + X|_+^{-(p+1)} \left\{ \prod_{j=1}^{p} |a_{jj}|^{p-j+1} \right\} dX.$$

The proof for the second part is parallel to that of the first part and hence omitted.

Theorem 1.33   *Let A and X be real upper triangular matrices where A is a constant matrix and X is composed of functionally independent real variables such that A and $A + X$ are nonsingular. Then, ignoring the sign,*

$$Y = (A + X)^{-1}(A - X) \Rightarrow$$

$$dY = 2^{\frac{p(p+1)}{2}} |A + X|_+^{-(p+1)} \left\{ \prod_{j=1}^{p} |a_{jj}|^j \right\} dX, \qquad (1.3.17)$$

and

$$Y = (A - X)(A + X)^{-1} \Rightarrow$$

$$dY = 2^{\frac{p(p+1)}{2}} |A + X|_+^{-(p+1)} \left\{ \prod_{j=1}^{p} |a_{jj}|^{p+1-j} \right\} dX \qquad (1.3.18)$$

where $|\cdot|_+$ *indicates that the absolute value is taken*

The proof is parallel to that of Theorem 1.32 and hence omitted.

Example 1.27 Verify Theorem 1.33 for

$$A = \begin{bmatrix} 1 & 0 & 0 \\ 0 & 1 & 2 \\ 0 & 0 & 3 \end{bmatrix}, \quad X = \begin{bmatrix} x_{11} & x_{12} & x_{13} \\ 0 & x_{22} & x_{23} \\ 0 & 0 & x_{33} \end{bmatrix}.$$

Solution

$$A + X = \begin{bmatrix} 1 + x_{11} & x_{12} & x_{13} \\ 0 & 1 + x_{22} & 2 + x_{23} \\ 0 & 0 & 3 + x_{33} \end{bmatrix} \Rightarrow$$

$$|A + X| = (1 + x_{11})(1 + x_{22})(3 + x_{33})$$

and

$$(A + X)^{-1} = \begin{bmatrix} \frac{1}{a} & \alpha & \beta \\ 0 & \frac{1}{b} & \gamma \\ 0 & 0 & \frac{1}{c} \end{bmatrix}$$

where

$$a = (1 + x_{11}), \ b = (1 + x_{22}), \ c = (3 + x_{33}), \ \alpha = -\frac{x_{12}}{ab}$$

$$\beta = \frac{x_{12}(2 + x_{23}) - x_{13}(1 + x_{22})}{abc}, \ \gamma = -\frac{(2 + x_{23})}{bc}.$$

But we have $Y = 2(A + X)^{-1}A - I$ and writing $U = (u_{ij}) = (A + X)^{-1}A$
we have

$$u_{11} = \frac{1}{a}, \ u_{22} = \frac{1}{b}, \ u_{33} = \frac{3}{c}$$

$$u_{12} = \alpha, \ u_{13} = 2\alpha + 3\beta, \ u_{23} = 3\gamma + \frac{2}{b}.$$

Take the u_{ij}'s in the order $u_{11}, u_{22}, u_{33}, u_{12}, u_{23}, u_{13}$ and the x_{ij}'s also in
the same order and form the matrix of partial derivatives. We have

$$\frac{\partial(u_{11}, u_{22}, u_{33})}{\partial(x_{11}, x_{22}, x_{33})} = \begin{bmatrix} -\frac{1}{a^2} & 0 & 0 \\ 0 & -\frac{1}{b^2} & 0 \\ 0 & 0 & -\frac{1}{c^2} \end{bmatrix}$$

and

$$\frac{\partial(u_{12}, u_{23}, u_{13})}{\partial(x_{12}, x_{23}, x_{13})} = \begin{bmatrix} -\frac{1}{ab} & 0 & 0 \\ 0 & -\frac{3}{bc} & 0 \\ * & * & -\frac{1}{ac} \end{bmatrix}$$

where a $*$ denotes the presence of a nonzero quantity. The absolute value of the determinant of the Jacobian matrix is then

$$\left|\frac{\partial U}{\partial X}\right| = \frac{3}{(abc)^4} = |A + X|^{-4} a_{11}^3 a_{22}^2 a_{33}.$$

Since there are 6 variables, 2^6 also comes in from $2(A + X)^{-1}A - I$. Thus one part of the theorem is verified. The other part can be verified in a similar fashion.

Theorem 1.34 *Let $X = \frac{T}{|T|}$ where X and T are $p \times p$ lower or upper triangular matrices of functionally independent real variables and with positive diagonal elements. Then, excluding the sign,*

$$X = \frac{T}{|T|} \Rightarrow dX = (p - 1)|T|^{-p(p+1)/2} dT. \qquad (1.3.19)$$

$$Y = X^{-1} = |T|T^{-1} \Rightarrow dY = (p - 1)|T|^{(p+1)(p-2)/2} dT. \qquad (1.3.20)$$

Proof   Note that if $|T|$ is defined as $|T|^{-1} = |X|^{\frac{1}{p-1}}$ where $|X|^{\frac{1}{p-1}}$ denotes the $(p - 1)$-th positive root of $|X|$ then the transformation is one-to-one. Consider the lower triangular case. Note that $|T| = t_{11}t_{22}...t_{pp}$. Thus $x_{jj} = t_{jj}/(t_{11}t_{22}...t_{pp})$. Hence

$$\frac{\partial x_{jj}}{\partial t_{jj}} = 0, \quad \frac{\partial x_{jj}}{\partial t_{ii}} = -\frac{t_{jj}}{t_{ii}}\frac{1}{|T|}, \quad i \neq j, \ i = 1, ..., p;$$

$$\frac{\partial x_{ij}}{\partial t_{ij}} = \frac{1}{|T|}, \quad i > j; \quad \frac{\partial x_{ij}}{\partial t_{ik}} = 0, \quad i > j, \ j \neq k.$$

When T is lower triangular take the x-variables in the order $x_{11}, x_{22}, ..., x_{pp}$, $x_{1j}, j = 2, ..., p$, $x_{2j}, j = 3, ..., p$, and so on and the t-variables also in the same order. Then $x_{1j}, j = 2, ..., p$, $x_{2j}, j = 3, ..., p$, and so on give a Jacobian matrix which is triangular with diagonal elements $1/|T|$ each and the determinant is therefore $|T|^{-p(p-1)/2}$ where $p(p - 1)/2$ is coming from $(p - 1) + (p - 2) + ... + 1$ diagonal elements of the $p - 1$ diagonal blocks. The determinant of the diagonal block of the Jacobian matrix corresponding to the variables $x_{11}, x_{22}, ..., x_{pp}$ is seen to be the following.

$$\left| \frac{\partial(x_{11}, ..., x_{pp})}{\partial(t_{11}, ..., t_{pp})} \right| = \frac{1}{|T|^p} \begin{vmatrix} 0 & -\frac{t_{11}}{t_{22}} & \cdots & -\frac{t_{11}}{t_{pp}} \\ -\frac{t_{22}}{t_{11}} & 0 & \cdots & -\frac{t_{22}}{t_{pp}} \\ \vdots & \vdots & \ddots & \vdots \\ -\frac{t_{pp}}{t_{11}} & -\frac{t_{pp}}{t_{22}} & \cdots & 0 \end{vmatrix}$$

$$= -\frac{(p-1)}{|T|^p} = -(p-1)|T|^{-p}.$$

Now multiply the two quantities $(p-1)|T|^{-p}$ and $|T|^{-p(p-1)/2}$ to obtain (1.3.19). The proof for the upper triangular case is parallel. For establishing (1.3.20), use (1.3.19) and the fact that $dY = |X|^{-(p+1)}dX$.

1.3.4   Jacobians in terms of eigenvalues

Here we consider more general nonlinear transformations involving one matrix where the Jacobians are available in terms of the eigenvalues of the matrix. The simplest of these is a transformation of a matrix to its square.

Theorem 1.35   *Let $X = (x_{ij})$ be a $p \times p$ positive definite matrix of functionally independent variables. Let $\lambda_1, ..., \lambda_p$ denote the eigenvalues of X. Then*

$$Y = X^2 \Rightarrow$$

$$dY = \left\{ \prod_{i=1}^{p} \prod_{j=1}^{p} (\lambda_i + \lambda_j) \right\} dX \quad \text{for a general } X,$$

$$= 2^p |X| \left\{ \prod_{i \neq j=1}^{p} (\lambda_i + \lambda_j) \right\} dX \quad \text{for a general } X, \quad (1.3.21)$$

$$= \left\{ \prod_{i \leq j=1}^{p} (\lambda_i + \lambda_j) \right\} dX \quad \text{for } X = X',$$

$$= 2^p |X| \left\{ \prod_{i < j=1}^{p} (\lambda_i + \lambda_j) \right\} dX \quad \text{for } X = X', \quad (1.3.22)$$

$$= \left\{ \prod_{i=1}^{p} \prod_{j=i}^{p} (t_{ii} + t_{jj}) \right\} dT$$

for $X = T = (t_{ij})$ lower or upper triangular with positive diagonal elements

$$= 2^p |T| \left\{ \prod_{i=1}^{p-1} \prod_{j=i+1}^{p} (t_{ii} + t_{jj}) \right\} dT. \tag{1.3.23}$$

Proof   We can write X^2 as XX. Thus by taking the differentials we have

$$Y = X^2 = XX \Rightarrow (dY) = (dX)X + X(dX).$$

Thus from Lemma 1.1

$$\text{vec}(dY) = (X' \otimes I + I \otimes X)\text{vec}(dX) \tag{1.3.24}$$

and when all the variables are functionally independent

$$dY = |X' \otimes I + I \otimes X| dX.$$

For evaluating the determinant we can use the following technique. The structure of the determinant is the following.

$$|X' \otimes I + I \otimes X| = \begin{vmatrix} x_{11}I + X & x_{21}I + 0 & \dots & x_{p1}I + 0 \\ x_{12}I + 0 & x_{22}I + X & \dots & x_{p2}I + 0 \\ \vdots & \vdots & \ddots & \vdots \\ x_{1p}I + 0 & x_{2p}I + 0 & \dots & x_{pp}I + X \end{vmatrix}. \tag{1.3.25}$$

Let P and P^{-1} be such that

$$PXP^{-1} = D = \text{diag}(\lambda_1, ..., \lambda_p).$$

Then by pre and postmultiplying with blocks of P and P^{-1} we can effectively replace X in the diagonal blocks by D in the determinant of (1.3.25). Then by interchanging rows and columns, without affecting the absolute value of the determinant, we can write the determinant in absolute value as follows:

$$|X' \otimes I + I \otimes X| = \begin{vmatrix} X' + \lambda_1 I & 0 & \dots & 0 \\ 0 & X' + \lambda_2 I & \dots & 0 \\ \vdots & \vdots & \ddots & \vdots \\ 0 & 0 & \dots & X' + \lambda_p I \end{vmatrix}$$

$$= \begin{vmatrix} D_1 & 0 & \dots & 0 \\ 0 & D_2 & \dots & 0 \\ \vdots & \vdots & \ddots & \vdots \\ 0 & 0 & \dots & D_p \end{vmatrix} \tag{1.3.26}$$

where
$$D_j = \text{diag}(\lambda_j + \lambda_1, \lambda_j + \lambda_2, ..., \lambda_j + \lambda_p), \ j = 1, ..., p.$$

If all the variables in X are functionally independent then this determinant is given by
$$\left\{ \prod_{j=1}^{p} (\lambda_1 + \lambda_j) \right\} ... \left\{ \prod_{j=1}^{p} (\lambda_p + \lambda_j) \right\}.$$

This establishes (1.3.21). If $X = X'$ then in the exterior product we take only dy_{11} from the set dy_{1j}, $j = 1, ..., p$, dy_{21} and dy_{22} from the set dy_{2j}, $j = 1, ..., p$, and so on. Thus the first element in D_1 is taken, the first two elements in D_2 are taken and so on. Again denoting the exterior products by the same symbols dY and dX we have in this case,

$$dY = \left\{ \prod_{j=1}^{1} (\lambda_1 + \lambda_j) \right\} ... \left\{ \prod_{j=1}^{p} (\lambda_p + \lambda_j) \right\} dX$$

$$= \left\{ \prod_{i \leq j=1}^{p} (\lambda_i + \lambda_j) \right\} dX.$$

This completes the proof for the symmetric case. Consider a lower triangular case. Let $Y = T^2$ where T is lower triangular then Y still remains lower triangular. Consider (1.3.24) after deleting all the elements corresponding to dy_{ij}, $i < j$. The coefficient matrix $(T' \otimes I + I \otimes T)$ is of the following form.

$$(T' \otimes I + I \otimes T)^* = \begin{bmatrix} t_{11}I + T_1 & t_{21}I + 0 & ... & t_{p1}I + 0 \\ 0 & t_{22}I + T_2 & ... & t_{p2}I + 0 \\ \vdots & \vdots & ... & \vdots \\ 0 & 0 & ... & t_{pp}I + T_p \end{bmatrix}$$

where a * indicates the resulting matrix after the deletion of the terms corresponding to dy_{ij}, $i < j$ and T_j is the submatrix of T obtained after the deletion of the first $j - 1$ rows and columns. Since $(T' \otimes I + I \otimes T)^*$ is now a triangular block matrix and the diagonal blocks remain the same for both upper and lower trinagular cases the determinant is the product of the determinants of the diagonal blocks. That is

$$dY = |t_{11}I + T_1| \ |t_{22}I + T_2| ... |t_{pp}I + T_p| dT$$

$$= \left\{ \prod_{j=1}^{p} (t_{11} + t_{jj}) \right\} \left\{ \prod_{j=2}^{p} (t_{22} + t_{jj}) \right\} ... \left\{ \prod_{j=p}^{p} (t_{pp} + t_{jj}) \right\} dT.$$

Simplifying this we have the result.

Remark 1.5   Note that the transformation $Y = X^2$ is not one-to-one. In general, $(-X)(-X) = X^2$. The theorem also holds when X is not positive definite and the eigenvalues not distinct. In this case $|\lambda_i - \lambda_j| \neq 0$ for all i and j. One likely situation where one can find a unique representation is when X is symmetric positive definite with distinct eigenvalues. In this case let P be an orthonormal matrix such that $PP' = I$, $P'YP = D$ $= \text{diag}(\lambda_1, ..., \lambda_p)$ with $\lambda_1 > ... > \lambda_p > 0$. Consider $D^{\frac{1}{2}} = \text{diag}(\lambda_1^{\frac{1}{2}}, ..., \lambda_p^{\frac{1}{2}})$. Then

$$\left(PD^{\frac{1}{2}}P'\right)\left(PD^{\frac{1}{2}}P'\right) = Y.$$

In the transformation $P'YP = D$ the matrix P is not unique but $P'YP$ is unique. Hence take $X = P\left((P'YP)^{\frac{1}{2}}\right)P'$. Thus in Theorem 1.35 the Jacobian should be interpreted as the Jacobian for a given or preselected representation $Y = X^2$ where X is a preselected function of Y.

Example 1.28   Verify Theorem 1.35 for $X = \begin{bmatrix} x_{11} & x_{12} \\ x_{21} & x_{22} \end{bmatrix}$.

Solution

$$Y = (y_{ij}) = X^2 = \begin{bmatrix} x_{11}^2 + x_{12}x_{21} & x_{11}x_{12} + x_{12}x_{22} \\ x_{21}x_{11} + x_{22}x_{21} & x_{21}x_{12} + x_{22}^2 \end{bmatrix}.$$

Taking the y_{ij}'s in the order $y_{11}, y_{12}, y_{21}, y_{22}$ and the x_{ij}'s also in the same order the matrix of partial derivatives is given by

$$\frac{\partial Y}{\partial X} = \begin{bmatrix} 2x_{11} & x_{21} & x_{12} & 0 \\ x_{12} & x_{11} + x_{22} & 0 & x_{12} \\ x_{21} & 0 & x_{11} + x_{22} & x_{21} \\ 0 & x_{21} & x_{12} & 2x_{22} \end{bmatrix}.$$

This determinant is given by

$$\left|\frac{\partial Y}{\partial X}\right| = 4\left(x_{11} + x_{22}\right)^2 |X|.$$

But if the eigenvalues of X are λ_1 and λ_2 then

$$x_{11} + x_{22} = \text{tr}(X) = \lambda_1 + \lambda_2$$

and

$$4\left(x_{11} + x_{22}\right)^2 = 2^2 \left(\lambda_1 + \lambda_2\right)\left(\lambda_2 + \lambda_1\right)$$

and hence the theorem is verified.

Remark 1.6   When X is skew symmetric $Y = X^2$ is symmetric since $(X^2)' = (XX)' = (X'X') = (-X)(-X) = X^2$. Hence the transformation $Y = X^2$ is not one-to-one in this case.

What happens if $Y = XX'$ instead of X^2? Note that XX' is symmetric and hence if $X \neq X'$ then it cannot be a one-to-one transformation. What about $Y = XAX$ where $A = A', X = X'$? Note that if A is at least positive semidefinite then we have a representation for A as $A = BB'$. But if B is singular then the transformation need not be one-to-one. From these considerations and from Theorem 1.35 we have the next result.

Theorem 1.36   *Let X and A be $p \times p$ symmetric positive definite matrices where A is a constant matrix and X is of functionally independent real variables. Let $\lambda_1, ..., \lambda_p$ be the eigenvalues of XA such that $\lambda_1 > ... > \lambda_p > 0$. Then*

$$Y = XAX \Rightarrow dY = \left\{\prod_{i \leq j}(\lambda_i + \lambda_j)\right\} dX. \qquad (1.3.27)$$

Proof   Let B be the symmetric square root of A so that $A = B^2, B = B'$. Then

$$Y = XAX = XBBX \Rightarrow$$
$$BYB = BXBBXB = (BXB)^2.$$

Let

$$U = BYB \text{ and } V = BXB.$$

Then by Theorem 1.35 and observing that the eigenvalues of BXB and $XBB = XA$ are one and the same

$$dU = \prod_{i \leq j}^{p}(\lambda_i + \lambda_j)dV.$$

But from Theorem 1.20 we have

$$dU = |B|^{p+1}dY \text{ and } dV = |B|^{p+1}dX$$

and hence the result.

For proving the next result we need the differential of a determinant which will be stated as a lemma.

Lemma 1.3   *Let $X = (x_{ij})$ be a $p \times p$ real nonsingular matrix with the inverse, the determinant and the cofactor of x_{ij} denoted by X^{-1}, $|X|$, $|X_{ij}|$ respectively. Then the differential of $|X|$, that is,*

$$d|X| = |X|\mathrm{tr}\left(X^{-1}(dX)\right).$$

Proof   Expanding the determinant in terms of the elements and their cofactors of the i-th row we have, for fixed i,

$$|X| = x_{i1}|X_{i1}| + \ldots + x_{ip}|X_{ip}|.$$

Then

$$\frac{\partial |X|}{\partial x_{ij}} = |X_{ij}| \quad \text{for a general } X,$$

$$= 2|X_{ij}| \quad \text{for } X = X',$$

$$= |X_{jj}| \text{ for } i = j.$$

Thus

$$d|X| = \sum_{i=1}^{p}\sum_{j=1}^{p} \frac{\partial |X|}{\partial x_{ij}} dx_{ij}$$

$$= \sum_{i=1}^{p}\sum_{j=1}^{p} |X_{ij}| dx_{ij}, \text{ for a general } X \tag{1.3.28}$$

$$= \sum_{i=1}^{p} |X_{ii}| dx_{ii} + 2\sum_{i>j} |X_{ij}| dx_{ij}, \; X = X'. \tag{1.3.29}$$

Consider the product of the transpose of the matrix of cofactors, denoted by $\mathrm{cof}(X')$, and the matrix of differentials (dx_{ij}) and take the trace of this product to get,

$$\mathrm{tr}\left(\begin{bmatrix} |X_{11}| & \cdots & |X_{p1}| \\ \vdots & \ddots & \vdots \\ |X_{1p}| & \cdots & |X_{pp}| \end{bmatrix} \begin{bmatrix} dx_{11} & \cdots & dx_{1p} \\ \vdots & \ddots & \vdots \\ dx_{p1} & \cdots & dx_{pp} \end{bmatrix}\right)$$

$$= \sum_{i=1}^{p}\sum_{j=1}^{p} |X_{ij}| dx_{ij} \quad \text{for a general } X \tag{1.3.30}$$

$$= \sum_{i=1}^{p} |X_{ii}| dx_{ii} + 2\sum_{i>j} |X_{ij}| dx_{ij} \quad \text{for } \; X = X'. \tag{1.3.31}$$

But $X^{-1} = \frac{1}{|X|}\text{cof}(X')$ and hence

$$d|X| = |X|\text{tr}\left(X^{-1}(dX)\right)$$

for all nonsingular X, that is, for both the cases $X = X'$ and general X.

Lemma 1.4   *Let $W = (w_{ij})$, $U = (u_{ij})$ be $p \times p$ matrices of real variables, then*

$$W = (\text{tr}(U))\, I - U \Rightarrow dW = (p-1)dU \tag{1.3.32}$$

ignoring the sign.

Proof

$$W = (\text{tr}(U))\, I - U \Rightarrow$$
$$w_{ii} = (u_{11} + ... + u_{pp}) - u_{ii}, \; i = 1, ..., p$$
$$w_{ij} = -u_{ij}, \; i \neq j \Rightarrow$$
$$dw_{ii} = (du_{11} + ... + du_{pp}) - du_{ii} = \sum_{j=1, j\neq i}^{p} du_{jj}$$

and

$$dw_{ij} = -du_{ij}, \; i \neq j. \tag{1.3.33}$$

Take the exterior products to get

$$dw_{11}...dw_{pp} = \pm(p-1)du_{11}...du_{pp}$$

and

$$\left\{ \prod_{i\neq j} dw_{ij} \right\} = \pm \left\{ \prod_{i\neq j} du_{ij} \right\}.$$

Hence the result.   Note that (1.3.33) holds irrespective of whether U is symmetric or not.

Theorem 1.37   *Let X be a $p \times p$ positive definite matrix of p^2 functionally independent real variables with its determinant and inverse denoted by $|X|$ and X^{-1} respectively. Let $\lambda_1, ..., \lambda_p$ be the eigenvalues of X and be distinct. Then, ignoring the sign,*

$$Y = |X|X^{-1} \Rightarrow dY = (p-1)|X|^{p(p-2)}dX$$

$$= (p-1)\left\{ \prod_{j=1}^{p} \lambda_j \right\}^{p(p-2)} dX.$$

Proof If $|X|$ is defined as $|X| = |Y|^{\frac{1}{p-1}}$ where $|Y|^{\frac{1}{p-1}}$ denotes the $(p-1)$-th positive root of $|Y|$ then the transformation is one-to-one. Taking the matrices of differentials we have

$$(dY) = (d|X|)X^{-1} + |X|d(X^{-1})$$
$$= |X|\left[tr(X^{-1}(dX))\right]X^{-1} - |X|X^{-1}(dX)X^{-1} \quad (1.3.34)$$

from (1.3.10) and Lemma 1.3. Postmultiply by X and let

$$(dV) = (dY)X, \ (dU) = X^{-1}(dX), \ (dW) = [tr(dU)]I - (dU).$$

Then (1.3.34) gives

$$(dV) = |X|(dW) \Rightarrow dV = |X|^{p^2}dW$$

by (1.2.14) since $|X|$ is scalar. By Lemma 1.4 and Theroem 1.18, and ignoring the sign,

$$dW = (p-1)dU, \ dU = |X|^{-p}dX, \ dV = |X|^p dY.$$

Thus

$$dY = |X|^{-p}dV = |X|^{-p}|X|^{p^2}dW$$
$$= |X|^{-p+p^2-p}(p-1)dX$$
$$= (p-1)|X|^{p(p-2)}dX$$
$$= (p-1)\left\{\prod_{j=1}^{p}\lambda_j\right\}^{p(p-2)}dX,$$

ignoring the sign. For writing dX in terms of dY use the fact that $|Y| = |X|^{p-1}$.

Corollary 1.37.1 *If X is as defined in Theorem 1.37 then, ignoring the sign,*

$$U = \frac{X}{|X|} \Rightarrow dU = (p-1)|X|^{-p^2}dX. \quad (1.3.35)$$

Proof

$$U = Y^{-1} \Rightarrow dU = |Y|^{-2p}dY$$

by using (1.3.9). Now substituting for dY and Y from Theorem 1.37 one has, ignoring the sign,

$$dU = ||X|X^{-1}|^{-2p}(p-1)|X|^{p(p-2)}dX$$
$$= (p-1)|X|^{-p^2}dX.$$

For writing dX in terms of dU use the fact that $|U| = |X|^{-(p-1)}$. Note that the case of triangular matrices is covered in Theorem 1.34.

Example 1.29   Illustrate Corollary 1.37.1 on the matrix

$$X = \begin{bmatrix} x_{11} & x_{12} \\ x_{21} & x_{22} \end{bmatrix}.$$

Solution
$$U = (u_{ij}) = \frac{X}{|X|}, \quad |X| = x_{11}x_{22} - x_{21}x_{12}.$$

Take the partial derivatives to get the Jacobian matrix

$$\frac{\partial U}{\partial X} = \frac{1}{|X|^2} \begin{bmatrix} -x_{12}x_{21} & x_{11}x_{21} & x_{11}x_{12} & -x_{11}^2 \\ -x_{12}x_{22} & x_{11}x_{22} & x_{12}^2 & -x_{12}x_{11} \\ -x_{21}x_{22} & x_{21}^2 & x_{11}x_{22} & -x_{21}x_{11} \\ -x_{22}^2 & x_{21}x_{22} & x_{12}x_{22} & -x_{12}x_{21} \end{bmatrix}.$$

The determinant is given by

$$\left|\frac{\partial U}{\partial X}\right| = \left[\frac{1}{|X|^2}\right]^4 \left[-x_{12}^2 x_{21}^2 |X|^2 + x_{11}^2 x_{22}^2 |X|^2 + 0\right]$$
$$= \frac{|X|^4}{|X|^8} = |X|^{-4}$$

and $p - 1 = 1$. This verifies the corollary.

Theorem 1.38   *Let the real matrix $X = (x_{ij}) = X' > 0$, that is, symmetric positive definite, with distinct eigenvalues $\lambda_1, ..., \lambda_p$. Then, ignoring the sign,*

$$Y = |X|X^{-1} \Rightarrow dY = (p-1)|X|^{\frac{1}{2}(p+1)(p-2)}dX$$
$$= (p-1)\left\{\prod_{j=1}^{p} \lambda_j\right\}^{\frac{1}{2}(p+1)(p-2)} dX.$$

Proof We will assume that the $(p-1)$-th positive root of $|Y|$ is $|X|$. Let $X^{\frac{1}{2}}$ be the symmetric square root of X. Pre and postmultiply (1.3.34) by $X^{\frac{1}{2}}$ and observe that

$$\text{tr}\left(X^{-1}(dX)\right) = \text{tr}\left(X^{-\frac{1}{2}}(dX)X^{-\frac{1}{2}}\right)$$

to get

$$X^{\frac{1}{2}}(dY)X^{\frac{1}{2}} = |X|\left[\text{tr}(X^{-\frac{1}{2}}(dX)X^{-\frac{1}{2}})\right]I - |X|X^{-\frac{1}{2}}(dX)X^{-\frac{1}{2}}. \quad (1.3.36)$$

Let

$$(dV) = X^{\frac{1}{2}}(dY)X^{\frac{1}{2}}, \ (dU) = X^{-\frac{1}{2}}(dX)X^{-\frac{1}{2}},$$

$$(dW) = \left[\text{tr}(dU)\right]I - (dU), \ (dV) = |X|(dW).$$

Then from Theorem 1.20 and Lemma 1.4

$$dV = |X|^{\frac{p+1}{2}}dY, \ dU = |X|^{-\frac{p+1}{2}}dX,$$

$$dW = (p-1)dU, \ dV = |X|^{\frac{p(p+1)}{2}}dW.$$

From these the result follows.

Corollary 1.38.1 *If X is as defined in Theorem 1.38 then*

$$U = \frac{X}{|X|} \Rightarrow dU = (p-1)|X|^{-p(p+1)/2}dX. \quad (1.3.37)$$

Proceed as in the proof of Corollary 1.37.1 to see the result.

Theorem 1.39 *Let $Y = AXA' + BXB'$ where A and B are $p \times p$ matrices of constants and X is a $p \times p$ matrix of functionally independent real variables such that $A \otimes A + B \otimes B$ is nonsingular. Then, ignoring the sign,*

$$Y = AXA' + BXB' \Rightarrow$$

$$dY = \left\{ \prod_{i=1}^{p} \prod_{j=1}^{p} |\lambda_i \lambda_j + \mu_i \mu_j| \right\} dX \quad \text{for a general } X$$

$$= \left\{ \prod_{i \leq j} |\lambda_i \lambda_j + \mu_i \mu_j| \right\} dX \quad \text{for } X = X'$$

where λ_j, μ_j, $j = 1, ..., p$ are the eigenvalues of A and B respectively.

Proof   Let the columns of Y be denoted by $y_{(1)}, ..., y_{(p)}$ and those of X by $x_{(1)}, ..., x_{(p)}$. Then from Lemma 1.1 we have

$$\text{vec}(Y) = (A \otimes A)\text{vec}(X) + (B \otimes B)\text{vec}(X).$$

That is,

$$\begin{bmatrix} y_{(1)} \\ \vdots \\ y_{(p)} \end{bmatrix} = (A \otimes A + B \otimes B) \begin{bmatrix} x_{(1)} \\ \vdots \\ x_{(p)} \end{bmatrix}. \tag{1.3.38}$$

Note that when $A \otimes A + B \otimes B$ is nonsingular the transformation in (1.3.38) is one-to-one. When X has p^2 functionally independent variables then from the Jacobian of the transformation we have

$$dY = |A \otimes A + B \otimes B|dX.$$

Let P and Q be orthogonal matrices such that $PP' = I, QQ' = I$,

$$P'AP = D_1 = \text{diag}(\lambda_1, ..., \lambda_p)$$

and

$$Q'BQ = D_2 = \text{diag}(\mu_1, ..., \mu_p).$$

But, for example,

$$A \otimes A = \begin{bmatrix} a_{11}A & \cdots & a_{1p}A \\ \vdots & \ddots & \vdots \\ a_{p1}A & \cdots & a_{pp}A \end{bmatrix}$$

$$= \bar{P} \begin{bmatrix} a_{11}D_1 & \cdots & a_{1p}D_1 \\ \vdots & \ddots & \vdots \\ a_{p1}D_1 & \cdots & a_{pp}D_1 \end{bmatrix} \bar{P}'$$

where $\bar{P}$ is the block diagonal matrix with p blocks of P each, that is,

$$\bar{P} = \begin{bmatrix} P & \cdots & 0 \\ \vdots & \ddots & \vdots \\ 0 & \cdots & P \end{bmatrix}.$$

One can rearrange the rows and columns, keeping the absolute value of the determinant the same, to get

$$\bar{P} \begin{bmatrix} a_{11}D_1 & \cdots & a_{1p}D_1 \\ \vdots & \ddots & \vdots \\ a_{p1}D_1 & \cdots & a_{pp}D_1 \end{bmatrix} \bar{P}' \to \bar{P} \begin{bmatrix} \lambda_1 A & \cdots & 0 \\ \vdots & \ddots & \vdots \\ 0 & \cdots & \lambda_p A \end{bmatrix} \bar{P}'$$

$$= \begin{bmatrix} \lambda_1 D_1 & \cdots & 0 \\ \vdots & \ddots & \vdots \\ 0 & \cdots & \lambda_p D_1 \end{bmatrix}.$$

Similar steps go through for $B \otimes B$ also and thus

$$A \otimes A + B \otimes B \to \begin{bmatrix} \lambda_1 D_1 + \mu_1 D_2 & \cdots & 0 \\ \vdots & \ddots & \vdots \\ 0 & \cdots & \lambda_p D_1 + \mu_p D_2 \end{bmatrix}. \qquad (1.3.39)$$

Thus

$$|A \otimes A + B \otimes B| = \left\{ \prod_{i=1}^{p} \prod_{j=1}^{p} (\lambda_i \lambda_j + \mu_i \mu_j) \right\}.$$

If $X = X'$ then all elements in $x_{(1)}$ are taken, the first element in $x_{(2)}$ is deleted, the first 2 elements in $x_{(3)}$ are deleted and so on. The effect of this on the coefficient matrix of $vec(X)$ in (1.3.38) is the following. Delete the first row and first column of all matrices in the second column block of the coefficient matrix, delete the first two rows and columns of all matrices in the third column block and so on. The effect in (1.3.39) will be the corresponding deletions in the second, third and so on of the diagonal blocks. Then the resulting determinant is $\left\{ \prod_{i \leq j} (\lambda_i \lambda_j + \mu_i \mu_j) \right\}$ and hence the result.

Example 1.30   Verify Theorem 1.39 for

$$A = \begin{bmatrix} 1 & 0 \\ 0 & 2 \end{bmatrix}, \quad B = \begin{bmatrix} 3 & 0 \\ 1 & 4 \end{bmatrix}, \quad X = \begin{bmatrix} x_{11} & x_{12} \\ x_{21} & x_{22} \end{bmatrix}.$$

Solution

The eigenvalues of A and B are respectively $\lambda_1 = 1, \lambda_2 = 2$; $\mu_1 = 3$, $\mu_2 = 4$.

$$AXA' = \begin{bmatrix} x_{11} & 2x_{12} \\ 2x_{21} & 4x_{22} \end{bmatrix}$$

$$BXB' = \begin{bmatrix} 9x_{11} & 3x_{11} + 12x_{12} \\ 3x_{11} + 12x_{21} & x_{11} + 4x_{21} + 4x_{12} + 16x_{22} \end{bmatrix}$$

$$AXA' + BXB' = \begin{bmatrix} 10x_{11} & 3x_{11} + 14x_{12} \\ 3x_{11} + 14x_{21} & x_{11} + 4x_{21} + 4x_{12} + 20x_{22} \end{bmatrix}.$$

That is,

$$\begin{bmatrix} y_{11} \\ y_{21} \\ y_{12} \\ y_{22} \end{bmatrix} = \begin{bmatrix} 10 & 0 & 0 & 0 \\ 3 & 14 & 0 & 0 \\ 3 & 0 & 14 & 0 \\ 1 & 4 & 4 & 20 \end{bmatrix} \begin{bmatrix} x_{11} \\ x_{21} \\ x_{12} \\ x_{22} \end{bmatrix}.$$

Then the Jacobian is

$$\begin{vmatrix} 10 & 0 & 0 & 0 \\ 3 & 14 & 0 & 0 \\ 3 & 0 & 14 & 0 \\ 1 & 4 & 4 & 20 \end{vmatrix} = (10)(14)(14)(20).$$

But from the formula

$$\begin{aligned} |J| &= (\lambda_1\lambda_1 + \mu_1\mu_1)(\lambda_1\lambda_2 + \mu_1\mu_2)(\lambda_2\lambda_1 + \mu_2\mu_1)(\lambda_2\lambda_2 + \mu_2\mu_2) \\ &= (\lambda_1^2 + \mu_1^2)(\lambda_1\lambda_2 + \mu_1\mu_2)^2(\lambda_2^2 + \mu_2^2) \\ &= (1+9)(2+12)^2(4+16) = (10)(14)(14)(20). \end{aligned}$$

The general polar coordinates transformations of Theorems 1.24 and 1.25 are used in many areas and more specifically in statistical distribution theory in connection with spherically symmetric and elliptically contoured distributions, see for example, Fang and Anderson (1990), and Fang and Zhang (1990). More results on the Jacobians of the types discussed in this section and the properties of matrices which are used in computing the Jacobians may be found in Bellman (1956, 1960), Deemer and Olkin (1951), Good (1981), Olkin (1953, 1959), Olkin and Roy (1954), Olkin and Sampson (1972), and Roy (1957). Some Jacobians in the singular case are useful in statistical applications, see for example Saw (1973, 1975).

Exercises

1.3.1   Prove Theorems 1.23, 1.24, 1.25, 1.26 and 1.28 by forming matrices of partial derivatives and then evaluating the determinants.

1.3.2   Show that the transformation $X = TT'$ or $X = T'T$ is one-to-one when $t_{jj} > 0$, $j = 1, ..., p$ where T is a $p \times p$ lower triangular matrix and X is real symmetric positive definite.

1.3.3 When the following Kronecker products are defined show that

$$A \otimes B \otimes C = (A \otimes B) \otimes C = A \otimes (B \otimes C) \qquad (i)$$
$$(A + B) \otimes (C + D) = A \otimes C + A \otimes D + B \otimes C + B \otimes D \qquad (ii)$$
$$(A \otimes B)(C \otimes D) = (AC) \otimes (BD) \qquad (iii)$$
$$(A \otimes B)' = A' \otimes B' \qquad (iv)$$
$$\text{tr}(A \otimes B) = (\text{tr}(A))(\text{tr}(B)) \qquad (v)$$
$$(A \otimes B)^{-1} = A^{-1} \otimes B^{-1} \qquad (vi)$$
$$a' \otimes b = ba' = b \otimes a' \text{ where } a \text{ and } b \text{ are vectors.} \qquad (vii)$$

1.3.4 Let the $p \times p$ real symmetric positive definite random matrix X have the density

$$f(X) = \begin{cases} \dfrac{|X|^{\alpha - \frac{p+1}{2}} e^{-\text{tr}(X)}}{\Gamma_p(\alpha)}, & X > 0, \ \text{Re}(\alpha) > \frac{p-1}{2} \\ \\ 0, \text{ elsewhere.} \end{cases}$$

Let $X = TT'$ where T is a lower triangular matrix with positive diagonal elements. Then show that t_{jj}, $j = 1, ..., p$ are gamma distributed, t_{ij}, $i > j$ are constant multiples of normal variables and all the t_{ij}'s are mutually independently distributed.

1.3.5 If a $p \times 1$ real random vector X has the density of the form $c\, h(X'X)$ where c is a normalizing constant and h is a positive function then by using Theorem 1.24 or Theorem 1.25 show that the density of $y = X'X$ is of the form

$$f(y) = \begin{cases} \dfrac{c\pi^{\frac{p}{2}} y^{\frac{p}{2}-1} h(y)}{\Gamma(\frac{p}{2})}, & y > 0 \\ \\ 0, \text{ elsewhere.} \end{cases}$$

1.3.6 If X has the density as given in Exercise 1.3.4 then evaluate the density of $Y = X^{-1}$.

1.3.7 If the real symmetric positive definite $p \times p$ matrix X has the density

$$f(X) = \begin{cases} \dfrac{\Gamma_p(\alpha+\beta)}{\Gamma_p(\alpha)\Gamma_p(\beta)} |X|^{\alpha - \frac{p+1}{2}} |I + X|^{-(\alpha+\beta)}, & X > 0, \\ \\ 0, \text{ elsewhere} \end{cases}$$

for $\text{Re}(\alpha) > \frac{p-1}{2}$, $\text{Re}(\beta) > \frac{p-1}{2}$ then evaluate the densities of

$$Y_1 = (I + X)^{-1} \text{ and } Y_2 = (I + X)^{-\frac{1}{2}} X (I + X)^{-\frac{1}{2}}.$$

1.3.8   Evaluate the integral

$$\int_{0<x_1<...<x_p<\infty} (x_1...x_p)^{\alpha - \frac{p+1}{2}} e^{-(x_1+...+x_p)} \left\{ \prod_{1=i>j}^{p} (x_i - x_j) \right\} dx_1...dx_p$$

by using Theorem 1.23.

1.3.9   By using Exercise 1.2.6 ((i)–(v)) of Section 1.2 prove the Theorems 1.27, 1.28 and 1.29.

1.3.10   Show that

$$B_p(\alpha, \beta) = \int_{0<X=X'<I} |X|^{\alpha - \frac{p+1}{2}} |I - X|^{\beta - \frac{p+1}{2}} dX$$

$$= \int_{Y=Y'>0} |Y|^{\alpha - \frac{p+1}{2}} |I + Y|^{-(\alpha+\beta)} dY$$

$$= \frac{\Gamma_p(\alpha)\Gamma_p(\beta)}{\Gamma_p(\alpha + \beta)}.$$

1.3.11   Let the $p \times p$ real matrix T be such that $T = \left(\frac{1}{2}(1 + \delta_{ij})t_{ij}\right)$, where $t_{ij} = t_{ji}$, $\delta_{ij} = 0$, $i \neq j$, $\delta_{jj} = 1$, for all i, $j = 1,...,p$. Let $X = (x_{ij})$ be a $p \times p$ real symmetric matrix. Then

$$\text{tr}(TX) = \sum_{i \leq j} t_{ij} x_{ij}.$$

If $X = X' > 0$ and $f(X)$ is a scalar function of X then the Laplace transform of f, denoted by $L_T(f)$, is given by

$$L_T(f) = \int_{X=X'>0} e^{-\text{tr}(TX)} f(X) dX.$$

Let $g(T) = L_T(f)$. Then prove the following.

(i) $g(T + A) = L_T \left(e^{-\text{tr}(AX)} f(X) \right);$

(ii)
$$L_T(1) = |T|^{-\frac{p+1}{2}} \Gamma_p \left(\frac{p+1}{2}\right);$$

(iii)
$$\int_{U>T} g(U) dU = L_T \left[|X|^{-\frac{p+1}{2}} \Gamma_p \left(\frac{p+1}{2}\right) f(X) \right];$$

(iv)
$$L_T \left[\left\{ \Gamma_p \left(\frac{p+1}{2}\right) |X|^{-\frac{p+1}{2}} \right\}^n f(X) \right]$$

$$= \int_{W_1 > T} \int_{W_2 > W_1} \cdots \int_{W_n > W_{n-1}} g(W_n) dW_1 ... dW_n$$

where $W_i = W_i' > 0$, $W_i - T > 0$, $W_{i+1} - W_i > 0$, $i = 1, ..., n-1$.
(The concept of Laplace transform is due to the French mathematician
P.S. Laplace (1749–1827)).

1.3.12 For the definitions as in Exercise 1.3.11 prove the following:

(i)
$$L_T \left(|X|^n \right) = |T|^{-n - \frac{p+1}{2}} \Gamma_p \left(n + \frac{p+1}{2}\right), \quad n > -1;$$

(ii)
$$L_T \left(|X - B|^\nu \right) = |T|^{-(\nu + \frac{p+1}{2})} e^{-\text{tr}(TB)} \Gamma_p \left(\nu + \frac{p+1}{2}\right)$$

for $B = B' > 0$, $X - B > 0$.

1.3.13 For real symmetric positive definite $p \times p$ matrices X and Y show
the following:

(i)
$$\lim_{b \to \infty} \left| I + \frac{XY}{b} \right|^{-b} = e^{-\text{tr}(XY)};$$

(ii)
$$\lim_{b \to \infty} \left| I - \frac{XY}{b} \right|^{-b} = e^{\text{tr}(XY)}.$$

Hint: Write determinants as products of eigenvalues.

1.3.14 If the $p \times p$ real symmetric positive definite matrix X has a type-2
matrix-variate beta density (see Appendix A) show that $(I + X^{-1})^{-1}$ has
a type-1 matrix-variate beta density.

1.3.15   If the $p \times p$ real symmetric positive definite matrix X has a type-1 beta density (see Appendix) show the following:

(i) $\left(X^{-1} - I\right)^{-1} \sim$ type-2 beta ;

(ii) $\left[I + (I - X)^{-\frac{1}{2}} X (I - X)^{-\frac{1}{2}}\right]^{-1} \sim$ type-1 beta ;

(iii) $\left[I + (I - X)^{-\frac{1}{2}} X (I - X)^{-\frac{1}{2}}\right]^{-\frac{1}{2}} (I - X)^{-\frac{1}{2}} X (I - X)^{-\frac{1}{2}}$

$$\times \left[I + (I - X)^{-\frac{1}{2}} X (I - X)^{-\frac{1}{2}}\right]^{-\frac{1}{2}} \sim \text{type-1 beta.}$$

1.3.16   For $x_j > 0$, $j = 1, ..., p$ and real, consider the transformation $y_1 = x_1 + ... + x_p, y_2 = x_1^2 + ... + x_p^2, ..., y_p = x_1^p + ... + x_p^p$. Show that

$$dy_1...dy_p = p! \left\{ \prod_{j=1}^{p} \prod_{k=j+1}^{p} |x_j - x_k| \right\} dx_1...dx_p.$$

1.3.17   Let X be a $p \times p$ matrix of p^2 functionally independent real variables. Let $S = \frac{1}{2}(X + X')$ and $A = \frac{1}{2}(X - X')$. Show that

$$dX = 2^{p(p-1)/2} dS \; dA.$$

1.3.18   By using Exercise 1.3.17 or otherwise show that for $\text{Re}(\alpha) > \frac{p-1}{2}$, $\text{Re}(\beta) > p$,

$$g(\alpha, \beta) = \int_X \frac{|X + X'|^{\alpha - \frac{p+1}{2}}}{|I + X|^{\alpha + \beta}} dX$$

$$= 2^{\frac{p(p-1)}{2} - p\beta} \pi^{\frac{p^2}{2}} \frac{\Gamma_p(\alpha) \Gamma_p \left(\beta - \frac{p+1}{2}\right)}{\Gamma_p \left(\frac{\alpha + \beta}{2}\right) \Gamma_p \left(\frac{\alpha + \beta + 1}{2}\right)}.$$

CHAPTER 2

Jacobians in orthogonal
and
related transformations

2.0   Introduction

In Chapter 1 we dealt with simple linear and smooth nonlinear transformations involving real matrices. Here we will look into more complicated types of transformations involving eigenvalues, decomposition into triangular, diagonal, orthogonal and semiorthogonal matrices as well as transformations involving submatrices, all in the real case.

2.1   Transformation of one matrix to components of two matrices

In Sections 2.1 and 2.2 we will mainly consider Jacobians when one matrix is transformed into components of two matrices. These component matrices may be triangular, diagonal, orthogonal and so on.

2.1.1   Decomposition with one matrix being triangular or diagonal

Here we consider a decomposition of a $p \times p$ matrix into two matrices where one is a diagonal matrix with p nonzero diagonal elements. Thus effectively p variables are removed. This can be achieved by suppressing p variables from the original matrix or by imposing p conditions on the elements of the original matrix. We will consider both of these situations here.

Theorem 2.1   *Let $X = (x_{jk})$ and W be $p \times p$ nonsingular lower triangular matrices of functionally independent real variables, W with unit diagonal elements and D a diagonal matrix with real and distinct diagonal* *elements* $\lambda_1, ..., \lambda_p$. *Then*

$$X = DW \Rightarrow dX = \left\{ \prod_{j=1}^{p} |\lambda_j|^{j-1} \right\} dD \ dW, \tag{a}$$

and

$$X = WD \Rightarrow dX = \left\{ \prod_{j=1}^{p} |\lambda_j|^{p-j} \right\} dD \ dW. \tag{b}$$

Proof   Take the x-variables in the order x_{jj}, $j = 1, ..., p$, x_{j1}, $j = 2, ..., p$, x_{j2}, $j = 3, ...$, and so on and the independent variables in the same order with λ_j, $j = 1, ..., p$ in place of the w_{jj}'s. Then the Jacobian matrix is triangular with diagonal elements having λ_2 appearing once, λ_3 appearing twice and so on when W is premultiplied by D, and λ_1 appearing $p - 1$ times and so on when W is postmultiplied by D. Hence the result.

   Note that in the notation of J we may write both (a) and (b) above as $J(X : D, W)$.

Corollary 2.1.1   *Let W and D be as in Theorem 2.1 with $\lambda_j > 0$ and $x_{jj} > 0$, $j = 1, ..., p$. Then*

$$Y = D^{\frac{1}{2}} W \Rightarrow dY = \left\{ 2^{-p} \prod_{j=1}^{p} (\lambda_j^{\frac{1}{2}})^{j-2} \right\} dD \ dW,$$

and

$$Y = WD^{\frac{1}{2}} \Rightarrow dY = \left\{ 2^{-p} \prod_{j=1}^{p} (\lambda_j^{\frac{1}{2}})^{p-1-j} \right\} dD \ dW$$

where $D^{\frac{1}{2}} = \mathrm{diag}(\lambda_1^{\frac{1}{2}}, ..., \lambda_p^{\frac{1}{2}})$.

Proof   Consider $Y = WD^{\frac{1}{2}}$. Then from Theorem 2.1

$$dY = \left\{ \prod_{j=1}^{p} (\lambda_j^{\frac{1}{2}})^{p-j} \right\} dD^{\frac{1}{2}} \ dW.$$

But

$$d\lambda_j^{\frac{1}{2}} = \frac{1}{2} \lambda_j^{-\frac{1}{2}} d\lambda_j.$$

That is,

$$dY = \left\{ \prod_{j=1}^{p} (\lambda_j^{\frac{1}{2}})^{p-j-1} \right\} dD \; dW.$$

Similar steps hold for the first part also.

Take the transpose of X in Theorem 2.1 and Corollary 2.1.1 and observe that $dX = dX'$ to get the corresponding results for the upper triangular case.

Theorem 2.2   *Let W and D be as defined in Theorem 2.1 with $\lambda_j > 0$, $x_{jj} > 0$, $j = 1, ..., p$. Then*

$$X = WDW' \Rightarrow dX = \left\{ \prod_{j=1}^{p} \lambda_j^{p-j} \right\} dD \; dW, \qquad (a)$$

$$X = D^{\frac{1}{2}}WW'D^{\frac{1}{2}} \Rightarrow dX = \left\{ \prod_{j=1}^{p} \lambda_j^{\frac{p-1}{2}} \right\} dD \; dW, \qquad (b)$$

$$X = W'DW \Rightarrow dX = \left\{ \prod_{j=1}^{p} \lambda_j^{j-1} \right\} dD \; dW, \qquad (c)$$

and

$$X = D^{\frac{1}{2}}W'WD^{\frac{1}{2}} \Rightarrow dX = \left\{ \prod_{j=1}^{p} \lambda_j^{\frac{p-1}{2}} \right\} dD \; dW. \qquad (d)$$

Proof   Let $Y = WD^{\frac{1}{2}}$. Then from Corollary 2.1.1 we have dY in terms of dD and dW. Now

$$X = WDW' = YY'.$$

From Theorem 1.28 of Chapter 1 for the lower triangular case we have

$$dX = 2^p \left\{ \prod_{j=1}^{p} (\lambda_j^{\frac{1}{2}})^{p+1-j} \right\} dY.$$

Now substituting for dY the result (a) follows. Similar steps hold for result (b). For establishing (c) and (d) use (1.3.12) of Chapter 1.

Theorem 2.3   *Let $Y = (y_{jk})$, $X = (x_{jk})$ be $p \times p$ matrices of functionally independent real variables with $y_{jj} > 0$, $x_{jj} = 1$, $j = 1,...,p$ and D a diagonal matrix with real distinct and positive diagonal elements $\lambda_1, ..., \lambda_p$. Then*

$$Y = DXD \Rightarrow dY = 2^p \left\{ \prod_{j=1}^{p} \lambda_j^{2p-1} \right\} dX\ dD, \qquad (a)$$

$$Y = DX \Rightarrow dY = \left\{ \prod_{j=1}^{p} \lambda_j^{p-1} \right\} dX\ dD, \qquad (b)$$

$$Y = XD \Rightarrow dY = \left\{ \prod_{j=1}^{p} \lambda_j^{p-1} \right\} dX\ dD, \qquad (c)$$

and

$$Y = DXD, X = X' \Rightarrow dY = \left\{ 2^p \prod_{j=1}^{p} \lambda_j^{p} \right\} dX\ dD. \qquad (d)$$

Proof   Consider (a). Note that $y_{jj} = \lambda_j^2$ since $x_{jj} = 1$, $j = 1,...,p$ and $y_{ij} = \lambda_i \lambda_j x_{ij}$, $i \neq j$. Then

$$\frac{\partial y_{jj}}{\partial \lambda_j} = 2\lambda_j, \quad \frac{\partial y_{ij}}{\partial x_{ij}} = \lambda_i \lambda_j, \ i \neq j.$$

Hence the Jacobian matrix is a diagonal matrix and the Jacobian is the product of all these diagonal elements. If these diagonal elements are arranged in the form of a matrix with the elements in the j-th diagonal block of the Jacobian matrix forming the j-th row of this new matrix then we have the following pattern:

$$\begin{bmatrix} 2\lambda_1 & \lambda_1\lambda_2 & \dots & \lambda_1\lambda_p \\ \lambda_2\lambda_1 & 2\lambda_2 & \dots & \lambda_2\lambda_p \\ \vdots & \vdots & \ddots & \vdots \\ \lambda_p\lambda_1 & \lambda_p\lambda_2 & \dots & 2\lambda_p \end{bmatrix}.$$

Then the result in (a) is obvious. When $X = X'$ we delete the elements above or below the leading diagonal in the above pattern. Hence (d) is obvious. For (b) and (c), the diagonal elements in the above pattern are unities and the j-th row consists of only λ_j's when X is premultiplied by

D, and the j-th column consists of only λ_j's when X is postmultiplied by D. Thus the results are obvious.

Example 2.1   Let X be a real $p \times p$ symmetric positive definite matrix having a matrix-variate gamma density with the parameters $\alpha, B = \frac{1}{2}I$ (see the definition from Appendix A; when $\alpha = n/2$, $n \geq p$ this case is the Wishart density $W_p(n, I)$). This density was introduced by the English statistician J. Wishart (1898-1956)). Let

$$R = (r_{ij}), \quad r_{ij} = \frac{x_{ij}}{(x_{ii}x_{jj})^{\frac{1}{2}}},$$

where $r_{jj} = 1$, $j = 1, ..., p$, $-1 < r_{ij} = r_{ji} < 1$, $i \neq j$. Note that $x_{jj} > 0$, $j = 1, ..., p$. Then

$$X = D^{\frac{1}{2}}RD^{\frac{1}{2}}, \quad D = \mathrm{diag}(x_{11}, ..., x_{pp}).$$

Show that (1) R and $x_{11}, x_{22}, ..., x_{pp}$ are independently distributed, (2) the density of R is given by

$$g(R) = \begin{cases} \frac{[\Gamma(\alpha)]^p}{\Gamma_p(\alpha)}|R|^{\alpha - \frac{p+1}{2}}, & R = R' > 0 \\ \\ 0, & \text{elsewhere.} \end{cases}$$

Solution   For the matrix-variate gamma, the density of X is given by

$$f(X) = \begin{cases} \frac{|X|^{\alpha - \frac{p+1}{2}}e^{-\frac{1}{2}\mathrm{tr}(X)}}{2^{\alpha p}\Gamma_p(\alpha)}, & X = X' > 0, \ \mathrm{Re}(\alpha) > \frac{p-1}{2} \\ \\ 0, & \text{elsewhere.} \end{cases}$$

Note that

$$|X| = |D| \, |R|$$

and

$$\mathrm{tr}(X) = \mathrm{tr}\left[D^{\frac{1}{2}}RD^{\frac{1}{2}}\right] = \mathrm{tr}[DR]$$
$$= x_{11} + x_{22} + ... + x_{pp}$$

observing that the diagonal elements of R are unities. From Theorem 2.3(d) for $\lambda_j = x_{jj}^{\frac{1}{2}}$, so that $d\lambda_j = \frac{1}{2}x_{jj}^{-\frac{1}{2}}dx_{jj}$,

$$f(X)dX = h(R, D)dRdD$$

where

$$h(R, D) = \frac{|D|^{\alpha - \frac{p+1}{2}} |R|^{\alpha - \frac{p+1}{2}} e^{-\frac{1}{2}(x_{11} + \ldots + x_{pp})} |D|^{\frac{p-1}{2}}}{2^{\alpha p} \Gamma_p(\alpha)}.$$

Thus the joint density of R and D is $h(R, D)$ which factorizes into functions of R and D. Hence R and D are independently distributed. Integrating out D we get the density of R. Note that $|D| = x_{11} \ldots x_{pp}$ and $0 < x_{jj} < \infty$, $j = 1, \ldots, p$. Integration over x_{jj} gives $2^{\alpha} \Gamma(\alpha)$ for each j. Thus the density of R reduces to $g(R)$. Hence the results.

For the next result we need a basic property of partitioned matrices. This will be stated as a lemma.

Lemma 2.1 *Let the $p \times p$ matrix A be partitioned into submatrices A_{11}, A_{12}, A_{21}, A_{22} where A_{11} is $r \times r$, $r < p$. Then*

$$A = \begin{bmatrix} A_{11} & A_{12} \\ A_{21} & A_{22} \end{bmatrix} \Rightarrow$$
$$|A| = |A_{11}| \, |A_{22} - A_{21} A_{11}^{-1} A_{12}|, \quad |A_{11}| \neq 0$$
$$= |A_{22}| \, |A_{11} - A_{12} A_{22}^{-1} A_{21}|, \quad |A_{22}| \neq 0.$$

The proof is trivial. For example, reduce A_{21} to a null matrix by adding $-A_{21} A_{11}^{-1}$ times the first r rows of A to the remaining $p - r$ rows if $|A_{11}| \neq 0$. Then the determinant remains the same and thus

$$\begin{vmatrix} A_{11} & A_{12} \\ A_{21} & A_{22} \end{vmatrix} = \begin{vmatrix} A_{11} & A_{12} \\ 0 & A_{22} - A_{21} A_{11}^{-1} A_{12} \end{vmatrix}.$$

Now the matrix on the right side determinant is block triangular and hence the determinant is the product of the determinants of the diagonal blocks which is $|A_{11}| \, |A_{22} - A_{21} A_{11}^{-1} A_{12}|$.

Theorem 2.4 *Let $X = (x_{jk})$, $W = (w_{jk})$ be lower triangular matrices of distinct real variables such that $x_{jj} > 0$, $w_{jj} > 0$ and $\sum_{k=1}^{j} w_{jk}^2 = 1$, $j = 1, \ldots, p$. Let $D = \text{diag}(\lambda_1, \ldots, \lambda_p)$, λ_j, $j = 1, \ldots, p$ be real positive and distinct. Let $D^{\frac{1}{2}} = \text{diag}(\lambda_1^{\frac{1}{2}}, \ldots, \lambda_p^{\frac{1}{2}})$. Then*

$$X = DW \Rightarrow dX = \left\{ \prod_{j=1}^{p} \lambda_j^{j-1} w_{jj}^{-1} \right\} dD \, dW, \qquad (a)$$

$$X = D^{\frac{1}{2}} W \Rightarrow dX = \left\{ 2^{-p} \prod_{j=1}^{p} (\lambda_j^{\frac{1}{2}})^{j-2} w_{jj}^{-1} \right\} dD \, dW, \qquad (b)$$

$$X = D^{\frac{1}{2}} WW'D^{\frac{1}{2}} \Rightarrow dX = \left\{ \prod_{j=1}^{p} \lambda_j^{\frac{p-1}{2}} w_{jj}^{p-j} \right\} dD \ dW, \qquad (c)$$

and

$$X = W'DW \Rightarrow dX = \left\{ \prod_{j=1}^{p} (\lambda_j w_{jj})^{j-1} \right\} dD \ dW. \qquad (d)$$

Proof   Note that from $X = DW$ we have

$$x_{ii} = \lambda_i w_{ii} = \lambda_i \left(1 - \sum_{j=1}^{i-1} w_{ij}^2 \right)^{\frac{1}{2}}$$

and

$$x_{ij} = \lambda_i w_{ij}, \quad i > j$$

which then yield

$$\frac{\partial x_{ij}}{\partial \lambda_i} = w_{ij}, \quad i > j, \qquad \frac{\partial x_{ij}}{\partial w_{ij}} = \lambda_i, \quad i > j$$

$$\frac{\partial x_{ii}}{\partial \lambda_i} = w_{ii} \quad \text{and} \quad \frac{\partial x_{ii}}{\partial w_{ij}} = -\lambda_i w_{ii}^{-1} w_{ij}.$$

Take the x-variables in the order $x_{j1}, x_{j2}, ..., x_{j\,j-1}, x_{jj}, \ j = 1, ..., p$ and the w-variables in the same order except that the w_{jj}'s are replaced by λ_j's. Then the matrix of partial derivatives can be seen to be a block diagonal matrix and the Jacobian is the product of the determinants of all these diagonal blocks. The j-th diagonal block is the following:

$$\begin{bmatrix} \lambda_j & 0 & \cdots & 0 & -\lambda_j w_{jj}^{-1} w_{j1} \\ 0 & \lambda_j & \cdots & 0 & -\lambda_j w_{jj}^{-1} w_{j2} \\ \vdots & \vdots & \ddots & \vdots & \vdots \\ 0 & 0 & \cdots & \lambda_j & -\lambda_j w_{jj}^{-1} w_{j\,j-1} \\ w_{j1} & w_{j2} & \cdots & w_{j\,j-1} & w_{jj} \end{bmatrix}'.$$

The determinant of this matrix can be evaluated by using Lemma 2.1 by taking $A_{11} = \text{diag}(\lambda_j, ..., \lambda_j)$ and $A_{22} = w_{jj}$. Then the determinant is given by

$$\lambda_j^{j-1} \left[w_{jj} + \frac{1}{w_{jj}} (w_{j1}, ..., w_{j\,j-1})(w_{j1}, ..., w_{j\,j-1})' \right]$$

$$= \lambda_j^{j-1} \left[\frac{w_{j1}^2 + \ldots + w_{jj}^2}{w_{jj}} \right] = \frac{\lambda_j^{j-1}}{w_{jj}}$$

since $\sum_{k=1}^{j} w_{jk}^2 = 1$. Now multiplying these determinants result (a) is established. For result (b), use the fact that $d\lambda_j^{\frac{1}{2}} = \frac{1}{2}\lambda_j^{-\frac{1}{2}}d\lambda_j$. For proving (c), use (b) and the first part of Theorem 1.28 of Chapter 1. For proving (d), note that in (b) $dX = dX'$. Now take the transpose in (b) and use the second part of Theorem 1.28 of Chapter 1. This completes the proofs.

Theorem 2.5　　*Let X, W and D be as defined in Theorem 2.4. Then*

$$X = WD \Rightarrow dX = \left\{ \prod_{j=1}^{p} \lambda_j^{p-j} w_{jj}^{-1} \right\} dD\ dW, \tag{a}$$

$$X = WD^{\frac{1}{2}} \Rightarrow dX = \left\{ 2^{-p} \prod_{j=1}^{p} (\lambda_j^{\frac{1}{2}})^{p-j-1} w_{jj}^{-1} \right\} dD\ dW, \tag{b}$$

$$X = WDW' \Rightarrow dX = \left\{ \prod_{j=1}^{p} (\lambda_j w_{jj})^{p-j} \right\} dD\ dW, \tag{c}$$

and

$$X = D^{\frac{1}{2}}W'WD^{\frac{1}{2}} \Rightarrow dX = \left\{ \prod_{j=1}^{p} \lambda_j^{\frac{p-1}{2}} w_{jj}^{j-1} \right\} dD\ dW. \tag{d}$$

The proof is parallel to that of Theorem 2.4 and hence omitted.

Example 2.2　　Let $R = (r_{ij})$, $r_{jj} = 1$, $j = 1, \ldots, p, -1 < r_{ij} = r_{ji} < 1$, $i \neq j$, have the density $g(R)$ of Example 2.1. Let $W = (w_{ij})$ be a real lower triangular matrix with $w_{jj} > 0$, $j = 1, \ldots, p$ and distinct such that $R = WW'$. Then (1) evaluate the density of W; (2) show that the sets $\{w_{j1}, \ldots, w_{j\,j-1}\}$, $j = 1, \ldots, p$ are independently distributed; (3) evaluate the density of w_{21}.

Solution　　From Example 2.1 the density of R, denoted by $f(R)$, is given by

$$f(R)dR = \begin{cases} \frac{[\Gamma(\alpha)]^p}{\Gamma_p(\alpha)} |R|^{\alpha - \frac{p+1}{2}} dR, & R = R' > 0 \\ \\ 0, & \text{elsewhere.} \end{cases}$$

Note that

$$|R| = |WW'| = \prod_{j=1}^{p} w_{jj}^2$$

and from Theorems 2.4 or 2.5 for $D = I$ or on evaluating the Jacobian directly from $R = WW'$, one has

$$dR = \left(\prod_{j=1}^{p} w_{jj}^{p-j} \right) dW.$$

Hence

$$f(R)dR = g(W)dW$$

where

$$g(W) = \frac{[\Gamma(\alpha)]^p}{\Gamma_p(\alpha)} |WW'|^{\alpha - \frac{p+1}{2}} \left\{ \prod_{j=1}^{p} w_{jj}^{p-j} \right\}.$$

That is,

$$g(W) = \begin{cases} \frac{[\Gamma(\alpha)]^p}{\Gamma_p(\alpha)} \left[\prod_{j=1}^{p} \left(w_{jj}^2 \right)^{\alpha - \frac{j+1}{2}} \right], & \sum_{k=1}^{j} w_{jk}^2 = 1, \ j = 1, ..., p \\ 0, & \text{elsewhere.} \end{cases}$$

Here $g(W)$ is the density of W and since it factorizes into $w_{jj}^2 = 1 - w_{j1}^2 - ... - w_{j\,j-1}^2$, $j = 1, ..., p$, the sets $\{w_{j1}, ..., w_{j\,j-1}\}, j = 1, ..., p$ are independently distributed. Note that $w_{11}^2 = 1$. The density of w_{21}, denoted by $g_{21}(w_{21})$, is given by

$$g_{21}(w_{21}) = \begin{cases} c \left(w_{22}^2 \right)^{\alpha - \frac{3}{2}}, & w_{21}^2 + w_{22}^2 = 1 \\ 0, & \text{elsewhere} \end{cases}$$

where c is a normalizing constant. But $w_{22}^2 = 1 - w_{21}^2$ with $-1 < w_{21} < 1$. By integrating out w_{21} with the help of a type-1 beta integral, we see that

$$c = \frac{\Gamma(\alpha)}{\Gamma\left(\frac{1}{2}\right) \Gamma\left(\alpha - \frac{1}{2}\right)}.$$

Hence the density of w_{21} is

$$g_{21}(w_{21}) = \begin{cases} \frac{\Gamma(\alpha)}{\Gamma(\frac{1}{2})\Gamma(\alpha - \frac{1}{2})} \left(1 - w_{21}^2 \right)^{\alpha - \frac{3}{2}}, & -1 < w_{21} < 1 \\ 0, & \text{elsewhere.} \end{cases}$$

The density of w_{22}, denoted by $g_{22}(w_{22})$ is given by

$$g_{22}(w_{22}) = \begin{cases} 2c\left(w_{22}^2\right)^{\alpha-1}\left(1 - w_{22}^2\right)^{-\frac{1}{2}}, & 0 < w_{22} < 1 \\ 0, & \text{elsewhere} \end{cases}$$

with the same c appearing above. Note also that the joint density of $w_{j1}^2, w_{j2}^2, ..., w_{jj-1}^2$ is a type-1 Dirichlet. Details of Dirichlet integrals and Dirichlet function are given in Section 5.1.8 of Chapter 5, see also Section 2.4.2 of this chapter. Dirichlet integrals were introduced and studied by the German mathematician G.P.L. Dirichlet (1805-1859).

Remark 2.1   It is easy to note that the transformations in Theorems 2.1–2.5 are one-to-one when the λ_j's and w_{jj}'s are real positive and distinct.

Theorem 2.6   *Let X, T, U be $p \times p$ matrices of functionally independent real variables where all the principal minors of X are nonzero, T is lower triangular and U is lower triangular with unit diagonal elements. Then, ignoring the sign,*

$$X = TU' \Rightarrow dX = \left\{ \prod_{j=1}^{p} |t_{jj}|^{p-j} \right\} dT \, dU, \tag{a}$$

and

$$X = T'U \Rightarrow dX = \left\{ \prod_{j=1}^{p} |t_{jj}|^{j-1} \right\} dT \, dU. \tag{b}$$

Proof   Take the differentials in $X = TU'$ to get

$$(dX) = (dT)U' + T(dU').$$

Denote the differentials by $x_{ij}^* = dx_{ij}$, $t_{ij}^* = dt_{ij}$, $u_{ij}^* = du_{ij}$ and note that $u_{jj}^* = 0$, $j = 1, ..., p$, $x_{ij}^* = 0 = t_{ij}^* = u_{ij}^*$ for $i < j$. Take the wedge product in the first row of (dX) to get

$$x_{11}^* ... x_{1p}^* = (t_{11}^* + 0)(t_{11}^* u_{21} + t_{11}u_{21}^*)$$
$$\times ...(t_{11}^* u_{p1} + t_{11}u_{p1}^*)$$
$$= t_{11}^* t_{11}^{p-1} u_{21}^* ... u_{p1}^*$$

deleting the higher powers of the differentials. From the second row we get

$$x_{21}^* \ldots x_{2p}^* = t_{21}^* t_{22}^* t_{22}^{p-2} u_{32}^* \ldots u_{p2}^*$$

deleting the higher powers, and so on. Thus the product of all these gives

$$dX = \left\{ t_{11}^{p-1} t_{22}^{p-2} \ldots t_{p-1\,p-1} \right\} dT \, dU.$$

This establishes (a). Similar steps hold in the proof for (b).

Remark 2.2   In the transformations $X = TU'$ and $X = T'U$ of Theorem 2.6, all the variables are free variables. Note that $-\infty < t_{ij} < \infty$, $i \geq j$, that is, including the case $i = j$, and $-\infty < u_{ij} < \infty$, $i > j$.

Example 2.3   Verify Theorem 2.6 for 3×3 matrices.

Solution   Let

$$X = \begin{bmatrix} x_{11} & x_{12} & x_{13} \\ x_{21} & x_{22} & x_{23} \\ x_{31} & x_{32} & x_{33} \end{bmatrix}, \quad T = \begin{bmatrix} t_{11} & 0 & 0 \\ t_{21} & t_{22} & 0 \\ t_{31} & t_{32} & t_{33} \end{bmatrix}, \quad U = \begin{bmatrix} 1 & 0 & 0 \\ u_{21} & 1 & 0 \\ u_{31} & u_{32} & 1 \end{bmatrix}.$$

Then

$$X = TU' = \begin{bmatrix} t_{11} & t_{11}u_{21} & t_{11}u_{31} \\ t_{21} & t_{21}u_{21} + t_{22} & t_{21}u_{31} + t_{22}u_{32} \\ t_{31} & t_{31}u_{21} + t_{32} & t_{31}u_{31} + t_{32}u_{32} + t_{33} \end{bmatrix}.$$

That is, $t_{11} = x_{11}$, $t_{21} = x_{21}$, $t_{31} = x_{31}$, $u_{21} = \frac{x_{12}}{x_{11}}$ and so on. Note that the t_{ij}'s and u_{ij}'s are uniquely determined in terms of the x_{ij}'s when the principal minors of X are nonzeros. Taking the matrices of partial derivatives we have

$$\frac{\partial (x_{11}, x_{21}, x_{31})}{\partial (t_{11}, t_{21}, t_{31})} = I$$

$$\frac{\partial (x_{12}, x_{22}, x_{32})}{\partial (u_{21}, t_{22}, t_{32})} = \begin{bmatrix} t_{11} & 0 & 0 \\ t_{21} & 1 & 0 \\ t_{31} & 0 & 1 \end{bmatrix}$$

$$\frac{\partial (x_{13}, x_{23}, x_{33})}{\partial (u_{31}, u_{32}, t_{33})} = \begin{bmatrix} t_{11} & 0 & 0 \\ t_{21} & t_{22} & 0 \\ t_{31} & t_{32} & 1 \end{bmatrix}$$

and the Jacobian matrix itself is triangular. Hence the Jacobian is the product of the determinants of the above matrices. That is, $t_{11}^2 t_{22}$. This verifies the theorem.

Theorem 2.7   *Let X, U, D be $p \times p$ matrices of functionally independent real variables where X is symmetric with the leading principal minors nonzero, U is lower triangular with unit diagonal elements and D is diagonal with distinct nonzero diagonal elements $\lambda_1, ..., \lambda_p$. Then*

$$X = UDU' \Rightarrow dX = \left\{ \prod_{j=1}^{p} |\lambda_j|^{p-j} \right\} dU \ dD. \qquad (a)$$

Proof   Note that the transformation is one-to-one. Take the differentials in $X = UDU'$ to get

$$(dX) = (dU)DU' + U(dD)U' + UD(dU'). \qquad (b)$$

Pre and postmultiply (b) by U^{-1} and U'^{-1} respectively to get

$$U^{-1}(dX)U'^{-1} = U^{-1}(dU)D + (dD) + D(dU')U'^{-1}. \qquad (c)$$

Let

$$(dY) = U^{-1}(dX)U'^{-1}, \ (dV) = U^{-1}(dU). \qquad (d)$$

Then

$$(dY) = (dV)D + (dD) + D(dV)'. \qquad (e)$$

From Theorem 1.20 of Chapter 1 we have $dY = dX$ since the determinant of U is unity. From Theorem 1.15 of Chapter 1 we have $dV = dU$. Note that $du_{11} = 0 \Rightarrow dv_{11} = 0$. Take the wedge product in (e). The product of the first row elements gives $\lambda_1^{p-1} dv_{21}...dv_{p1} \ d\lambda_1$, and so on. Thus the product of all the elements gives the final result.

Example 2.4   Let X be a $p \times p$ real symmetric positive definite random matrix having a matrix-variate real gamma density

$$f(X) = \frac{|X|^{\alpha - \frac{p+1}{2}} e^{-\text{tr}(X)}}{\Gamma_p(\alpha)}, \ X > 0, \ \text{Re}(\alpha) > \frac{p-1}{2}$$

and $f(X) = 0$ elsewhere. Let $X = UDU'$ where $U = (u_{jk})$, $u_{jk} = 0$, $j < k$, $u_{jj} = 1$, is a lower triangular matrix with unit diagonal elements and

$D = \text{diag}(\lambda_1, ..., \lambda_p)$ where the λ_j's are positive and distinct. Show that (1) the λ_j's are independently distributed; (2) $\left\{u_{jk}^2, j = k+1, ..., p\right\}, k = 1, ..., p-1$ are mutually independently distributed; (3) $\left\{u_{jk}^2, j = k+1, ..., p\right\}$ has a type-2 Dirichlet distribution for each k.

Solution Consider the transformation

$$X = UDU' \Rightarrow dX = \left\{\prod_{j=1}^{p} |\lambda_j|^{p-j}\right\} dU dD.$$

Note that since the diagonal elements of U are unities

$$|X| = |UDU'| = |D| = \lambda_1...\lambda_p$$

and

$$\text{tr}(X) = \text{tr}(UDU') = \text{tr}(DU'U)$$
$$= \lambda_1(1 + u_{21}^2 + ... + u_{p1}^2) + \lambda_2(1 + u_{32}^2 + ... + u_{p2}^2) + ... + \lambda_p.$$

Then the joint density factorizes into functions of $\lambda_1, ..., \lambda_p$ and hence the λ_j's are independently distributed. Integrating out $\lambda_1, ..., \lambda_p$, we have the marginal density of U given by

$$g(U) = C \prod_{k=1}^{p-1} \left(1 + \sum_{j=k+1}^{p} u_{jk}^2\right)^{-(\alpha - \frac{p+1}{2} + p - k)}$$

where C is a normalizing constant. The sets $\left\{u_{jk}^2, j = k+1, ..., p\right\}$, $k = 1, ..., p-1$ are independently distributed since $g(U)$ factorizes . For a specific k the marginal density of $\left\{u_{jk}^2, \ j = k+1, ..., p\right\}$ is given by

$$g_k = C_k \left(1 + \sum_{j=k+1}^{p} u_{jk}^2\right)^{-(\alpha - \frac{p+1}{2} + p - k)} , \ 0 < u_{jk}^2 < \infty$$

where C_k is the normalizing constant. This has the structure of a type-2 Dirichlet density and hence $u_{jk}^2, j = k+1, ..., p$ is a type-2 Dirichlet with the parameters $(1, 1, ..., 1; \alpha - \frac{p+1}{2})$. Hence the result.

Theorem 2.8 *Let X be a $p \times p$ matrix of functionally independent real variables and with real distinct positive eigenvalues $\lambda_1 > ... > \lambda_p > 0$. Let*

Q be a nonsingular matrix such that $X = QDQ^{-1}$, $D = \text{diag}(\lambda_1, ..., \lambda_p)$. Let $Y = F(X) = QD_{f(\lambda)}Q^{-1}$ where $D_{f(\lambda)} = \text{diag}(f(\lambda_1), ..., f(\lambda_p))$ with $f(x)$ being a differentiable function of x such that $f(\lambda_i) - f(\lambda_j) \neq 0$, for all i and j, $i \neq j$, $f'(\lambda_j) \neq 0$, $j = 1, ..., p$. Then, ignoring the sign,

$$X = QDQ^{-1} \Rightarrow$$

$$dX = \left\{ \prod_{i<j} (\lambda_i - \lambda_j)^2 \right\} dD \wedge [Q^{-1}(dQ)],$$

for a general X and

$$dX = \left\{ \prod_{i<j} |\lambda_i - \lambda_j| \right\} dD \wedge [Q^{-1}(dQ)], \qquad (a)$$

for $X = X'$, where $\wedge [Q^{-1}(dQ)]$ denotes the wedge product in $Q^{-1}(dQ)$, and

$$Y = F(X) = QD_{f(\lambda)}Q^{-1} \Rightarrow$$

$$dY = \left\{ \prod_{i<j} \left| \frac{f(\lambda_i) - f(\lambda_j)}{\lambda_i - \lambda_j} \right|^2 \right\} \left\{ \prod_{j=1}^{p} |f'(\lambda_j)| \right\} dX,$$

for a general X and

$$dY = \left\{ \prod_{i<j} \left| \frac{f(\lambda_i) - f(\lambda_j)}{\lambda_i - \lambda_j} \right| \right\} \left\{ \prod_{j=1}^{p} |f'(\lambda_j)| \right\} dX, \qquad (b)$$

for $X = X'$, where $f'(x)$ denotes the derivative of $f(x)$ with respect to x.

Proof   Taking the differentials we have

$$I = Q^{-1}Q \Rightarrow Q^{-1}(dQ) = -(dQ^{-1})Q.$$

Let

$$(dG) = Q^{-1}(dQ).$$
$$X = QDQ^{-1} \Rightarrow$$
$$(dX) = (dQ)DQ^{-1} + Q(dD)Q^{-1} + QD(dQ^{-1}). \qquad (c)$$

Pre and postmultiply (c) by Q^{-1} and Q respectively to get

$$Q^{-1}(dX)Q = (dG)D + (dD) - D(dG). \qquad (d)$$

Let $(\mathrm{d}\mathbf{W}) = Q^{-1}(\mathrm{d}\mathbf{X})Q$. Then from Theorem 1.18 of Chapter 1, we have $\mathrm{d}X = \mathrm{d}W$. Note that from (d) we have

$$\mathrm{d}w_{ij} = (\lambda_i - \lambda_j)\mathrm{d}g_{ij}, \ i \neq j$$
$$= \mathrm{d}\lambda_i, \ i = j.$$

Taking the wedge product, one has the result in (a). Note that when W is not symmetric both $(\lambda_i - \lambda_j)$ and $(\lambda_j - \lambda_i)$, for fixed i and j, $i \neq j$, enter into the product giving $-(\lambda_i - \lambda_j)^2$, and when W is symmetric $(\lambda_i - \lambda_j)$ appears only once. Now consider

$$Y = QD_{f(\lambda)}Q^{-1} \Rightarrow Q^{-1}YQ = D_{f(\lambda)}. \tag{e}$$

Using the results from (d) and (a) we have, for a general X,

$$\mathrm{d}Y = \left\{ \prod_{i<j} (f(\lambda_i) - f(\lambda_j))^2 \right\} \mathrm{d}D_{f(\lambda)} \wedge \left[Q^{-1}(\mathrm{d}\mathbf{Q})\right],$$

and for $X = X'$,

$$\mathrm{d}Y = \left\{ \prod_{i<j} |f(\lambda_i) - f(\lambda_j)| \right\} \mathrm{d}D_{f(\lambda)} \wedge \left[Q^{-1}(\mathrm{d}\mathbf{Q})\right]. \tag{f}$$

But by taking the product of the differentials we have

$$\mathrm{d}D_{f(\lambda)} = \left\{ \prod_{j=1}^{p} |f'(\lambda_j)| \right\} \mathrm{d}D. \tag{g}$$

Take $\mathrm{d}D \wedge \left[Q^{-1}(\mathrm{d}\mathbf{Q})\right]$ from (a) and substitute in (f) to establish the result (b).

Remark 2.3 X written as a function of Q and D as in Theorem 2.8 is evidently not a one-to-one transformation. Note that if Q is multiplied by any scalar QDQ^{-1} still remains the same. Hence $\mathrm{d}X$ written in terms of $\mathrm{d}D$ and $\mathrm{d}Q$ should be interpreted as the Jacobian for a specific Q. But whichever Q is taken $\wedge \left[Q^{-1}(\mathrm{d}\mathbf{Q})\right]$ remains the same.

An explicit form of the wedge product of the type $\wedge \left[Q^{-1}(\mathrm{d}\mathbf{Q})\right]$ will be evaluated later on. As a corollary, one can get the Jacobian when transforming X to X^n by taking $f(\lambda)$ in Theorem 2.8 as λ^n. Then

$$\prod_{j=1}^{p} f'(\lambda_j) = \prod_{j=1}^{p} (n\lambda_j^{n-1})$$

$$= n^p \left\{ \prod_{j=1}^{p} \lambda_j \right\}^{n-1} = n^p |X|^{n-1}.$$

Corollary 2.8.1   *Let X be a $p \times p$ matrix of functionally independent real variables and with real distinct and positive eigenvalues $\lambda_1, ..., \lambda_p$. Then, ignoring the sign,*

$$Y = X^n \Rightarrow$$

$$dY = n^p |X|^{n-1} \left\{ \prod_{i<j} \left(\frac{\lambda_i^n - \lambda_j^n}{\lambda_i - \lambda_j} \right)^2 \right\} dX,$$

for a general X, and

$$dY = n^p |X|^{n-1} \left\{ \prod_{i<j} \left| \frac{\lambda_i^n - \lambda_j^n}{\lambda_i - \lambda_j} \right| \right\} dX, \quad \text{for } X = X'$$

where $|X| = |Y|^{\frac{1}{n}}$ is the n-th positive root of $|Y|$.

Example 2.5   Let X be a $p \times p$ real symmetric positive definite random matrix with distinct and positive eigenvalues and having the density

$$f(X) = \frac{|X|^{\alpha - \frac{p+1}{2}} e^{-\text{tr}(X)}}{\Gamma_p(\alpha)}, \quad X = X' > 0, \ \text{Re}(\alpha) > \frac{p-1}{2},$$

evaluate the density of $Y = X^m$, $m > 0$, where $Y^{\frac{1}{m}}$ denotes the m-th positive definite symmetric root of Y.

Solution   From Theorem 2.8

$$Y = X^m \Rightarrow dY = m^p |X|^{m-1} \left\{ \prod_{j<k} \left| \frac{\lambda_j^m - \lambda_k^m}{\lambda_j - \lambda_k} \right| \right\} dX$$

where the λ_j's are the eigenvalues of X. Hence the eigenvalues of Y, denoted by μ_j's, are such that $\mu_j = \lambda_j^m$. Consider the m-th symmetric root of Y. Then $|X| = |Y|^{\frac{1}{m}}$. Hence the density of Y, denoted by $g(Y)$ is given by

$$g(Y) = |Y|^{\frac{a}{m} - \frac{p+1}{2m} - \frac{m-1}{m}} e^{-\text{tr}\left(Y^{\frac{1}{m}}\right)}$$

$$\times m^{-p} \left\{ \prod_{j<k} \left| \frac{\mu_j - \mu_k}{\mu_j^{\frac{1}{m}} - \mu_k^{\frac{1}{m}}} \right| \right\}^{-1}, \quad Y = Y' > 0$$

and $g(Y) = 0$ elsewhere.

2.1.2   Decomposition with one matrix being skew symmetric

For skew symmetric matrices the following properties are well known. If X is skew symmetric, that is, $X' = -X$ then the diagonal elements of X are zeros and no eigenvalue of X is equal to -1. Then $|X + I| \neq 0$ and $(X + I)^{-1}$ exists. Any nonsingular matrix Y of p^2 functionally independent real elements can be represented in terms of a skew symmetric matrix X and a lower triangular matrix T in the form $Y = T\left[2(X + I)^{-1} - I\right]$. When X is skew symmetric $Z = 2(X + I)^{-1} - I$ is orthogonal, that is, $ZZ' = I$. If X is skew symmetric with the elements as functionally independent variables then there are $p(p-1)/2$ variables in X since the diagonal elements are zeros. If T is lower triangular with functionally independent elements then there are $p(p+1)/2$ variables in T. Thus combined, there are $\frac{p(p-1)}{2} + \frac{p(p+1)}{2} = p^2$ variables in X and T together.

$$Z = 2(I + X)^{-1} - I \Rightarrow$$
$$X = 2(I + Z)^{-1} - I.$$

Then Z is uniquely determined by X and vice versa. By solving the equation $Y = TZ$, where T is lower triangular, one can see that every element t_{jk}, $j \geq k$ and x_{jk}, $j < k$ can be written as a function of the elements of Y under the following two situations: (1) $t_{jj} > 0$, $j = 1, ..., p$, $-\infty < t_{jk} < \infty$, $j > k$, $-\infty < x_{jk} < \infty$, $j < k$, that is, all the diagonal elements of T are positive, the nonzero nondiagonal elements of T as well as the nonzero elements of X are arbitrary, and (2) $-\infty < t_{jk} < \infty$, $j \geq k$ and x_{jk}'s are restricted in such a way that there is a unique choice of Z.

Note that since $ZZ' = I$, Z being orthonormal, if any column of Z is multiplied by -1 the orthonormality still holds. One way of selecting a unique orthonormal matrix is to restrict one element each from each row and column or the diagonal elements to be of a specific sign. But when X is skew symmetric it can be shown that when the first column of Z is multiplied by -1, the resulting determinant $|Z_1 + I| = 0$ where Z_1 is Z with the first column multiplied by -1. It can be shown that $|Z + I|$ enters into the Jacobian of the transformation $Y = TZ$. Let G be a diagonal matrix with the first diagonal element -1 and all other diagonal elements 1. Any matrix postmultiplied by G will result in the first column of that matrix multiplied by -1. Then $Z_1 = ZG$. That is,

$$|Z_1 + I| = |ZG + I| = |2(I + X)^{-1}G - G + I|$$
$$= |I + X|^{-1}|I + G - XG + X|.$$

Note that the first row of $I + G - XG + X$ is a null vector since X is skew symmetric and hence

$$|I + G - XG + X| = 0 \Rightarrow |Z_1 + I| = 0.$$

Thus only $p - 1$ restrictions are sufficient to determine Z uniquely. For example, restrict the first row elements, except the first element, that is, the elements in the $(1, 2), (1, 3), ..., (1, p)$ positions of Z or of $(I + X)^{-1}$ to be negative.

Note that if all the diagonal elements of T are restricted to be positive, that is, $t_{jj} > 0$, $j = 1, ..., p$ and t_{jk}, $j > k$, x_{jk}, $j < k$ arbitrary then not every point in the p^2-demensional Y-space is described by T and X in the transformation $Y = TZ$. Take for example $p = 1$. Then $Y = TZ$ gives $y_{11} = t_{11}$. Note that $-\infty < y_{11} < \infty$ whereas $0 < t_{11} < \infty$. It is not difficult to see that every point in the Y-space is described by $t_{jj} > 0$, $j = 1, ..., p - 1$, $-\infty < t_{pp} < \infty$, $-\infty < t_{jk} < \infty$, $j > k$, $-\infty < x_{jk} < \infty$, $j < k$. Thus the following will be two unique choices of T and Z in the transformation $Y = TZ$: (i) $t_{jj} > 0$, $j = 1, ..., p - 1$, $-\infty < t_{pp} < \infty$, $-\infty < t_{jk} < \infty$, $j > k$, $-\infty < x_{jk} < \infty$, $j < k$ or (ii) $-\infty < t_{jk} < \infty$, $j \geq k$, $-\infty < x_{1k}^* < 0$, $k = 2, ..., p$, $-\infty < x_{jk}^* < \infty$, $j < k$, $j \neq 1$ where x_{jk}^*'s are the elements of $(I + X)^{-1}$. We will make use of these properties for the next theorem.

Theorem 2.9   *Let Y, X and T be $p \times p$ matrices where Y and T are nonsingular, X is skew symmetric and T is lower triangular of functionally independent real variables y_{ij}'s, x_{ij}'s and t_{ij}'s respectively. Let $t_{jj} > 0$, $j = 1, ..., p - 1$, $-\infty < t_{jk} < \infty$, $j > k$, $-\infty < t_{pp} < \infty$, $-\infty < x_{jk} < \infty$, $j < k$*

or $-\infty < t_{jk} < \infty$, $j \geq k$ *and the first row elements, except the first one,*
of $(I + X)^{-1}$ *are negative. Then the unique representation*

$$Y = T\left[2(X + I)^{-1} - I\right] = T(I - X)(I + X)^{-1} \Rightarrow$$

$$dY = 2^{p(p-1)/2}\left\{\prod_{j=1}^{p}|t_{jj}|^{p-j}\right\}|X + I|^{-(p-1)}dT\ dX.$$

Proof   Take the differentials to get

$$(dY) = (dT)\left[2(X + I)^{-1} - I\right] + T\left[2(d(X + I)^{-1})\right]$$

since $(d(I)) = 0$. But from (1.3.10) of Chapter 1

$$\left(d(X + I)^{-1}\right) = -(X + I)^{-1}(dX)(X + I)^{-1}$$

$$= -\frac{1}{4}(Z + I)(dX)(Z + I)$$

where

$$Z = 2(X + I)^{-1} - I,\quad ZZ' = I.$$

Thus

$$(dY) = (dT)Z - \frac{1}{2}T\left[Z + I\right](dX)\left[Z + I\right]. \tag{a}$$

Pre and postmultiply (a) by T^{-1} and Z' respectively to get

$$T^{-1}(dY)Z' = T^{-1}(dT) - \frac{1}{2}(Z + I)(dX)(Z' + I). \tag{b}$$

The Jacobian of the transformation of T and X going to Y is equal to the
Jacobian of $(dT), (dX)$ going to (dY). Now treat $(dT), (dX)$ and (dY) as
variables and everything else as constants. Let

$$(dU) = T^{-1}(dY)Z',\quad (dV) = T^{-1}(dT)$$

and

$$(dW) = (Z + I)(dX)(Z' + I).$$

From Theorem 1.18 of Chapter 1, excluding the sign,

$$dU = |T|^{-p}|Z'|^p dY = |T|^{-p}dY \Rightarrow$$

$$dY = |T|^p dU = \left\{\prod_{j=1}^{p}|t_{jj}|^p\right\}dU. \tag{c}$$

Note that $|Z'| = \pm 1$ since Z is orthogonal. Since X is skew symmetric one has from Theorem 1.21 of Chapter 1

$$dW = |Z + I|^{p-1}dX = 2^{p(p-1)}|X + I|^{-(p-1)}dX. \qquad (d)$$

From Theorem 1.15 of Chapter 1

$$dV = \left\{ \prod_{j=1}^{p} |t_{jj}|^{-j} \right\} dT. \qquad (e)$$

Now (b) is the same as

$$(\mathrm{dU}) = (\mathrm{dV}) - \frac{1}{2}(\mathrm{dW}) \Rightarrow U = V - \frac{1}{2}W. \qquad (f)$$

Let $U = (u_{ij})$, $V = (v_{ij})$, $W = (w_{ij})$. Then since T is lower triangular $t_{ij} = 0$, $i < j$ and thus V is lower triangular, and since X is skew symmetric $x_{jj} = 0$ for all j and $x_{ij} = -x_{ji}$ for $i \neq j$ and thus W is skew symmetric. Thus we have

$$u_{ii} = v_{ii}, \quad u_{ij} = -\frac{1}{2}w_{ij}, \quad i < j$$

$$u_{ij} = v_{ij} + \frac{1}{2}w_{ij}, \quad i > j.$$

Take the u-variables in the order u_{ii}, $i = 1, ..., p$, u_{ij}, $i < j$, u_{ij}, $i > j$ and the v-variables and w-variables in the order v_{ii}, $i = 1, ..., p$, w_{ij}, $i < j$, v_{ij}, $i > j$. Then the Jacobian matrix is of the following form:

$$\begin{bmatrix} I & 0 & 0 \\ 0 & -\frac{1}{2}I & 0 \\ 0 & \frac{1}{2}I & I \end{bmatrix}$$

and the determinant, in absolute value, is $(\frac{1}{2})^{p(p-1)/2}$. That is,

$$dU = 2^{-p(p-1)/2}dV \; dW.$$

Now substitute for dU, dW and dV from (c), (d) and (e) respectively to obtain the result. In terms of J the above transformations can be summarized as follows:

$$J(Y : X, T) = J(Y : U)J(U : V, W)J(V : T)J(W : X).$$

Remark 2.4 When evaluating an integral by using Theorem 2.9, integrate t_{jj}'s over $0 \leq t_{jj} < \infty$, $j = 1, ..., p-1$, t_{pp} over $-\infty < t_{pp} < \infty$, t_{jk}'s over $-\infty < t_{jk} < \infty$, $j > k$ and x_{jk}'s over $-\infty < x_{jk} < \infty$, $j < k$ or integrate over $-\infty < t_{jk} < \infty$, $j \geq k$ and x_{jk}'s such that the first row elements of $(I + X)^{-1}$, except the first element, are negative and other elements arbitrary, see also Exercises 2.1.8 to 2.1.12 at the end of this section.

Verification   Let us verify Theorem 2.9 by evaluating a known integral by using this theorem. Let Y be a 3×3 matrix of functionally independent real variables. Let $Y = TZ$ where T is a 3×3 lower triangular matrix and $Z = 2(I + X)^{-1} - I$ with $X' = -X$ a 3×3 skew symmetric matrix. Consider the integral

$$A = \int_Y e^{-\text{tr}(YY')} dY = \int_Y e^{-\sum_{j,k} v_{jk}^2} dY$$
$$= \prod_{j,k} \left\{ \int_{-\infty}^{\infty} e^{-v_{jk}^2} dy_{jk} \right\} = \pi^{p^2/2} = \pi^{9/2}.$$

By using Theorem 2.9

$$A = \int_T 2^{p(p-1)/2} \left\{ \prod_{j=1}^{p} |t_{jj}|^{p-j} \right\} e^{-\text{tr}(TT')} dT$$
$$\times |I + X|^{-(p-1)} dX.$$

For $p = 3$ and $X' = -X = -(x_{jk})$, $x_{jj} = 0$, $j = 1, 2, 3$, and by direct computation

$$|I + X|^{-(p-1)} = \left(1 + x_{12}^2 + x_{13}^2 + x_{23}^2\right)^{-2},$$
$$\prod_{j=1}^{p} |t_{jj}|^{p-j} = |t_{11}|^2 |t_{22}|, \quad 2^{p(p-1)/2} = 2^3,$$

and

$$\text{tr}(TT') = t_{11}^2 + t_{22}^2 + t_{33}^2 + t_{21}^2 + t_{31}^2 + t_{32}^2.$$

Consider the situation where $t_{jj} > 0$, $j = 1, 2$, $-\infty < t_{33} < \infty$, $-\infty < t_{jk} < \infty$, $j > k$, $-\infty < x_{jk} < \infty$, $j < k$. Then

$$2^3 \int_T e^{-\text{tr}(TT')} \left\{ \prod_{j=1}^{3} |t_{jj}|^{3-j} \right\} dT = 2 \left\{ \prod_{j=1}^{2} 2 \int_0^{\infty} t_{jj}^{3-j} e^{-t_{jj}^2} dt_{jj} \right\}$$
$$\times \int_{-\infty}^{\infty} e^{-t_{33}^2} dt_{33} \left\{ \prod_{j>k=1}^{3} \int_{-\infty}^{\infty} e^{-t_{jk}^2} dt_{jk} \right\}$$

$$= 2\Gamma\left(\frac{3}{2}\right)\Gamma(1)\Gamma\left(\frac{1}{2}\right)\left[\Gamma\left(\frac{1}{2}\right)\right]^3$$

$$= \pi^{5/2}.$$

$$\int_X |I+X|^{-(p-1)}dX = \int_{-\infty}^{\infty}\int_{-\infty}^{\infty}\int_{-\infty}^{\infty}(1+x_{12}^2+x_{13}^2+x_{23}^2)^{-2}dx_{12}dx_{13}dx_{23}$$

$$= \int_0^{\infty}\int_0^{\infty}\int_0^{\infty}u_{12}^{-\frac{1}{2}}u_{13}^{-\frac{1}{2}}u_{23}^{-\frac{1}{2}}$$

$$\times (1+u_{12}+u_{13}+u_{23})^{-2}du_{12}du_{13}du_{23}$$

$$= \frac{\left[\Gamma\left(\frac{1}{2}\right)\right]^4}{\Gamma(2)},$$

evaluating by using a type-2 Dirichlet integral. Now $\pi^{5/2}\pi^2 = \pi^{9/2}$ and hence the result is verified.

Now consider the situation where $-\infty < t_{jk} < \infty$, $j \geq k$ and the first row elements of $(I+X)^{-1}$, except the first element, are negative. Then the integral over T gives

$$\int_T = \left(\int_{-\infty}^{\infty}|t_{11}|^2e^{-t_{11}^2}dt_{11}\right)\left(\int_{-\infty}^{\infty}|t_{22}|e^{-t_{22}^2}dt_{22}\right)$$

$$\times \left(\int_{-\infty}^{\infty}e^{-t_{33}^2}dt_{33}\right)\left(\prod_{j>k=1}^{3}\int_{-\infty}^{\infty}e^{-t_{jk}^2}dt_{jk}\right)$$

$$= \frac{1}{2}\left[\Gamma\left(\frac{1}{2}\right)\right]^5.$$

Write

$$(I+X)^{-1} = \begin{pmatrix} 1 & w \\ -w' & I+X_1 \end{pmatrix}^{-1}$$

where $w = (x_{12}, x_{13})$, w' is its transpose and

$$I+X_1 = \begin{pmatrix} 1 & x_{23} \\ -x_{23} & 1 \end{pmatrix}.$$

Then taking the inverse by using properties of partitioned matrices we have

$$(I+X)^{-1} = \begin{pmatrix} \alpha^{-1} & \beta \\ -\beta' & \gamma \end{pmatrix}$$

where
$$\alpha = 1 + w(I + X_1)^{-1}w', \quad \beta' = -\alpha^{-1}(I + X_1)^{-1}w'.$$

Put
$$(I + X_1)^{-1}w' = u' \Rightarrow |I + X_1|^{-1}dw = du$$

and
$$1 + w(I + X_1)^{-1}w' = 1 + u(I + X_1)u'$$
$$= 1 + uu' = 1 + u_{12}^2 + u_{13}^2 > 0$$

since $uX_1u' = 0$ due to X_1 being skew symmetric. Then the condition that the first row elements, except the first element, are negative implies that the elements in β are negative which means that $u_{12} > 0$, $u_{13} > 0$ since $\alpha > 0$, observing that

$$\beta = -\alpha^{-1}(u_{12}, u_{13}).$$

But
$$|I + X| = |I + X_1| \left[1 + w(I + X_1)^{-1}w'\right]$$

and
$$\int |I + X|^{-2}dX = \int |I + X_1|^{-2} \left[1 + w(I + X_1)^{-1}w'\right]^{-2} dX$$
$$= \int_{X_1} |I + X_1|^{-1}dX_1$$
$$\times \int_{u_{12}>0, u_{13}>0} \left(1 + u_{12}^2 + u_{13}^2\right)^{-2} du_{12}du_{13}.$$

Now there are no more restrictions on the elements of X_1. Hence

$$\int_{X_1} |I + X_1|^{-1}dX_1 = \int_{-\infty}^{\infty} \left(1 + x_{23}^2\right)^{-1} dx_{23}$$
$$= \frac{\Gamma\left(\frac{1}{2}\right)\Gamma\left(\frac{1}{2}\right)}{\Gamma(1)} = \pi$$

and

$$\int_{u_{12}>0, u_{13}>0} \left(1 + u_{12}^2 + u_{13}^2\right)^{-2} du_{12}du_{13} = \frac{1}{2^2}\frac{\Gamma\left(\frac{1}{2}\right)\Gamma\left(\frac{1}{2}\right)\Gamma(1)}{\Gamma(2)}$$
$$= \frac{\pi}{2^2}.$$

Multiply the T-integral and the X-integral to obtain $2^{-3}\pi^{9/2}$ which verifies the result.

Remark 2.5   When X is skew symmetric and D is diagonal with distinct nonzero diagonal elements then it can be shown that a symmetric matrix Y can be uniquely represented in terms of X and D as

$$Y = \left[2(I + X)^{-1} - I\right] D \left[2(I + X)^{-1} - I\right]'$$

provided the following conditions are satisfied: (i) Y does not belong to the set of symmetric matrices which constitutes a set of measure zero in the $p(p+1)/2$-dimensional space and, (ii) the elements in any row of $(I+X)^{-1}$, except the first element, are of a specific sign, for example, the elements of the first row, except the first element, are negative. Under these conditions we will state the next theorem.

Theorem 2.10   *Let Y, X and D be $p \times p$ matrices of functionally independent real variables where Y is symmetric with distinct and nonzero eigenvalues, X is skew symmetric with the elements of the first row of $(I+X)^{-1}$, except the first element, negative and $D = \mathrm{diag}(\lambda_1, ..., \lambda_p)$, with $\lambda_1 > ... > \lambda_p$. Then, excluding the sign,*

$$Y = \left[2(I + X)^{-1} - I\right] D \left[2(I + X)^{-1} - I\right]' \Rightarrow$$

$$dY = 2^{p(p-1)/2}|I + X|^{-(p-1)} \left\{ \prod_{1=i<j}^{p} |\lambda_i - \lambda_j| \right\} dX \, dD.$$

Proof   Let $Z = 2(I + X)^{-1} - I$. Take the differentials to get

$$(d\mathbf{Y}) = 2\left(d(\mathbf{I} + \mathbf{X})^{-1}\right) DZ' + Z(d\mathbf{D})Z'$$
$$+ ZD\left[2(d(\mathbf{I} + \mathbf{X})^{-1})\right]'. \qquad (a)$$

Proceed as in the proof of Theorem 2.9, and observing that $(d\mathbf{X}') = -(d\mathbf{X})$ since $X' = -X$, to get

$$Z'(d\mathbf{Y})Z = -\frac{1}{2}(Z' + I)(d\mathbf{X})(Z + I)D + (d\mathbf{D})$$
$$+ \frac{1}{2}D(I + Z')(d\mathbf{X})(I + Z).$$

Put

$$(d\mathbf{U}) = Z'(d\mathbf{Y})Z, \quad (d\mathbf{W}) = (d\mathbf{D})$$
$$(d\mathbf{V}) = (Z' + I)(d\mathbf{X})(Z + I).$$

Then (a) gives

$$(\mathrm{d}\mathbf{U}) = -\frac{1}{2}(\mathrm{d}\mathbf{V})D + \frac{1}{2}D(\mathrm{d}\mathbf{V}) + (\mathrm{d}\mathbf{W}). \qquad (b)$$

But since $(\mathrm{d}Y)$ is symmetric and Z is orthogonal

$$\mathrm{d}U = |Z|^{p+1}\mathrm{d}Y = \mathrm{d}Y \qquad (c)$$

excluding the sign. Since X is skew symmetric we have

$$\mathrm{d}V = |I + Z|^{p-1}\mathrm{d}X = 2^{p(p-1)}|I + X|^{-(p-1)}\mathrm{d}X, \qquad (d)$$

and

$$\mathrm{d}W = \mathrm{d}D. \qquad (e)$$

In terms of J the above transformations can be summarized as follows:

$$J(Y : X, D) = J(Y : U)J(U : V, W)J(W : D)J(V : X).$$

From (b) we have

$$\mathrm{d}u_{ii} = \mathrm{d}w_{ii}, \quad \mathrm{d}u_{ij} = \frac{1}{2}(\lambda_i - \lambda_j)\mathrm{d}v_{ij}, \ i < j.$$

Take the u-variables in the order u_{ii}, $i = 1, ..., p$, u_{ij}, $i < j$ and the w and v-variables in the order w_{ii}, $i = 1, ..., p$, v_{ij}, $i < j$. Then the matrix of partial derivatives is of the following form:

$$\begin{bmatrix} I & 0 \\ 0 & M \end{bmatrix}$$

where M is a diagonal matrix with the diagonal elements $\frac{1}{2}(\lambda_i - \lambda_j)$, $i < j$. There are $p(p-1)/2$ elements. Hence the determinant of the above matrix, in absolute value, is $(\frac{1}{2})^{p(p-1)/2}\left\{\prod_{i<j}|\lambda_i - \lambda_j|\right\}$. That is,

$$\mathrm{d}U = 2^{-p(p-1)/2}\left\{\prod_{i<j}|\lambda_i - \lambda_j|\right\}\mathrm{d}V \ \mathrm{d}D. \qquad (f)$$

Hence from $(c), (d)$ and (e), excluding the sign,

$$\mathrm{d}Y = 2^{p(p-1)/2}|I + X|^{-(p-1)}\left\{\prod_{i<j}|\lambda_i - \lambda_j|\right\}\mathrm{d}X \ \mathrm{d}D.$$

Remark 2.6   When integrating over the skew symmetric matrix X using the transformation in Theorem 2.10, under the unique choice for $Z = 2(I + X)^{-1} - I$, observe that

$$2^{p(p-1)/2} \int_X |I + X|^{-(p-1)} dX = \frac{\pi^{\frac{p^2}{2}}}{\Gamma_p\left(\frac{p}{2}\right)}.$$

Note that the λ_j's are to be integrated out over $\infty > \lambda_1 > \lambda_2 > \ldots > \lambda_p$ and X over a unique choice of Z.

Remark 2.7   Theorem 2.10 can also be derived as a particular case in Section 2.3 when considering the singular value decomposition for the symmetric case.

Verification   Let us verify Theorem 2.10 by evaluating a known integral directly as well as by using the theorem. Consider a 3×3 real symmetric positive definite matrix Y with distinct eigenvalues $\lambda_1 > \lambda_2 > \lambda_3 > 0$. From a matrix-variate gamma integral

$$\int_Y e^{-\text{tr}(Y)} dY = \Gamma_p\left(\frac{p+1}{2}\right) = \Gamma_3(2) = \frac{\pi^2}{2}.$$

If Theorem 2.10 is used then we have to evaluate two integrals, one over X and one over $\lambda_1, \lambda_2, \lambda_3$. Since X is restricted by keeping the first row elements of $(I + X)^{-1}$, except the first one, negative we have

$$2^{\frac{p(p-1)}{2}} \int_X |I + X|^{-(p-1)} dX = \frac{\pi^{\frac{p(p+1)}{4}}}{\prod_{j=1}^{p} \Gamma\left(\frac{j}{2}\right)} = 2\pi^2$$

for $p = 3$, see also Exercise 2.1.10 at the end of this section. Let

$$A = \int_{\lambda_1 > \lambda_2 > \lambda_3 > 0} (\lambda_1 - \lambda_2)(\lambda_1 - \lambda_3)(\lambda_2 - \lambda_3) e^{-(\lambda_1 + \lambda_2 + \lambda_3)} d\lambda_1 d\lambda_2 d\lambda_3.$$

Make the transformations $u = \lambda_1 - \lambda_2$, $v = \lambda_2 - \lambda_3$. Then

$$A = \int_{u=0}^{\infty} \int_{v=0}^{\infty} \int_{\lambda_3=0}^{\infty} u(u+v)v e^{-(u+2v+3\lambda_3)} du \, dv \, d\lambda_3$$

$$= \frac{1}{4}.$$

But $\left(\frac{1}{4}\right)\left(2\pi^2\right) = \frac{\pi^2}{2}$ which verifies the result.

In Theorem 2.10, a $p \times p$ skew symmetric matrix X and a diagonal matrix D created $p + p(p-1)/2 = p(p+1)/2$ elements in the symmetric $p \times p$ matrix Y. If Y is not symmetric then the p^2 elements in Y can be created by a skew symmetric matrix and a lower or upper triangular matrix, that is, $p(p-1)/2 + p(p+1)/2 = p^2$. We will consider a theorem of this type next.

Theorem 2.11   *Let X and T be $p \times p$ nonsingular matrices of functionally independent real variables where X is skew symmetric and $T = (t_{ij})$, $t_{ij} = 0$, $i < j$ is lower triangular with t_{jj}, $j = 1, .., p$ real and distinct. Consider the transformation $Y = HTH'$ where*

$$H = (I + X)^{-1}(I - X) = 2(I + X)^{-1} - I = (I - X)(I + X)^{-1}$$

with the first row elements of $(I + X)^{-1}$, except the first one, all negative. Then

$$Y = HTH' \Rightarrow$$

$$dY = 2^{\frac{p(p-1)}{2}} |I + X|^{-(p-1)} \left\{ \prod_{i<j} |t_{ii} - t_{jj}| \right\} dT\, dX.$$

Proof   Take the differentials in $Y = HTH'$ to get

$$(dY) = (dH)TH' + H(dT)H' + HT(dH'). \tag{a}$$

Since $X = -X'$, $H'H = I = HH'$ and thus H is an orthogonal matrix. Pre and postmultiply (a) by H' and H respectively to get

$$H'(dY)H = H'(dH)T + (dT) + T(dH')H. \tag{b}$$

By taking the differentials in $H'H = I$, we note that $H'(dH)$ is a skew symmetric matrix. Make the following substitutions:

$$(dW) = H'(dY)H, \quad (dU) = H'(dH). \tag{c}$$

Then since H is orthogonal, $dW = dY$ ignoring the sign. Let us simplify $H'(dH)$. From (1.3.10) of Chapter 1

$$(dH) = (d\,[2(I + X)^{-1} - I])$$
$$= -2(I + X)^{-1}(dX)(I + X)^{-1}.$$

Note that

$$H' = (I - X)^{-1}(I + X)$$

and hence

$$H'(\mathrm{d}H) = -2(I - X)^{-1}(I + X)(I + X)^{-1}(\mathrm{d}X)(I + X)^{-1}$$
$$= -2(I - X)^{-1}(\mathrm{d}X)(I + X)^{-1}$$
$$= -2\left[(I + X)^{-1}\right]'(\mathrm{d}X)(I + X)^{-1}.$$

Thus since $(\mathrm{d}X)$ is skew symmetric

$$(\mathrm{d}U) = H'(\mathrm{d}H) \Rightarrow$$

$$dU = 2^{\frac{p(p-1)}{2}}|I + X|^{-(p-1)}\mathrm{d}X \tag{d}$$

ignoring the sign, see also Theorem 1.21 of Chapter 1. From (b) and (c) we have

$$(\mathrm{d}W) = (\mathrm{d}U)T + (\mathrm{d}T) + T(\mathrm{d}U)'$$
$$= (\mathrm{d}U)T + (\mathrm{d}T) - T(\mathrm{d}U)$$

since $(\mathrm{d}U)$ is skew symmetric. Note that T is lower triangular and $(\mathrm{d}U)$ is of the form

$$(\mathrm{d}U)T = \begin{bmatrix} 0 & -du_{21} & -... & -du_{p1} \\ du_{21} & 0 & -... & -du_{p2} \\ \vdots & \vdots & ... & \vdots \\ du_{p1} & du_{p2} & ... & 0 \end{bmatrix} \begin{bmatrix} t_{11} & 0 & ... & 0 \\ t_{21} & t_{22} & ... & 0 \\ \vdots & \vdots & ... & \vdots \\ t_{p1} & t_{p2} & ... & t_{pp} \end{bmatrix}.$$

Then the elements are the following.

$$dw_{ii} = [-du_{i+1\,i}t_{i+1\,i} - ... - du_{pi}t_{pi}] + [dt_{ii}]$$
$$\qquad + [t_{i1}du_{i1} + ... + t_{i\,i-1}du_{i\,i-1}]$$
$$dw_{ij} = [-du_{ji}t_{jj} - ... - du_{pi}t_{pj}] + [t_{i1}du_{j1} + ... + t_{ii}du_{ji}],\ i < j$$
$$\qquad = [du_{ij}t_{jj} + ... + du_{i\,i-1}t_{j\,i-1}] + [dt_{ij}]$$
$$\qquad + [-t_{i1}du_{j1} - ... - t_{ii}du_{ji}],\ i > j.$$

Taking $w_{ij}^* = dw_{ij}$, $u_{ij}^* = du_{ij}$, $t_{ij}^* = dt_{ij}$ as the new variables and everything else fixed, the matrix of partial derivatives of w_{ii}^*'s with respect to the t_{ii}^*'s gives the identity matrix and with respect to the t_{ij}^*'s gives a null matrix and so on. The Jacobian matrix J can be found to have the following structure:

$$
\begin{array}{cccc}
 & t_{ii}^* & t_{ij}^* & u_{ij}^* \\
w_{ii}^* & I & 0 & ** \\
w_{ij}^*,\ i<j & 0 & 0 & G \\
w_{ij}^*,\ i>j & 0 & I & **
\end{array}
\quad \Rightarrow J = \begin{bmatrix} I & 0 & ** \\ 0 & 0 & G \\ 0 & I & ** \end{bmatrix} \tag{e}
$$

where ** denotes the presence of some nonnull matrix and G has the diagonal elements $(t_{ii} - t_{jj})$ for $i < j$ and it can be reduced to a triangular form when taking the determinant. Since J is a triangular block matrix its determinant is given by

$$|J| = \left\{ \prod_{i<j} |t_{ii} - t_{jj}| \right\}. \tag{f}$$

Now from $(f), (d)$ and the fact that $dW = dY$ the theorem is established since

$$J(Y : X, T) = J(Y : W)J(W : H, T)J(H : X).$$

Remark 2.8 Note that the structure of the Jacobian matrix in (e) above remains the same even if T is upper triangular and hence Theorem 2.11 holds for both lower and upper triangular T.

Remark 2.9 The transformation $Y = HTH'$ is not unique; this can be verified by checking the case $p = 2$.

Another result containing a symmetric matrix and an orthogonal matrix, which can be written in terms of a skew symmetric matrix, is given in Section 2.2 as Theorem 2.22.

Parametrization in terms of skew symmetric matrices, the Jacobians associated with such decompositions, and many other Jacobians can be found in the collected papers of Hsu, see for example, Chung (1983), Deemer and Olkin (1951) and Njoroge (1988).

Exercises

2.1.1 If $X = f(Y)$ and $Z = h(W)$ then show that whenever the Jacobians are defined $J(X, Z : Y, W) = J(X : Y)J(Z : W)$.

2.1.2 Let X and Y be $p \times p$ symmetric matrices of functionally independent real variables and $u = \text{tr}(X)$. Then show that

$$X = uY \Rightarrow dX = u^{\frac{1}{2}p(p+1)-1}dudY \quad \text{and} \quad \text{tr}(Y) = 1.$$

2.1.3   Let X, T, U be $p \times p$ matrices of functionally independent real variables. Let all the principal minors of X be nonzero, T and U be upper triangular, T with unit diagonal elements. Then evaluate the Jacobians for the transformations $Y = TU'$ and $Z = T'U$.

2.1.4   Let X be a $p \times p$ symmetric positive definite matrix of functionally independent real random variables with X^2 having the density

$$f(X^2) = C|I + X^2|^{-(\beta + \frac{p+1}{2})}, \ \mathrm{Re}(\beta) > \frac{p-1}{2}$$

and $f(X^2) = 0$ elsewhere, where C is a normalizing constant. Evaluate C and the density of X, the symmetric positive definite square root of X^2.

2.1.5   Let X be a $p \times p$ symmetric positive definite matrix of functionally independent real random variables having the density

$$f(X) = \frac{|X|^{\alpha - \frac{p+1}{2}} e^{-\mathrm{tr}(X)}}{\Gamma_p(\alpha)}, \ X = X' > 0, \ \mathrm{Re}(\alpha) > \frac{p-1}{2}$$

and $f(X) = 0$ elsewhere. Let $D^{\frac{1}{2}} Y D^{\frac{1}{2}} = X$ where $D = \mathrm{diag}(\lambda_1, ..., \lambda_p)$ with the λ_j's real, distinct and positive such that Y has diagonal elements unities. Evaluate the density of Y.

2.1.6   Let $X = (x_{ij})$ be a $p \times p$ symmetric matrix of functionally independent real variables such that $x_{jj} = 1, \ j = 1, ..., p, \ -1 < x_{ij} < 1, \ i \neq j$. Let $X = WW'$ where $W = (w_{ij})$ is a lower triangular matrix with $w_{jj} > 0$, $j = 1, ..., p$ and distinct. Then show that

$$dX = \left(\prod_{j=1}^{p} w_{jj}^{p-j} \right) dW.$$

2.1.7   Evaluate dX in terms of dW if $W = (w_{ij})$ of Exercise 2.1.6 is upper triangular with $w_{jj} > 0, \ j = 1, ..., p$ and distinct, and $X = WW'$.

2.1.8   Let T, X be $p \times p$ matrices of functionally independent real variables where T is lower triangular and X skew symmetric. Show that Y can be uniquely represented as

$$Y = T \left[2(I + X)^{-1} - I \right]$$

if $|T^{-1}Y + I| \neq 0$.

2.1.9   Show that the condition for unique representation in Exercise 2.1.8 is also equivalent to either (*i*) the first row elements of $(I + X)^{-1}$, except the first element, are all negative , or (*ii*) the elements in the following row vector are all positive

$$\frac{x'(I + X_1)^{-1}x}{1 + x'(I + X_1)^{-1}x}$$

where

$$X = \begin{pmatrix} 0 & x' \\ -x & X_1 \end{pmatrix},$$

with the $p \times p$ skew symmetric matrix X partitioned into submatrices where x is $(p - 1) \times 1$ and X_1 is $(p - 1) \times (p - 1)$.

2.1.10   Consider the differential elements coming from the transformation in Theorem 2.9 where X is $p \times p$ skew symmetric and T is $p \times p$ triangular. Let

$$\gamma_p = \int_X |I + X|^{-(p-1)} dX.$$

Then show that if each of the independent variables in X is freely varying over the real line then

$$\gamma_p = \frac{\pi^{p(p+1)/4}}{2^{(p-1)(p-2)/2} \prod_{j=1}^{p} \Gamma\left(\frac{j}{2}\right)}$$

and if the independent variables are such that the first row elements of $(I + X)^{-1}$, except the first element, are negative, that is, when the representation $Y = T\left[2(I + X)^{-1} - I\right]$ is unique, then

$$\gamma_p = \frac{\pi^{p(p+1)/4}}{2^{p(p-1)/2} \prod_{j=1}^{p} \Gamma\left(\frac{j}{2}\right)}.$$

2.1.11   Verify the results in Exercise 2.1.10 for $p = 2$, and $p = 3$, by direct evaluation of γ_p.

2.1.12   Establish the results in Exercise 2.1.10 by using Theorem 2.9, evaluating the integral over T, and the result

$$\int_Y e^{-\text{tr}(YY')} dY = \pi^{\frac{p^2}{2}}.$$

2.1.13   For distinct real variables $\lambda_1, ..., \lambda_p$ show that

$$\int_{\infty > \lambda_1 > ... > \lambda_p > -\infty} \left\{ \prod_{j<k} |\lambda_j - \lambda_k| \right\} e^{-(\lambda_1^2 + ... + \lambda_p^2)} d\lambda_1 ... d\lambda_p$$

$$= 2^{-\frac{p(p-1)}{4}} \prod_{j=1}^{p} \Gamma\left(\frac{j}{2}\right).$$

2.1.14   For distinct real and positive variables $\lambda_1, ..., \lambda_p$ show that

$$\int_{\lambda_1 > ... > \lambda_p > 0} \left\{ \prod_{j<k} |\lambda_j - \lambda_k| \right\} e^{-(\lambda_1 + ... + \lambda_p)} d\lambda_1 ... d\lambda_p = \frac{\Gamma_p\left(\frac{p+1}{2}\right) \Gamma_p\left(\frac{p}{2}\right)}{\pi^{\frac{p^2}{2}}}.$$

2.1.15   Show that the result

$$\int_{-\infty}^{\infty} ... \int_{-\infty}^{\infty} \exp\left[-\beta \left(\frac{1}{2} \sum_{i=1}^{N} x_i^2 - \sum_{i<j} \ln|x_i - x_j| \right) \right] dx_1 ... dx_N$$

$$= (\pi)^{\frac{N}{2}} \beta^{-\frac{N}{2} - \beta \frac{N(N-1)}{4}} \left[\Gamma\left(1 + \frac{\beta}{2}\right) \right]^{-N} \prod_{j=1}^{N} \Gamma\left(1 + \frac{\beta}{2}j\right)$$

holds for $\beta = 1, 2, 4$. Does it hold for other values of β ?

2.2   Decomposition with one matrix being orthogonal or an element of the Stiefel manifold

Here we consider transformations of the type where a matrix X is transformed to two matrices where one may be an orthogonal or semiorthogonal matrix belonging to an *orthogonal group* or *Stiefel manifold*, respectively.

Definition 2.1   Stiefel manifold   Let A be a $q \times p$, $q \geq p$ matrix with real elements such that $A'A = I_p$, that is, the p columns of A are *orthonormal vectors*. The set of all such matrices A is known as the *Stiefel manifold*, denoted by $V_{p,q}$. That is, for all $q \times p$ matrices A,

$$V_{p,q} = \{A : A'A = I_p\}. \tag{2.1}$$

(This manifold was introduced and studied by the Swiss mathematician E.L. Stiefel (1909-)). The equation $A'A = I_p$ imposes $p(p+1)/2$ conditions on the elements of A. Thus the number of free elements in A is $pq - p(p+1)/2$.

If $p = q$ then A is an orthogonal matrix. The set of such orthogonal matrices form a group. This group is known as the *orthogonal group* of $p \times p$ matrices.

Definition 2.2   Orthogonal group   Let B be a $p \times p$ matrix with real elements such that $B'B = I_p$. The set of all B is called an orthogonal group, denoted by $O_{(p)}$. That is, for all $p \times p$ matrices A,

$$O_{(p)} = \{B : B'B = I_p\}. \tag{2.2}$$

Note that $B'B = I_p$ imposes $p(p+1)/2$ conditions and hence the number of free elements in B is only $p^2 - p(p+1)/2 = p(p-1)/2$.

Theorem 2.12   *Let U be a $p \times p$ orthogonal matrix of functionally independent real variables, that is, $UU' = I = U'U$, with the diagonal elements or the elements in the first row of U all positive. Let $(\mathrm{d}U)$ be the matrix of differentials. Consider a transformation of U going to V through the relation $(\mathrm{d}V) = U'(\mathrm{d}U)$. Then, ignoring the sign,*

$$(\mathrm{d}V) = U'(\mathrm{d}U) \Rightarrow$$
$$\mathrm{d}V = \wedge_{i=1}^{p-1} \wedge_{j=i+1}^{p} u_i'(\mathrm{d}u_j) \tag{a}$$
$$= 2^{p(p-1)/2}|I + X|^{-(p-1)}\mathrm{d}X \tag{b}$$

ignoring the sign, where X is a skew symmetric matrix such that the first row elements of $(I + X)^{-1}$, except the first element, are negative.

Proof   Let the columns of U be denoted by $u_1, u_2, ..., u_p$. Since the columns are orthonormal we have $u_i' u_i = 1$, $u_i' u_j = 0$, $i \neq j$. Then

$$u_i' u_i = 1 \Rightarrow u_i'(du_i) + (du_i')u_i = 0$$
$$2u_i'(du_i) = 0 \Rightarrow u_i'(du_i) = 0$$

since $u_i'(du_i)$ is a scalar quantity and hence its transpose is itself.

$$u_i' u_j = 0 \Rightarrow u_i'(du_j) = -u_j'(du_i), \quad i \neq j.$$

Then $U'(dU)$ is a skew symmetric matrix. That is,

$$(dV) = U'(dU)$$

$$= \begin{bmatrix} u_1' \\ u_2' \\ \vdots \\ u_p' \end{bmatrix} [(du_1), ..., (du_p)] = \begin{bmatrix} 0 & -u_1'(du_2) & ... & -u_1'(du_p) \\ u_1'(du_2) & 0 & ... & -u_2'(du_p) \\ \vdots & \vdots & ... & \vdots \\ u_1'(du_p) & u_2'(du_p) & ... & 0 \end{bmatrix}.$$

Then there are only $(p - 1) + (p - 2) + ... + 1 = p(p - 1)/2$ functionally independent variables in V. Take these as $v_{21}, ..., v_{p1}, v_{32}, ..., v_{p2}, ..., v_{p\,p-1}$ and the u-variables as $u_{12}, ..., u_{1p}, u_{23}, ..., u_{2p}, ..., u_{p-1\,p}$. Then dV is the wedge product of the elements below the leading diagonal in the matrix $U'(dU)$, see also $(1.2.7)$–$(1.2.9)$ of Chapter 1:

$$dV = \wedge_{i=1}^{p-1} \wedge_{j=i+1}^{p} u_i'(du_j).$$

This establishes (a). For establishing (b) take a skew symmetric matrix X considered in Theorems 2.9 and 2.10. Then

$$Z = 2(I + X)^{-1} - I$$

is orthonormal such that $ZZ' = I$. Further the matrix of differentials in Z is given by

$$(dZ) = -2(I + X)^{-1}(dX)(I + X)^{-1}$$
$$= -\frac{1}{2}(I + Z)(dX)(I + Z).$$

Thus

$$Z'(\mathrm{d}Z) = -\frac{1}{2}(I + Z')(\mathrm{d}X)(I + Z)$$

and the wedge product is evaluated in Theorem 2.9 which establishes (b).

For practical situations the most convenient representation may be the one in terms of a skew symmetric matrix. As pointed out in Remark 2.6,

$$\int_X 2^{p(p-1)/2}|I + X|^{-(p-1)}\mathrm{d}X = \frac{\pi^{\frac{p^2}{2}}}{\Gamma_p\left(\frac{p}{2}\right)}$$

where $Z = 2(I+X)^{-1} - I$ is uniquely chosen, that is, for example, the first row elements of $(I + X)^{-1}$, except the first element, are negative. Hence one can expect

$$\int_U \wedge[U'(\mathrm{d}U)] = \int_U \wedge[(\mathrm{d}U)U'] = \frac{\pi^{\frac{p^2}{2}}}{\Gamma_p\left(\frac{p}{2}\right)}$$

where U is orthonormal and uniquely chosen. This result will be established as a corollary to Theorem 2.15 later on.

Theorem 2.13   *Let X be a $p \times p$ symmetric matrix of functionally independent real variables and with distinct and nonzero eigenvalues $\lambda_1 > \lambda_2 > ... > \lambda_p$ and let $D = \mathrm{diag}(\lambda_1, ..., \lambda_p)$, $\lambda_j \neq 0$, $j = 1, ..., p$. Let V be a unique $p \times p$ orthogonal matrix $V'V = I = VV'$ such that $X = VDV'$. Then, ignoring the sign,*

$$\mathrm{d}X = \left\{ \prod_{i=1}^{p-1} \prod_{j=i+1}^{p} |\lambda_i - \lambda_j| \right\} \mathrm{d}D \; \mathrm{d}G, \quad (\mathrm{d}G) = V'(\mathrm{d}V).$$

Proof   Take the differentials in $X = VDV'$ to get

$$(\mathrm{d}X) = (\mathrm{d}V)DV' + V(\mathrm{d}D)V' + VD(\mathrm{d}V'). \tag{a}$$

Pre and postmultiply (a) by V' and V, respectively, to get

$$V'(\mathrm{d}X)V = V'(\mathrm{d}V)D + (\mathrm{d}D) + D(\mathrm{d}V')V.$$

Let

$$(\mathrm{d}Y) = V'(\mathrm{d}V) \text{ and } (\mathrm{d}W) = V'(\mathrm{d}X)V \text{ for fixed } V.$$

Then observing that $|V| = \pm 1$ since V is orthonormal, and ignoring the sign,

$$dW = dX \text{ and } (dW) = (dY)D + (dD) + D(dY'). \tag{b}$$

Taking the differentials in $V'V = I$, one has $V'(dV) = -(dV')V$, that is, Y is skew symmetric. Then

$$J(X : D, V) = J(X : W)J(W : Y, D)J(Y : V).$$

Letting $W = (w_{ij})$, one has from (b)

$$dw_{ii} = d\lambda_i, \quad dw_{ij} = (\lambda_j - \lambda_i)dy_{ij}, \quad i > j = 1, ..., p.$$

Then $\{\prod_{i=1}^{p} dw_{ii}\} = dD$ and from Theorem 2.12, ignoring the sign,

$$\left\{ \prod_{i>j} dw_{ij} \right\} = \left\{ \prod_{i>j} |\lambda_j - \lambda_i| \right\} \left\{ \prod_{i>j} dy_{ij} \right\}$$

$$= \left\{ \prod_{i<j} |\lambda_i - \lambda_j| \right\} dG \, dV$$

and hence the result.

Remark 2.10   In general, $X = VDV'$, $D = \text{diag}(\lambda_1, ..., \lambda_p)$, $VV' = I$ is not a one-to-one transformation since X determines 2^p matrices $V = (\pm v_1, ..., \pm v_p)$ where $v_1, ..., v_p$ are the columns of V, such that $X = VDV'$. This transformation can be shown to be unique if one element each from each row and column are of a specified sign, for example, the diagonal elements are positive. Once this is done we are considering only 2^p-th part of the orthogonal group $O(p)$. Hence if we are integrating with respect to dG, $(dG) = V'(dV)$ over the full orthogonal group $O(p)$, the result must be divided by 2^p to get the result for a unique transformation $X = VDV'$.

Example 2.6   Let a $p \times p$ real symmetric positive definite matrix X have the real matrix-variate type-1 beta density

$$f(X) = C|X|^{\alpha - \frac{p+1}{2}}|I - X|^{\beta - \frac{p+1}{2}},$$

for $0 < X < I$, $\text{Re}(\alpha) > \frac{p-1}{2}$, $\text{Re}(\beta) > \frac{p-1}{2}$ and $f(X) = 0$ elsewhere, where C is the normalizing constant. Let X have distinct eigenvalues $\lambda_1, ..., \lambda_p$. Evaluate the joint density of $\lambda_1, ..., \lambda_p$.

Solution   Let $V'XV = \text{diag}(\lambda_1, ..., \lambda_p)$, $VV' = I$. Then $0 < \lambda_j < 1$, $j = 1, ..., p$,

$$|X| = \lambda_1...\lambda_p, \quad |I - X| = (1 - \lambda_1)(1 - \lambda_2)...(1 - \lambda_p)$$

and from Theorems 2.12 and 2.13

$$dX = \prod_{j<k} |\lambda_j - \lambda_k| dD \, dG, \quad (dG) = V'(dV).$$

The joint density of the λ_j's and G is given by

$$g(D, G)dD \, dG = h_1(D)dD \, dG$$

where

$$h_1(D) = CC_1 (\lambda_1...\lambda_p)^{\alpha - \frac{p+1}{2}} \left[\prod_{j=1}^{p} (1 - \lambda_j) \right]^{\beta - \frac{p+1}{2}} \left[\prod_{j<k} |\lambda_j - \lambda_k| \right].$$

where $C_1 = 2^{-p} \int_V dG$, and the integral is over the full orthogonal group, see Remark 2.10 for the presence of 2^{-p}. The explicit form of C_1 is given in Corollary 2.15.2 later on.

Theorem 2.14   *Let X be a $p \times n$, $n \geq p$, matrix of rank p and let $X = TU_1'$ where T is a $p \times p$ lower triangular matrix with distinct nonzero diagonal elements and U_1 is a unique $n \times p$ semiorthogonal matrix, $U_1'U_1 = I_p$, all are of functionally independent real variables, see Theorem 2.9 and the preceeding discussion for the unique choices of T and U_1. Let U_2 be an $n \times (n - p)$ semiorthogonal matrix such that U_1 augmented with U_2 is a full orthogonal matrix. That is, $U = (U_1 \quad U_2)$, $U'U = I_n$, $U_2'U_2 = I_{n-p}$, $U_1'U_2 = 0$. Let u_j be the j-th column of U and (du_j) its differential. Then, ignoring the sign,*

$$X = TU_1' \Rightarrow$$

$$dX = \left\{ \prod_{j=1}^{p} |t_{jj}|^{n-j} \right\} dT dU_1,$$

where

$$dU_1 = \wedge_{j=1}^{p} \wedge_{k=j+1}^{n} u_k'(du_j).$$

Proof   Note that

$$U = (U_1 \quad U_2) \Rightarrow U'U = \begin{pmatrix} U_1' \\ U_2' \end{pmatrix} (U_1 \quad U_2)$$

$$= \begin{pmatrix} U_1'U_1 & U_1'U_2 \\ U_2'U_1 & U_2'U_2 \end{pmatrix} = \begin{pmatrix} I_p & 0 \\ 0 & I_{n-p} \end{pmatrix}.$$

Take the differentials in $X = TU_1'$ to get

$$(dX) = (dT)U_1' + T(dU_1'). \tag{a}$$

Multiply on the right of (a) by U to get

$$(dX)U = (dT)U_1'(U_1 \quad U_2) + T(dU_1')(U_1 \quad U_2)$$
$$= [(dT) + T(dU_1')U_1 \quad T(dU_1')U_2] \tag{b}$$

since $U_1'U_1 = I_p$, $U_1'U_2 = 0$. Make the substitutions

$$(dW) = (dX)U, \quad (dY) = (dU_1')U_1$$
$$(dS) = (dU_1')U_2, \quad (dH) = T(dS).$$

Now (b) becomes

$$(dW) = [(dT) + T(dY) \quad (dH)]. \tag{c}$$

Note that

$$U_1'U_1 = I_p \Rightarrow U_1'(dU_1) + (dU_1')U_1 = 0$$

which means that $(dU_1')U_1$ is a skew symmetric matrix. Hence $T(dY)$ is of the form

$$T(dY) = \begin{bmatrix} t_{11} & 0 & \cdots & 0 \\ t_{21} & t_{22} & \cdots & 0 \\ \vdots & \vdots & \cdots & \vdots \\ t_{p1} & t_{p2} & \cdots & t_{pp} \end{bmatrix} \begin{bmatrix} 0 & dy_{21} & \cdots & dy_{p1} \\ -dy_{21} & 0 & \cdots & dy_{p2} \\ \vdots & \vdots & \cdots & \vdots \\ -dy_{p1} & -dy_{p2} & \cdots & 0 \end{bmatrix}.$$

By straight multiplication, deleting the higher powers of the differentials, and remembering that the variables are only $dy_{21}, ..., dy_{p1}, dy_{32}, ..., dy_{p2}, ..., dy_{p\,p-1}$, the exterior product of

$$T(dY) \quad \text{gives} \quad \left\{ \prod_{j=1}^{p-1} |t_{jj}|^{p-j} \right\} dY,$$

ignoring the sign. Thus

$$(dT) + T(dY) \quad \text{gives} \quad \left\{ \prod_{j=1}^{p} |t_{jj}|^{p-j} \right\} dY \, dT, \tag{d}$$

ignoring the sign, and

$$dY = \wedge_{j=1}^{p} \wedge_{k=j+1}^{p} u'_k(du_j). \tag{e}$$

Now consider $(dH) = T(dS)$. Since (dS) is a $p \times (n-p)$ matrix, we have from Theorem 1.13 of Chapter 1

$$dH = |T|^{n-p} dS = \left\{ \prod_{j=1}^{p} |t_{jj}|^{n-p} \right\} dS, \tag{f}$$

$$dX = dW, \tag{g}$$

ignoring the sign. The wedge product in (dS) is the following:

$$dS = \wedge_{j=1}^{p} \wedge_{k=p+j}^{n} u'_k(du_j). \tag{h}$$

Hence from $(c) - (h)$ one has, ignoring the sign,

$$dX = dW = \left\{ \prod_{j=1}^{p} |t_{jj}|^{n-j} \right\} dY \, dT \, dS.$$

Substituting back one has

$$dX = \left\{ \prod_{j=1}^{p} |t_{jj}|^{n-j} \right\} dT \, \wedge_{j=1}^{p} \wedge_{k=j+1}^{n} u'_k(du_j)$$

which establishes the result. In terms of J, the above transformations can be written as follows:

$$J(X:T,U) = J(X:W)J(W:T,Y,H)$$
$$\times J(H:S)J(S:U)J(Y:U).$$

Remark 2.11   If the triangular matrix T is restricted to the one with positive diagonal elements, that is, $t_{jj} > 0$, $j = 1,...,p$ then while integrating over T using Theorem 2.14, the result must be multiplied by 2^p. Without the factor 2^p, the t_{jj}'s must be integrated over $-\infty < t_{jj} < \infty$, $j = 1,...,p$. If the expression to be integrated contains both T and U then restrict $t_{jj} > 0$, $j = 1,...,p$ and integrate U over the full Stiefel manifold.

If the rows of U are $u_1, ..., u_p$ then $\pm u_1, ..., \pm u_p$ give 2^p choices. Similarly $t_{jj} > 0$, $t_{jj} < 0$ give 2^p choices. But there are not 2^{2p} choices in $X = TU$. There are only 2^p choices. Hence either integrate out the t_{jj}'s over $-\infty < t_{jj} < \infty$ and a unique U or over $0 < t_{jj} < \infty$ and the U over the full Stiefel manifold.

Example 2.7   Let X be a $p \times n$, $n \geq p$ random matrix having an np-variate real Gaussian density

$$f(X) = \frac{e^{-\frac{1}{2}\text{tr}(V^{-1}XX')}}{(2\pi)^{\frac{np}{2}}|V|^{\frac{n}{2}}}, \quad V = V' > 0.$$

Evaluate the density of $S = XX'$.

Solution   Consider the transformation $X = TU_1'$ as in Theorem 2.14. Then $XX' = TU_1'U_1T' = TT'$ and

$$dX = \left\{ \prod_{j=1}^{p} |t_{jj}|^{n-j} \right\} dT \ \wedge_{j=1}^{p} \wedge_{k=j+1}^{n} u_k'(du_j).$$

Then the joint density of T and U_1 is given by

$$g(U_1, T)dT \ dU_1 = \frac{e^{-\frac{1}{2}\text{tr}(V^{-1}TT')}}{(2\pi)^{\frac{np}{2}}|V|^{\frac{n}{2}}} \left\{ \prod_{j=1}^{p} |t_{jj}|^{n-j} \right\} dT \ dU_1.$$

By integrating out U_1, one gets the density of T as

$$g_T(T)dT = Ce^{-\frac{1}{2}\text{tr}(V^{-1}TT')} \left\{ \prod_{j=1}^{p} |t_{jj}|^{n-j} \right\} dT$$

where C is a normalizing constant. Let $t_{jj} > 0$, $j = 1, ..., p$ and

$$S = TT' \Rightarrow dS = 2^p \left\{ \prod_{j=1}^{p} t_{jj}^{p+1-j} \right\} dT,$$

with $t_{jj} > 0$, $j = 1, ..., p$, see also Theorem 1.28 of Chapter 1. Substituting for dT in terms of dS and observing that $|S| = |TT'| = \prod_{j=1}^{p} t_{jj}^2$, we get the density of S, denoted by $h(S)$, as

$$h(S) = C_1|S|^{\frac{n}{2}-\frac{p+1}{2}}e^{-\frac{1}{2}\text{tr}(V^{-1}S)}, \quad S = S' > 0.$$

Since $\int_S h(S)\mathrm{d}S = 1$, integrating out S by using a matrix-variate gamma integral one has

$$C_1 = \left[2^{\frac{np}{2}}\Gamma_p(\frac{n}{2})|V|^{\frac{n}{2}}\right]^{-1}.$$

Theorem 2.15 *Let X_1 be an $n \times p$, $n \geq p$ matrix of rank p of functionally independent real variables and let $X_1 = U_1 T_1$ where T_1 is a real $p \times p$ upper triangular matrix with distinct nonzero diagonal elements and U_1 is a unique real $n \times p$ semiorthogonal matrix, that is, $U_1' U_1 = I_p$. Let $U = (U_1 \quad U_2)$ such that $U'U = I_n$, $U_2'U_2 = I_{n-p}$, $U_1'U_2 = 0$. Let u_j be the j-th column of U and $(\mathrm{d}u_j)$ its differential. Then, ignoring the sign,*

$$X_1 = U_1 T_1 \Rightarrow$$

$$\mathrm{d}X_1 = \left\{\prod_{j=1}^{p} |t_{jj}|^{n-j}\right\} \mathrm{d}T_1 \, \mathrm{d}U_1,$$

where

$$\mathrm{d}U_1 = \wedge_{j=1}^{p} \wedge_{k=j+1}^{n} u_k'(\mathrm{d}u_j).$$

The proof is parallel to that in Theorem 2.14 since by taking the transposes of the matrices in Theorem 2.14 we get the situation in Theorem 2.15. Note that in this case $U_1'(\mathrm{d}U_1)$ will be a skew symmetric matrix.

Let X_1, T_1 and U_1 be as defined in Theorem 2.15. Then by using the results in Theorem 2.15, we can evaluate the surface area or the volume content of the Stiefel manifold $V_{p,n}$, see also Definition 2.1. That is, what is the total integral of the wedge product $\wedge_{j=1}^{p} \wedge_{k=j+1}^{n} u_k'(\mathrm{d}u_j)$ over the Stiefel manifold:

$$\int_{V_{p,n}} \wedge_{j=1}^{p} \wedge_{k=j+1}^{n} u_k'(\mathrm{d}u_j) \ ?$$

This will be stated as a corollary.

Corollary 2.15.1 *If X_1, T_1 and U_1 are as defined in Theorem 2.15 then the surface area of the full Stiefel manifold $V_{p,n}$ or the total integral of the wedge product $\wedge_{j=1}^{p} \wedge_{k=j+1}^{n} u_k'(\mathrm{d}u_j)$ over $V_{p,n}$ is given by*

$$\int_{V_{p,n}} \wedge_{j=1}^{p} \wedge_{k=j+1}^{n} u_k'(\mathrm{d}u_j) = \frac{2^p \pi^{pn/2}}{\Gamma_p(\frac{n}{2})} \tag{a}$$

where

$$\Gamma_p(\alpha) = \pi^{\frac{p(p-1)}{4}}\Gamma(\alpha)\Gamma\left(\alpha - \frac{1}{2}\right)...\Gamma\left(\alpha - \frac{p-1}{2}\right), \quad \mathrm{Re}(\alpha) > \frac{p-1}{2}$$

and $\mathrm{Re}(.)$ *denotes the real part of* (.).

Proof   Note that since X_1 is $n \times p$, the sum of squares of the np variables in X_1 is given by

$$\mathrm{tr}(X_1'X_1) = \sum_{ij} x_{ij}^2.$$

Let $\int_{X_1}$ denote the integral over all the variables in X_1. Then

$$\int_{X_1} e^{-\mathrm{tr}(X_1'X_1)}\mathrm{d}X_1 = \int_{-\infty}^{\infty} ... \int_{-\infty}^{\infty} e^{-(\sum_{ij} x_{ij}^2)}\left\{\prod_{ij} \mathrm{d}x_{ij}\right\}$$

$$= \pi^{np/2} \tag{b}$$

by direct evaluation of the exponential integrals. Make the transformation as in Theorem 2.15:

$$X_1 = U_1 T_1 \Rightarrow (X_1'X_1) = T_1'T_1$$

since $U_1'U_1 = I_p$ and

$$\mathrm{tr}(X_1'X_1) = \mathrm{tr}(T_1'T_1) = \sum_{i \leq j} t_{ij}^2.$$

Note that $\mathrm{d}X_1$ is available from Theorem 2.15. Now the left side of (b) reduces to the following:

$$\int_{X_1} e^{-\mathrm{tr}(X_1'X_1)}\mathrm{d}X_1 = \int_{T_1} \left\{\prod_{j=1}^{p} |t_{jj}|^{n-j}\right\} e^{-(\sum_{i \leq j} t_{ij}^2)}\mathrm{d}T_1$$

$$\times \int_{V_{p,n}} \wedge_{j=1}^{p} \wedge_{k=j+1}^{n} u_k'(\mathrm{d}u_j). \tag{c}$$

But for $0 < t_{jj} < \infty$, $-\infty < t_{ij} < \infty$, $i < j$ and U_1 unrestricted, see also Remark 2.11,

$$\int_{T_1} \left\{\prod_{j=1}^{p} |t_{jj}|^{n-j}\right\} e^{-(\sum_{i \leq j} t_{ij}^2)}\mathrm{d}T_1 = 2^{-p}\Gamma_p\left(\frac{n}{2}\right) \tag{d}$$

observing that for $j = 1, ..., p$ the p integrals

$$\int_0^\infty |t_{jj}|^{n-j} e^{-t_{jj}^2} dt_{jj} = 2^{-1}\Gamma\left(\frac{n}{2} - \frac{j-1}{2}\right), \ n > j - 1,$$

and each of the $p(p-1)/2$ integrals

$$\int_{-\infty}^\infty e^{-t_{ij}^2} dt_{ij} = \sqrt{\pi}, \ i < j.$$

Thus from (a) to (d) the result follows.

Remark 2.12   Note that in deriving the result (a) above, there is no restriction imposed on U_1 or the integral is over the full Stiefel manifold. Hence if an integral containing a unique U_1 is evaluated by using Corollary 2.5.1 then the final result must be divided by 2^p. This remark applies to Corollary 2.15.2 also.

When $n = p$, $V_{p,n}$ is the full orthogonal group $O_{(p)}$. Thus one has the following as a particular case:

Corollary 2.15.2   If X_1, T_1 and U_1 are as defined in Theorem 2.15 with $n = p$, then the volume content of the full orthogonal group $O_{(p)}$ is given by

$$\int_{O_{(p)}} \wedge [U_1'(dU_1)] = \frac{2^p \pi^{\frac{p^2}{2}}}{\Gamma_p(\frac{p}{2})}.$$

Since U in Theorem 2.14 is a semiorthogonal matrix, the representation

$$X = TU_1', \ U_1'U_1 = I_p \Rightarrow XX' = TT'$$

which is free of U_1. Select T with $t_{jj} > 0$, $j = 1, ..., p$ and an unrestricted U_1. If XX' is denoted by A then from $(1.3.11)$ of Chapter 1 one can write, for $t_{jj} > 0$, $j = 1, ..., p$,

$$A = XX' = TT' \Rightarrow dA = 2^p \left\{ \prod_{j=1}^p t_{jj}^{p+1-j} \right\} dT$$

$$\Rightarrow dT = 2^{-p} \left\{ \prod_{j=1}^p t_{jj}^{-p-1+j} \right\} dA. \tag{a}$$

Then we can consider a transformation of X going to U_1 and A and state the following theorem.

Theorem 2.16   *Let X, T and U_1 be as defined in Theorem 2.14 with $t_{jj} > 0$, $j = 1, ..., p$ and an unrestricted U_1. Further, let $A = XX' = TT'$. Then, ignoring the sign,*

$$X = TU_1' \text{ and } A = XX' = TT' \Rightarrow$$
$$dX = 2^{-p}|A|^{\frac{n}{2} - \frac{p+1}{2}} dA \ dU_1,$$

where

$$dU_1 = \wedge_{j=1}^{p} \wedge_{k=j+1}^{n} u_k'(du_j).$$

Proof   From (a) of Corollary 2.15.2, take dT and substitute for dT in terms of dA in Theorem 2.14 to obtain the result.

By taking the transposes of X, T, U we get the corresponding result for an upper triangular T and an $n \times p$ matrix X.

Remark 2.13   The dU_1 appearing in Theorem 2.16 is from an unrestricted U_1. If this theorem is used to evaluate an integral, then U_1 must be integrated out over the full Stiefel manifold or

$$\int_{U_1} dU_1 = \frac{2^p \pi^{\frac{np}{2}}}{\Gamma_p\left(\frac{n}{2}\right)}.$$

One of the earliest applications of the transformations considered in Theorems 2.15 and 2.16 appears in statistical distribution theory in connection with rectangular coordinates transformations in sampling distributions, see Mahalanobis, Bose and Roy (1937) and Roy (1957). Some discussion of Stiefel manifolds can be found in Atiyah and Todd (1960).

Example 2.8   Let X be a 2×2 real symmetric positive definite matrix with eigenvalues $1 > \lambda_2 > \lambda_1 > 0$. Evaluate $\int_0^I |X|^\alpha dX$, $\mathrm{Re}(\alpha) \geq 0$.

Solution   For the sake of illustration we will evaluate this by using different methods.

Method 1   Consider it as a matrix-variate type-1 beta integral, that is,

$$\int_0^I |X|^\alpha dX = \int_0^I |X|^{\alpha + \frac{3}{2} - \frac{3}{2}} |I - X|^{\frac{3}{2} - \frac{3}{2}} dX$$

$$= \frac{\Gamma_2(\alpha + \frac{3}{2})\Gamma_2(\frac{3}{2})}{\Gamma_2(\alpha + 3)} = \frac{\pi}{(\alpha + 1)(\alpha + 2)(2\alpha + 3)}.$$

Hence for $\alpha = 0$

$$\int_0^I dX = \frac{\pi}{3!}.$$

Method 2   Consider an orthogonal transformation $X = VDV'$, $VV' = I$, $D = \text{diag}(\lambda_1, \lambda_2)$. Then from Theorem 2.13 we have

$$\int_0^I |X|^\alpha dX = \int_{0 < \lambda_1 < \lambda_2 < 1} (\lambda_1 \lambda_2)^\alpha (\lambda_2 - \lambda_1) d\lambda_1 d\lambda_2$$

$$\times \left(2^{-2} \int_V dG \right), \quad (dG) = V'(dV),$$

see also Remark 2.10. Also $2^{-2} \int_V dG$ is available from Corollary 2.15.2 as

$$\frac{\pi^2}{\Gamma_2(1)} = \pi.$$

But

$$\int_{0 < \lambda_1 < \lambda_2 < 1} (\lambda_1 \lambda_2)^\alpha (\lambda_2 - \lambda_1) d\lambda_1 d\lambda_2 = \int_0^1 \lambda_1^\alpha \left[\int_{\lambda_1}^1 \lambda_2^\alpha (\lambda_2 - \lambda_1) d\lambda_2 \right] d\lambda_1$$

$$= \frac{1}{(\alpha + 1)(\alpha + 2)(2\alpha + 3)}.$$

Hence

$$\int_0^I |X|^\alpha dX = \frac{\pi}{(\alpha + 1)(\alpha + 2)(2\alpha + 3)}.$$

Theorem 2.17   *Let X be a $p \times p$ symmetric positive definite matrix of functionally independent real variables and with distinct eigenvalues $\lambda_1 > \lambda_2 > ... > \lambda_p > 0$. Let $X = U D_\lambda U'$ where $D_\lambda = \text{diag}(\lambda_1, ..., \lambda_p)$ and U is an orthonormal matrix, $U'U = I = UU'$, with positive diagonal elements. Let $f(x)$ be a differentiable function of the scalar variable x with $f'(\lambda_j) \neq 0$, $j =$*

$1, ..., p$ and $F(D_\lambda) = \text{diag}\,(f(\lambda_1), ..., f(\lambda_p))$ *such that* $f(\lambda_i) - f(\lambda_j) \neq 0$ *for all* i *and* j, $i \neq j$. *Let* $F(X) = U F(D_\lambda) U'$. *Then, ignoring the sign,*

$$X = U D_\lambda U' \text{ and } Y = F(X) = U F(D_\lambda) U' \Rightarrow$$

$$dY = \left\{ \prod_{i=1}^{p-1} \prod_{j=i+1}^{p} \left| \frac{f(\lambda_i) - f(\lambda_j)}{\lambda_i - \lambda_j} \right| \right\} \left\{ \prod_{j=1}^{p} |f'(\lambda_j)| \right\} dX$$

where for example $f'(x)$ *denotes the derivative of* $f(x)$ *with respect to* x.

Proof   Take the differentials in Y to get

$$(d\mathbf{Y}) = (d\mathbf{U}) F(D_\lambda) U' + U \left(dF(\mathbf{D}_\lambda) \right) U' + U F(D_\lambda)(d\mathbf{U}').$$

Pre and postmultiply by U' and U respectively to get

$$U'(d\mathbf{Y}) U = U'(d\mathbf{U}) F(D_\lambda) + (dF(\mathbf{D}_\lambda)) + F(D_\lambda)(d\mathbf{U}') U.$$

Let

$$(d\mathbf{W}) = U'(d\mathbf{Y}) U, \ \ (d\mathbf{G}) = U'(d\mathbf{U}).$$

Thus

$$J(Y : U, F(D_\lambda)) = J(Y : W) J(W : G, F(D_\lambda)) J(G : U).$$

Then from Theorem 2.13 we have the following, ignoring the sign:

$$dY = dW$$

$$= \left\{ \prod_{i=1}^{p-1} \prod_{j=i+1}^{p} |f(\lambda_i) - f(\lambda_j)| \right\} dG \, dF(D_\lambda) \qquad (a)$$

and

$$dX = \left\{ \prod_{i=1}^{p-1} \prod_{j=i+1}^{p} |\lambda_i - \lambda_j| \right\} dG \, dD_\lambda. \qquad (b)$$

But

$$dF(D_\lambda) = \left\{ \prod_{j=1}^{p} |f'(\lambda_j)| \right\} dD_\lambda. \qquad (c)$$

Replace $dF(D_\lambda)$ in (a) by the one from (c) and then substitute for $dG \, dD_\lambda$ from (b) to establish the result.

Remark 2.14 The transformation $X = UD_\lambda U'$, $UU' = I$ is unique if one element each from each row and column of U are of specified sign; for example if the diagonal elements are positive.

Theorem 2.18 *Let X be a $p \times p$ symmetric positive definite matrix of functionally independent real variables and with distinct eigenvalues. Let $Y = \left(\frac{n}{2}\right)^{\frac{1}{2}} \ln\left(\frac{1}{n}X\right)$ in the sense*

$$X = ne^{[(\frac{2}{n})^{\frac{1}{2}}Y]} = n\left[I + \left(\frac{2}{n}\right)^{\frac{1}{2}}Y + \left(\frac{2}{n}\right)\frac{Y^2}{2!} + \cdots\right].$$

Then, ignoring the sign,

$$dX = \left[(2n)^{\frac{p(p+1)}{4}}e^{\mathrm{tr}((2/n)^{\frac{1}{2}}Y)}\left\{\prod_{i=1}^{p-1}\prod_{j=i+1}^{p}\left|\frac{f(\lambda_i) - f(\lambda_j)}{\lambda_i - \lambda_j}\right|\right\}\right]dY$$

where $f(\lambda_j) = e^{\lambda_j}$, the λ_j's being the eigenvalues of $\left(\frac{2}{n}\right)^{\frac{1}{2}}Y$.

Proof Let $Z = \left(\frac{2}{n}\right)^{\frac{1}{2}}Y$ and $U'ZU = D = \mathrm{diag}(\lambda_1, ..., \lambda_p)$, $\lambda_1, ..., \lambda_p$ are the eigenvalues of Z, $U'U = I = UU'$. Note that

$$\frac{X}{n} = \left[I + Z + \frac{Z^2}{2!} + \cdots\right].$$

Then

$$W = U'\left(\frac{X}{n}\right)U = I + D + \frac{D^2}{2!} + \cdots$$

$$= \mathrm{diag}\left(e^{\lambda_1}, ..., e^{\lambda_p}\right).$$

Thus

$$\frac{X}{n} = U\mathrm{diag}\left(e^{\lambda_1}, ..., e^{\lambda_p}\right)U' \qquad (a)$$

and

$$\left(\frac{2}{n}\right)^{\frac{1}{2}}Y = U\mathrm{diag}(\lambda_1, ..., \lambda_p)U'. \qquad (b)$$

Now compare (a) and (b) with the statement of Theorem 2.17 to see that

$$d\left(\frac{X}{n}\right) = \left\{\prod_{i=1}^{p-1}\prod_{j=i+1}^{p}\left|\frac{f(\lambda_i) - f(\lambda_j)}{\lambda_i - \lambda_j}\right|\right\}$$

$$\times \left\{\prod_{j=1}^{p}|f'(\lambda_j)|\right\}d\left(\left(\frac{2}{n}\right)^{\frac{1}{2}}Y\right)$$

where

$$f(\lambda_j) = e^{\lambda_j}$$

$$d\left(\frac{X}{n}\right) = n^{-\frac{p(p+1)}{2}}dX \tag{c}$$

$$d\left(\left(\frac{2}{n}\right)^{\frac{1}{2}}Y\right) = \left(\frac{2}{n}\right)^{\frac{p(p+1)}{4}}dY \tag{d}$$

$$\left\{\prod_{j=1}^{p}|f'(\lambda_j)|\right\} = e^{\lambda_1+\ldots+\lambda_p} = e^{\mathrm{tr}\left((2/n)^{\frac{1}{2}}Y\right)}.$$

Now substitute for dX and dY from (c) and (d) to obtain the result.

Theorem 2.19   *Let X be a $p \times p$ nonsingular matrix of p^2 functionally independent real variables. Let $X = UTU'$ where T is a lower or upper triangular matrix of real, distinct and nonzero diagonal elements t_{jj}'s and U is an orthonormal matrix, that is, $U'U = I = UU'$. Then, ignoring the sign,*

$$X = UTU' \Rightarrow$$

$$dX = \left\{\prod_{i=1}^{p-1}\prod_{j=i+1}^{p}|t_{ii} - t_{jj}|\right\} dT \, dG, \ (dG) = U'(dU).$$

Proof   Take the differentials in $X = UTU'$ to get

$$(dX) = (dU)TU' + U(dT)U' + UT(dU').$$

Pre and postmultiply by U' and U, respectively, to get

$$U'(dX)U = U'(dU)T + (dT) + T(dU')U.$$

Let

$$(dY) = U'(dX)U, \ (dG) = U'(dU).$$

Thus

$$J(X:T,U) = J(X:Y)J(Y:G,T)J(G:U).$$

Then since U is orthonormal and G is skew symmetric

$$dY = |U|^{2p}dX = dX$$

and

$$dG = \wedge[U'(dU)]. \tag{a}$$

Thus

$$(dY) = (dG)T + (dT) - T(dG). \tag{b}$$

Let $dy_{ij} = y_{ij}^*$, $dt_{ij} = t_{ij}^*$, $g_{ij}^* = dg_{ij}$ and let T be a lower triangular matrix. Then the variables in (b) are y_{ij}^*, $i, j = 1, ..., p$, t_{ij}^*, $i \geq j = 1, ..., p$ and g_{ij}, $i > j = 1, ..., p$. Write the y_{ij}^*'s in terms of the t_{ij}^*'s and g_{ij}^*'s and consider the matrix of partial derivatives. We get the following structure.

	$t_{ii}^*, i = 1, ..., p$	$g_{ij}^*, i > j = 1, ..., p$	$t_{ij}^*, i > j = 1, ..., p$
$y_{ii}^*, \ i = 1, ..., p$	I	A	0
$y_{ij}^*, \ i < j$	0	B	0
$y_{ij}^*, \ i > j$	0	C	I^*

where the diagonal elements of the $\frac{p(p-1)}{2} \times \frac{p(p-1)}{2}$ matrix B are $(t_{ii} - t_{jj})$, $i < j$. Note that I^* is an identity matrix of order $\frac{p(p-1)}{2}$ and I is an identiy matrix of order p. By using the last $\frac{p(p-1)}{2}$ columns, one can wipe out C and then by using the last $\frac{p(p-1)}{2}$ rows, one can wipe out all the elements below the leading diagonal of B. Then

$$\begin{vmatrix} I & A & 0 \\ 0 & B & 0 \\ 0 & C & I^* \end{vmatrix} = \begin{vmatrix} I & A & 0 \\ 0 & B^* & 0 \\ 0 & 0 & I^* \end{vmatrix}$$

$$= \left\{ \prod_{i<j} (t_{ii} - t_{jj}) \right\} \tag{c}$$

where B^* is a triangular matrix with the diagonal elements $(t_{ii} - t_{jj})$ for $i < j$. Thus from (a) and (c) the result is established for lower triangular T. But note that $X' = UT'U'$ where T' is upper triangular and hence the result holds for upper triangular T as well.

Remark 2.15   Note that $X = UTU'$, $UU' = I$ and T lower triangular is not a one-to-one transformation.

Theorem 2.20   Let X, T, U be as defined in Theorem 2.19. Then, ignoring the sign,

$$Y = UT^2U' \Rightarrow$$

$$dY = 2^p \left\{ \prod_{j=1}^{p} |t_{jj}| \right\} \left\{ \prod_{i=1}^{p-1} \prod_{j=i+1}^{p} |t_{ii} - t_{jj}||t_{ii} + t_{jj}|^2 \right\}$$
$$\times dG \; dT, \quad (dG) = U'(dU).$$

Proof Let $X = UTU'$. Then $Y = X^2$. The eigenvalues of X are $t_{ii}, i = 1, ..., p$. From Theorem 1.35 of Chapter 1, ignoring the sign,

$$dY = 2^p |X| \left\{ \prod_{i<j} |\lambda_i + \lambda_j|^2 \right\} dX$$

$$= 2^p \left\{ \prod_{j=1}^{p} |t_{jj}| \right\} \left\{ \prod_{i<j} |t_{ii} + t_{jj}|^2 \right\} dX$$

since $|X| = |T|$ and $\lambda_j = t_{jj}$, $j = 1, ..., p$. Now substituting for dX from Theorem 2.19 the result follows.

Note that in a $p \times p$ orthogonal matrix of functionally independent real variables, we have $p^2 - p(p+1)/2 = p(p-1)/2$ variables and in a $p \times p$ symmetric matrix, we have $p(p+1)/2$ variables. Thus combined, we have $p(p-1)/2 + p(p+1)/2 = p^2$ variables. Hence if a $p \times p$ matrix of p^2 functionally independent real variables can be represented in terms of a symmetric matrix and an orthogonal matrix, we have the following result.

Theorem 2.21 *Let Z be a $p \times p$ matrix of p^2 functionally independent real variables and with positive eigenvalues, Y a $p \times p$ real symmetric positive definite matrix and V a $p \times p$ orthogonal matrix, that is, $V'V = I$. Let $U'YU = \text{diag}(\lambda_1, ..., \lambda_p)$ where the λ_j's are distinct, $UU' = I$. Then, ignoring the sign,*

$$Z = YV \Rightarrow dZ = \left\{ \prod_{i<j} (\lambda_i + \lambda_j) \right\} dY \wedge [(dV)V'].$$

Proof Taking the differentials of $Z = YV$ we get

$$(dZ) = (dY)V + Y(dV). \tag{a}$$

Postmultiply by V' to get

$$(dZ)V' = (dY) + Y(dV)V'. \qquad (b)$$

Let U be an orthogonal matrix such that $UU' = I$, $U'YU = \text{diag}(\lambda_1, ..., \lambda_p)$, where $\lambda_1, ..., \lambda_p$ are the eigenvalues of Y. Pre and postmultiply (b) by U' and U to get

$$U'(dZ)V'U = U'(dY)U + DU'(dV)V'U. \qquad (c)$$

Let

$$(dW) = U'(dZ)V'U, \quad (dY^*) = U'(dY)U$$

and

$$(dV^*) = U'(d\bar{V})U, \quad (d\bar{V}) = (dV)V'.$$

Since $(dV)V'$ is skew symmetric, $(d\bar{V})$ and (dV^*) are skew symmetric, and U and V are orthogonal. Thus, ignoring the sign, we have

$$dV^* = |U|^{p-1}d\bar{V} = d\bar{V}$$

and

$$dW = |U'|^p|V'U|^p dZ = dZ.$$

From (c) we have

$$(dW) = (dY^*) + D(dV^*) \Rightarrow \qquad (d)$$
$$dw_{ii} = dy_{ii}^*, \ i = 1, ..., p;$$
$$dw_{ij} = dy_{ij}^* + \lambda_i dv_{ij}^*, \ i < j;$$
$$dw_{ji} = dy_{ij}^* - \lambda_j dv_{ij}^*, \ i < j.$$

The matrix of partial derivatives has the following form:

$$\begin{pmatrix} & y_{ii}^*, \ i = 1, ..., p & y_{ij}^*, \ i < j & v_{ij}^*, \ i < j \\ w_{ii}, \ i = 1, ..., p & I & 0 & 0 \\ w_{ij}, \ i < j & 0 & I & A \\ w_{ji}, \ i < j & 0 & I & B \end{pmatrix}$$

where A is a diagonal matrix with λ_1 repeated $p - 1$ times, λ_2 repeated $p - 2$ times and so on, and B is a diagonal matrix with diagonal elements $\lambda_2, ..., \lambda_p, \lambda_3, ..., \lambda_p, ..., \lambda_p$, each multiplied by (-1). But the determinant

$$\begin{vmatrix} I & A \\ I & B \end{vmatrix} = |B - A| = (-1)^{\frac{p(p-1)}{2}} \left\{ \prod_{i<j} (\lambda_i + \lambda_j) \right\}.$$

The wedge product $\wedge\,[(\mathrm{d}\mathbf{V})V']$ is available from Theorem 2.12. This completes the proof.

Theorem 2.22 *In Theorem 2.21, if $V = 2(I+X)^{-1} - I$ where X is a skew symmetric matrix then, ignoring the sign,*

$$Z = Y[2(I+X)^{-1} - I] \Rightarrow$$

$$\mathrm{d}Z = 2^{p(p-1)/2}|I+X|^{-(p-1)}\left\{\prod_{i<j}(\lambda_i + \lambda_j)\right\}\mathrm{d}Y\;\mathrm{d}X.$$

Proof Consider $(\mathrm{d}\mathbf{V}^*)$ in the proof of Theorem 2.21. Writing

$$V = 2(I+X)^{-1} - I$$

we have

$$(\mathrm{d}\mathbf{V}^*) = -\left(\frac{1}{2}\right)U'(I+V)(\mathrm{d}X)(I+V')U \Rightarrow$$

$$\begin{aligned}
\mathrm{d}V^* &= \left|\frac{1}{\sqrt{2}}U'(I+V)\right|^{p-1}\mathrm{d}X \\
&= 2^{-p(p-1)/2}|I+V|^{p-1}\mathrm{d}X \\
&= 2^{-p(p-1)/2}|2(I+X)^{-1}|^{p-1}\mathrm{d}X \\
&= 2^{p(p-1)/2}|I+X|^{-(p-1)}\mathrm{d}X.
\end{aligned}$$

Substituting for $\mathrm{d}V^*$ above and $\mathrm{d}W$ from the proof of Theorem 2.21 the result follows.

Remark 2.16 If a function of Z is integrated out using Theorem 2.21 or Theorem 2.22 then the integral cannot be taken over Y directly due to the presence of the λ_j's which are also functions of Y. Hence if Y is written as $Y = UDU'$ where $D = \mathrm{diag}(\lambda_1, ..., \lambda_p)$, $UU' = I$ then the theorems hold for any selection of U. Hence if Y is integrated out over D and U, then U is integrated over the full orthogonal group $O_{(p)}$. That is,

$$\int_U \mathrm{d}G = \frac{2^p \pi^{\frac{p^2}{2}}}{\Gamma_p\left(\frac{p}{2}\right)}\;,\;(\mathrm{d}\mathbf{G}) = U'(\mathrm{d}U).$$

Note that the orthogonal matrix V in Theorem 2.21 and the skew symmetric matrix X in Theorem 2.22 are uniquely chosen and hence the factor 2^p will be absent in the corresponding integrals. That is,

$$\int_V dG_1 = \int_X 2^{p(p-1)/2}|I + X|^{-(p-1)}dX = \frac{\pi^{\frac{p^2}{2}}}{\Gamma_p\left(\frac{p}{2}\right)}, \quad (dG_1) = V'(dV).$$

Many more Jacobians of the types discussed in this section as well as their extensions from the real to the complex case are studied in the many papers of Khatri and the references therein, see for example, Khatri (1965a).

Exercises

2.2.1   Let $D = \text{diag}(\lambda_1, ..., \lambda_p)$, $0 < \lambda_1 < ... < \lambda_p < \infty$. Show that

$$\int_{0 < \lambda_1 < ... < \lambda_p < \infty} \left[\prod_{j>k}(\lambda_j - \lambda_k)\right] e^{-\text{tr}(D)}dD = \frac{\Gamma_p\left(\frac{p+1}{2}\right)\Gamma_p\left(\frac{p}{2}\right)}{\pi^{\frac{p^2}{2}}}.$$

2.2.2   For the same D in Execise 2.2.1 and for $\text{Re}(\alpha) > \frac{p-1}{2}$ show that

$$\int_{0 < \lambda_1 < ... < \lambda_p < \infty} \left(\prod_{j=1}^{p} \lambda_j\right)^{\alpha - \frac{p+1}{2}} \left(\prod_{j>k}(\lambda_j - \lambda_k)\right) e^{-\text{tr}(D)}dD$$

$$= \frac{\Gamma_p(\alpha)\Gamma_p\left(\frac{p}{2}\right)}{\pi^{\frac{p^2}{2}}}.$$

2.2.3   For $D = \text{diag}(\lambda_1, ..., \lambda_p)$, $0 < \lambda_j < 1$, $j = 1, ..., p$, $\text{Re}(\alpha) > \frac{p-1}{2}$, $\text{Re}(\beta) > \frac{p-1}{2}$, show that

$$\int_{0 < \lambda_1 < ... < \lambda_p < 1} \left(\prod_{j=1}^{p} \lambda_j\right)^{\alpha - \frac{p+1}{2}} \left(\prod_{j=1}^{p}(1 - \lambda_j)\right)^{\beta - \frac{p+1}{2}} \left(\prod_{j>k}(\lambda_j - \lambda_k)\right) dD$$

$$= \frac{\Gamma_p(\alpha)\Gamma_p(\beta)\Gamma_p\left(\frac{p}{2}\right)}{\Gamma_p(\alpha + \beta)\pi^{\frac{p^2}{2}}}.$$

2.2.4 Verify the result in Exercise 2.2.2 for $p = 2$ by carrying out the integration on the left side.

2.2.5 Let O_p be the set of $p \times p$ orthogonal matrices and let $h(\cdot)$ be a real valued function of $(\cdot)$. Then show that the functional equation $h(V_1)h(V_2) = h(V_1 V_2)$ implies $h(V) = 1$ when $h(\cdot)$ is positive, where $V_1 \in O_p$, $V_2 \in O_p$. See also Olkin and Sampson (1972).

2.2.6 Let Y be a $p \times p$ symmetric positive definite matrix of functionally independent real variables and with eigenvalues $\lambda_1 > ... > \lambda_p > 0$. Show that

$$\int_Y e^{-\text{tr}(YY')} \left\{ \prod_{i<j} (\lambda_i + \lambda_j) \right\} dY = 2^{-p} \Gamma_p \left(\frac{p}{2} \right).$$

2.2.7 Let $\lambda_1 > ... > \lambda_p > 0$ be real variables and $D = \text{diag}(\lambda_1, ..., \lambda_p)$. Show that

$$\int_D e^{-\text{tr}(D^2)} \left\{ \prod_{i<j} |\lambda_i^2 - \lambda_j^2| \right\} dD = \frac{[\Gamma_p \left(\frac{p}{2} \right)]^2}{2^p \pi^{\frac{p^2}{2}}}.$$

2.2.8 Let Z be a $p \times p$ matrix of p^2 functionally independent real variables. By using the decompositions in Theorems 2.21 and 2.22, evaluate the integral

$$\int_Z e^{-\text{tr}(ZZ')} dZ$$

by evaluating the integrals over the components Y and V when using Theorem 2.21, and Y and X when using Theorem 2.22.

2.2.9 For a 4×4 matrix X such that $X = X' > 0$ and $I - X > 0$, show that

$$\int_X dX = \frac{2\pi^4}{7!5}.$$

2.2.10 For a $p \times p$ matrix Y of p^2 functionally independent real variables with positive eigenvalues, show that

$$\int_Y |Y'Y|^{\alpha - \frac{p+1}{2}} e^{-\text{tr}(Y'Y)} dY = 2^{-p} \frac{\Gamma_p \left(\alpha - \frac{1}{2} \right) \pi^{\frac{p^2}{2}}}{\Gamma_p \left(\frac{p}{2} \right)}.$$

2.3 Singular value decomposition

Definition 2.3 **Singular values** Let X be a real $n \times p$, $n \geq p$, matrix of rank r. Then there exist three matrices U, D and V where U is $n \times r$ with $U'U = I_r$, $D = \text{diag}(\mu_1, ..., \mu_r)$ and V is $p \times r$ with $V'V = I_r$ such that $X = UDV'$. Then the diagonal elements $\mu_1, ..., \mu_r$ are called the singular values of X.

If $n = p$ and $r = p$ then the square matrix X has rank p and the elements of X can represent p^2 functionally independent real variables. Thus we have the following situation.

2.3.1 Square matrices of full rank

Any real $p \times p$ matrix X can be decomposed into the form

$$X = UDV', \ U'U = I, \ V'V = I, \ D = \text{diag}(\mu_1, ..., \mu_p)$$

where U and V are orthonormal matrices and D is a diagonal matrix with the elements $\mu_1, ..., \mu_p$. These $\mu_1, ..., \mu_p$ are the singular values of X. Note that

$$XX' = UDD'U' = UD^2U', \ U'U = I$$

and

$$X'X = VDD'V' = VD^2V', \ V'V = I$$

and thus $\lambda_1 = \mu_1^2, ..., \lambda_p = \mu_p^2$ are the eigenvalues of XX' or of $X'X$. Let us assume that the singular values $\mu_1, ..., \mu_p$ are distinct. If the elements in X are p^2 functionally independent real variables then these p^2 variables are split into $p(p-1)/2$ variables in U and V each and p variables in D. What will be the wedge product dX in terms of dU, dV and dD?

Theorem 2.23 *Let X be a $p \times p$ matrix of p^2 functionally independent real variables. Let $D = \text{diag}(\mu_1, ..., \mu_p)$, $\mu_1 > \mu_2 > ... > \mu_p$, μ_j real, $j = 1, ..., p$, and let U and V be orthonormal matrices such that $X = UDV'$. Then, ignoring the sign,*

$$X = UDV' \Rightarrow$$

$$dX = \left\{ \prod_{i<j} |\mu_i^2 - \mu_j^2| \right\} dG \ dH \ dD,$$

where $(\mathrm{dG}) = U'(\mathrm{dU})$ *and* $(\mathrm{dH}) = (\mathrm{dV}')V$.

Proof   Take the differentials in $X = UDV'$ to get

$$(\mathrm{dX}) = (\mathrm{dU})DV' + U(\mathrm{dD})V' + UD(\mathrm{dV}').$$

Pre and postmultiply by U' and V to get

$$U'(\mathrm{dX})V = U'(\mathrm{dU})D + (\mathrm{dD}) + D(\mathrm{dV}')V.$$

Since $U'U = I$ and $V'V = I$, $U'(\mathrm{dU})$ and $(\mathrm{dV}')V$ are skew symmetric matrices. Let $(\mathrm{dG}) = (\mathrm{d}g_{ij}) = U'(\mathrm{dU})$ and $(\mathrm{dH}) = (\mathrm{d}h_{ij}) = (\mathrm{dV}')V$. Then

$$(\mathrm{dG}) = \begin{bmatrix} 0 & -\mathrm{d}g_{21} & \cdots & -\mathrm{d}g_{p1} \\ \mathrm{d}g_{21} & 0 & \cdots & -\mathrm{d}g_{p2} \\ \vdots & \vdots & \cdots & \vdots \\ \mathrm{d}g_{p1} & \mathrm{d}g_{p2} & \cdots & 0 \end{bmatrix}, (\mathrm{dH}) = \begin{bmatrix} 0 & \cdots & \mathrm{d}h_{p1} \\ -\mathrm{d}h_{21} & \cdots & \mathrm{d}h_{p2} \\ \vdots & \cdots & \vdots \\ -\mathrm{d}h_{p1} & \cdots & 0 \end{bmatrix}.$$

Let $(\mathrm{dY}) = U'(\mathrm{dX})V$. Since (dX) is a nonsymmetric matrix and since U and V are orthonormal matrices we have $\mathrm{d}Y = \mathrm{d}X$, ignoring the sign, as $|U| = \pm 1$ and $|V| = \pm 1$, see also Theorem 1.18 of Chapter 1. Let us compute the wedge product in (dY).

$$\mathrm{d}y_{ii} = \mathrm{d}\mu_i$$
$$\mathrm{d}y_{ij} = \mu_j \mathrm{d}g_{ij} - \mu_i \mathrm{d}h_{ij}, \; i > j$$
$$= -\mu_j \mathrm{d}g_{ji} + \mu_i \mathrm{d}h_{ji}, \; i < j.$$

Thus if the elements are taken in the order $\mathrm{d}y_{11}, ..., \mathrm{d}y_{pp}$, $(\mathrm{d}y_{12}, \mathrm{d}y_{21})$, $..., (\mathrm{d}y_{1p}, \mathrm{d}y_{p1})$, $(\mathrm{d}y_{23}, \mathrm{d}y_{32})$, $..., (\mathrm{d}y_{p-1\,p}, \mathrm{d}y_{p\,p-1})$ then the corresponding matrix of partial derivatives is a triangular block matrix with the first diagonal block the identity matrix and the remaining blocks as

$$\begin{bmatrix} -\mu_2 & \mu_1 \\ \mu_1 & -\mu_2 \end{bmatrix}, \begin{bmatrix} -\mu_3 & \mu_1 \\ \mu_1 & -\mu_3 \end{bmatrix}, ..., \begin{bmatrix} -\mu_p & \mu_1 \\ \mu_1 & -\mu_p \end{bmatrix}, \begin{bmatrix} -\mu_3 & \mu_2 \\ \mu_2 & -\mu_3 \end{bmatrix}, ...$$

and thus the determinant is $(\mu_2^2 - \mu_1^2) ...(\mu_p^2 - \mu_1^2)(\mu_3^2 - \mu_2^2) ...(\mu_p^2 - \mu_2^2)...$, ignoring the sign. If the wedge products are taken then remember to use the fact $\mathrm{d}g_{ij}\mathrm{d}h_{ij} = -\mathrm{d}h_{ij}\mathrm{d}g_{ij}$. Thus, ignoring the sign,

$$\mathrm{d}X = \left\{ \prod_{i=1}^{p-1} \prod_{j=i+1}^{p} |\mu_i^2 - \mu_j^2| \right\} \mathrm{d}D \wedge [U'(\mathrm{dU})] \wedge [(\mathrm{dV}')V].$$

But the wedge products $\wedge[U'(\mathrm{d}U)]$ and $\wedge[(\mathrm{d}V')V]$ are available from Theorem 2.12 and thus the result is established.

Remark 2.17 The orthonormal matrices U and V are not unique. If any column of U and the corresponding column of V are multiplied by (-1), UDV' still remains the same. Hence the $\mathrm{d}X$ written up in Theorem 2.23 is for a preselected transformation $X = UDV'$. There are 2^p choices for which D will remain the same. Hence if $\wedge[U'(\mathrm{d}U)]$ and $\wedge[(\mathrm{d}V')V]$ are integrated out over the full orthogonal group then the final combined result is to be divided by 2^p.

Example 2.9 Let X be a 2×2 positive definite matrix of 4 functionally independent real variables. Let X have distinct singular values $0 < \mu_2 < \mu_1 < 1$. Evaluate $\int_0^I |X|^\alpha \mathrm{d}X$ for $\mathrm{Re}(\alpha) \geq 0$.

Solution In Example 2.8 we have evaluated this integral when X is symmetric positive definite. It is the nonsymmetric case here. We will evaluate this by using two different methods for the sake of illustration.

Method 1 Let Y be a 2×2 symmetric positive definite matrix with distinct eigenvalues and let V be an orthonormal matrix such that $X = YV$, $V'V = I$. Note that $XX' = YY' = Y^2$ since $Y = Y'$. Let $0 < \lambda_2 < \lambda_1$ be the eigenvalues of Y. Observe that $0 < X < I \Rightarrow 0 < XX' < I$ $\Rightarrow 0 < \lambda_2^2 < \lambda_1^2 < 1$ or $0 < \lambda_2 < \lambda_1 < 1$ or $0 < Y < I$. When $X = YV$ we have $\mathrm{d}X = (\lambda_1 + \lambda_2)\mathrm{d}Y\ \mathrm{d}G$, with $(\mathrm{d}G) = (\mathrm{d}V)V'$ from Theorem 2.21. Let

$$A(\alpha) = \int_{0<X<I} |X|^\alpha \mathrm{d}X$$
$$= \int_{0<Y<I} (\lambda_1\lambda_2)^\alpha (\lambda_1 + \lambda_2)\mathrm{d}Y \int_V \mathrm{d}G.$$

But from Theorem 2.13

$$\mathrm{d}Y = (\lambda_1 - \lambda_2)\mathrm{d}D\ \mathrm{d}H$$

where $D = \mathrm{diag}(\lambda_1, \lambda_2)$, $(\mathrm{d}H) = U'(\mathrm{d}U)$, $U'YU = D$. Thus

$$A(\alpha) = \int_D \phi(D)\mathrm{d}D \int_U \mathrm{d}H \int_V \mathrm{d}G$$

where

$$\int_D \phi(D)\mathrm{d}D = \int_{0<\lambda_2<\lambda_1<1} (\lambda_1\lambda_2)^\alpha \left(\lambda_1^2 - \lambda_2^2\right) \mathrm{d}\lambda_1 \mathrm{d}\lambda_2.$$

From Corollary 2.15.2 for $p = 2$ we have, see also Remarks 2.12 and 2.16,

$$\int_U dH \int_V dG = 2^{-p} \left[\frac{2^p \pi^{p^2/2}}{\Gamma_p \left(\frac{p}{2} \right)} \right]^2 = 4(\pi)(\pi)$$

and

$$\int_D \phi(D) dD = \int_0^1 \lambda_2^\alpha \left[\int_{\lambda_2}^1 \lambda_1^\alpha \left(\lambda_1^2 - \lambda_2^2 \right) d\lambda_1 \right] d\lambda_2$$

$$= \frac{1}{(\alpha+1)(\alpha+2)(\alpha+3)}.$$

Hence

$$\int_{0 < X < I} |X|^\alpha dX = \frac{4\pi^2}{(\alpha+1)(\alpha+2)(\alpha+3)}.$$

Method 2   Let U and V be as defined in Theorem 2.23. Then $U'XV = \text{diag}(\mu_1, \mu_2)$ and $dX = \left(\mu_1^2 - \mu_2^2 \right) dG\, dH$ where $(dH) = U'(dU)$ and $(dG) = (dV')V$. But $|X| = \mu_1\mu_2$. Then

$$A(\alpha) = \int_{0 < \mu_2 < \mu_1 < 1} (\mu_1\mu_2)^\alpha \left(\mu_1^2 - \mu_2^2 \right) d\mu_1 d\mu_2 \int_U dH \int_V dG.$$

Now proceeding as in Method 1 the result follows.

One can have a unique representation of U and V in terms of skew symmetric matrices. Let X_1 and X_2 be skew symmetric matrices. Then we can write

$$U = 2(I + X_1)^{-1} - I, \quad V = 2(I + X_2)^{-1} - I.$$

The representations are unique if the elements of the first row, except the first element, of $(I + X_1)^{-1}$ and $(I + X_2)^{-1}$ are negative. In terms of this parametrization we can restate Theorem 2.23 as follows:

Theorem 2.24   *Let*

$$U = 2(I + X_1)^{-1} - I \quad and \quad V = 2(I + X_2)^{-1} - I$$

where X_1 and X_2 are skew symmetric matrices such that the first row elements, except the first element, of $(I + X_1)^{-1}$ and $(I + X_2)^{-1}$ are negative. Let X be a $p \times p$ matrix of p^2 functionally independent real variables such that $X = UDV'$, $D = \text{diag}(\mu_1, ..., \mu_p)$ where the μ_j's are real, distinct and nonzero. Then, ignoring the sign,

$$X = \left[2(I + X_1)^{-1} - I \right] D \left[2(I + X_2)^{-1} - I \right]' \Rightarrow$$

$$dX = 2^{p(p-1)}|I + X_1|^{-(p-1)}|I + X_2|^{-(p-1)}$$

$$\times \left\{ \prod_{i=1}^{p-1} \prod_{j=i+1}^{p} |\mu_i^2 - \mu_j^2| \right\} dD \; dX_1 \; dX_2.$$

Proof   From Theorem 2.23 we have dX in terms of the wedge products $\wedge[U'(dU)]$ and $\wedge[(dV')V]$. Let us evaluate these in terms of dX_1 and dX_2. Taking the differentials in $U = 2(I + X_1)^{-1} - I$ we have

$$(dU) = -2(I + X_1)^{-1}(dX_1)(I + X_1)^{-1}$$
$$= -\frac{1}{2}(I + U)(dX_1)(I + U),$$

see also (1.3.10) of Chapter 1. Hence

$$U'(dU) = -\frac{1}{2}(I + U')(dX_1)(I + U)$$
$$= -\left[\frac{1}{\sqrt{2}}(I + U)\right]'(dX_1)\left[\frac{1}{\sqrt{2}}(I + U)\right]$$

since $U'(I + U) = U' + I$. From Theorem 1.21 of Chapter 1 the wedge product is given by

$$\wedge[U'(dU)] = 2^{-p(p-1)/2}|I + U|^{p-1}dX_1$$
$$= 2^{p(p-1)/2}|I + X_1|^{-(p-1)}dX_1.$$

A similar expression for $\wedge[(dV')V]$ is also available in terms of dX_2. Now substituting these in Theorem 2.23 the result is established.

Remark 2.18   Even if U and V are uniquely parametrized in terms of X_1 and X_2 there are still 2^p choices of U and V to give rise to the same D, see Remark 2.17. Hence if an integral is evaluated by using Theorem 2.24 then the final result must be multiplied by 2^p if the integrals over X_1 and X_2 are taken as

$$\int_{X_j} 2^{p(p-1)/2}|I + X_j|^{-(p-1)}dX_j = \frac{\pi^{\frac{p^2}{2}}}{\Gamma_p\left(\frac{p}{2}\right)}, \quad j = 1, 2.$$

2.3.2   Rectangular matrices of full rank

In Definition 2.3, if the rank of X is p whereas $n \geq p$, then we can have a situation of np functionally independent real variables represented by the elements of X, $np - p(p+1)/2$ variables in U, p variables in D and $p^2 - p(p+1)/2$ variables in V, that is,

$$np = \left[np - \frac{p(p+1)}{2} \right] + p + \left[p^2 - \frac{p(p+1)}{2} \right].$$

What will be the wedge product dX in terms of the wedge products dU, dD and dV?

Theorem 2.25   *Let X be an $n \times p$, $n \geq p$ matrix of full rank p and of np functionally independent real variables. Let U_1 be an element of the Stiefel manifold $V_{p,n}$ that is, U_1 is an $n \times p$, $n \geq p$ semiorthonormal matrix such that $U_1'U_1 = I$. Let V be a $p \times p$ orthonormal matrix and $D = \mathrm{diag}(\mu_1, ..., \mu_p)$ be a diagonal matrix with the diagonal elements being the singular values $\mu_1, ..., \mu_p$ of X which are real distinct and nonzero such that $X = U_1 D V'$. Then, ignoring the sign,*

$$X = U_1 D V', \quad D = \mathrm{diag}(\mu_1, ..., \mu_p), \quad U_1'U_1 = I, \quad V'V = I \Rightarrow$$

$$dX = |\mu_1...\mu_p|^{n-p} \left\{ \prod_{i=1}^{p-1} \prod_{j=i+1}^{p} |\mu_i^2 - \mu_j^2| \right\} dD \; dG \; dU_1$$

where

$$dG = \wedge_{j=1}^{p-1} \wedge_{k=j+1}^{p} (dv_j')v_k, \quad dU_1 = \wedge_{j=1}^{p} \wedge_{k=j+1}^{n} u_k'(du_j).$$

Proof   Let $U = (U_1 \;\; U_2)$, that is, U_1 augmented with U_2 where U_2 is $n \times (n-p)$ containing $n - p$ orthonormal vectors so that U is a full orthonormal matrix. Let $D_1 = \binom{D}{0}$ where 0 is a null matrix. Then

$$UD_1V' = (U_1 \;\; U_2)\binom{D}{0}V' = U_1 D V'$$

and

$$X = UD_1V' \Rightarrow$$
$$(dX) = (dU)D_1V' + U(dD_1)V' + UD_1(dV') \Rightarrow$$
$$U'(dX)V = U'(dU)D_1 + (dD_1) + D_1(dV')V$$

$$= \left[\begin{array}{c} U_1'(\mathrm{d}\mathbf{U_1})D + (\mathrm{d}D) + D(\mathrm{d}\mathbf{V'})V \\ U_2'(\mathrm{d}\mathbf{U_1})D \end{array} \right].$$

Let

$$(\mathrm{d}\mathbf{Y}) = U'(\mathrm{d}\mathbf{X})V \Rightarrow \mathrm{d}Y = \mathrm{d}X,$$

ignoring the sign, since U and V are orthonormal matrices. Consider the wedge product in $U_1'(\mathrm{d}\mathbf{U_1})D + (\mathrm{d}D) + D(\mathrm{d}\mathbf{V'})V$. Remember that $(\mathrm{d}\mathbf{U_1})$ is now the matrix of variables and that there are np variables. If u_i denotes the i-th column of U_1 then u_i is $n \times 1$, $i = 1, ..., p$. If $(\mathrm{d}u_i)$ denotes the differential of this vector, then writing $U_1'(\mathrm{d}\mathbf{U_1}) = (z_{ij})$, we have $z_{ij} = u_i'(\mathrm{d}u_j)$. If $u_i = \begin{pmatrix} u_{i1} \\ u_{i2} \end{pmatrix}$ where u_{i1} is the vector of the first p components of u_i and u_{i2} the vector of the remaining components is then

$$z_{ij} = u_{i1}'(\mathrm{d}u_{j1}) + u_{i2}'(\mathrm{d}u_{j2}).$$

Let the matrices on the right side be denoted by

$$A = (u_{i1}'(\mathrm{d}u_{j1})) \quad \text{and} \quad B = (u_{i2}'(\mathrm{d}u_{j2})).$$

Then proceeding as in the proof of Theorem 2.23, the wedge product in $AD + (\mathrm{d}D) + D(\mathrm{d}\mathbf{V'})V$ would have yielded the factor $\prod_{i<j} (\mu_i^2 - \mu_j^2)$. The remaining $(n-p) \times p$ variables are in B. Note that, in BD, the j-th column of B is multiplied by μ_j. Thus when the wedge product in BD is taken, μ_j appears $n - p$ times for each $j = 1, ..., p$. Hence the wedge product

$$\wedge [U_1'(\mathrm{d}\mathbf{U_1})D + (\mathrm{d}D) + D(\mathrm{d}\mathbf{V'})V]$$

$$= |\mu_1 ... \mu_p|^{n-p} \left\{ \prod_{i=1}^{p-1} \prod_{j=i+1}^{p} |\mu_i^2 - \mu_j^2| \right\} \mathrm{d}D \wedge [U_1'(\mathrm{d}\mathbf{U_1})] \wedge [(\mathrm{d}\mathbf{V'})V]$$

where

$$\wedge [U_1'(\mathrm{d}\mathbf{U_1})] = \wedge_{j=1}^{p-1} \wedge_{k=j+1}^{p} u_k'(\mathrm{d}u_j)$$

and

$$\wedge [(\mathrm{d}\mathbf{V'})V] = \wedge_{j=1}^{p-1} \wedge_{k=j+1}^{p} (\mathrm{d}v_j')v_k.$$

Now consider the wedge product in $U_2'(\mathrm{d}\mathbf{U_1})D$. If the columns of U_2 are denoted by $\alpha_1, ..., \alpha_{n-p}$ and those of U_1 by $u_1, ..., u_p$ then

$$U_2'(\mathrm{d}\mathbf{U_1})D = \left[\begin{array}{ccc} \mu_1 \alpha_1'(\mathrm{d}u_1) & ... & \mu_p \alpha_1'(\mathrm{d}u_p) \\ \vdots & ... & \vdots \\ \mu_1 \alpha_{n-p}'(\mathrm{d}u_1) & ... & \mu_p \alpha_{n-p}'(\mathrm{d}u_p) \end{array} \right].$$

Note that $U_2'(\mathrm{d}U_1)D$ contains the differentials $\mathrm{d}u_{ij}$'s and α_k's, $k = 1, ..., n-p$. Thus the wedge product coming from here, excluding the μ_j's which are already considered, is of the form

$$\wedge_{j=1}^{p} \wedge_{k=1}^{n-p} \alpha_k'(\mathrm{d}u_j).$$

Writing α_j's, $j = 1, ..., n-p$ as u_j, $j = p+1, ..., n$ and combining the wedge products the result follows.

Example 2.10   Let the $n \times p$ real random matrix $X = (x_{jk})$ have a real np-variate Gaussian density given by

$$f(X) = \frac{e^{-\frac{1}{2}\mathrm{tr}(W^{-1}X'X)}}{(2\pi)^{np/2}|W|^{n/2}}, \quad W = W' > 0, \quad -\infty < x_{jk} < \infty.$$

Evaluate the density of $X'X$ if X has distinct and nonzero singular values $\mu_1, ..., \mu_p$, $\mu_1 > ... > \mu_p$.

Solution   Let U be an $n \times p$ semiorthonormal matrix and V a $p \times p$ orthonormal matrix such that

$$X = UD_\mu V', \quad D_\mu = \mathrm{diag}(\mu_1, ..., \mu_p).$$

Then

$$\mathrm{d}X = |\mu_1...\mu_p|^{n-p} \left\{ \prod_{i<j} |\mu_i^2 - \mu_j^2| \right\} \mathrm{d}D_\mu \mathrm{d}G \, \mathrm{d}H$$

where

$$\mathrm{d}G = \wedge_{j=1}^{p} \wedge_{k=j+1}^{n} u_k'(\mathrm{d}u_j)$$

and

$$\mathrm{d}H = \wedge_{j=1}^{p-1} \wedge_{k=j+1}^{p} v_k'(\mathrm{d}v_j).$$

Note that

$$X'X = VD_\mu^2 V', \quad D_\mu^2 = D_\lambda = \mathrm{diag}(\lambda_1, ..., \lambda_p), \quad \lambda_j = \mu_j^2.$$

Since U is absent in the transformed $f(X)$, we get the joint density of D_μ and V by integrating out U. From Corollary 2.15.1 we have, see also Remark 2.17,

$$\int_U \mathrm{d}G = \frac{2^p \pi^{np/2}}{\Gamma_p(n/2)}.$$

Then the joint density of D_μ and V, denoted by $g(D_\mu, V)$, is available as

$$g(D_\mu, V)dD_\mu dV = \frac{2^p e^{-\frac{1}{2}\text{tr}(W^{-1}VD_\mu^2 V')}}{2^{np/2}\Gamma_p(n/2)|W|^{n/2}}$$

$$\times (\mu_1...\mu_p)^{n-p} \left\{ \prod_{i<j} |\mu_i^2 - \mu_j^2| \right\} dD_\mu dV.$$

But since $\mu_j^2 = \lambda_j$, $dD_\mu = 2^{-p}(\lambda_1...\lambda_p)^{-1/2}dD_\lambda$. Let $S = VD_\lambda V'$. Then from Theorem 2.13 we have

$$dS = \left\{ \prod_{i<j} |\lambda_i - \lambda_j| \right\} dD_\lambda dV.$$

Note that

$$VD_\mu^2 V' = S, \quad (\mu_1...\mu_p)^{n-p} = |S|^{(n-p)/2}, \quad \text{and} \quad (\lambda_1...\lambda_p)^{-\frac{1}{2}} = |S|^{-\frac{1}{2}}.$$

Substituting these we have the density of S, denoted by $h(S)$, given by

$$h(S)dS = \begin{cases} \dfrac{|S|^{\frac{n}{2}-\frac{p+1}{2}}\exp\{-\frac{1}{2}\text{tr}(W^{-1}S)\}}{2^{np/2}\Gamma_p(n/2)|W|^{n/2}}dS, & W = W' > 0, \ S = S' > 0 \\ 0, & \text{elsewhere.} \end{cases}$$

Remark 2.19 The transformations in Theorems 2.23 and 2.25 can be shown to be unique if the singular values are real distinct and nonzero and the diagonal elements of V are positive.

Exercises

2.3.1 Let X be a $p \times p$ symmetric positive definite matrix of functionally independent real variables with distinct eigenvalues $\lambda_1 > \lambda_2 > ... > \lambda_p > 0$. Let $X = HDH'$ where H is an orthonormal matrix and $D = \text{diag}(\lambda_1, ..., \lambda_p)$. Then show that

$$dX = \left\{ \prod_{j<k} |\lambda_j - \lambda_k| \right\} dDdP$$

where

$$dD = \prod_{j=1}^{p} d\lambda_j, \quad dP = \prod_{j>k} h'_j(dh_k)$$

and h_j is the j-th column vector of H with (dh_k) the differential of the k-th column of H.

2.3.2 Let X be a $p \times q$ matrix of rank $p, q \ge p$, $T = (t_{jk})$ a $p \times p$ lower triangular matrix with $t_{jj} > 0$, $j = 1, ..., p$ and L a $q \times p$ matrix satisfying $L'L = I_p$ where the matrices are of functionally independent real variables. Then show that

$$X = TL' \Rightarrow dX = \left(\prod_{j=1}^{p} t_{jj}^{p-j} \right) dTd\hat{L}$$

where $d\hat{L} = \prod_{k=1}^{p} \prod_{j=k+1}^{q} l'_j(dl_k)$, $l_j = j$-th column of $\hat{L} = (\, L \quad L_1\,) \in O_q$, $(dl_k) = $ the differential of the k-th column of L.

2.3.3 Let X be a $p \times q$ matrix, $q \ge p$, S a $p \times p$ symmetric positive definite matrix and L a $q \times p$ matrix with $L'L = I_p$, all are of functionally independent real variables. Then show that

$$X = S^{\frac{1}{2}} L' \Rightarrow dX = 2^{-p} |S|^{\frac{1}{2}(q-p-1)} dSd\hat{L}$$

where $d\hat{L}$ is defined in Exercise 2.3.2 and $S^{\frac{1}{2}}$ is the symmetric square root of S.

2.3.4 Using the same notations as in Exercise 2.3.3 show that

$$\int_{L'L=I_p} d\hat{L} = \frac{2^p \pi^{\frac{pq}{2}}}{\Gamma_p\left(\frac{q}{2}\right)}$$

and for $p = q$

$$\int_{O_p} d\hat{H} = \frac{2^p \pi^{\frac{p^2}{2}}}{\Gamma_p\left(\frac{p}{2}\right)}$$

where O_p is the full orthogonal group. Define the normalized orthogonal measures as

$$dH = \frac{\Gamma_p\left(\frac{p}{2}\right)}{2^p \pi^{\frac{p^2}{2}}} d\hat{H} = \frac{\Gamma_p\left(\frac{p}{2}\right)}{2^p \pi^{\frac{p^2}{2}}} \prod_{j>k} h'_j(dh_k)$$

where h_j is the j-th column vector of H, and

$$dL = \frac{\Gamma_p\left(\frac{q}{2}\right)}{2^p \pi^{\frac{pq}{2}}} d\hat{L}$$

$$= \frac{\Gamma_p\left(\frac{q}{2}\right)}{2^p \pi^{\frac{pq}{2}}} \left(\prod_{j>k} l'_j(dl_k) \right) \left(\prod_{j=1}^{p} \prod_{k=1}^{q-p} b'_k(dl_j) \right)$$

where l_j is the j-th column of L and b_k the k-th column of B, $(L \quad B) \in O_q$. Show that $d\hat{H} = \wedge(H'(dH))$ is invariant under simultaneous translations $H \to H_1 H H_2$, $H_1, H_2 \in O_p$.

2.3.5 If the density function of an $m \times n$, $n \geq m$ real random matrix X is $f(XX')$ then show that the density function of $S = XX'$ is expressed as

$$g(S) = \frac{\pi^{\frac{mn}{2}}}{\Gamma_m\left(\frac{n}{2}\right)} |S|^{\frac{n}{2} - \frac{m+1}{2}} f(S).$$

(For more situations of this type see Fang and Anderson (1990))

2.3.6 Let X be a 2×2 positive definite matrix of 4 functionally independent real variables with singular values μ_1 and μ_2, $0 < \mu_2 < \mu_1 < 1$. Evaluate the integral $\int_0^I |X|^\alpha dX$, $\text{Re}(\alpha) \geq 0$ by using Theorem 2.24.

2.3.7 Redo Exercise 2.3.6 using the same procedure when X is symmetric positive definite.

2.4 Simultaneous transformations involving many matrices

In the previous sections we considered transformations of the type where one matrix is written as a function of one or more matrices. Here we consider simultaneous transformations involving many matrices.

2.4.1 Transformations involving two matrices

Consider the case of two matrices, that is, two matrices written as functions of two or more matrices. Such problems abound in statistical distribution theory. As a simple case, consider two real $p \times p$ matrices X_1 and X_2 transformed to two $p \times p$ matrices Y_1 and Y_2 by the relation

$$Y_1 = X_1, \quad Y_2 = X_1 + X_2. \tag{2.4.1}$$

When X_1 and X_2 are not symmetric, we have $p^2 + p^2 = 2p^2$ functionally independent variables in X_1 and X_2 together. Then the transformation (2.4.1) can be treated as a transformation of the combined set of $2p^2$ variables in X_1 and X_2 to the $2p^2$ variables in Y_1 and Y_2. The wedge products can be taken as

$$\wedge [(d\mathbf{X_1}), (d\mathbf{X_2})] = dX_1 \, dX_2 \quad \text{and} \quad \wedge [(d\mathbf{Y_1}), (d\mathbf{Y_2})] = dY_1 \, dY_2, \tag{2.4.2}$$

respectively. What is the relationship between $dX_1 dX_2$ and $dY_1 dY_2$? The matrix of partial derivatives in (2.4.1) will be of the structure

$$
\begin{array}{cc}
 & X_1 \quad X_2 \\
\begin{array}{c} Y_1 \\ Y_2 \end{array} &
\begin{array}{cc} I & 0 \\ I & I \end{array}
\end{array}
$$

and then the determinant is unity. Thus it is easy to see that in (2.4.1) we have

$$dX_1 \, dX_2 = dY_1 \, dY_2.$$

Theorem 2.26 *Let X_1 and X_2 be $p \times p$ symmetric positive definite matrices of functionally independent real variables. Then*

$$Y_1 = X_2^{-\frac{1}{2}} X_1 X_2^{-\frac{1}{2}} \text{ and } Y_2 = X_2$$
$$\Rightarrow dY_1 \, dY_2 = |X_2|^{-\frac{p+1}{2}} dX_1 \, dX_2$$

where $X_2^{-\frac{1}{2}}$ is the symmetric square root of X_2^{-1}.

Proof Since Y_2 does not contain X_1, the matrix of partial derivatives is of the following form:

$$
\begin{array}{cc}
 & X_1 \quad X_2 \\
\begin{array}{c} Y_1 \\ Y_2 \end{array} &
\begin{array}{cc} G & H \\ 0 & I \end{array}
\end{array}
$$

where G and I are $\frac{p(p+1)}{2} \times \frac{p(p+1)}{2}$ matrices. Since the Jacobian matrix is a triangular block matrix, the Jacobian is $|G| \, |I| = |G|$. But $|G|$ is the Jacobian of Y_1 as a function of X_1 with X_2 fixed. From Theorem 1.20 of Chapter 1,

$$dY_1 = |X_2|^{-(p+1)/2} dX_1$$

for fixed X_2. Hence

$$dY_1 \, dY_2 = |X_2|^{-(p+1)/2} dX_1 \, dX_2.$$

Remark 2.20   If Y_2 or Y_1 contains only one of the matrices X_1 and X_2, then we can evaluate the Jacobian by using the results in the previous sections of this chapter and those of Chapter 1.

Theorem 2.27   *Let X_1 and X_2 be as in Theorem 2.26. Then*

$$Y_1 = (X_1 + X_2)^{-1/2} X_1 (X_1 + X_2)^{-1/2}$$

and

$$Y_2 = X_1 + X_2 \Rightarrow dY_1 \, dY_2 = |X_1 + X_2|^{-(p+1)/2} dX_1 \, dX_2.$$

Proof   Let $U_1 = X_1$ and $U_2 = X_1 + X_2$. Then $dU_1 \, dU_2 = dX_1 \, dX_2$. Thus

$$Y_1 = U_2^{-1/2} U_1 U_2^{-1/2} \text{ and } Y_2 = U_2.$$

Then from Theorem 2.26 the result follows.

In terms of J we can write this as follows, where $J(Y_1, Y_2 : X_1, X_2)$ means the Jacobian when Y_1 and Y_2 are written as functions of X_1 and X_2.

$$J(Y_1, Y_2 : X_1, X_2) = J(Y_1, Y_2 : U_1, U_2) J(U_1, U_2 : X_1, X_2).$$

Example 2.11   Let X_1 and X_2 be $p \times p$ real symmetric positive definite random matrices having gamma densities

$$f_j(X_j) = \frac{|B|^{\alpha_j}}{\Gamma_p(\alpha_j)} |X_j|^{\alpha_j - \frac{p+1}{2}} e^{-\mathrm{tr}(BX_j)}$$

for $X_j = X_j' > 0$, $B = B' > 0$, $\text{Re}(\alpha_j) > \frac{p-1}{2}$, $j = 1, 2$ and $f_j(X_j) = 0$ elsewhere. Let X_1 and X_2 be independently distributed. Evaluate the density of

$$Y_1 = (X_1 + X_2)^{-\frac{1}{2}} X_1 (X_1 + X_2)^{-\frac{1}{2}}.$$

Solution   Due to statistical independence, the joint density of X_1 and X_2, denoted by $f(X_1, X_2)$, is given by

$$\begin{aligned} f(X_1, X_2) dX_1 dX_2 &= f_1(X_1) f_2(X_2) dX_1 dX_2 \\ &= \frac{|B|^{\alpha_1 + \alpha_2}}{\Gamma_p(\alpha_1) \Gamma_p(\alpha_2)} |X_1|^{\alpha_1 - \frac{p+1}{2}} |X_2|^{\alpha_2 - \frac{p+1}{2}} \\ &\quad \times e^{-\text{tr}(B(X_1 + X_2))} dX_1 dX_2. \end{aligned}$$

Consider the transformation

$$Y_1 = (X_1 + X_2)^{-\frac{1}{2}} X_1 (X_1 + X_2)^{-\frac{1}{2}} \text{ and } Y_2 = X_1 + X_2.$$

Then from Theorem 2.27

$$dY_1 dY_2 = |X_1 + X_2|^{-\frac{(p+1)}{2}} dX_1 dX_2 = |Y_2|^{-\frac{(p+1)}{2}} dX_1 dX_2.$$

Note that

$$Y_1 = Y_2^{-\frac{1}{2}} X_1 Y_2^{-\frac{1}{2}} \Rightarrow Y_2^{\frac{1}{2}} Y_1 Y_2^{\frac{1}{2}} = X_1$$

and

$$X_2 = Y_2 - X_1 = Y_2^{\frac{1}{2}} (I - Y_1) Y_2^{\frac{1}{2}}.$$

Substituting for X_1, X_2, dX_1, dX_2 the joint density of Y_1 and Y_2, denoted by $g(Y_1, Y_2)$, is given by

$$\begin{aligned} g(Y_1, Y_2) dY_1 dY_2 &= \frac{|B|^{\alpha_1 + \alpha_2}}{\Gamma_p(\alpha_1) \Gamma_p(\alpha_2)} |Y_2|^{\alpha_1 + \alpha_2 - \frac{p+1}{2}} e^{-\text{tr}(BY_2)} \\ &\quad \times |Y_1|^{\alpha_1 - \frac{p+1}{2}} |I - Y_1|^{\alpha_2 - \frac{p+1}{2}} dY_1 dY_2. \end{aligned}$$

Integrating out Y_2 by using a gamma integral we have the density of Y_1, denoted by $g_1(Y_1)$ as follows:

$$g_1(Y_1) = \frac{\Gamma_p(\alpha_1 + \alpha_2)}{\Gamma_p(\alpha_1) \Gamma_p(\alpha_2)} |Y_1|^{\alpha_1 - \frac{p+1}{2}} |I - Y_1|^{\alpha_2 - \frac{p+1}{2}}$$

for $0 < Y_1 < I$, $\text{Re}(\alpha_j) > \frac{p-1}{2}$, $j = 1, 2$ and $g_1(Y_1) = 0$ elsewhere. This is a type-1 matrix-variate real beta density.

In the previous theorems one of Y_1 and Y_2 is a linear function of X_1 and X_2. Now we will consider a case where both are nonlinear functions.

Theorem 2.28 *Let X, D be $p \times p$ matrices of functionally independent real variables such that $D = \text{diag}(\lambda_1, ..., \lambda_p)$, $\lambda_1 > \lambda_2 > ... > \lambda_p > 0$. Then*

$$Y_1 = XDX' \text{ and } Y_2 = XX' \Rightarrow \tag{2.4.3}$$

$$dY_1 \, dY_2 = 2^p |X|^{p+2} \left\{ \prod_{i<j} |\lambda_i - \lambda_j| \right\} dX \, dD.$$

Proof Taking differentials we have

$$(dY_1) = (dX)DX' + X(dD)X' + XD(dX'); \tag{a}$$
$$(dY_2) = (dX)X' + X(dX'). \tag{b}$$

Pre and postmultiply (a) and (b) by X^{-1} and X'^{-1} to get

$$X^{-1}(dY_1)X'^{-1} = X^{-1}(dX)D + (dD) + D(dX')X'^{-1}; \tag{c}$$
$$X^{-1}(dY_2)X'^{-1} = X^{-1}(dX) + (dX')X'^{-1}. \tag{d}$$

Let

$$(dU) = X^{-1}(dX) \Rightarrow dU = |X|^{-p}dX;$$
$$(dW) = X^{-1}(dY_1)X'^{-1} \Rightarrow dW = |X|^{-(p+1)}dY_1;$$
$$(dV) = X^{-1}(dY_2)X'^{-1} \Rightarrow dV = |X|^{-(p+1)}dY_2.$$

From (c) and (d) we have

$$(dW) = (dU)D + (dD) + D(dU'); \tag{e}$$
$$(dV) = (dU) + (dU'). \tag{f}$$

That is,

$$dw_{ii} = d\lambda_i + 2\lambda_i du_{ii};$$
$$dw_{ij} = \lambda_j du_{ij} + \lambda_i du_{ji}, \ i \neq j;$$
$$dv_{ii} = 2du_{ii};$$
$$dv_{ij} = du_{ij} + du_{ji}, \ i \neq j.$$

Let us examine the table of partial derivatives.

	$d\lambda_i,\ \forall i$	$du_{ii},\ \forall i$	$du_{ij},\ i<j$	$du_{ij},\ i>j$
$dw_{ii},\ \forall i$	I	A	0	0
$dv_{ii},\ \forall i$	0	$2I$	0	0
$dw_{ij},\ i<j$	0	0	B	C
$dv_{ij},\ i<j$	0	0	I	I

It is a triangular block matrix and hence the determinant is

$$|2I|\begin{vmatrix} B & C \\ I & I \end{vmatrix} = 2^p|B-C| = 2^p\left\{\prod_{i>j}(\lambda_i-\lambda_j)\right\}. \tag{g}$$

That is, $J(W,V:D,U)$ gives

$$dW\ dV = \left\{\prod_{i>j}|\lambda_i-\lambda_j|\right\}dD\ dU.$$

Substituting for dU, dV, and dW the result follows. In terms of J the above transformations can be written as

$$J(Y_1,Y_2:X,D) = J(Y_1,Y_2:W,V)$$
$$\times\ J(W,V:U,D)J(U:X).$$

Remark 2.21   The Jacobians in the following transformations can be shown to be the same when X and $D = D_\lambda$ are as defined in Theorem 2.28.

(i) $Y_1 = XD_\lambda X',\ Y_2 = XX',\ D_\lambda = \text{diag}(\lambda_1,...,\lambda_p)$;

(ii) $Y_1 = XD_\lambda X',\ Y_1 + Y_2 = XX'$;

(iii) $Y_1 = XD_\nu X',\ Y_2 = XD_\mu X',\ D_\mu = \text{diag}(\mu_1,...,\mu_p)$,
$$\nu_j^2 + \mu_j^2 = 1,\ j = 1,...,p.$$

It can also be noted that $\lambda_1 > ... > \lambda_p > 0$ are the roots of the determinantal equation

$$|Y_1 - \lambda Y_2| = 0,\quad |X| \neq 0.$$

Remark 2.22   The transformation in (2.4.3) with X a square matrix is unique when $x_{jj} > 0$, $j = 1,...,p$, X nonsingular and the λ_j's are distinct and nonzero. Hence if an integral is evaluated by using Theorem 2.28 and by integrating out over an arbitrary nonsingular matrix X, then the final result must be divided by 2^p.

2.4.2   Transformations involving many matrices

In Section 2.4.1 we considered simultaneous transformations involving two matrices. Here we will generalize to k matrices. As a simple case, consider k real $p \times p$ matrices $X_1, ..., X_k$ transformed to the $p \times p$ matrices $Y_1, ..., Y_k$ by the relation

$$
\begin{aligned}
Y_1 &= X_1, \\
Y_2 &= X_1 + X_2, \\
&\ \ \vdots \\
Y_k &= X_1 + ... + X_k.
\end{aligned}
\tag{2.4.4}
$$

The wedge products can be taken as

$$\wedge [(\mathrm{d}\mathbf{X_1}), ..., (\mathrm{d}\mathbf{X_k})] = \mathrm{d}X_1 ... \mathrm{d}X_k \text{ and } \wedge [(\mathrm{d}\mathbf{Y_1}), ..., (\mathrm{d}\mathbf{Y_k})] = \mathrm{d}Y_1 ... \mathrm{d}Y_k$$

respectively. What is the relationship between $\mathrm{d}X_1 ... \mathrm{d}X_k$ and $\mathrm{d}Y_1 ... \mathrm{d}Y_k$? The matrix of partial derivatives in (2.4.4) will then be a triangular block matrix, see also Section 2.4.1, and the determinant is unity. Thus it is easy to see that in (2.4.4) we have

$$\mathrm{d}X_1 ... \mathrm{d}X_k = \mathrm{d}Y_1 ... \mathrm{d}Y_k.$$

Now we consider a simple nonlinear transformation which is applicable when dealing with the matrix-variate Dirichlet densities in statistical distribution theory.

Theorem 2.29   *Let $X_1, ..., X_k$ be $p \times p$ symmetric positive definite matrices of functionally independent real variables such that $0 < X_j < I$, $j = 1, ..., k$, that is, all the eigenvalues of X_j are between 0 and 1. Let*

$$
\begin{aligned}
Y_1 &= X_1 \\
Y_2 &= (I - X_1)^{\frac{1}{2}} X_2 (I - X_1)^{\frac{1}{2}} \\
Y_3 &= (I - X_1)^{\frac{1}{2}} (I - X_2)^{\frac{1}{2}} X_3 (I - X_2)^{\frac{1}{2}} (I - X_1)^{\frac{1}{2}} \\
&\ \ \vdots \\
Y_k &= (I - X_1)^{\frac{1}{2}} ... (I - X_{k-1})^{\frac{1}{2}} X_k (I - X_{k-1})^{\frac{1}{2}} ... (I - X_1)^{\frac{1}{2}}, \quad (2.4.5)
\end{aligned}
$$

where $(I - X_j)^{\frac{1}{2}}$ *is the symmetric square root of* $I - X_j$. *Then*

$$dY_1...dY_k = |I - X_1|^{\frac{1}{2}(p+1)(k-1)}|I - X_2|^{\frac{1}{2}(p+1)(k-2)}$$
$$\times ...|I - X_{k-1}|^{\frac{1}{2}(p+1)}dX_1...dX_k.$$

Proof   Since the transformation in (2.4.5) is of a triangular formation, we can compute dY_j in terms of dX_j with $X_1, ..., X_{j-1}$ fixed for $j = 1, ..., k$. Now from Theorem 1.20 of Chapter 1, we have

$$dY_j = |I - X_1|^{\frac{1}{2}(p+1)}...|I - X_{j-1}|^{\frac{1}{2}(p+1)}dX_j$$

for fixed $X_1, ..., X_{j-1}$, and for $j = 1, ..., k$. Multiplying these, the result follows.

Note that the transformation in (2.4.5) can also be written as

$$X_1 = Y_1$$
$$X_2 = (I - Y_1)^{-\frac{1}{2}}Y_2(I - Y_1)^{-\frac{1}{2}}$$
$$X_3 = (I - Y_1 - Y_2)^{-\frac{1}{2}}Y_3(I - Y_1 - Y_2)^{-\frac{1}{2}}$$
$$\vdots$$
$$X_k = (I - Y_1 - ... - Y_{k-1})^{-\frac{1}{2}}Y_k(I - Y_1 - ... - Y_{k-1})^{-\frac{1}{2}}. \quad (2.4.6)$$

Theorem 2.30   *Let* X_j, $j = 1, ..., k$ *be symmetric positive definite matrices of functionally independent real variables such that* $0 < X_j < I$, $j = 1, ..., k$ *and* $0 < X_1 + ... + X_k < I$. *Let* $X_0 = X_1 + ... + X_k$. *Then*

$$Y_j = (I - X_0)^{-\frac{1}{2}}X_j(I - X_0)^{-\frac{1}{2}}, \, j = 1, ..., k \Rightarrow \quad (2.4.7)$$
$$dY_1...dY_k = |I - X_0|^{-(k+1)(\frac{p+1}{2})}dX_1...dX_k$$

or

$$dX_1...dX_k = |I + Y_1 + ... + Y_k|^{-(k+1)(\frac{p+1}{2})}dY_1...dY_k.$$

Proof   From (2.4.7) one has

$$I + Y_1 + ... + Y_k = (I - X_0)^{-1}.$$

Let $U_1 = Y_1, ..., U_{k-1} = Y_{k-1}$ and $U_k = I + Y_1 + ... + Y_k$. Then

$$dY_1...dY_k = dU_1...dU_k.$$

Note that

$$X_j = U_k^{-\frac{1}{2}} U_j U_k^{-\frac{1}{2}}, \; j = 1, ..., k - 1$$

$$X_k = I - U_k^{-\frac{1}{2}}(I + U_1 + ... + U_{k-1})U_k^{-\frac{1}{2}}.$$

The matrix of partial derivatives is of the following form:

	U_1	U_2	$\cdots$	U_{k-1}	U_k
X_1	G_1	0	$\cdots$	0	G_1^*
X_2	0	G_2	$\cdots$	0	G_2^*
$\vdots$	$\vdots$	$\vdots$	$\ddots$	$\vdots$	$\vdots$
X_{k-1}	0	0	$\cdots$	G_{k-1}	G_{k-1}^*
X_k	$-G_1$	$-G_2$	$\cdots$	$-G_{k-1}$	$-G^* - G_1^* - ... - G_{k-1}^*$

where, for example, G_j is the partial derivative matrix of the elements in X_j with respect to the elements in U_j, $j = 1, ..., k-1$, G_j^* is that of $U_k^{-\frac{1}{2}} U_j U_k^{-\frac{1}{2}}$ with respect to the elements in U_k and G^* is that of U_k^{-1} with respect to U_k. The determinant of the above matrix configuration is

$$|G_1| \, |G_2| \, ... |G_{k-1}| \, |G^*|$$

ignoring the sign. But, for example, $|G_j|$ is the Jacobian coming from the transformation

$$X_j = U_k^{-\frac{1}{2}} U_j U_k^{-\frac{1}{2}}$$

with U_k fixed. That is, from Theorem 1.20 of Chapter 1

$$dX_j = |U_k|^{-\frac{p+1}{2}} dU_j, \; j = 1, ..., k - 1$$

and from Theorem 1.27 of Chapter 1

$$dX_k = |G^*| dU_k = |U_k|^{-(p+1)} dU_k.$$

Thus

$$dX_1...dX_k = |U_k|^{-(k+1)(\frac{p+1}{2})} dU_1...dU_k.$$

Now substitute for the U_j's in terms of the Y_j's to complete the proof.

Example 2.12 Under the transformation in Theorem 2.30, show that a real type-1 Dirichlet distribution goes to a type-2 Dirichlet distribution.

Solution   The real type-1 matrix-variate Dirichlet density is given by

$$f(X_1, ..., X_k) = C\left\{\prod_{j=1}^{k} |X_j|^{\alpha_j - \frac{p+1}{2}}\right\} |I - X_1 - ... - X_k|^{\alpha_{k+1} - \frac{p+1}{2}}$$

for $0 < X_j < I$, $j = 1, ..., k$, $0 < X_1 + ... + X_k < I$, $\mathrm{Re}(\alpha_j) > \frac{p-1}{2}$, $j = 1, ..., k+1$ where C is the normalizing constant. Consider the transformation in Theorem 2.30 and from the proof thereof note that

$$X_0 = X_1 + ... + X_k, \quad (I - X_0)^{-1} = I + Y_1 + ... + Y_k$$

and

$$X_j = (I + Y_1 + ... + Y_k)^{-\frac{1}{2}} Y_j (I + Y_1 + ... + Y_k)^{-\frac{1}{2}}.$$

Substituting for the X_j's, $I - X_0$ and $dX_1...dX_k$, we have the joint density of $Y_1, ..., Y_k$ given by

$$g(Y_1, ..., Y_k) = C\left\{\prod_{j=1}^{k} |Y_j|^{\alpha_j - \frac{p+1}{2}}\right\} |I + Y_1 + ... + Y_k|^{-(\alpha_1 + ... + \alpha_{k+1})}$$

for $Y_j = Y_j' > 0$, $j = 1, ..., k$. This is the real type-2 matrix-variate Dirichlet density and hence the result.

Theorem 2.31   Let $X_1, ..., X_k$ be symmetric positive definite $p \times p$ matrices of functionally independent real variables. Let

$$X_0 = X_1 + ... + X_k$$

and

$$Y_j = (I + X_0)^{-\frac{1}{2}} X_j (I + X_0)^{-\frac{1}{2}}, \quad j = 1, ..., k.$$

Then

$$dY_1...dY_k = |I + X_0|^{-(k+1)\left(\frac{p+1}{2}\right)} dX_1...dX_k$$

or

$$dX_1...dX_k = |I - Y_1 - ... - Y_k|^{-(k+1)\left(\frac{p+1}{2}\right)} dY_1...dY_k.$$

Proof   Note that

$$I - Y_1 - ... - Y_k = (I + X_0)^{-1}.$$

Now the steps are parallel to those used in the proof of Theorem 2.30 and hence omitted.

Theorem 2.32   Let $X_0, X_1, ..., X_k$ be $p \times p$ real symmetric positive definite matrices. Let

$$Y_0 = X_0 + X_1 + ... + X_k$$

and

$$Y_j = Y_0^{-\frac{1}{2}} X_j Y_0^{-\frac{1}{2}}, \ j = 1, ..., k.$$

Then

$$\mathrm{d}X_0 \ \mathrm{d}X_1...\mathrm{d}X_k = |Y_0|^{k\left(\frac{p+1}{2}\right)}\mathrm{d}Y_0 \ \mathrm{d}Y_1...\mathrm{d}Y_k$$

or

$$\mathrm{d}Y_0 \ \mathrm{d}Y_1...\mathrm{d}Y_k = |X_0 + X_1 + ... + X_k|^{-k\left(\frac{p+1}{2}\right)}\mathrm{d}X_0 \ \mathrm{d}X_1...\mathrm{d}X_k.$$

Proof   Note that

$$X_j = Y_0^{\frac{1}{2}} Y_j Y_0^{\frac{1}{2}}, \ j = 1, ..., k$$

and

$$X_0 = Y_0^{\frac{1}{2}}[I - Y_1 - ... - Y_k]Y_0^{\frac{1}{2}}.$$

Now proceeding as in the proof of Theorem 2.30 the result follows.

Example 2.13   Let the $p \times p$ real matrices $X_j, \ j = 0, 1, ..., k$ be independently gamma distributed with the densities

$$f_j(X_j) = [\Gamma_p(\alpha_j)]^{-1} |X_j|^{\alpha_j - \frac{p+1}{2}} e^{-\mathrm{tr}(X_j)}$$

for $X_j = X_j' > 0$, $\mathrm{Re}(\alpha_j) > \frac{p-1}{2}$, $j = 0, 1, ..., k$. Let

$$Y_0 = X_0 + X_1 + ... + X_k \text{ and } Y_j = Y_0^{-\frac{1}{2}} X_j Y_0^{-\frac{1}{2}}, \ j = 1, ..., k.$$

Show that $Y_1, ..., Y_k$ have a real type-1 matrix-variate Dirichlet distribution.

Solution   The joint density of $X_0, ..., X_k$, denoted by $f(X_0, X_1, ..., X_k)$ is given by the product of the $f_j(X_j)$'s. Make the transformation as in Theorem 2.32. Then

$$\mathrm{d}Y_0\mathrm{d}Y_1...\mathrm{d}Y_k = |Y_0|^{-k\left(\frac{p+1}{2}\right)}\mathrm{d}X_0...\mathrm{d}X_k.$$

Also

$$X_j = Y_0^{\frac{1}{2}} Y_j Y_0^{\frac{1}{2}}, \quad j = 1, ..., k$$

and

$$X_0 = Y_0^{\frac{1}{2}} (I - Y_1 - ... - Y_k) Y_0^{\frac{1}{2}}.$$

Substitute these values to get the joint density of $Y_0, Y_1, ..., Y_k$. Now integrate out Y_0 by using a gamma integral to get $\Gamma_p(\alpha_0 + \alpha_1 + ... + \alpha_k)$. Then the joint density of $Y_1, ..., Y_k$, denoted by $h(Y_1, ..., Y_k)$ is given by

$$h(Y_1, ..., Y_k) = \frac{\Gamma_p(\alpha_0 + \alpha_1 + ... + \alpha_k)}{\Gamma_p(\alpha_0)...\Gamma_p(\alpha_k)} \left\{ \prod_{j=1}^{k} |Y_j|^{\alpha_j - \frac{p+1}{2}} \right\}$$

$$\times |I - Y_1 - ... - Y_k|^{\alpha_0 - \frac{p+1}{2}}$$

for $0 < Y_j < I$, $j = 1, ..., k$, $0 < Y_1 + ... + Y_k < I$, $\mathrm{Re}(\alpha) > \frac{p-1}{2}$, $j = 0, 1, ..., k$ and $h(Y_1, ..., Y_k) = 0$ elsewhere, which is a real type-1 Dirichlet density. Hence the result.

Exercises

2.4.1   Show that the transformation $Y_1 = XDX'$ and $Y_2 = XX'$ where $D = \mathrm{diag}(\lambda_1, ..., \lambda_p)$, $\lambda_1 > ... > \lambda_p$ and all the matrices are of functionally independent real variables is one-to-one when $x_{jj} > 0$, $j = 1, ..., p$.

2.4.2   Let $\lambda_1 > ... > \lambda_p > 0$ be the roots of the determinantal equation $|Y_1 - \lambda Y_2| = 0$ and $|X| \neq 0$. Then show that the Jacobian in the following transformations

(i) $Y_1 = XD_\lambda X', \quad Y_2 = XX', \quad D_\lambda = \mathrm{diag}(\lambda_1, ..., \lambda_p);$

(ii) $Y_1 = XD_\lambda X', \quad Y_1 + Y_2 = XX'$

where all the matrices are of functionally independent real variables, is given by

$$dY_1 dY_2 = 2^p |X|^{p+2} \left\{ \prod_{i<j} |\lambda_i - \lambda_j| \right\} dX dD.$$

2.4.3   Let $Y_1 = XD_\nu X'$, $Y_2 = XD_\mu X'$, $D_\nu = \text{diag}(\nu_1, ..., \nu_p)$, $D_\mu = \text{diag}(\mu_1, ..., \mu_p)$ such that $\nu_j^2 + \mu_j^2 = 1$, $j = 1, ..., p$, $\nu_1 > ... > \nu_p > 0$ and the μ_j's are real and distinct, where all the matrices are of functionally independent real variables. Show that $dY_1 dY_2$ is given by the same expression as that in Exercise 2.4.2.

2.4.4   Let X_j, $j = 1, ..., k$ have a real matrix-variate type-1 Dirichlet distribution. Then under the transformation listed in Theorem 2.30 what is the distribution of $Y_1, ..., Y_k$?

2.4.5   Let $X_j, j = 1, ..., k$ have a real type-2 matrix-variate Dirichlet distribution. Then under the transformation listed in Theorem 2.31 what is the distribution of Y_j, $j = 1, ..., k$?

2.4.6   By using Theorem 2.28 or otherwise show that

$$\int_X \int_D \exp\{-\text{tr}\,[X(I + D)X']\}\,\psi(X, D)dD\,dX = \left[\Gamma_p\left(\frac{p+1}{2}\right)\right]^2$$

with

$$\psi(X, D) = 2^p |X|^{p+2} \left\{\prod_{i<j} |\lambda_i - \lambda_j|\right\}$$

where X is a $p \times p$ nonsingular matrix of p^2 functionally independent real variables and $D = \text{diag}(\lambda_1, ..., \lambda_p)$ with the λ_j's real and $\lambda_1 > ... > \lambda_p > 0$.

2.5   Transformations involving submatrices

In some problems of statistical distribution theory, one needs transformations involving submatrices of a given matrix. One result of this type will be discussed in this section. We need some results from matrix theory which will be stated as lemmas.

Lemma 2.2   *Let X be a symmetric positive definite matrix of functionally independent real variables, partitioned into*

$$X = \begin{bmatrix} X_{11} & X_{12} \\ X_{21} & X_{22} \end{bmatrix}$$

where X_{11} is $m \times m$ and X_{22} is $n \times n$, with $m \leq n$ so that $X_{12} = X'_{21}$ is $m \times n$. Then there exists a nonsingular diagonal block matrix

$$G = \begin{bmatrix} G_1 & 0 \\ 0 & G_2 \end{bmatrix}$$

where G_1 is $m \times m$ and G_2 is $n \times n$ such that

$$GXG' = \begin{bmatrix} I_m & D & 0 \\ D & I_m & 0 \\ 0 & 0 & I_\delta \end{bmatrix}, \quad \delta = n - m \tag{2.5.1}$$

where I_m denotes an identity matrix of order m and D is a diagonal matrix with the diagonal elements $\lambda_1, ..., \lambda_m$ where $\lambda_1^2, ..., \lambda_m^2$ are the roots of the determinantal equation

$$|X_{12}X_{22}^{-1}X_{21} - \theta X_{11}| = 0.$$

Thus writing

$$G^{-1} = \begin{bmatrix} P & 0 \\ 0 & Q \end{bmatrix},$$

one has

$$X = \begin{bmatrix} P & 0 \\ 0 & Q \end{bmatrix} \begin{bmatrix} I_m & D & 0 \\ D & I_m & 0 \\ 0 & 0 & I_\delta \end{bmatrix} \begin{bmatrix} P' & 0 \\ 0 & Q' \end{bmatrix}. \tag{2.5.2}$$

Note that since X is symmetric, $(m + n) \times (m + n)$, and of functionally independent real variables, there are only $\frac{1}{2}(m + n)(m + n + 1)$ distinct elements x_{ij}'s in X. But P, Q and D have a total of $m^2 + n^2 + m$ elements. The difference is

$$m^2 + n^2 + m - \frac{1}{2}(m + n)(m + n + 1) = \frac{1}{2}\delta(\delta - 1), \quad \delta = n - m.$$

Thus if $m = n$ then the total numbers of distinct elements on both sides of (2.5.2) are equal. If $m \neq n$ then one can reduce the number of elements on the right by $\frac{1}{2}\delta(\delta - 1)$ by using the following procedure. Write

$$Q = \begin{bmatrix} Q_{11} & Q_{12} \\ Q_{21} & Q_{22} \end{bmatrix}$$

and postmultiply Q_{12} and Q_{22} by an orthogonal matrix Δ such that $Q_{22}\Delta$ becomes a triangular matrix with zeros below the leading diagonal thereby reducing the number of elements in Q_{22} by $\frac{1}{2}\delta(\delta - 1)$. Note that the final product remains the same, which may be seen by multiplying them out. That is,

$$
X = \begin{bmatrix} P & 0 & 0 \\ 0 & Q_{11} & Q_{12} \\ 0 & Q_{21} & Q_{22} \end{bmatrix} \begin{bmatrix} I_m & D & 0 \\ D & I_m & 0 \\ 0 & 0 & I_\delta \end{bmatrix} \begin{bmatrix} P' & 0 & 0 \\ 0 & Q'_{11} & Q'_{21} \\ 0 & Q'_{12} & Q'_{22} \end{bmatrix}
$$

$$
= \begin{bmatrix} P & 0 & 0 \\ 0 & Q_{11} & Q_{12}\Delta \\ 0 & Q_{21} & Q_{22}\Delta \end{bmatrix} \begin{bmatrix} I_m & D & 0 \\ D & I_m & 0 \\ 0 & 0 & I_\delta \end{bmatrix} \begin{bmatrix} P' & 0 & 0 \\ 0 & Q'_{11} & Q'_{21} \\ 0 & \Delta'Q'_{12} & \Delta'Q'_{22} \end{bmatrix}.
$$

This result will be used to establish the next theorem.

Lemma 2.3 Let a nonsingular $(m+n) \times (m+n)$ matrix A and its inverse A^{-1} be partitioned into submatrices as follows:

$$
A = \begin{bmatrix} A_{11} & A_{12} \\ A_{21} & A_{22} \end{bmatrix}, \quad |A_{11}| \neq 0, \quad |A_{22}| \neq 0, \quad A^{-1} = \begin{bmatrix} A^{11} & A^{12} \\ A^{21} & A^{22} \end{bmatrix}
$$

where A_{11}, A^{11} are $m \times m$, A_{22}, A^{22} are $n \times n$, A_{12}, A^{12} are $m \times n$ and A_{21}, A^{21} are $n \times m$. Then the following results hold.

$$
\begin{aligned}
|A| &= |A_{11}| |A_{22} - A_{21} A_{11}^{-1} A_{12}| \\
&= |A_{22}| |A_{11} - A_{12} A_{22}^{-1} A_{21}|,
\end{aligned} \tag{i}
$$

$$
A^{11} = \left(A_{11} - A_{12} A_{22}^{-1} A_{21} \right)^{-1},
$$

$$
A_{11} = \left(A^{11} - A^{12} \left(A^{22} \right)^{-1} A^{21} \right)^{-1},
$$

$$
A^{22} = \left(A_{22} - A_{21} A_{11}^{-1} A_{12} \right)^{-1},
$$

$$
A_{22} = \left(A^{22} - A^{21} \left(A^{11} \right)^{-1} A^{12} \right)^{-1}. \tag{ii}
$$

$$
|A^{11}| = |A_{11} - A_{12} A_{22}^{-1} A_{21}|^{-1} = \frac{|A_{22}|}{|A|},
$$

$$
|A^{22}| = \frac{|A_{11}|}{|A|}. \tag{iii}
$$

Theorem 2.33 Let $Q = \begin{bmatrix} Q_{11} & Q_{12} \\ Q_{21} & T \end{bmatrix}$ where Q_{11} is $m \times m$, Q_{12} is $m \times \delta$, $m \leq n$, $\delta = n - m$ and T is $\delta \times \delta$ and upper triangular with

$t_{jj} > 0$, $j = 1, ..., \delta$, $t_{ij} = 0$, $i > j$. *Let the $m \times m$ matrix P and the $(m + n) \times (m + n)$ symmetric matrix X be as defined in Lemma 2.2, all the matrices being of functionally independent real variables. Consider the transformation*

$$X = \begin{bmatrix} P & 0 \\ 0 & Q \end{bmatrix} \Phi \begin{bmatrix} P' & 0 \\ 0 & Q' \end{bmatrix} \qquad (2.5.3)$$

where

$$\Phi = \begin{bmatrix} I_m & D & 0 \\ D & I_m & 0 \\ 0 & 0 & I_\delta \end{bmatrix}$$

is as defined in (2.5.1). Then

$$dX = 2^{m+n} |P|^{n+1} |Q|^{m+1} \left(\prod_{j=1}^{\delta} t_{jj}^{j-1} \right)$$

$$\times \left(\prod_{j=1}^{m} |\lambda_j|^\delta \right) \left(\prod_{i<j}^{m} |\lambda_i^2 - \lambda_j^2| \right) dP \, dQ \, dD.$$

Proof   Take differentials on both sides of (2.5.3) to obtain

$$X^* = \begin{bmatrix} P^* & 0 \\ 0 & Q^* \end{bmatrix} \Phi \begin{bmatrix} P' & 0 \\ 0 & Q' \end{bmatrix}$$

$$+ \begin{bmatrix} P & 0 \\ 0 & Q \end{bmatrix} \Phi \begin{bmatrix} P^* & 0 \\ 0 & Q^* \end{bmatrix}'$$

$$+ \begin{bmatrix} P & 0 \\ 0 & Q \end{bmatrix} \begin{bmatrix} 0 & D^* & 0 \\ D^* & 0 & 0 \\ 0 & 0 & 0 \end{bmatrix} \begin{bmatrix} P' & 0 \\ 0 & Q' \end{bmatrix} \qquad (2.5.4)$$

where

$$X^* = (dX), \quad P^* = (dP), \quad Q^* = (dQ),$$
$$D^* = (dD) = \text{diag}(d\lambda_1, ..., d\lambda_m) = \text{diag}(\lambda_1^*, ..., \lambda_m^*),$$
$$Q^* = \begin{bmatrix} Q_{11}^* & Q_{12}^* \\ Q_{21}^* & T^* \end{bmatrix}, \quad Q_{11}^* = (dQ_{11}), \quad Q_{12}^* = (dQ_{12}),$$
$$Q_{21}^* = (dQ_{21}), \quad \text{and} \quad T^* = (dT).$$

Pre and postmultiply (2.5.4) by $\begin{bmatrix} P^{-1} & 0 \\ 0 & Q^{-1} \end{bmatrix}$ and $\begin{bmatrix} P^{-1} & 0 \\ 0 & Q^{-1} \end{bmatrix}'$. Then one has

$$Y^* = \begin{bmatrix} P^{-1} & 0 \\ 0 & Q^{-1} \end{bmatrix} X^* \begin{bmatrix} P^{-1} & 0 \\ 0 & Q^{-1} \end{bmatrix}' \qquad (2.5.5)$$

$$
= \begin{bmatrix} P^{-1}P^* & 0 \\ 0 & Q^{-1}Q^* \end{bmatrix} \Phi + \Phi \begin{bmatrix} P^{-1}P^* & 0 \\ 0 & Q^{-1}Q^* \end{bmatrix}'
$$

$$
+ \begin{bmatrix} 0 & D^* & 0 \\ D^* & 0 & 0 \\ 0 & 0 & 0 \end{bmatrix}. \tag{2.5.6}
$$

Treating (2.5.5) and (2.5.6) as transformations in differentials or treating differentials as new variables and everything else as constants, one has dY^* in terms of dX^*, dP^*, dQ^*, and dD^*, that is, dY in terms of dX, dP, dQ, and dD. From (2.5.5), using Theorem 1.20 of Chapter 1 and the fact that $J(Y : X) = J(Y^* : X^*)$, one has

$$
dY^* = \begin{vmatrix} P^{-1} & 0 \\ 0 & Q^{-1} \end{vmatrix}^{m+n+1} dX^*
$$

$$
= |P|^{-(m+n+1)}|Q|^{-(m+n+1)}dX^* \Rightarrow
$$

$$
dX^* = |P|^{m+n+1}|Q|^{m+n+1}dY^*. \tag{2.5.7}
$$

Let

$$
A^* = (dA) = P^{-1}P^*,
$$

$$
B^* = (dB) = Q^{-1}\begin{bmatrix} Q_{11}^* & Q_{12}^* \\ Q_{21}^* & T^* \end{bmatrix} = \begin{bmatrix} B_{11}^* & B_{12}^* \\ B_{21}^* & B_{22}^* \end{bmatrix}.
$$

Then from Theorem 1.13 of Chapter 1

$$
dA^* = |P|^{-m}dP^*
$$

and

$$
d\begin{pmatrix} B_{11}^* \\ B_{21}^* \end{pmatrix} = |Q|^{-m}d\begin{pmatrix} Q_{11}^* \\ Q_{21}^* \end{pmatrix}. \tag{2.5.8}
$$

From (2.5.6) we have

$$
Y^* = \begin{bmatrix} A^* & 0 & 0 \\ 0 & B_{11}^* & B_{12}^* \\ 0 & B_{21}^* & B_{22}^* \end{bmatrix} \Phi + \Phi \begin{bmatrix} A^* & 0 & 0 \\ 0 & B_{11}^* & B_{12}^* \\ 0 & B_{21}^* & B_{22}^* \end{bmatrix}'
$$

$$
+ \begin{bmatrix} 0 & D^* & 0 \\ D^* & 0 & 0 \\ 0 & 0 & 0 \end{bmatrix}
$$

$$
= \begin{bmatrix} Y_{11}^* & Y_{12}^* & Y_{13}^* \\ Y_{21}^* & Y_{22}^* & Y_{23}^* \\ Y_{31}^* & Y_{32}^* & Y_{33}^* \end{bmatrix} \tag{2.5.9}
$$

where

$$
Y_{11}^* = A^* + A'^*, \quad Y_{12}^* = A^*D + DB_{11}'^* + D^*,
$$

$$
Y_{13}^* = DB_{21}'^*, \quad Y_{22}^* = B_{11}^* + B_{11}'^*,
$$

$$Y_{23}^* = B_{12}^* + B_{21}^{\prime*}, \quad Y_{33}^* = B_{22}^* + B_{22}^{\prime*}. \tag{2.5.10}$$

For computing the Jacobian directly from (2.5.9) there is a small problem. In Q^* we had T^* triangular but when Q^* is premultiplied by Q^{-1} this property is lost and B_{22}^* is no longer triangular. Hence not all variables in B_{12}^* and B_{22}^* are free variables. But from (2.5.10) note that A^*, D^*, B_{11}^* appear only in Y_{11}^*, Y_{12}^* and Y_{22}^*. Hence by using these three submatrices one can evaluate $dY_{11}^* dY_{12}^* dY_{22}^*$ in terms of dA^*, dD^* and dB_{11}^*. For convenience write

$$Y_{11}^* = U = (u_{ij}), \quad Y_{12}^* = V = (v_{ij}), \quad Y_{22}^* = W = (w_{ij}).$$

Note that

$$u_{ii} = 2a_{ii}^*; \quad v_{ii} = \lambda_i a_{ii}^* + \lambda_i b_{ii}^* + \lambda_i^*;$$
$$w_{ii} = 2b_{ii}^*; \quad u_{ij} = a_{ij}^* + a_{ji}^* \ (i < j);$$
$$v_{ij} = \lambda_j a_{ij}^* + \lambda_i b_{ij}^* \ (i \neq j); \quad w_{ij} = b_{ij}^* + b_{ji}^* \ (i < j).$$

Let us examine the matrices of partial derivatives treating $a_{ij}^*, b_{ij}^*, \lambda_j^*$ as variables.

	a_{ii}^*	b_{ii}^*	λ_i^*
u_{ii}	$2I$	0	0
w_{ii}	0	$2I$	0
v_{ii}	L	M	I

$$\begin{vmatrix} 2I & 0 & 0 \\ 0 & 2I & 0 \\ L & M & I \end{vmatrix} = 2^{2m}$$

and note that L and M do not affect the value of the determinant. For $i < j = 1, ..., m$,

	a_{ij}^*	b_{ji}^*	a_{ji}^*	b_{ij}^*
u_{ij}	I	0	I	0
w_{ij}	0	I	0	I
v_{ji}	0	C_1	C_2	0
v_{ij}	C_1	0	0	C_2

where C_1 is a diagonal matrix with the elements $\lambda_2, \lambda_3, ..., \lambda_m; \lambda_3, ..., \lambda_m;$ ..., $\lambda_{m-1}, \lambda_m; \lambda_m$ and C_2 is a diagonal matrix with the elements λ_1 repeated $m - 1$ times, λ_2 repeated $m - 2$ times, ..., λ_{m-1}. Thus the determinant, in absolute value, is the following:

$$\begin{vmatrix} I & 0 & I & 0 \\ 0 & I & 0 & I \\ 0 & C_1 & C_2 & 0 \\ C_1 & 0 & 0 & C_2 \end{vmatrix} = \left| \begin{pmatrix} C_2 & 0 \\ 0 & C_2 \end{pmatrix} - \begin{pmatrix} 0 & C_1 \\ C_1 & 0 \end{pmatrix} \right|$$

$$= |C_2^2 - C_1^2| = \prod_{i<j}^{m} |\lambda_i^2 - \lambda_j^2|.$$

Thus

$$dY_{11}^* dY_{12}^* dY_{22}^* = 2^{2m} \left\{ \prod_{i<j}^{m} |\lambda_i^2 - \lambda_j^2| \right\} dA^* dD^* dB_{11}^*. \qquad (2.5.11)$$

From Y_{13}^* using Theorem 1.13 of Chapter 1 we have

$$dY_{13}^* = |D|^\delta dB_{21}^* = \left(\prod_{i=1}^{m} |\lambda_i| \right)^\delta dB_{21}^*. \qquad (2.5.12)$$

Now consider the equation connecting B_{12}^*, B_{22}^* to Q_{12}^*, T^*. That is,

$$\begin{pmatrix} B_{12}^* \\ B_{22}^* \end{pmatrix} = Q^{-1} \begin{pmatrix} Q_{12}^* \\ T^* \end{pmatrix}.$$

Since the higher order differentials are zeros when the wedge product is taken and since we have counted dB_{21}^* already in (2.5.12), note that dY_{23}^* leads to $d\tilde{Y}_{23}^*$ where $\tilde{Y}_{23}^* = B_{12}^*$. Hence we need to consider only the following equations for computing the balance of the Jacobian:

$$\tilde{Y}_{23}^* = B_{12}^*,$$
$$Y_{33}^* = B_{22}^* + B_{22}^{*'}.$$

As stated earlier, not all variables in B_{12}^* and B_{22}^* are free variables. Let

$$Q = \begin{bmatrix} Q_{11} & Q_{12} \\ Q_{21} & T \end{bmatrix}, \quad Q^{-1} = \begin{bmatrix} Q^{11} & Q^{12} \\ Q^{21} & Q^{22} \end{bmatrix}$$

where Q_{11} and Q^{11} are $m \times m$, Q_{12} and Q^{12} are $m \times \delta$, Q_{21} and Q^{21} are $\delta \times m$ and T and Q^{22} are $\delta \times \delta$. Then

$$Q^{-1} \begin{bmatrix} Q_{12}^* \\ T^* \end{bmatrix} = \begin{bmatrix} Q^{11} & Q^{12} \\ Q^{21} & Q^{22} \end{bmatrix} \begin{bmatrix} Q_{12}^* \\ T^* \end{bmatrix}$$
$$= \begin{bmatrix} Q^{11} Q_{12}^* + Q^{12} T^* \\ Q^{21} Q_{12}^* + Q^{22} T^* \end{bmatrix}.$$

Thus

$$d \begin{pmatrix} \tilde{Y}_{23}^* \\ Y_{33}^* \end{pmatrix} = d \begin{pmatrix} Q^{11} Q_{12}^* + Q^{12} T^* \\ Q^{21} Q_{12}^* + Q_{12}^{*'} Q^{21} + Q^{22} T^* + T^{*'} Q^{22'} \end{pmatrix}.$$

Let

$$\tilde{Y}_{23}^* = (\alpha_{ij}^*), \quad Y_{33}^* = (\beta_{ij}^*),$$
$$Q_{12}^* = (r_{ij}^*), \quad T^* = (t_{ij}^*) \text{ and } Q^{-1} = (q^{ij}).$$

Thus writing α_{ij}^*, $i = 1, ..., m$, $j = 1, ..., \delta$; β_{ij}^*, $i \le j = 1, ..., \delta$ in terms of the r_{ij}^*, $i = 1, ..., m$, $j = 1, ..., \delta$ and the t_{ij}^*, $i \le j = 1, ..., \delta$ we have the following equations:

$$\alpha_{11}^* = q^{11}r_{11}^* + q^{12}r_{21}^* + ... + q^{1m}r_{m1}^* + q^{1\,m+1}t_{11}^*$$
$$\alpha_{21}^* = q^{21}r_{11}^* + q^{22}r_{21}^* + ... + q^{2m}r_{m1}^* + q^{2\,m+1}t_{11}^*$$

$$\vdots \qquad \vdots$$

$$\alpha_{m1}^* = q^{m\,1}r_{11}^* + ... + q^{m\,m}r_{m1}^* + q^{m\,m+1}t_{11}^*$$
$$\alpha_{12}^* = q^{1\,1}r_{12}^* + ... + q^{1\,m}r_{m2}^* + q^{1\,m+1}t_{12}^* + q^{1\,m+2}t_{22}^*$$

$$\vdots \qquad \vdots$$

$$\alpha_{m2}^* = q^{m\,1}r_{12}^* + ... + q^{m\,m}r_{m2}^* + q^{m\,m+1}t_{12}^* + q^{m\,m+2}t_{22}^*$$

$$\vdots \qquad \vdots$$

$$\beta_{11}^* = 2\left(q^{m+1\,1}r_{11}^* + q^{m+1\,2}r_{21}^* + ... + q^{m+1\,m}r_{m1}^*\right)$$
$$\quad + 2\left(q^{m+1\,m+1}t_{11}^*\right)$$
$$\beta_{12}^* = q^{m+1\,1}r_{12}^* + ... + q^{m+1\,m}r_{m2}^*$$
$$\quad + q^{m+2\,1}r_{11}^* + ... + q^{m+2\,m}r_{m1}^*$$
$$\quad + q^{m+1\,m+1}t_{12}^* + q^{m+1\,m+2}t_{22}^* + q^{m+2\,m+1}t_{11}^*$$

$$\vdots \qquad \vdots$$

$$\beta_{1\delta}^* = q^{m+1\,1}r_{1\delta}^* + ... + q^{m+1\,m}r_{m\delta}^*$$
$$\quad + q^{m+\delta\,1}r_{11}^* + ... + q^{m+\delta\,m}r_{m1}^*$$
$$\quad + q^{m+1\,m+1}t_{1\delta}^* + ... + q^{m+1\,m+\delta}t_{\delta\delta}^* + q^{m+\delta\,1}t_{11}^*$$
$$\beta_{22}^* = 2\left(q^{m+2\,1}r_{12}^* + ... + q^{m+2\,m}r_{m2}^*\right)$$
$$\quad + 2\left(q^{m+2\,m+1}t_{12}^* + q^{m+2\,m+2}t_{22}^*\right)$$

$$\vdots \qquad \vdots$$

$$\beta_{2\delta}^* = q^{m+2\,1}r_{1\delta}^* + ... + q^{m+2\,\delta}r_{m\delta}^*$$
$$\quad + q^{m+\delta\,1}r_{12}^* + ... + q^{m+\delta\,m}r_{m2}^*$$
$$\quad + q^{m+2\,m+1}t_{1\delta}^* + ... + q^{m+2\,m+\delta}t_{\delta\delta}^*$$
$$\quad + q^{m+\delta\,m+1}t_{12}^* + q^{m+\delta\,m+2}t_{22}^*$$

$$\vdots \qquad \vdots$$

$$\beta_{\delta\delta}^* = 2\left(q^{m+\delta\,1}r_{1\delta}^* + ... + q^{m+\delta\,m}r_{m\delta}^*\right)$$
$$+ 2\left(q^{m+\delta\,m+1}t_{1\delta}^* + ... + q^{m+\delta\,m+\delta}t_{\delta\delta}^*\right).$$

The Jacobian matrix can be put into a triangular block matrix form by the following technique. Arrange the variables α_{ij}^*'s and β_{ij}^*'s in the following order:

$$\alpha_{11}^*, \alpha_{21}^*, ..., \alpha_{m1}^*, \beta_{11}^*; \alpha_{12}^*, \alpha_{22}^*, ..., \alpha_{m2}^*, \beta_{12}^*, \beta_{22}^*; ...; \alpha_{1\delta}^*, ..., \alpha_{m\delta}^*, \beta_{1\delta}^*, ..., \beta_{\delta\delta}^*,$$

and the r_{ij}^*'s and t_{ij}^*'s in the order

$$r_{11}^*, r_{21}^*, ..., r_{m1}^*, t_{11}^*; r_{12}^*, ..., r_{m2}^*, t_{12}^*, t_{22}^*; ...; r_{1\delta}^*, ..., r_{m\delta}^*, t_{1\delta}^*, ..., t_{\delta\delta}^*.$$

Then in the triangular block Jacobian matrix the diagonal blocks are the following matrices. We will use the notation $\frac{\partial(x,y,...,z)}{\partial(\alpha,\beta,...,\gamma)}$ to denote the matrix obtained by differentiating each element of $x, y, ..., z$ partially with respect to each element of $\alpha, \beta, ..., \gamma$. Then we have

$$\frac{\partial\left(\alpha_{11}^*, \alpha_{21}^*, ..., \alpha_{m1}^*, \beta_{11}^*\right)}{\partial\left(r_{11}^*, r_{21}^*, ..., r_{m1}^*, t_{11}^*\right)} = \bar{B}_{m+1},$$

$$\frac{\partial\left(\alpha_{12}^*, \alpha_{22}^*, ..., \alpha_{m2}^*, \beta_{12}^*, \beta_{22}^*\right)}{\partial\left(r_{12}^*, r_{22}^*, ..., r_{m2}^*, t_{12}^*, t_{22}^*\right)} = \bar{B}_{m+2}, ...,$$

$$\frac{\partial\left(\alpha_{1\delta}^*, ..., \alpha_{m\delta}^*, \beta_{1\delta}^*, ..., \beta_{\delta\delta}^*\right)}{\partial\left(r_{1\delta}^*, ..., r_{m\delta}^*, t_{1\delta}^*, ..., t_{\delta\delta}^*\right)} = \bar{B}_{m+\delta}$$

where

$$\bar{B}_{m+j} = \begin{bmatrix} q^{11} & q^{12} & \cdots & q^{1\,m+j} \\ \vdots & \vdots & \ddots & \vdots \\ 2q^{m+j\,1} & 2q^{m+j\,2} & \cdots & 2q^{m+j\,m+j} \end{bmatrix}$$

and

$$B_{m+j} = \begin{bmatrix} q^{11} & q^{12} & \cdots & q^{1\,m+j} \\ \vdots & \vdots & \ddots & \vdots \\ q^{m+j\,1} & q^{m+j\,2} & \cdots & q^{m+j\,m+j} \end{bmatrix}.$$

Thus the determinant is $2^\delta |B_{m+1}| |B_{m+2}| ... |B_{m+\delta}|$. Consider a partitioning of Q^{-1} with the first leading submatrix equal to B_{m+j}. That is,

$$Q^{-1} = \begin{bmatrix} \tilde{Q}^{11} & \tilde{Q}^{12} \\ \tilde{Q}^{21} & \tilde{Q}^{22} \end{bmatrix}, \quad Q = \begin{bmatrix} \tilde{Q}_{11} & \tilde{Q}_{12} \\ \tilde{Q}_{21} & \tilde{Q}_{22} \end{bmatrix}$$

with $\tilde{Q}^{11} = B_{m+j}$. Then from Lemma 2.3(iii) we have

$$|B_{m+j}| = \frac{\left|\tilde{Q}_{22}\right|}{|Q|}.$$

Note that $\tilde{Q}_{22}$ is T^* with the first j rows and columns removed. Thus

$$\left|\tilde{Q}_{22}\right| = t_{j+1\,j+1}...t_{\delta\delta}.$$

Hence we have

$$|B_{m+1}|\,...\,|B_{m+\delta}| = |Q|^{-\delta}t_{22}^1...t_{\delta\delta}^{\delta-1}$$

$$= \left\{\prod_{j=1}^{\delta} t_{jj}^{j-1}\right\} |Q|^{-\delta}.$$

That is,

$$d\tilde{Y}_{23}^* dY_{33}^* = 2^\delta |Q|^{-\delta} \left\{\prod_{j=1}^{\delta} t_{jj}^{j-1}\right\} dQ_{12}^* dT^*. \qquad (2.5.13)$$

From (2.5.11), (2.5.12) and (2.5.13) we have

$$dY^* = dY_{11}^* dY_{12}^* dY_{13}^* dY_{22}^* dY_{23}^* dY_{33}^*$$

$$= 2^{2m} \left\{\prod_{i<j}^{m} |\lambda_i^2 - \lambda_j^2|\right\} \left\{\prod_{i=1}^{m} |\lambda_i|^\delta\right\}$$

$$\times\ 2^\delta |Q|^{-\delta} \left\{\prod_{j=1}^{\delta} t_{jj}^{j-1}\right\}$$

$$\times\ dA^* dD^* dB_{11}^* dB_{21}^* dQ_{12}^* dT^*. \qquad (2.5.14)$$

From (2.5.8)

$$dA^* dB_{11}^* dB_{21}^* dQ_{12}^* dT^* = |P|^{-m}|Q|^{-m} dP^*$$

$$\times\ dQ_{11}^* dQ_{21}^* dQ_{12}^* dT^*$$

$$= |P|^{-m}|Q|^{-m} dP^* dQ^*. \qquad (2.5.15)$$

Substituting for dY^* in (2.5.7) from (2.5.13) and (2.5.14) and using the fact that $J(Y : X) = J((dY) : (dX))$ the result is established.

More Jacobians on partitioned matrices are given in Roy (1957).

Exercises

2.5.1   Let X be a $p \times p$ symmetric positive definite matrix of functionally independent real variables. Let X be decomposed as

$$X = H \begin{bmatrix} \lambda_1 & 0' \\ 0 & W \end{bmatrix} H'$$

where λ_1 is the largest eigenvalue of X, $H = (h_{ij})$ is an orthonormal matrix with the first column vector preselected, and W is a $(p-1) \times (p-1)$ real symmetric matrix such that $\lambda_1 I_{p-1} > W > 0$. Then show that

$$dX = \frac{|\lambda_1 I_{p-1} - W|}{\left(1 - \sum_{i=2}^{p} h_{ij}^2\right)^{\frac{1}{2}}} d\lambda_1 \left\{ \prod_{i=2}^{p} dh_{ij} \right\} dW.$$

2.5.2   Show that

$$\int_{\sum_{j=1}^{p} \left(\frac{x_j}{a_j}\right)^{m_j} \leq 1} \left\{ \prod_{j=1}^{p} x_j^{n_j - 1} dx_j \right\} = \frac{\prod_{j=1}^{p} \Gamma\left(\frac{n_j}{m_j}\right) a_j^{n_j}}{\Gamma\left(\sum_{j=1}^{p} \frac{n_j}{m_j} + 1\right) \left(\prod_{j=1}^{p} m_j\right)}$$

for $x_j > 0$, $n_j, m_j, a_j > 0$, $j = 1, ..., p$.

2.5.3   As a special case of Exercise 2.5.2 or otherwise, evaluate the integral

$$\int_{\sum_{j=1}^{p} \left(\frac{x_j}{a}\right)^2 \leq 1} \left\{ \prod_{j=1}^{p} dx_j \right\}.$$

2.5.4   Consider the Vandermonde matrix $V_p = (\beta_{jk})$, $\beta_{jk} = \beta_j^{k-1}$, $j, k = 1, ..., p$. Let $|V_p|$ denote the absolute value of the determinant of V_p. If $\beta_1, ..., \beta_p$ are independently and identically distributed as real standard normal variables and if E denotes the expected value then show that

$$E|V_p|^m = \int_{-\infty}^{\infty} \cdots \int_{-\infty}^{\infty} \frac{|V_p|^m}{(2\pi)^{p/2}} e^{-\frac{1}{2}(\beta_1^2 + ... + \beta_p^2)} d\beta_1 ... d\beta_p$$

$$= \frac{\prod_{j=1}^{p} \Gamma\left(1 + \frac{jm}{2}\right)}{\left[\Gamma\left(1 + \frac{m}{2}\right)\right]^p}.$$

(Vandermonde matrix and Vandermonde determinant are due to the French mathematician A.T. Vandermonde (1735–1796)).

2.5.5   Show that

$$\int_{\infty > y_1 > \ldots > y_p > -\infty} \left\{ \prod_{j > k} (y_j - y_k))^m \right\} e^{-\frac{1}{2}(v_1^2 + \ldots + v_p^2)} dy_1 \ldots dy_p$$

$$= \frac{1}{p!} (2\pi)^{p/2} \frac{\prod_{j=1}^p \Gamma \left(1 + \frac{jm}{2} \right)}{\left[\Gamma \left(1 + \frac{m}{2} \right) \right]^p}.$$

CHAPTER 3

Jacobians in the complex case

3.0 Introduction

In Chapters 1 and 2 we dealt with matrices where the elements are either real constants or real variables. Here we consider Jacobians of matrix transformations where the elements of the matrices are complex quantities. When the matrices are real, we will use the same notations as in Chapters 1 and 2. In the complex case, the matrix variable X will be denoted by $\tilde{X}$ to indicate that the elements in X are complex variables so that the statements of theorems in Chapters 3 and 4 will not be mixed up with those in Chapters 1 and 2. The complex conjugate of a matrix A will be denoted by $\bar{A}$ and the conjugate transpose by $\bar{A}'$ or by A^*. The determinant of A will be denoted by $\det A$, $\det(A)$ or $|A|$. The absolute value of a scalar a will also be denoted by $|a|$. The wedge product of differentials in $\tilde{X}$ will be denoted by $d\tilde{X}$ and the matrix of differentials by $(d\tilde{\mathbf{X}})$ with $\tilde{X}$ written in bold.

Some statistical concepts in the real and complex cases will be listed in the Appendix. It is assumed that the reader is familiar with the basic properties of real and complex matrices. Some properties of complex matrices will be listed here for convenience.

Definition 3.1   Semiunitary and unitary matrices   A $p \times n$ matrix U is said to be semiunitary if $U\bar{U}' = UU^* = I_p$ for $p < n$ or $\bar{U}'U = U^*U = I_n$ for $n < p$. When $n = p$ and $UU^* = I_p$ then U is called a unitary matrix.

Definition 3.2   A hermitian or a skew hermitian matrix   If $A = \bar{A}' = A^*$ then A is said to be hermitian and if $A = -\bar{A}' = -A^*$ then it is skew hermitian. (These matrices are named after the French mathematician C. Hermite (1822-1901)).

Lemma 3.1   *Consider a $p \times p$ matrix A and $(2p) \times (2p)$ matrices B and*

171

C where

$$A = A_1 + iA_2, \ B = \begin{bmatrix} A_1 & A_2 \\ -A_2 & A_1 \end{bmatrix}, \ C = \begin{bmatrix} A_1 & -A_2 \\ A_2 & A_1 \end{bmatrix}$$

where A_1, A_2 are real matrices and $i = \sqrt{-1}$. Then for $|A_1| \neq 0$

$$|\det(A)| = |\det(B)|^{\frac{1}{2}} = |\det(C)|^{\frac{1}{2}}.$$

Proof Let $\det(A) = a + ib$ where a and b are real scalars. Then the absolute value is available as $[(a + ib)(a - ib)]^{\frac{1}{2}}$. If $\det(A_1 + iA_2)$ is $a + ib$ then $a - ib = \det(A_1 - iA_2)$. Hence

$$(a + ib)(a - ib) = \det(A_1 + iA_2)\det(A_1 - iA_2)$$
$$= \begin{vmatrix} A_1 + iA_2 & 0 \\ 0 & A_1 - iA_2 \end{vmatrix}.$$

Adding the last p columns to the first p columns and then adding the last p rows to the first p rows we have

$$\begin{vmatrix} A_1 + iA_2 & 0 \\ 0 & A_1 - iA_2 \end{vmatrix} = \begin{vmatrix} 2A_1 & A_1 - iA_2 \\ A_1 - iA_2 & A_1 - iA_2 \end{vmatrix}.$$

Using similar steps we have

$$\begin{vmatrix} 2A_1 & A_1 - iA_2 \\ A_1 - iA_2 & A_1 - iA_2 \end{vmatrix} = \begin{vmatrix} 2A_1 & A_1 - iA_2 \\ -iA_2 & \frac{1}{2}A_1 - \frac{1}{2}iA_2 \end{vmatrix}$$
$$= \begin{vmatrix} 2A_1 & -iA_2 \\ -iA_2 & \frac{1}{2}A_1 \end{vmatrix}$$
$$= |A_1|\,|A_1 + A_2 A_1^{-1} A_2|$$
$$= \begin{vmatrix} A_1 & A_2 \\ -A_2 & A_1 \end{vmatrix}$$
$$= \begin{vmatrix} A_1 & -A_2 \\ A_2 & A_1 \end{vmatrix}$$

by evaluating as the determinant of partitioned matrices. Thus the absolute value of $\det(A)$ is given by

$$|\det(A)| = \{|A_1|\,|A_1 + A_2 A_1^{-1} A_2|\}^{\frac{1}{2}}$$
$$= |\det(B)|^{\frac{1}{2}} = |\det(C)|^{\frac{1}{2}}.$$

This establishes the result.

A matrix $\tilde{X}$ with complex elements can always be written as $\tilde{X} = X_1 + iX_2, i = \sqrt{-1}$ where X_1 and X_2 are real matrices. Let us examine the wedge product of the differentials in $\tilde{X}$. In general, there are p^2 real variables in X_1 and another p^2 real variables in X_2 when $\tilde{X}$ is a $p \times p$ matrix. Thus $\tilde{X}$ is a function of $2p^2$ real variables and the wedge product of the differentials will be denoted by the following:

Notation 3.1
$$d\tilde{X} = dX_1 dX_2 \tag{3.0.1}$$
where dX_1 is the wedge product in (dX_1) and dX_2 that in (dX_2). In this notation an empty product is interpreted as unity. That is, when the matrix $\tilde{X}$ is real then X_2 is null and $d\tilde{X} = dX_1$. If $\tilde{X}$ is a hermitian matrix, then X_1 is symmetric and X_2 is skew symmetric, and in this case

$$dX_1 = \wedge_{j \geq k} dx_{jk1} \text{ and } dX_2 = \wedge_{j > k} dx_{jk2} \tag{3.0.2}$$

where $X_1 = (x_{jk1})$ and $X_2 = (x_{jk2})$. If $\tilde{Y}$ is a scalar function of $\tilde{X} = X_1 + iX_2$ then $\tilde{Y}$ can be written as $\tilde{Y} = Y_1 + iY_2$ where Y_1 and Y_2 are real. Thus if $\tilde{Y} = F(\tilde{X})$ it is a transformation of (X_1, X_2) to (Y_1, Y_2) or where (Y_1, Y_2) is written as a function of (X_1, X_2) then we will use the following notation for the Jacobian in the complex case.

Notation 3.2 Jacobians in the complex case $J(Y_1, Y_2 : X_1, X_2)$: Jacobian of the transformation where Y_1 and Y_2 are written as functions of X_1 and X_2 or where $\tilde{Y} = Y_1 + iY_2$ is a function of $\tilde{X} = X_1 + iX_2$.

Notation 3.3 $A^{\frac{1}{2}}$: square root of a matrix in the complex case
If a complex matrix A is hermitian positive definite then there exists a nonsingular hermitian matrix C such that $A = C^2$. This C will be written as $C = A^{\frac{1}{2}}$.

We need the following result regarding the absolute value of a determinant. This will be stated as a lemma.

Lemma 3.2   *Let G be a matrix with complex elements. Consider the square of the absolute value of the determinant of G. Then*

$$|\det(G)|^2 = |\det(G\bar{G}')| = |\det(GG^*)| \tag{3.0.3}$$

where $\bar{G}' = G^*$ is the conjugate transpose of G.

This result is evident by observing the following: Let $G = G_1 + iG_2$, where G_1 and G_2 are real matrices and $i = \sqrt{-1}$. Let $\det(G) = a + ib$. Then the square of the absolute value of this determinant is $|\det(G)|^2 = (a+ib)(a-ib) = a^2 + b^2$. Note that $\bar{G} = G_1 - iG_2$ and hence $\det(\bar{G}) = a - ib$. Thus

$$\det(G)\det(\bar{G}) = (a + ib)(a - ib)$$
$$= (a^2 + b^2) = |\det(G)|^2.$$

But for any matrix B, $|\det(B)| = |\det(B')|$ and hence we have

$$|\det(G)\det(\bar{G})| = |\det(G)\det(\bar{G}')|$$
$$= |\det(G\bar{G}')| = |\det(GG^*)|.$$

Example 3.1   Let $\tilde{x}$ and $\tilde{y}$ be scalar complex variables such that $\tilde{y} = a\tilde{x}$ where $a = 2 + i3$. Evaluate $d\tilde{y}$ in terms of $d\tilde{x}$.

Solution   Let

$$\tilde{x} = x_1 + ix_2,\ \tilde{y} = y_1 + iy_2,\ \tilde{x}^* = x_1 - ix_2,\ a^* = \bar{a} = 2 - i3,\ i = \sqrt{-1}$$

where x_1, x_2, y_1, y_2 are real. Then

$$\tilde{y} = y_1 + iy_2$$
$$= (2 + i3)\tilde{x} = (2 + i3)(x_1 + ix_2) \Rightarrow$$
$$y_1 = 2x_1 - 3x_2 \text{ and } y_2 = 3x_1 + 2x_2.$$

The matrix of partial derivatives is given by

$$\frac{\partial(y_1, y_2)}{\partial(x_1, x_2)} = \begin{bmatrix} 2 & -3 \\ 3 & 2 \end{bmatrix}$$

and hence the determinant is 13, that is,

$$\begin{vmatrix} 2 & -3 \\ 3 & 2 \end{vmatrix} = 2^2 + 3^2 = |a|^2$$

where $|a|$ is the absolute value of a and thus

$$d\tilde{y} = |a|^2 d\tilde{x}.$$

Note that if a is real, say $a = 2$, then $d\tilde{y} = 4d\tilde{x} = a^2 d\tilde{x}$. If x is real then $dy = adx$ irrespective of whether a is real or complex as long as $a \neq 0$. If $\tilde{y} = a^*\tilde{x}$ then also $d\tilde{y} = |a|^2 d\tilde{x}$. If $\tilde{y} = a\tilde{x}^*$ or $\tilde{y} = a^*\tilde{x}^*$ then $d\tilde{y} = -|a|^2 d\tilde{x}$. Note that $d\tilde{x}^* = -d\tilde{x}$ for a complex variable and $d\tilde{X}^* = (-1)^{pq} d\tilde{X}$ if $\tilde{X}$ is a $p \times q$ matrix of complex variables.

3.1 Jacobians of linear transformations in the complex case

First we consider the simplest type of linear transformations where a vector
is premultiplied by a scalar constant or a constant matrix. Then we will
examine more complicated types. These results will be stated as theorems.

Theorem 3.1 *Let $\tilde{X}$ and $\tilde{Y}$ be $p \times 1$ vectors of p functionally independent
complex variables each, A a nonsingular matrix of constants, α a nonzero
scalar and C a $p \times 1$ vector of constants. Then, ignoring the sign,*

$$\tilde{Y} = A\tilde{X} + C \Rightarrow d\tilde{Y} = |\det(A)|^2 d\tilde{X}$$
$$= |\det(AA^*)| d\tilde{X}, \qquad (a)$$
$$\tilde{Y} = \alpha\tilde{X} + C \Rightarrow d\tilde{Y} = |\alpha|^{2p} d\tilde{X}, \qquad (b)$$
$$\tilde{Y}' = \tilde{X}'A' + C' \Rightarrow d\tilde{Y} = |\det(A)|^2 d\tilde{X}, \qquad (c)$$
$$= |\det(AA^*)| d\tilde{X}$$

and

$$\tilde{Y}^* = \tilde{X}^* A^* + C^* \Rightarrow d\tilde{Y}^* = (-1)^p |\det(AA^*)| d\tilde{X}. \qquad (d)$$

Proof Let $\tilde{X} = X_1 + iX_2$, where $i = \sqrt{-1}$ and X_1 and X_2 are real $p \times 1$
vectors. Let $\tilde{Y} = Y_1 + iY_2$ where Y_1 and Y_2 are real. Since C is a constant
$d(A\tilde{X} + C) = d(A\tilde{X})$ and hence we may ignore C. $\tilde{Y} = A\tilde{X} \Rightarrow Y_1 = AX_1$
and $Y_2 = AX_2$ if A is real. This transformation is such that the $2p$ real
variables in (Y_1, Y_2) are written as functions of the $2p$ real variables in
(X_1, X_2). Let

$$X_1' = (x_{11}, ..., x_{p1}), \quad X_2' = (x_{12}, ..., x_{p2}),$$
$$Y_1' = (y_{11}, ..., y_{p1}), \text{ and } Y_2' = (y_{12}, ..., y_{p2}).$$

Then the Jacobian is the determinant of the following matrix of partial
derivatives:

$$\frac{\partial(Y_1, Y_2)}{\partial(X_1, X_2)} = \frac{\partial(y_{11}, ..., y_{p1}, y_{12}, ..., y_{p2})}{\partial(x_{11}, ..., x_{p1}, x_{12}, ..., x_{p2})}.$$

Note that

$$\frac{\partial Y_1}{\partial X_1} = \frac{\partial(y_{11}, ..., y_{p1})}{\partial(x_{11}, ..., x_{p1})} = A,$$
$$\frac{\partial Y_1}{\partial X_2} = 0, \; \frac{\partial Y_2}{\partial X_1} = 0, \; \frac{\partial Y_2}{\partial X_2} = A.$$

Thus the Jacobian is

$$J = \begin{vmatrix} A & 0 \\ 0 & A \end{vmatrix} = [\det(A)]^2.$$

If A is complex then let $A = A_1 + iA_2$ where A_1 and A_2 are real. Then

$$\begin{aligned}
Y = Y_1 + iY_2 &= (A_1 + iA_2)(X_1 + iX_2) \\
&= (A_1X_1 - A_2X_2) + i(A_1X_2 + A_2X_1) \Rightarrow \\
Y_1 &= A_1X_1 - A_2X_2
\end{aligned}$$

and

$$Y_2 = A_1X_2 + A_2X_1.$$

Then

$$\frac{\partial Y_1}{\partial X_1} = A_1, \quad \frac{\partial Y_1}{\partial X_2} = -A_2$$

$$\frac{\partial Y_2}{\partial X_1} = A_2, \quad \frac{\partial Y_2}{\partial X_2} = A_1.$$

Thus the Jacobian is

$$J = \begin{vmatrix} A_1 & -A_2 \\ A_2 & A_1 \end{vmatrix}$$

$$= |\det(A)|^2 = |\det(AA^*)|$$

from Lemmas 3.1 and 3.2, which establishes the result. By putting $A = \alpha I$, where I is the identity matrix of order p, (b) follows. By taking the transposes and observing that $d\tilde{X} = d\tilde{X}'$ and $d\tilde{Y} = d\tilde{Y}'$ the result (c) follows. (d) follows by noting that $d\tilde{Y}^* = dY_1(-1)^p dY_2 = (-1)^p d\tilde{Y}$.

Example 3.2   Let the $p \times 1$ complex random vector $\tilde{X}$ have a nonsingular complex normal density, that is, $\tilde{X} \sim \tilde{N}_p(\tilde{\mu}, \tilde{\Sigma})$. Evaluate $M_{\tilde{X}}(T)$ the moment generating function (m.g.f.) of $\tilde{X}$.

Solution   Let T be a $p \times 1$ parametric vector so that the m.g.f. is

$$M_{\tilde{X}}(T) = E\left[e^{\text{Re}(T^*\tilde{X})}\right]$$

where E denotes the expected value and $\text{Re}(\cdot)$ denotes the real part of $(\cdot)$. Integrating over the density of $\tilde{X}$ we have

$$\begin{aligned}
M_{\tilde{X}}(T) &= e^{\text{Re}(T^*\tilde{\mu})} E\left[e^{\text{Re}(T^*(\tilde{X}-\tilde{\mu}))}\right] \\
&= \frac{e^{\text{Re}(T^*\tilde{\mu})}}{\pi^p |\det(\tilde{\Sigma})|} \\
&\quad \times \int_{\tilde{X}} \exp\left\{\text{Re}[T^*(\tilde{X} - \tilde{\mu}) - (\tilde{X} - \tilde{\mu})^* \tilde{\Sigma}^{-1}(\tilde{X} - \tilde{\mu})]\right\} d\tilde{X}.
\end{aligned}$$

Since $\tilde{\Sigma}$ is hermitian positive definite there exists a nonsingular C such that $\tilde{\Sigma} = CC^*$ and then $\tilde{\Sigma}^{-1} = (C^*)^{-1}C^{-1}$. Let

$$\tilde{U} = C^{-1}(\tilde{X} - \tilde{\mu}) \Rightarrow d\tilde{U} = |\det(CC^*)|^{-1}d\tilde{X}$$

from Theorem 3.1. Then

$$M_{\tilde{X}}(T) = \frac{e^{\mathrm{Re}(T^*\tilde{\mu})}}{\pi^p} \int_{\tilde{U}} \exp\left\{-\mathrm{Re}[\tilde{U}^*\tilde{U} - T^*C\tilde{U}]\right\} d\tilde{U}$$

$$= e^{\mathrm{Re}(T^*\tilde{\mu} + \frac{1}{4}T^*\tilde{\Sigma}T)} \int_{\tilde{V}} \frac{\exp\{-(\tilde{V}^*\tilde{V})\}}{\pi^p} d\tilde{V}$$

where $\tilde{V} = \tilde{U} - \frac{1}{2}C^*T$. Note that for $\tilde{V}' = (\tilde{v}_1, ..., \tilde{v}_p)$

$$\tilde{V}^*\tilde{V} = |\tilde{v}_1|^2 + ... + |\tilde{v}_p|^2$$

where $|\tilde{v}_j|$ is the *absolute value* of $\tilde{v}_j$. But for a scalar variable $\tilde{z} = x + iy$ where x and y are real

$$\int_{\tilde{z}} e^{-|\tilde{z}|^2} d\tilde{z} = \int_{-\infty}^{\infty}\int_{-\infty}^{\infty} e^{-(x^2+y^2)} dx dy$$
$$= \pi.$$

Hence

$$\int_{\tilde{V}} \frac{e^{-(\tilde{V}^*\tilde{V})}}{\pi^p} d\tilde{V} = 1$$

and thus

$$M_{\tilde{X}}(T) = e^{\mathrm{Re}(T^*\tilde{\mu} + \frac{1}{4}T^*\tilde{\Sigma}T)}$$

where $\tilde{\Sigma} = \Sigma_1 + i\Sigma_2$ with Σ_1 real symmetric positive definite and Σ_2 real skew symmetric.

Theorem 3.2   Let $\tilde{X}$ and $\tilde{Y}$ be $p \times q$ matrices of pq functionally independent complex variables each. Let A and B be $p \times p$ and $q \times q$ nonsingular matrices of constants, C a $p \times q$ matrix of constants and α a nonzero scalar quantity. Then, ignoring the sign,

$$\tilde{Y} = A\tilde{X} + C \Rightarrow d\tilde{Y} = |\det(A\bar{A}')|^q d\tilde{X}, \qquad (a)$$
$$\tilde{Y} = \tilde{X}B + C \Rightarrow d\tilde{Y} = |\det(B\bar{B}')|^p d\tilde{X}, \qquad (b)$$
$$\tilde{Y} = A\tilde{X}B + C \Rightarrow$$
$$d\tilde{Y} = |\det(A\bar{A}')|^q |\det(B\bar{B}')|^p d\tilde{X}, \qquad (c)$$

and

$$\tilde{Y} = \alpha \tilde{X} + C \Rightarrow d\tilde{Y} = |\alpha|^{2pq} d\tilde{X}. \qquad (d)$$

Proof   Since C is a constant $\mathrm{d}(A\tilde{X} + C) = \mathrm{d}(A\tilde{X})$ and hence we may ignore C. Let $\tilde{Y} = Y_1 + iY_2$ and $\tilde{X} = X_1 + iX_2$. The result in (a) can be considered as a repetition of (a) of Theorem 3.1 on the q columns of $\tilde{X}$. The Jacobian matrix is a block diagonal matrix with q diagonal blocks, each equal to the Jacobian matrix in (a) of Theorem 3.1. Thus, denoting $A = A_1 + iA_2$, the determinant is

$$J = \begin{vmatrix} A_1 & -A_2 \\ A_2 & A_1 \end{vmatrix}^q.$$

The result in (b) can be considered as a repetition of (c) of Theorem 3.1 on the p rows of $\tilde{X}$. Hence the Jacobian in this case, denoting $B = B_1 + iB_2$, is given by

$$J = \begin{vmatrix} B_1 & B_2 \\ -B_2 & B_1 \end{vmatrix}^p.$$

For establishing (c) write $\tilde{Y} = A\tilde{U}$ where $\tilde{U} = \tilde{X}B$, and apply the results in (a) and (b). For proving (d) use the fact that each of the pq elements in $\tilde{X}$ is multiplied by the scalar α, and use Example 3.1.

Example 3.3   Let V and W be $q \times q$ and $p \times p$ hermitian positive definite matrices of constants, that is, $V = V^* > 0$, $W = W^* > 0$, M a $p \times q$ matrix of constants and $\tilde{X}$ a $p \times q$ matrix of pq functionally independent complex random variables. Then show that

$$f(\tilde{X}) = \pi^{-(pq)} |\det(V)|^{-p} |\det(W)|^{-q}$$
$$\times \exp\left\{ -\mathrm{tr}\left[V^{-1}(\tilde{X} - M)^* W^{-1}(\tilde{X} - M) \right] \right\}$$

is a density function.

Solution   Evidently $f(\tilde{X}) \geq 0$ for all $\tilde{X}$ and it remains to prove that the total integral is unity. Since V and W are hermitian positive definite there exist nonsingular matrices A and B such that

$$W^{-1} = A^* A \text{ and } V^{-1} = B^* B.$$

Let $\tilde{U} = A(\tilde{X} - M)B^*$ then from (c) of Theorem 3.2

$$d\tilde{U} = |\det(AA^*)|^q |\det(B^*B)|^p d(\tilde{X} - M)$$
$$= |\det(W)|^{-q}|\det(V)|^{-p}d\tilde{X}.$$

Hence

$$\int_{\tilde{X}} f(\tilde{X})d\tilde{X} = \pi^{-(pq)} \int_{\tilde{U}} e^{-\mathrm{tr}(\tilde{U}^*\tilde{U})}d\tilde{U}.$$

But

$$\mathrm{tr}\left(\tilde{U}^*\tilde{U}\right) = \sum_{j=1}^{p}\sum_{k=1}^{q} |\tilde{u}_{jk}|^2$$

where $\tilde{U} = (\tilde{u}_{jk})$. Note that

$$\int_{\tilde{u}_{jk}} e^{-|\tilde{u}_{jk}|^2} d\tilde{u}_{jk} = \int_{-\infty}^{\infty}\int_{-\infty}^{\infty} e^{-(u_{jk1}^2+u_{jk2}^2)} du_{jk1} du_{jk2}$$
$$= \pi$$

where u_{jk1} and u_{jk2} denote the real and the imaginary parts of $\tilde{u}_{jk}$ respectively. There are pq such $\tilde{u}_{jk}$'s and hence the result. This density is known as the *matrix-variate nonsingular complex normal density*.

Theorem 3.3   *Let $\tilde{Y}, \tilde{X}, A, B, C$ be $p \times p$ lower triangular matrices where $\tilde{Y}$ and $\tilde{X}$ are matrices of $p(p+1)/2$ functionally independent complex variables, A, B and C are matrices of constants, A and B nonsingular, and let α be a nonzero complex number. Then, ignoring the sign,*

$$\tilde{Y} = \tilde{X} + \tilde{X}' + C \Rightarrow d\tilde{Y} = 2^{2p}d\tilde{X}, \tag{a}$$
$$\Rightarrow d\tilde{Y} = 2^p d\tilde{X} \text{ if the } x_{jj}\text{'s are real ;} \tag{b}$$

$$\tilde{Y} = A\tilde{X} + C \Rightarrow d\tilde{Y} = \left\{\prod_{j=1}^{p} |a_{jj}|^{2j}\right\} d\tilde{X}, \tag{c}$$

$$\Rightarrow d\tilde{Y} = \left\{\prod_{j=1}^{p} |a_{jj}|^{2j-1}\right\} d\tilde{X} \text{ if the } a_{jj}\text{'s and } x_{jj}\text{'s are real ;} \tag{d}$$

$$\tilde{Y} = \tilde{X}B + C \Rightarrow d\tilde{Y} = \left\{\prod_{j=1}^{p} |b_{jj}|^{2(p-j+1)}\right\} d\tilde{X}, \tag{e}$$

$$\Rightarrow d\tilde{Y} = \left\{\prod_{j=1}^{p} |b_{jj}|^{2(p-j)+1}\right\} d\tilde{X} \text{ if the } b_{jj}\text{'s and } x_{jj}\text{'s are real ;} \tag{f}$$

and

$$\tilde{Y} = \alpha\tilde{X} + C \Rightarrow d\tilde{Y} = |\alpha|^{p(p+1)}d\tilde{X}. \tag{g}$$

Proof   Results $(a), (b)$ and (g) are available from Example 3.1. Let

$$\tilde{Y} = Y_1 + iY_2, \quad \tilde{X} = X_1 + iX_2, \quad A = A_1 + iA_2, \quad B = B_1 + iB_2,$$
$$Y_1 = (y_{jk1}), \quad Y_2 = (y_{jk2}), \quad X_1 = (x_{jk1}), \quad X_2 = (x_{jk2}),$$
$$A_1 = (a_{jk1}), \quad A_2 = (a_{jk2}), \quad B_1 = (b_{jk1}), \quad \text{and } B_2 = (b_{jk2})$$

where Y_1, Y_2, X_1, X_2, A_1, A_2, B_1 and B_2 are all real. Note that since C is a constant we may omit it when considering the differentials. When $\tilde{Y} = A\tilde{X}$ we have $Y_1 = A_1X_1 - A_2X_2$ and $Y_2 = A_1X_2 + A_2X_1$. The matrix of partial derivative of Y_1 with respect to X_1, that is $\frac{\partial Y_1}{\partial X_1}$, can be seen to be a lower triangular matrix with a_{jj1} repeated j times, $j = 1, ..., p$, on the diagonal. Let this matrix be denoted by G_1. Let G_2 be a matrix of the same structure with a_{jj2}'s on the diagonal. Then the Jacobian matrix is given by

$$\frac{\partial (Y_1, Y_2)}{\partial (X_1, X_2)} = \begin{bmatrix} G_1 & -G_2 \\ G_2 & G_1 \end{bmatrix}.$$

From Lemma 3.1 the determinant is available as $|\det(G)|^2$ where $G = G_1 + iG_2$. Since G is triangular the absolute value of the determinant is given by

$$|\det(G)|^2 = \prod_{j=1}^{p} \left[|a_{jj}|^2\right]^j.$$

This establishes (c). If the x_{jj}'s and a_{jj}'s are real then note that the x_{jk}'s for $j > k$ contribute a_{jj} twice that is, corresponding to x_{jk1} and x_{jk2}, whereas the a_{jj}'s appear only once corresponding to the x_{jj1}'s since the x_{jj2}'s are zeros. This establishes (d). If $\tilde{Y} = \tilde{X}B$ and if a matrix H_1 is defined corresponding to G_1 then note that the b_{jj1}'s appear $p - j + 1$ times on the diagonal for $j = 1, ..., p$. Results (e) and (f) are established by using similar steps as in the case of (c) and (d).

Remark 3.1   The upper triangular cases are available by taking the transposes in Theorem 3.3 and then using the fact that $d\tilde{Y} = d\tilde{Y}'$. This will be stated as the next theorem without proof.

Theorem 3.4   *Let $\tilde{Y}, \tilde{X}, A, B, C$ be $p \times p$ upper triangular matrices where $\tilde{Y}$ and $\tilde{X}$ are matrices of $p(p+1)/2$ functionally independent complex variables, A, B and C are matrices of constants, A and B nonsingular, and let*

α be a nonzero complex number. Then, ignoring the sign,

$$\tilde{Y} = \tilde{X} + \tilde{X}' + C \Rightarrow d\tilde{Y} = 2^{2p}d\tilde{X}, \qquad (a)$$

$$\Rightarrow d\tilde{Y} = 2^p d\tilde{X} \text{ if the } x_{jj}\text{'s are real ;} \qquad (b)$$

$$\tilde{Y} = A\tilde{X} \Rightarrow d\tilde{Y} = \left\{ \prod_{j=1}^{p} |a_{jj}|^{2(p-j+1)} \right\} d\tilde{X}, \qquad (c)$$

$$\Rightarrow d\tilde{Y} = \left\{ \prod_{j=1}^{p} |a_{jj}|^{2(p-j)+1} \right\} d\tilde{X} \text{ if the } a_{jj}\text{'s and } x_{jj}\text{'s are real ; } (d)$$

$$\tilde{Y} = \tilde{X}B + C \Rightarrow d\tilde{Y} = \left\{ \prod_{j=1}^{p} |b_{jj}|^{2j} \right\} d\tilde{X}, \qquad (e)$$

$$\Rightarrow d\tilde{Y} = \left\{ \prod_{j=1}^{p} |b_{jj}|^{2j-1} \right\} d\tilde{X} \text{ if the } b_{jj}\text{'s and } x_{jj}\text{'s are real ; } \qquad (f)$$

and

$$\tilde{Y} = \alpha\tilde{X} + C \Rightarrow d\tilde{Y} = |\alpha|^{p(p+1)}d\tilde{X}. \qquad (g)$$

Remark 3.2   If $\tilde{X}$ is lower or upper triangular with $p(p+1)/2$ function-ally independent complex variables then $\tilde{U} = \tilde{X} + \tilde{X}^*$ is no longer a one-to-one transformation unless the diagonal elements of $\tilde{X}$ are real to start with, because $\tilde{X} + \tilde{X}^*$ is hermitian. Thus in $\tilde{U}$ there are only $p + 2[p(p-1)/2] = p^2$ real variables whereas in $\tilde{X}$ there are $2[p(p+1)/2] = p(p+1)$ real variables. If the diagonal elements of $\tilde{X}$ are real then $\tilde{U} = \tilde{X} + \tilde{X}^* \Rightarrow d\tilde{U} = 2^p d\tilde{X}$.

Example 3.4   Verify the result (c) of Theorem 3.3 for $C = 0, \tilde{X} = X_1 + iX_2$ where

$$A = \begin{bmatrix} 1 + 2i & 0 \\ 2 - i & 1 - i \end{bmatrix}$$

and

$$\tilde{X} = \begin{bmatrix} \tilde{x}_{11} & 0 \\ \tilde{x}_{21} & \tilde{x}_{22} \end{bmatrix} = \begin{bmatrix} x_{111} + ix_{112} & 0 \\ x_{211} + ix_{212} & x_{221} + ix_{222} \end{bmatrix}.$$

Solution   Let

$$\tilde{Y} = \begin{bmatrix} \tilde{y}_{11} & 0 \\ \tilde{y}_{21} & \tilde{y}_{22} \end{bmatrix} = A\tilde{X}.$$

Then

$$\tilde{y}_{11} = (1 + 2i)(x_{111} + ix_{112})$$
$$= x_{111} - 2x_{112} + i(2x_{111} + x_{112}),$$
$$\tilde{y}_{21} = (2 - i)(x_{111} + ix_{112}) + (1 - i)(x_{211} + ix_{212})$$
$$= (2x_{111} + x_{112} + x_{211} + x_{212}) + i(-x_{111} + 2x_{112} - x_{211} + x_{212}),$$
$$\tilde{y}_{22} = (1 - i)(x_{221} + ix_{222})$$
$$= x_{221} + x_{222} + i(-x_{221} + x_{222}).$$

If

$$\tilde{Y} = Y_1 + iY_2 = (y_{jk1}) + i(y_{jk2})$$

then

$$y_{111} = x_{111} - 2x_{112}, \quad y_{211} = 2x_{111} + x_{112} + x_{211} + x_{212},$$
$$y_{221} = x_{221} + x_{222}, \quad y_{112} = 2x_{111} + x_{112},$$
$$y_{212} = -x_{111} + 2x_{112} - x_{211} + x_{212}, \text{ and } y_{222} = -x_{221} + x_{222}.$$

$$\frac{\partial(Y_1, Y_2)}{\partial(X_1, X_2)} = \begin{bmatrix} B_1 & B_2 \\ -B_2 & B_1 \end{bmatrix}$$

where

$$\frac{\partial Y_1}{\partial X_1} = B_1 = \begin{bmatrix} 1 & 0 & 0 \\ 2 & 1 & 0 \\ 0 & 0 & 1 \end{bmatrix}$$

$$\frac{\partial Y_1}{\partial X_2} = B_2 = \begin{bmatrix} -2 & 0 & 0 \\ 1 & 1 & 0 \\ 0 & 0 & 1 \end{bmatrix}.$$

Then from Lemmas 3.1 and 3.2 the determinant is given by

$$J = \det[(B_1 + iB_2)(B_1 + iB_2)^*]$$
$$= (1^2 + (-2)^2)(1^2 + 1^2)(1^2 + 1^2) = 20$$

where

$$B_1 + iB_2 = \begin{bmatrix} 1 - 2i & 0 & 0 \\ 2 + i & 1 + i & 0 \\ 0 & 0 & 1 + i \end{bmatrix}$$

and

$$(B_1 + iB_2)^* = \begin{bmatrix} 1+2i & 2-i & 0 \\ 0 & 1-i & 0 \\ 0 & 0 & 1-i \end{bmatrix}.$$

But

$$\prod_{j=1}^{p} |a_{jj}|^{2j} = |a_{11}|^2 |a_{22}|^4$$

$$= |1 + 2i|^2 |1 - i|^2 = (\sqrt{5})^2 (\sqrt{2})^4 = 20.$$

Thus the result is verified.

Theorem 3.5   *Let $\tilde{X}$ be a $p \times p$ hermitian matrix of functionally indepen-dent complex variables and A be a nonsingular matrix of constants. Then, ignoring the sign,*

$$\tilde{Y} = A\tilde{X}A^* \Rightarrow d\tilde{Y} = |\det(A)|^{2p} d\tilde{X}$$
$$= |\det(AA^*)|^p d\tilde{X}.$$

Proof   Since A is nonsingular it can be written as a product of elementary matrices. Let $E_1, ..., E_k$ be elementary matrices such that

$$A = E_k E_{k-1} ... E_1 \Rightarrow A^* = E_1^* E_2^* ... E_k^*.$$

For example let E_1 be such that the j-th row of an identity matrix is multiplied by a scalar $c = a + ib$ where a and b are real and $i = \sqrt{-1}$. Then $E_1 \tilde{X} E_1^*$ means that the j-th row of $\tilde{X}$ is multiplied by $a + ib$ and the j-th column of $\tilde{X}$ is multiplied by $a - ib$. Let $\tilde{U}_1 = E_1 \tilde{X} E_1^*$, $\tilde{U}_2 = E_2 \tilde{U}_1 E_2^*$,..., $\tilde{U}_k = \tilde{Y} = E_k \tilde{U}_{k-1} E_k^*$. Then the Jacobian of $\tilde{Y}$ written as a function of $\tilde{X}$ is given by

$$J(\tilde{Y} : \tilde{X}) = J(\tilde{Y} : \tilde{U}_{k-1}) ... J(\tilde{U}_1 : \tilde{X}).$$

Let us evaluate $d\tilde{U}_1$ in terms of $d\tilde{X}$ by direct computation. Since $\tilde{X}$ is hermitian its diagonal elements are real and the elements above the leading diagonal are the complex conjugates of those below the leading diagonal, and $\tilde{U}_1$ is also of the same structure as $\tilde{X}$. Let $\tilde{U}_1 = U + iV$ and $\tilde{X} = Z + iW$ where $U = (u_{jk})$, $V = (v_{jk})$, $Z = (z_{jk})$, $W = (w_{jk})$ are all real and the diagonal elements of V and W are zeros. Take the u_{jj}'s and z_{jj}'s separately. The matrix of partial derivatives of $u_{11}, ..., u_{pp}$ with respect to $z_{11}, ..., z_{pp}$

is a diagonal matrix with the j-th element $a^2 + b^2$ and all other elements unities. The remaining variables produce a $\frac{p(p-1)}{2} \times \frac{p(p-1)}{2}$ matrix of the following type

$$\frac{\partial (U_0, V_0)}{\partial (Z_0, W_0)} = \begin{bmatrix} A_1 & B_1 \\ -B_1 & A_1 \end{bmatrix}$$

where U_0, V_0, Z_0, W_0 mean that the diagonal elements are deleted, A_1 is a diagonal matrix with $p - 1$ of the diagonal elements equal to a and the remaining unities and B_1 is a diagonal matrix such that corresponding to every a in A_1 there is a b or $-b$ with $j - 1$ of them equal to $-b$ and $p - j$ of them equal to b. The configuration of the partial derivatives for $p = 3$ with $j = 2$ is the following:

	z_{11}	z_{22}	z_{33}	z_{21}	z_{31}	z_{32}	w_{21}	w_{31}	w_{32}
u_{11}	1	0	0	0	0	0	0	0	0
u_{22}	0	$a^2 + b^2$	0	0	0	0	0	0	0
u_{33}	0	0	1	0	0	0	0	0	0
u_{21}	0	0	0	a	0	0	$-b$	0	0
u_{31}	0	0	0	0	1	0	0	0	0
u_{32}	0	0	0	0	0	a	0	0	b
v_{21}	0	0	0	b	0	0	a	0	0
v_{31}	0	0	0	0	0	0	0	1	0
v_{32}	0	0	0	0	0	$-b$	0	0	a

From Lemmas 3.1 and 3.2 the determinant is $|\det(A_1 + iB_1)|^2$. That is,

$$\left| \frac{\partial (U_0, V_0)}{\partial (Z_0, W_0)} \right| = \left| \begin{matrix} A_1 & B_1 \\ -B_1 & A_1 \end{matrix} \right| = |\det(A_1 + iB_1)|^2$$
$$= |\det(A_1 + iB_1)(A_1 + iB_1)^*)|$$
$$= (a^2 + b^2)^{p-1}.$$

Thus

$$d\tilde{U}_1 = (a^2 + b^2)^p \, d\tilde{X} = |\det (E_1 E_1^*)|^p d\tilde{X}.$$

Note that interchanges of rows and columns can produce only a change in the sign in the determinant, the addition of a row (column) to another row (column) does not change the determinant and elementary matrices of the type E_1 will produce $|\det (E_1 E_1^*)|^p$ in the Jacobian. Thus by computing $J(\tilde{U}_1 : \tilde{X})$, $J(\tilde{U}_2 : \tilde{U}_1)$ etc we have

$$d\tilde{Y} = |\det (AA^*)|^p d\tilde{X}.$$

Hence the result.

Remark 3.3 If $\tilde{X}$ is skew hermitian then the diagonal elements are purely imaginary, that is, the real parts are zeros. It is easy to note that the structure of the Jacobian matrix for a transformation of the type $\tilde{Y} = A\tilde{X}A^*$, where $\tilde{X} = -\tilde{X}^*$, remains the same as that in the hermitian case of Theorem 3.5. The roles of (u_{jj}, z_{jj})'s and (v_{jj}, w_{jj})'s are interchanged. Thus the next theorem will be stated without proof.

Theorem 3.6 *Let $\tilde{X}$ be a $p \times p$ skew hermitian matrix of functionally independent complex variables. Let A be a nonsingular matrix of constants. Then, ignoring the sign,*

$$\tilde{Y} = A\tilde{X}A^* \Rightarrow \mathrm{d}\tilde{Y} = |\det(AA^*)|^p\mathrm{d}\tilde{X}.$$

Remark 3.4 All the nonsingular linear transformations in Theorems 3.1–3.6 are easily seen to be one-to-one.

Exercises

3.1.1 Verify Theorem 3.1 for $p = 3$ and

$$A = \begin{pmatrix} 1 & 0 & 1-i \\ 2+i & 3 & 1+i \\ 1 & 2 & 4 \end{pmatrix}, \quad C = 0.$$

3.1.2 Verify Theorem 3.2 for $p = 3, q = 2$, $C = 0$, the same A as in Exercise 3.1.1 and

$$B = \begin{pmatrix} 1 & 1+i \\ 2 & 3+2i \end{pmatrix}.$$

3.1.3 Verify Theorem 3.3(c) for $p = 2$, $C = 0$,

$$A = \begin{pmatrix} 1-i & 0 \\ 2+i & 2+3i \end{pmatrix}, \quad \text{and for } A = \begin{pmatrix} 1 & 0 \\ 2+i & 2 \end{pmatrix}$$

when the x_{jj}'s are real.

3.1.4   Let $\tilde{X}$ be a $p \times p$ hermitian positive definite matrix of functionally independent complex variables and B be a $p \times p$ matrix of constants such that $B = B^* > 0$. Show that

$$\int_{\tilde{X}} e^{-\operatorname{tr}(B\tilde{X})} d\tilde{X} = |\det(B)|^{-2p} \pi^{\frac{p(p-1)}{2}} (p-1)!(p-2)!...1!, \ p \geq 1.$$

3.1.5   If $\tilde{X}$ is a $p \times p$ hermitian positive definite matrix of functionally independent complex variables evaluate the integral

$$\int_{0 < \tilde{X} < I} |\tilde{X}|^{\alpha} d\tilde{X} \text{ for } \operatorname{Re}(\alpha) \geq 0.$$

3.2   Smooth nonlinear transformations in the complex case

Some simple nonlinear transformations will be considered here. These are transformations which become linear transformations in the differentials so that the Jacobian of the original transformation becomes the Jacobian of the linear transformation where the matrices of differentials are treated as the new variables and everything else as constants.

Theorem 3.7   *Let $\tilde{X}$ be a $p \times p$ hermitian positive definite matrix of functionally independent complex variables. Let $\tilde{T}$ be $p \times p$ lower triangular and $\tilde{Q}$ be $p \times p$ upper triangular matrices of functionally independent complex variables with real and positive diagonal elements. Then*

$$\tilde{X} = \tilde{T}\tilde{T}^* \Rightarrow \mathrm{d}\tilde{X} = 2^p \left\{ \prod_{j=1}^{p} t_{jj}^{2(p-j)+1} \right\} \mathrm{d}\tilde{T}$$

and

$$\tilde{X} = \tilde{Q}\tilde{Q}^* \Rightarrow \mathrm{d}\tilde{X} = 2^p \left\{ \prod_{j=1}^{p} t_{jj}^{2(j-1)+1} \right\} \mathrm{d}\tilde{Q}.$$

Proof   When the diagonal elements of the triangular matrices are real and positive there exist unique representations $\tilde{X} = \tilde{T}\tilde{T}^*$ and $\tilde{X} = \tilde{Q}\tilde{Q}^*$. Let $\tilde{X} = X_1 + iX_2$ and $\tilde{T} = T_1 + iT_2$ where $\tilde{X} = (\tilde{x}_{jk})$, $\tilde{T} = (\tilde{t}_{jk})$, $\tilde{t}_{jk} = 0$, $j < k$, $X_1 = (x_{jk1})$, $X_2 = (x_{jk2})$, $T_1 = (t_{jk1})$, $T_2 = (t_{jk2})$. Note that X_1 is symmetric positive definite and X_2 is skew symmetric. The diagonal elements of X_2 and T_2 are zeros. Hence when considering the Jacobian we should take $\tilde{x}_{jj}$, $j = 1, ..., p$ and $\tilde{x}_{jk}$, $j > k$ separately.

$$\tilde{X} = \tilde{T}\tilde{T}^* \Rightarrow X_1 + iX_2 = (T_1 + iT_2)(T_1' - iT_2') \Rightarrow$$
$$X_1 = T_1 T_1' + T_2 T_2' \text{ and } X_2 = T_2 T_1' - T_1 T_2'$$

with $t_{jj1} = t_{jj}$, $t_{jj2} = 0$, $j = 1, ..., p$. Note that

$$\frac{\partial x_{jj1}}{\partial t_{jj1}} = 2t_{jj1} = 2t_{jj}, \ j = 1, ..., p.$$

Now consider the x_{jk1}'s for $j > k$. It is easy to note that

$$\frac{\partial (X_{10}, X_{20})}{\partial (T_{10}, T_{20})} = \begin{bmatrix} U & V \\ W & Y \end{bmatrix}$$

where a zero indicates that the x_{jj1}'s are removed and the derivatives are taken with respect to the t_{jk1}'s and t_{jk2}'s for $j > k$. U and Y are lower triangular matrices with t_{jj} repeated $p - j$ times along the diagonal and V is of the same form as U but with $t_{jj2} = 0$ along the diagonal and the t_{jk1}'s replaced by the t_{jk2}'s. For example, take the x_{jk1}'s in the order $x_{211}, x_{311}, ..., x_{p11}, x_{321}, ..., x_{p\,p-11}$ and the t_{jk1}'s also in the same order. Then we get the $\frac{p(p-1)}{2} \times \frac{p(p-1)}{2}$ matrix

$$U = \frac{\partial X_{10}}{\partial T_{10}} = \begin{bmatrix} t_{11} & 0 & \cdots & 0 \\ * & t_{11} & \cdots & 0 \\ \vdots & \vdots & \ddots & 0 \\ * & * & \cdots & t_{p-1\,p-1} \end{bmatrix}$$

where the $*$'s indicate the presence of elements some of which may be zeros. Since U and V are lower triangular with the diagonal elements of V being zeros, one can make W null by adding suitable combinations of the rows of $(U \;\; V)$. This will not alter the lower triangular nature or the diagonal elements of Y. Then the determinant is given by

$$\begin{vmatrix} U & V \\ W & Y \end{vmatrix} = |U| \, |Y| = \prod_{j=1}^{p} t_{jj}^{2(p-j)}.$$

Multiply with the $2t_{jj}$'s for $j = 1, ..., p$ to establish the result.

The proof in the case of $\tilde{X} = \tilde{Q}\tilde{Q}^*$ is similar but in this case it can be seen that the triangular matrices corresponding to U and Y will have t_{jj} repeated $j - 1$ times along the diagonal for $j = 1, ..., p$.

Example 3.5   Show that for $\tilde{X} = \tilde{X}^* > 0$, $p \times p$, and $\operatorname{Re}(\alpha) > p - 1$,

$$\tilde{\Gamma}_p(\alpha) = \int_{\tilde{X} > 0} |\det(\tilde{X})|^{\alpha - p} e^{-\operatorname{tr}(\tilde{X})} d\tilde{X}$$

$$= \pi^{\frac{p(p-1)}{2}} \Gamma(\alpha) \Gamma(\alpha - 1) ... \Gamma(\alpha - p + 1).$$

Solution   Let $\tilde{T} = (\tilde{t}_{jk})$, $\tilde{t}_{jk} = 0$, $j < k$ be a lower triangular matrix with real and positive diagonal elements $t_{jj} > 0$, $j = 1, ..., p$ such that $\tilde{X} = \tilde{T}\tilde{T}^*$. Then from Theorem 3.7

$$d\tilde{X} = 2^p \left\{ \prod_{j=1}^{p} t_{jj}^{2(p-j)+1} \right\} d\tilde{T}.$$

Note that

$$\text{tr}(\tilde{X}) = \text{tr}(\tilde{T}\tilde{T}^*)$$
$$= t_{11}^2 + t_{22}^2 + \dots + t_{pp}^2$$
$$+ \left|\tilde{t}_{21}\right|^2 + \dots + \left|\tilde{t}_{p1}\right|^2 + \dots + \left|\tilde{t}_{p\,p-1}\right|^2$$

and

$$|\det(\tilde{X})|^{\alpha-p}d\tilde{X} = 2^p \left\{ \prod_{j=1}^p t_{jj}^{2\alpha-2j+1} \right\} d\tilde{T}.$$

The integral over $\tilde{X}$ splits into p integrals over the t_{jj}'s and $p(p-1)/2$ integrals over the $\tilde{t}_{jk}$'s, $j > k$. Note that $0 < t_{jj} < \infty$, $-\infty < t_{jk1} < \infty$, $-\infty < t_{jk2} < \infty$ where t_{jk1} and t_{jk2} denote the real and the imaginary parts of $\tilde{t}_{jk}$ respectively. But

$$2 \int_0^\infty t_{jj}^{2\alpha-2j+1} e^{-t_{jj}^2} dt_{jj} = \Gamma(\alpha-j+1), \ \ \text{Re}(\alpha) > j-1,$$

for $j = 1, \dots, p \Rightarrow \text{Re}(\alpha) > p-1$ and

$$\int_{\tilde{t}_{jk}} e^{-|\tilde{t}_{jk}|^2} d\tilde{t}_{jk} = \int_{-\infty}^\infty \int_{-\infty}^\infty e^{-(t_{jk1}^2 + t_{jk2}^2)} dt_{jk1} dt_{jk2}$$
$$= \pi \Rightarrow$$
$$\prod_{j>k} (\pi) = \pi^{\frac{p(p-1)}{2}}.$$

Hence the result.

Definition 3.3 $\tilde{\Gamma}_p(\alpha)$: **complex matrix-variate gamma**   It is defined as stated in Example 3.5. We will write with a tilde over Γ to distinguish it from the matrix-variate gamma in the real case.

Example 3.6   Show that

$$f(\tilde{X}) = \frac{|\det(B)|^\alpha |\det(\tilde{X})|^{\alpha-p} e^{-\text{tr}(B\tilde{X})}}{\tilde{\Gamma}(\alpha)}$$

for $B = B^* > 0$, $\tilde{X} = \tilde{X}^* > 0$, $\text{Re}(\alpha) > p-1$ and $f(\tilde{X}) = 0$ elsewhere, is a density function for $\tilde{X}$ where B is a constant matrix, with $\tilde{\Gamma}(\alpha)$ as given in Definition 3.3.

Solution   Evidently $f(\tilde{X}) \geq 0$ for all $\tilde{X}$ and it remains to show that the total integral is unity. Since B is hermitian positive definite there exists a nonsingular C such that $B = C^*C$. Then

$$\text{tr}(B\tilde{X}) = \text{tr}(C^*C\tilde{X}) = \text{tr}(C\tilde{X}C^*).$$

Hence from Theorem 3.5

$$\tilde{Y} = C\tilde{X}C^* \Rightarrow d\tilde{Y} = |\det(CC^*)|^p d\tilde{X},$$

and

$$\tilde{X} = C^{-1}\tilde{Y}C^{*-1} \Rightarrow |\det(\tilde{X})| = |\det(CC^*)|^{-1}|\det(\tilde{Y})|.$$

Then

$$\int_{\tilde{X}>0} f(\tilde{X})d\tilde{X} = \int_{\tilde{Y}>0} \frac{|\det(\tilde{Y})|^{\alpha-p}e^{-\text{tr}(\tilde{Y})}}{\tilde{\Gamma}(\alpha)} d\tilde{Y}.$$

But from Example 3.5 the right side is unity for $\text{Re}(\alpha) > p - 1$. This density $f(\tilde{X})$ is known as the *complex matrix-variate gamma density with the parameters α and B.*

When $\tilde{X}$ is nonsingular we have a unique $\tilde{X}^{-1}$ such that $\tilde{X}\tilde{X}^{-1} = I$ where I is the identity matrix. Taking the differentials and writing the matrices of differentials one has

$$\tilde{X}\tilde{X}^{-1} = I \Rightarrow (d\tilde{X})\tilde{X}^{-1} + \tilde{X}(d\tilde{X}^{-1}) = 0$$
$$\Rightarrow (d\tilde{X}^{-1}) = -\tilde{X}^{-1}(d\tilde{X})\tilde{X}^{-1}. \qquad (3.2.1)$$

As in the real case of Chapter 1 it is easy to note that the Jacobian of a smooth nonlinear transformation is the Jacobian of the linear transformation of the matrices of differentials. For example, if $\tilde{X}$ is nonsingular and if $\tilde{Y} = \tilde{X}^{-1}$ then for evaluating the Jacobian one needs to consider only (3.2.1) treating the matrices $(d\tilde{Y})$ and $(d\tilde{X})$ as variables and everything else, including $\tilde{X}$, as constants. Then one can apply Theorems 3.2, 3.5 and 3.6 to write down the Jacobian. Hence the following theorem will be stated without proof.

Theorem 3.8   *Let $\tilde{X}$ be a $p \times p$ nonsingular matrix of functionally independent complex variables. Then, ignoring the sign,*

$$\tilde{Y} = \tilde{X}^{-1} \Rightarrow$$
$$d\tilde{Y} = |\det(\tilde{X}\tilde{X}^*)|^{-2p}d\tilde{X} \text{ for a general } \tilde{X}, \qquad (3.2.2)$$
$$= |\det(\tilde{X}\tilde{X}^*)|^{-p}d\tilde{X} \text{ for } \tilde{X} = \tilde{X}^* \text{ or } \tilde{X} = -\tilde{X}^*. \qquad (3.2.3)$$

Remark 3.5 The result is different when $\tilde{X}$ is triangular. This can be written up from equation (3.2.1) by using Theorems 3.3 and 3.4.

Example 3.7 If $\tilde{x}$ is a scalar complex variable, $\tilde{x} \neq 0$ and if $\tilde{y} = \tilde{x}^{-1} = \frac{1}{\tilde{x}}$ evaluate $d\tilde{y}$ in terms of $d\tilde{x}$ from first principles.

Solution Let $\tilde{y} = y_1 + iy_2$ and $\tilde{x} = x_1 + ix_2$ where y_1, y_2, x_1, x_2 are real.

$$\tilde{y} = \frac{1}{\tilde{x}} = \frac{1}{x_1 + ix_2} = \frac{x_1 - ix_2}{x_1^2 + x_2^2} \Rightarrow$$
$$y_1 = \frac{x_1}{u}, \ y_2 = \frac{-x_2}{u}, \ u = x_1^2 + x_2^2.$$

The partial derivatives are

$$\frac{\partial y_1}{\partial x_1} = \frac{x_2^2 - x_1^2}{u^2}, \ \frac{\partial y_1}{\partial x_2} = -\frac{2x_1 x_2}{u^2},$$
$$\frac{\partial y_2}{\partial x_1} = \frac{2x_1 x_2}{u^2} \ \text{and} \ \frac{\partial y_2}{\partial x_2} = \frac{x_2^2 - x_1^2}{u^2}.$$

The determinant of the matrix of partial derivatives is therefore

$$\left| \frac{\partial (y_1, y_2)}{\partial (x_1, x_2)} \right| = \frac{1}{u^2} = |\tilde{x}\tilde{x}^*|^{-2}.$$

Then

$$d\tilde{y} = |\tilde{x}\tilde{x}^*|^{-2} d\tilde{x}.$$

Note that if $\tilde{x} = \tilde{x}^*$ then $\tilde{x} = x_1$, and from (3.2.3) $d\tilde{y} = |x_1^2|^{-1} dx_1 = x_1^{-2} dx_1$, ignoring the sign. Remember that an empty product is interpreted as unity and hence in this case $d\tilde{y} = dy_1$. If $\tilde{x} = -\tilde{x}^*$ then $\tilde{x} = ix_2$ and from (3.2.3) $d\tilde{y} = x_2^{-2} dx_2$, ignoring the sign.

Example 3.8 If $\tilde{X}_1$ and $\tilde{X}_2$ are $p \times p$ statistically independent hermitian positive definite matrices of complex random variables having complex matrix-variate gamma distributions as in Example 3.6 with parameters (α_1, B) and (α_2, B) respectively with $B = B^* > 0$, evaluate the density of

$$\tilde{Y} = \left(\tilde{X}_1 + \tilde{X}_2 \right)^{-\frac{1}{2}} \tilde{X}_1 \left(\tilde{X}_1 + \tilde{X}_2 \right)^{-\frac{1}{2}}.$$

Solution   The joint density of $\tilde{X}_1$ and $\tilde{X}_2$, denoted by $f(\tilde{X}_1, \tilde{X}_2)$, is given by

$$f(\tilde{X}_1, \tilde{X}_2)\mathrm{d}\tilde{X}_1\mathrm{d}\tilde{X}_2 = a|\det(\tilde{X}_1)|^{\alpha_1-p}|\det(\tilde{X}_2)|^{\alpha_2-p}$$
$$\times e^{-\mathrm{tr}(B(\tilde{X}_1+\tilde{X}_2))}\mathrm{d}\tilde{X}_1\mathrm{d}\tilde{X}_2, \ \tilde{X}_j = \tilde{X}_j^* > 0, \ j = 1,2$$

and $f(\tilde{X}_1, \tilde{X}_2) = 0$ elsewhere, where

$$a = \frac{|\det(B)|^{\alpha_1+\alpha_2}}{\tilde{\Gamma}(\alpha_1)\tilde{\Gamma}(\alpha_2)}.$$

Let $\tilde{Y}_j = B^{\frac{1}{2}}\tilde{X}_jB^{\frac{1}{2}}$, $j = 1,2$. This will get rid of B, and $\tilde{Y}$ is invariant. Let $\tilde{U} = \tilde{Y}_1 + \tilde{Y}_2$ for fixed $\tilde{Y}_2$. Then $\mathrm{d}\tilde{U} = \mathrm{d}\tilde{Y}_1$ and $\tilde{Y}_2 = \tilde{U} - \tilde{Y}_1$. The joint density of $\tilde{U}$ and $\tilde{Y}_1$ is then

$$f(\tilde{Y}_1, \tilde{U} - \tilde{Y}_1)\mathrm{d}\tilde{U}\mathrm{d}\tilde{Y}_1 = a|\det(\tilde{Y}_1)|^{\alpha_1-p}|\det(\tilde{U} - \tilde{Y}_1)|^{\alpha_2-p}$$
$$\times e^{-\mathrm{tr}(\tilde{U})}\mathrm{d}\tilde{U}\mathrm{d}\tilde{Y}_1, \ \tilde{U} - \tilde{Y}_1 = (\tilde{U} - \tilde{Y}_1)^* > 0$$
$$= a|\det(\tilde{Y}_1)|^{\alpha_1-p}|\det(\tilde{U})|^{\alpha_2-p}e^{-\mathrm{tr}(\tilde{U})}$$
$$\times \left|\det\left(I - \tilde{U}^{-\frac{1}{2}}\tilde{Y}_1\tilde{U}^{-\frac{1}{2}}\right)\right|^{\alpha_2-p}\mathrm{d}\tilde{U}\mathrm{d}\tilde{Y}_1.$$

Put $\tilde{Y} = \tilde{U}^{-\frac{1}{2}}\tilde{Y}_1\tilde{U}^{-\frac{1}{2}}$ for fixed $\tilde{U}$ then

$$\mathrm{d}\tilde{Y} = |\det(\tilde{U}\tilde{U}^*)|^{-\frac{p}{2}}\mathrm{d}\tilde{Y}_1$$
$$= |\det(\tilde{U})|^{-p}\mathrm{d}\tilde{Y}_1.$$

Substitute for $\tilde{Y}_1$ and $\mathrm{d}\tilde{Y}_1$. Now the joint density of $\tilde{Y}$ and $\tilde{U}$, denoted by $g(\tilde{U}, \tilde{Y})$, is

$$g(\tilde{U}, \tilde{Y})\mathrm{d}\tilde{U}\mathrm{d}\tilde{Y} = a|\det(\tilde{U})|^{\alpha_1+\alpha_2-p}e^{-\mathrm{tr}(\tilde{U})}$$
$$\times |\det(\tilde{Y})|^{\alpha_1-p}|\det(I - \tilde{Y})|^{\alpha_2-p}\mathrm{d}\tilde{U}\mathrm{d}\tilde{Y}.$$

The marginal density or the density of $\tilde{Y}$, denoted by $h(\tilde{Y})$, is available by integrating out $\tilde{U}$, noting that $\tilde{U} = \tilde{U}^* > 0$. Integrating out $\tilde{U}$ by using Example 3.5 we have

$$h(\tilde{Y}) = \frac{\tilde{\Gamma}_p(\alpha_1 + \alpha_2)}{\tilde{\Gamma}_p(\alpha_1)\tilde{\Gamma}_p(\alpha_2)}|\det(\tilde{Y})|^{\alpha_1-p}|\det(I - \tilde{Y})|^{\alpha_2-p} \qquad (3.2.4)$$

for $\mathrm{Re}(\alpha_1) > p-1$, $\mathrm{Re}(\alpha_2) > p-1$, $0 < \tilde{Y} < I$, $\tilde{Y} = \tilde{Y}^* > 0$ and $h(\tilde{Y}) = 0$ elsewhere. The density $h(\tilde{Y})$ is known as the *complex type-1 matrix-variate* *beta density with the parameters* (α_1, α_2). *The notation* $0 < \tilde{Y} < I$ *means that* $\tilde{Y} = \tilde{Y}^* > 0$ *and* $I - \tilde{Y} = (I - \tilde{Y})^* > 0$.

Theorem 3.9   *Let* $\tilde{X}$ *and* A *be* $p \times p$ *nonsingular matrices where* $\tilde{X}$ *is a matrix of functionally independent complex variables and* A *is a constant matrix such that* $A + \tilde{X}$ *is nonsingular. Then, ignoring the sign,*

$$\tilde{Y} = (A + \tilde{X})^{-1}(A - \tilde{X}) \text{ or } (A - \tilde{X})(A + \tilde{X})^{-1} \Rightarrow$$
$$d\tilde{Y} = 2^{2p^2} |\det(AA^*)|^p \left| \det\left((A + \tilde{X})(A + \tilde{X})^* \right) \right|^{-2p} d\tilde{X}$$
for a general $\tilde{X}$;
$$d\tilde{Y} = 2^{p^2} \left| \det\left((I + \tilde{X})(I + \tilde{X})^* \right) \right|^{-p} d\tilde{X}$$
for $A = I, \tilde{X} = \tilde{X}^*$.

Proof   Note that $(A + \tilde{X})^{-1}(A - \tilde{X}) = 2(A + \tilde{X})^{-1}A - I$ and the transformation is one-to-one. Let $\tilde{U} = (A + \tilde{X})^{-1}$. Then from Theorem 3.8

$$d\tilde{U} = \left| \det\left((A + \tilde{X})(A + \tilde{X})^* \right) \right|^{-2p} d\tilde{X}$$

for a general $\tilde{X}$ and

$$d\tilde{U} = \left| \det\left((A + \tilde{X})(A + \tilde{X})^* \right) \right|^{-p} d\tilde{X}$$

for $A + \tilde{X} = (A + \tilde{X})^*$ or $A + \tilde{X} = -(A + \tilde{X})^*$. Note that $\tilde{Y} = 2\tilde{U}A - I$ and when $A = I$ and $\tilde{X} = \tilde{X}^*$ then $\tilde{U} = \tilde{U}^*$. Then from Theorem 3.2

$$d\tilde{Y} = 2^{2p^2} |\det(AA^*)|^p d\tilde{U}$$

for a general $\tilde{U}$ and

$$d\tilde{Y} = 2^{p^2} d\tilde{U} \text{ for } A = I, \ \tilde{X} = \tilde{X}^*.$$

Substituting for $d\tilde{U}$ in terms of $d\tilde{X}$ the result follows.

Theorem 3.10   *Let* $\tilde{X} = (\tilde{x}_{jk})$ *be a* $p \times p$ *hermitian positive definite matrix of functionally independent complex variables with diagonal elements unities, that is,* $\tilde{x}_{jj} = 1$, $j = 1, ..., p$. *Let* $\tilde{T} = (\tilde{t}_{jk})$ *be a lower triangular* *matrix with real and positive diagonal elements, that is,* $\tilde{t}_{jk} = 0$, $j < k$ *and* $\tilde{t}_{jj} = t_{jj} > 0$, $j = 1, ..., p$. *Then*

$$\tilde{X} = \tilde{T}\tilde{T}^* \text{ with } \sum_{k=1}^{j} \tilde{t}_{jk}\tilde{t}_{jk}^* = 1 = \sum_{k=1}^{j} |\tilde{t}_{jk}|^2, \ j = 1, ..., p \Rightarrow$$

$$d\tilde{X} = \left\{ \prod_{j=1}^{p} t_{jj}^{2(p-j)} \right\} d\tilde{T}, \ t_{11} = 1$$

and

$$\tilde{X} = \tilde{T}^*\tilde{T} \text{ with } \sum_{j=k}^{p} \tilde{t}_{jk}^*\tilde{t}_{jk} = 1 = \sum_{j=k}^{p} |\tilde{t}_{jk}|^2, \ k = 1, ..., p \Rightarrow$$

$$d\tilde{X} = \left\{ \prod_{j=1}^{p} t_{jj}^{2(j-1)} \right\} d\tilde{T}, \ t_{pp} = 1.$$

Proof   Note that the transformation is one-to-one. Let $\tilde{X} = X_1 + iX_2$, $\tilde{T} = T_1 + iT_2$, $i = \sqrt{-1}$ where X_1, X_2, T_1, T_2 are real matrices. The diagonal elements of X_2 and T_2 are zeros and the diagonal elements of X_1 are unities. Also

$$\tilde{X} = \tilde{T}\tilde{T}^* \Rightarrow X_1 = T_1T_1' + T_2T_2' \text{ and } X_2 = T_2T_1' - T_1T_2'.$$

The submatrices of the Jacobian matrix coming from $\frac{\partial X_1}{\partial T_1}$ and $\frac{\partial X_2}{\partial T_2}$ are each triangular with t_{11} repeated $p-1$ times, t_{22} repeated $p-2$ times and so on. The submatrices $\frac{\partial X_1}{\partial T_2}$ and $\frac{\partial X_2}{\partial T_1}$ can be made null by operating with the rows when evaluating the determinant. Thus the Jacobian is the product of the diagonal elements in $\frac{\partial X_1}{\partial T_1}$ and $\frac{\partial X_2}{\partial T_2}$ observing that $t_{11} = 1$ since $t_{11} > 0$ and $t_{11}^2 = 1$. Hence the result. The second part can be established in a similar fashion. For the second part, note that $t_{pp} = 1$.

Example 3.9   Let $\tilde{X} = (\tilde{x}_{jk})$ be $p \times p$ hermitian positive definite matrix with functionally independent complex random variables such that $\tilde{x}_{jj} = 1$, $j = 1, ..., p$ and having the density

$$f(\tilde{X}) = \begin{cases} \{[\Gamma(n)]^p/\tilde{\Gamma}_p(n)\}|\det(\tilde{X})|^{n-p}, \ \tilde{X} = \tilde{X}^* > 0, \ n \geq p \\ \\ 0 \text{ elsewhere.} \end{cases}$$

Let $\tilde{X} = \tilde{T}\tilde{T}^*$ where $\tilde{T} = (\tilde{t}_{jk})$ is lower triangular with $\sum_{k=1}^{j} |\tilde{t}_{jk}|^2 = 1$, $j = 1, ..., p$ and with real and positive diagonal elements $t_{jj} > 0$, $j = 1, ..., p$. Show that $\tilde{u}_j = (\tilde{t}_{j1}, \tilde{t}_{j2}, ..., \tilde{t}_{j\,j-1})$, $j = 2, ..., p$ are mutually independently distributed and $\{|\tilde{t}_{jk}|^2, k = 1, ..., j-1\}$ have a complex type-1 Dirichlet distribution.

Solution   Let the density of $\tilde{T}$ be denoted by $g(\tilde{T})$. Then from Theorem 3.10

$$g(\tilde{T}) = \frac{[\Gamma(n)]^p}{\tilde{\Gamma}_p(n)} |\det(\tilde{X})|^{n-p} \left\{ \prod_{j=2}^{p} t_{jj}^{2(p-j)} \right\}$$

$$= \frac{[\Gamma(n)]^p}{\tilde{\Gamma}_p(n)} \left\{ \prod_{j=1}^{p} (t_{jj}^2)^{n-p+p-j} \right\}, \quad t_{11}^2 = 1$$

$$= \frac{[\Gamma(n)]^p}{\tilde{\Gamma}_p(n)} \prod_{j=2}^{p} \left[1 - |\tilde{t}_{j1}|^2 - ... - |\tilde{t}_{j\,j-1}|^2 \right]^{n-j}.$$

Since the density factorizes into functions of $\tilde{u}_j$ we have $\tilde{u}_j$, $j = 2, ..., p-1$ are mutually independently distributed. Then the marginal density of $\tilde{u}_j$ is $c_j [1 - |\tilde{t}_{j1}|^2 - ... - |\tilde{t}_{j\,j-1}|^2]^{n-j}$ where c_j is a normalizing constant. This is a special case of a type-1 Dirichlet for $\{|\tilde{t}_{jk}|^2, k = 1, ..., j-1\}$. Hence the result.

Theorem 3.11   *Let $\tilde{X}$ be a $p \times p$ nonsingular lower triangular matrix of functionally independent complex variables and A a nonsingular lower triangular matrix of constants such that $(A + \tilde{X})$ is nonsingular. Then, ignoring the sign,*

$$\tilde{Y} = (A + \tilde{X})^{-1}(A - \tilde{X}) \Rightarrow$$

$$d\tilde{Y} = 2^{p(p+1)} \left| \det\left((A + \tilde{X})(A + \tilde{X})^* \right) \right|^{-(p+1)}$$

$$\times \left\{ \prod_{j=1}^{p} |a_{jj}|^{2(p-j+1)} \right\} d\tilde{X} \text{ if all elements are complex ;}$$

$$d\tilde{Y} = 2^{p^2} \left\{ \prod_{j=1}^{p} |a_{jj} + x_{jj}|^{-2p} \right\} \left\{ \prod_{j=1}^{p} |a_{jj}|^{2(p-j)+1} \right\} d\tilde{X}$$

if all the diagonal elements of $\tilde{X}$ and A are real and others complex and

$$\tilde{Y} = (A - \tilde{X})(A + \tilde{X})^{-1} \Rightarrow$$

$$d\tilde{Y} = 2^{p(p+1)} \left| \det\left((A + \tilde{X})(A + \tilde{X})^* \right) \right|^{-(p+1)}$$

$$\times \left\{ \prod_{j=1}^{p} |a_{jj}|^{2j} \right\} d\tilde{X} \text{ if all elements are complex; and}$$

$$d\tilde{Y} = 2^{p^2} \left\{ \prod_{j=1}^{p} |a_{jj} + x_{jj}|^{-2p} \right\} \left\{ \prod_{j=1}^{p} |a_{jj}|^{2j-1} \right\} d\tilde{X}$$

if the diagonal elements of $\tilde{X}$ and A are all real.

Proof   Observe that the transformation is one-to-one and

$$\tilde{Y} = (A + \tilde{X})^{-1}(A - \tilde{X}) = 2(A + \tilde{X})^{-1}A - I.$$

Taking the differentials and using the fact that $d(A + \tilde{X}) = d\tilde{X}$ one has

$$(d\tilde{Y}) = -(A + \tilde{X})^{-1}(d\tilde{X})(A + \tilde{X})^{-1}(2A).$$

Let

$$(d\tilde{U}) = -(A + \tilde{X})^{-1}(d\tilde{X})(A + \tilde{X})^{-1}.$$

Then from Theorem 3.3, and ignoring the sign,

$$d\tilde{U} = |\det\left((A + \tilde{X})(A + \tilde{X})^* \right)|^{-(p+1)} d\tilde{X}$$

for a general A and $\tilde{X}$ and

$$d\tilde{U} = \left\{ \prod_{j=1}^{p} |a_{jj} + x_{jj}|^{-2p} \right\} d\tilde{X}$$

if the diagonal elements of A and $\tilde{X}$ are all real. Now

$$(d\tilde{Y}) = (d\tilde{U})(2A)$$

and using Theorem 3.3 once again the result follows. Note that

$$\prod_{j=1}^{p} 2^{2(p-j)+1} = 2^{p^2}.$$

The second part as well as the next theorem can be established in a similar fashion.

Theorem 3.12 *Let $\tilde{X}$ be a nonsingular upper triangular matrix of functionally independent complex variables and A a nonsingular upper triangular matrix of complex constants such that $(A + \tilde{X})$ is nonsingular. Then, ignoring the sign,*

$$\tilde{Y} = (A + \tilde{X})^{-1}(A - \tilde{X}) \Rightarrow$$

$$d\tilde{Y} = 2^{p(p+1)} \left| \det \left((A + \tilde{X})(A + \tilde{X})^* \right) \right|^{-(p+1)}$$

$$\times \left\{ \prod_{j=1}^{p} |a_{jj}|^{2j} \right\} d\tilde{X} \text{ when all elements are complex ;}$$

$$d\tilde{Y} = 2^{p^2} \left\{ \prod_{j=1}^{p} |a_{jj} + x_{jj}|^{-2p} \right\} \left\{ \prod_{j=1}^{p} |a_{jj}|^{2j-1} \right\} d\tilde{X}$$

when the diagonal elements of X and A are all real ; and

$$\tilde{Y} = (A - \tilde{X})(A + \tilde{X})^{-1} \Rightarrow$$

$$d\tilde{Y} = 2^{p(p+1)} \left| \det \left((A + \tilde{X})(A + \tilde{X})^* \right) \right|^{-(p+1)}$$

$$\times \left\{ \prod_{j=1}^{p} |a_{jj}|^{2(p+1-j)} \right\} d\tilde{X} \text{ when all elements are complex ;}$$

$$d\tilde{Y} = 2^{p^2} \left\{ \prod_{j=1}^{p} |a_{jj} + x_{jj}|^{-2p} \right\} \left\{ \prod_{j=1}^{p} |a_{jj}|^{2(p-j)+1} \right\} d\tilde{X}$$

when the diagonal elements of $\tilde{X}$ and A are all real.

Since the Jacobians in the complex case can be worked out by using steps parallel to those in the real case there are not many papers as such on the Jacobians in the complex case. When scalar functions of matrices in the complex case are used in various applied areas the required Jacobians are usually computed along the way. Most of the applications can be found in statistics and engineering problems. Some of the sample references are the following, and more could be found thereon. For engineering applications, see for example Biyari and Lindsey (1991, 1995), Divsalar, Simon and Shahshahani (1990), Huang and Campbell (1991) and Mehta (1967); for applications in physics, see for example Bronk (1965) and Mehta (1967); for applications in statistics, see for example Conradie and Gupta (1987), Conradie and Troskie (1984), Goodman (1963), Hannan

(1970), Hayakawa (1972), Kabe (1966), Khatri (1970), Krishnaiah (1976), Mitra (1970), Shaman (1980), Srivastava (1965), and Wooding (1956).

Exercises

3.2.1   For $p \times p$ matrix $\tilde{X} = \tilde{X}^* > 0$, $0 < \tilde{X} < I$ show that

$$\int_{0 < \tilde{X} < I} |\det(\tilde{X})|^{\alpha - p} |\det(I - \tilde{X})|^{\beta - p} d\tilde{X} = \frac{\tilde{\Gamma}_p(\alpha)\tilde{\Gamma}_p(\beta)}{\tilde{\Gamma}_p(\alpha + \beta)}$$

for $\mathrm{Re}(\alpha) > p - 1$, $\mathrm{Re}(\beta) > p - 1$. (The integrand divided by the right side is the complex type-1 matrix-variate beta density.)

3.2.2   For $p \times p$ matrix $\tilde{X} = \tilde{X}^* > 0$ show that

$$\int_{\tilde{X}} |\det(\tilde{X})|^{\alpha - p} |\det(I + \tilde{X})|^{-(\alpha + \beta)} d\tilde{X} = \frac{\tilde{\Gamma}_p(\alpha)\tilde{\Gamma}_p(\beta)}{\tilde{\Gamma}_p(\alpha + \beta)}$$

for $\mathrm{Re}(\alpha) > p - 1$, $\mathrm{Re}(\beta) > p - 1$. (The integrand divided by the right side is the complex type-2 matrix-variate beta density.)

3.2.3   If $\tilde{X}_1$ and $\tilde{X}_2$ are $p \times p$ hermitian positive definite matrices which are independently distributed with the densities

$$f_j\left(\tilde{X}_j\right) = \frac{|\det(\tilde{X}_j)|^{\alpha_j - p} e^{-\mathrm{tr}(\tilde{X}_j)}}{\tilde{\Gamma}_p(\alpha_j)}$$

for $\tilde{X}_j = \tilde{X}_j^* > 0$, $\mathrm{Re}(\alpha_j) > p - 1$ and $f_j\left(\tilde{X}_j\right) = 0$ elsewhere, $j = 1, 2$, show that (1) $\tilde{Y} = \tilde{X}_1 + \tilde{X}_2$ is complex matrix-variate gamma with the parameter $\alpha_1 + \alpha_2$, (2) $\tilde{U} = \tilde{X}_2^{-\frac{1}{2}} \tilde{X}_1 \tilde{X}_2^{-\frac{1}{2}}$ has a complex type-2 matrix-variate beta density.

3.2.4   Extend Theorems 1.23 and 1.26 of Chapter 1 to the complex case.

3.2.5   Extend Theorems 1.24 and 1.25 of Chapter 1 to the complex case.

3.2.6   If $\tilde{X}$ is a $p \times p$ lower or upper triangular matrix of functionally independent complex variables and if $\tilde{Y} = \tilde{X}^{-1}$ evaluate $d\tilde{Y}$ in terms of $d\tilde{X}$.

3.2.7   Let $\tilde{X} = (\tilde{x}_{jk})$ be a $p \times p$ hermitian positive definite matrix of functionally independent complex variables such that $\tilde{x}_{jj} = 1$, $j = 1, ..., p$. Show that

$$\int_{\tilde{X}} d\tilde{X} = \pi^{\frac{p(p-1)}{2}} \frac{(p-2)!(p-3)!...1!}{[(p-1)!]^{p-1}}.$$

3.2.8   Show that

$$\int_{\tilde{X} > B} |\det(\tilde{X} - B)|^{\nu} e^{-\operatorname{tr}(T\tilde{X})} d\tilde{X} = |\det(\tilde{T})|^{-(\nu+p)} e^{-\operatorname{tr}(TB)} \tilde{\Gamma}_p(\nu + p)$$

for $\operatorname{Re}(\nu) > -1$ where $\tilde{X} = \tilde{X}^* > 0$, $B = B^* > 0$, $T = T^* > 0$, $\tilde{X} - B > 0$, and all $p \times p$ matrices.

3.2.9   For $\tilde{X} = \tilde{X}^* > 0$, $B = B^* > 0$ and $\operatorname{Re}(\alpha) > p - 1$ show that

$$|\det(B)|^{-\alpha} = \frac{1}{\tilde{\Gamma}_p(\alpha)} \int_{\tilde{X} > 0} |\tilde{X}|^{\alpha-p} e^{-\operatorname{tr}(B\tilde{X})} d\tilde{X}$$

where all the matrices are $p \times p$.

3.2.10   For the same conditions as in Exercise 3.2.9 and for $C = C^* > 0$ show that

$$|\det(B)|^{-\alpha} e^{-\operatorname{tr}(CB)} = \frac{1}{\tilde{\Gamma}_p(\alpha)} \int_{\tilde{X} > C} |\det(\tilde{X} - C)|^{\alpha-p} e^{-\operatorname{tr}(B\tilde{X})} d\tilde{X}.$$

3.3   Transformations involving determinants and diagonal matrices

Transformations considered in this section are those where one matrix is written as a function of two or more components. These components may be determinants, diagonal matrices, triangular matrices and so on.

Theorem 3.13   *Let $\tilde{T}$ be a $p \times p$ nonsingular lower or upper triangular matrix of functionally independent complex variables. Let $|\det(\tilde{T})|$ denote the absolute value of its determinant. Then, ignoring the sign,*

$$\tilde{X} = \frac{\tilde{T}}{|\det(\tilde{T})|} \Rightarrow d\tilde{X} = (p-1)|\det(\tilde{T})|^{-p(p+1)} d\tilde{T}.$$

Proof   Interpreting $|\det(\tilde{T})|^{-1}$ as a specific $(p-1)$-th root of $|\det(\tilde{X})|$ one has a one-to-one transformation. Let $\tilde{T}$ be lower triangular, that is,

$$\tilde{T} = (\tilde{t}_{jk}), \ \tilde{t}_{jk} = t_{jk1} + it_{jk2}, \ \tilde{t}_{jk} = 0 \text{ for } j < k$$

where t_{jk1} and t_{jk2} are real variables and $i = \sqrt{-1}$. Then

$$|\det(\tilde{T})| = \left[(t_{111}^2 + t_{112}^2) \cdots (t_{pp1}^2 + t_{pp2}^2) \right]^{\frac{1}{2}}.$$

Note that every element in $\tilde{x}_{jk}$, $j > k$ contains t_{jjm}'s, $m = 1, 2$, $j = 1, ..., p$. Let $\tilde{x}_{jk} = x_{jk1} + ix_{jk2}$. Separate the x_{jkm}'s for $j \neq k$ from the x_{jjm}'s and then consider the matrices of partial derivatives. For $j > k$,

$$\left(\frac{\partial x_{jk1}}{\partial t_{jk1}} \right) = \text{diag} \left(\alpha^{-1}, ..., \alpha^{-1} \right) = \left(\frac{\partial x_{jk2}}{\partial t_{jk2}} \right)$$

where $\alpha = |\det(\tilde{T})|$.

$$\left(\frac{\partial x_{jj1}}{\partial t_{jj1}} \right) = \left(\frac{t_{jj2}^2}{\alpha \left(t_{jj1}^2 + t_{jj2}^2 \right)} \right), \ \left(\frac{\partial x_{jj1}}{\partial t_{jj2}} \right) = \left(\frac{-t_{jj1}t_{jj2}}{\alpha \left(t_{jj1}^2 + t_{jj2}^2 \right)} \right).$$

For $j \neq k$,

$$\left(\frac{\partial x_{jj1}}{\partial t_{kk1}} \right) = \left(\frac{-t_{jj1}t_{kk1}}{\alpha \left(t_{kk1}^2 + t_{kk2}^2 \right)} \right), \ \left(\frac{\partial x_{jj1}}{\partial t_{kk2}} \right) = \left(\frac{-t_{jj1}t_{kk2}}{\alpha \left(t_{kk1}^2 + t_{kk2}^2 \right)} \right).$$

Same types of matrices are obtained for the derivatives of x_{jj2} with respect to t_{jj1}, t_{jj2}, t_{kkm}, $m = 1, 2$. For $j, k = 1, ..., p$, let

$$A = \left(\frac{\partial x_{jj1}}{\partial t_{kk1}} \right), \ B = \left(\frac{\partial x_{jj1}}{\partial t_{kk2}} \right),$$

$$C = \left(\frac{\partial x_{jj2}}{\partial t_{kk1}} \right), \ D = \left(\frac{\partial x_{jj2}}{\partial t_{kk2}} \right).$$

The Jacobian matrix is of the following form

$$\begin{bmatrix} \alpha^{-1}I & 0 & * & * \\ 0 & \alpha^{-1}I & * & * \\ 0 & 0 & A & B \\ 0 & 0 & C & D \end{bmatrix}$$

where $*$ indicates the presence of a nonnull matrix. Hence the determinant is

$$|\alpha^{-1}I| \times |\alpha^{-1}I| \times \begin{vmatrix} A & B \\ C & D \end{vmatrix} = \alpha^{-p(p-1)} \begin{vmatrix} A & B \\ C & D \end{vmatrix}.$$

After taking out α^{-2p} from the determinant on the right, the Jacobian becomes

$$J = \alpha^{-p(p+1)} \begin{vmatrix} A_1 & B_1 \\ C_1 & D_1 \end{vmatrix}$$

where A_1, B_1, C_1, D_1 are respectively A, B, C, D with α removed from the denominators of every element. By taking out $(t_{jj1}^2 + t_{jj2}^2)$, $j = 1, ..., p$ from the denominators and t_{jjm}, $j = 1, ..., p$, $m = 1, 2$ from the various rows and columns we have

$$\begin{vmatrix} A_1 & B_1 \\ C_1 & D_1 \end{vmatrix} = \alpha^{-4} \left\{ \prod_{j=1}^{p} t_{jj1}^2 t_{jj2}^2 \right\} \begin{vmatrix} A_2 & B_2 \\ C_2 & D_2 \end{vmatrix}$$

where B_2 and C_2 have all the elements equal to -1, A_2 and D_2 have all the nondiagonal elements equal to -1, A_2 has the diagonal elements t_{jj2}^2 / t_{jj1}^2, $j = 1, ..., p$ and D_2 has the diagonal elements t_{jj1}^2 / t_{jj2}^2, $j = 1, ..., p$. By adding -1 times the first column to all other columns and then taking out

$$\prod_{j=1}^{p} (t_{jj1}^2 + t_{jj2}^2)^2 \quad \text{and} \quad \prod_{j=1}^{p} (t_{jj1}^{-2} t_{jj2}^{-2}),$$

one has a determinant of the form

$$\begin{vmatrix} A_3 & B_3 \\ C_3 & D_3 \end{vmatrix}, \quad D_3 = I_{2p-1}, \quad A_3 = \frac{t_{112}^2}{t_{111}^2 + t_{112}^2}, \quad B_3 = (-1, ..., -1)$$

and C_3 has one element the same as that of A_3, the remaining elements being $t_{jjm}^2 / (t_{jj1}^2 + t_{jj2}^2)$, $j = 2, ..., p$, $m = 1, 2$. Evaluating the determinant by using the formula $|D_3| |A_3 - B_3 D_3^{-1} C_3|$, one gets $-(p-1)$. The procedure is the same for the upper triangular case.

Theorem 3.14 Let $\tilde{T}$ be a $p \times p$ nonsingular lower or upper triangular matrix of functionally independent complex variables. Let $|\det(\tilde{T})|$ denote the absolute value of its determinant. Then, ignoring the sign,

$$\tilde{X} = |\det(\tilde{T})| \tilde{T}^{-1} \Rightarrow d\tilde{X} = (p-1) |\det(\tilde{T})|^{(p+1)(p-2)} d\tilde{T}.$$

Proof Interpreting $|\det(\tilde{T})|$ as a specific $(p-1)$-th root of $|\det(\tilde{X})|$ one has a one-to-one transformation. Let $\tilde{U} = (\tilde{u}_{jk}) = \frac{\tilde{T}}{|\det(\tilde{T})|}$. Then $\tilde{X} = \tilde{U}^{-1}$.

From (3.2.1) and Theorem 3.3

$$d\tilde{X} = \left\{ \prod_{j=1}^{p} |\tilde{u}_{jj}|^{-2(p+1)} \right\} d\tilde{U}.$$

Substituting for $d\tilde{U}$ and $|\tilde{u}_{jj}|$ from Theorem 3.13 the result follows.

Example 3.10   Verify Theorem 3.14 for a 2×2 matrix.

Solution   Let

$$\tilde{T} = \begin{bmatrix} x_1 + ix_2 & 0 \\ y_1 + iy_2 & z_1 + iz_2 \end{bmatrix}$$

where x_1, x_2, y_1, y_2, z_1, z_2 are real and $i = \sqrt{-1}$. Then $\det(\tilde{T}) = (x_1 + ix_2)(z_1 + iz_2)$ and $|\det(\tilde{T})| = \left[(x_1^2 + x_2^2)(z_1^2 + z_2^2) \right]^{\frac{1}{2}}$.

$$\tilde{T}^{-1} = \begin{bmatrix} \frac{1}{x_1+ix_2} & 0 \\ \frac{-(y_1+iy_2)}{(x_1+ix_2)(z_1+iz_2)} & \frac{1}{z_1+iz_2} \end{bmatrix}$$

$$= \begin{bmatrix} \frac{x_1-ix_2}{x_1^2+x_2^2} & 0 \\ \frac{-(y_1+iy_2)(x_1-ix_2)(z_1-iz_2)}{(x_1^2+x_2^2)(z_1^2+z_2^2)} & \frac{z_1-iz_2}{z_1^2+z_2^2} \end{bmatrix}.$$

Hence

$$\tilde{U} = \begin{bmatrix} u_1 + iv_1 & 0 \\ u_2 + iv_2 & u_3 + iv_3 \end{bmatrix} = |\det(\tilde{T})|\tilde{T}^{-1} \Rightarrow$$

$$u_1 = \frac{x_1(z_1^2 + z_2^2)^{1/2}}{(x_1^2 + x_2^2)^{1/2}}, \quad v_1 = -\frac{x_2(z_1^2 + z_2^2)^{1/2}}{(x_1^2 + x_2^2)^{1/2}}$$

$$u_2 = -\frac{y_1(x_1z_1 - x_2z_2) + y_2(x_2z_1 + x_1z_2)}{(x_1^2 + x_2^2)^{1/2}(z_1^2 + z_2^2)^{1/2}}$$

$$v_2 = -\frac{y_2(x_1z_1 - x_2z_2) - y_1(x_2z_1 + x_1z_2)}{(x_1^2 + x_2^2)^{1/2}(z_1^2 + z_2^2)^{1/2}}$$

$$u_3 = \frac{z_1(x_1^2 + x_2^2)^{1/2}}{(z_1^2 + z_2^2)^{1/2}}, \quad v_3 = -\frac{z_2(x_1^2 + x_2^2)^{1/2}}{(z_1^2 + z_2^2)^{1/2}}.$$

Consider the matrix of partial derivatives

$$\frac{\partial(u_2, v_2, u_1, u_3, v_1, v_3)}{\partial(y_1, y_2, x_1, z_1, x_2, z_2)}.$$

Since

$$\frac{\partial(u_1, u_3)}{\partial(y_1, y_2)} = 0, \quad \frac{\partial(v_1, v_3)}{\partial(y_1, y_2)} = 0$$

we need not evaluate the derivatives of (u_2, v_2) with respect to x_1, z_1, x_2, z_2. Thus the determinant of the Jacobian matrix reduces to

$$\left| \frac{\partial(u_2, v_2)}{\partial(y_1, y_2)} \right| \times \left| \frac{\partial(u_1, u_3, v_1, v_3)}{\partial(x_1, z_1, x_2, z_2)} \right|$$

of which, evidently, $\left| \frac{\partial(u_2, v_2)}{\partial(y_1, y_2)} \right| = 1$. Consider the determinant $\left| \frac{\partial(u_1, u_3, v_1, v_3)}{\partial(x_1, z_1, x_2, z_2)} \right|$.
Take out $|\det(\tilde{T})|$ from each denominator to get $|\det(\tilde{T})|^{-4}$. By rearranging rows and columns the remaining determinant reduces to the form

$$(x_1^2 + x_2^2)^2 (z_1^2 + z_2^2)^2 \begin{vmatrix} z_1^2 + z_2^2 & -1 & 0 & 0 \\ 0 & 0 & -1 & 0 \\ 0 & 0 & 1 & -1 \\ -1 & 0 & 0 & 0 \end{vmatrix} = -(x_1^2 + x_2^2)^2 (z_1^2 + z_2^2)^2.$$

Thus the absolute value of the Jacobian is unity. From the formula

$$(p - 1)|\det(\tilde{T})|^{(p+1)(p-2)} = 1.$$

Thus the theorem is verified.

Lemma 3.3   Let $\tilde{X} = (\tilde{x}_{jk}) = (x_{jk1} + ix_{jk2}) = X_1 + iX_2$, $i = \sqrt{-1}$ be a $p \times p$ nonsingular matrix of functionally independent complex variables. Let $|\tilde{X}|$, $|\tilde{X}_{jk}|$, $\tilde{X}^{-1}$ denote the determinant, cofactor of $\tilde{x}_{jk}$ and inverse respectively. Then

$$d|\tilde{X}| = \text{tr} \left[(|\tilde{X}_{jk}|)'(d\tilde{X}) \right] = |\tilde{X}| \text{tr} \left[\tilde{X}^{-1}(d\tilde{X}) \right]$$

where $(|\tilde{X}_{jk}|)$ and $(d\tilde{X})$ denote the matrices of cofactors and differentials respectively.

Proof   The determinant can be expanded as

$$|\tilde{X}| = \tilde{x}_{j1}|\tilde{X}_{j1}| + ... + \tilde{x}_{jp}|\tilde{X}_{jp}|$$

for any j with

$$0 = \tilde{x}_{k1}|\tilde{X}_{j1}| + ... + \tilde{x}_{kp}|\tilde{X}_{jp}|$$

for $j \neq k$. If all the elements in $\tilde{X}$ are functionally independent then

$$\frac{\partial |\tilde{X}|}{\partial x_{jk1}} = |\tilde{X}_{jk}| \text{ and } \frac{\partial |\tilde{X}|}{\partial x_{jk2}} = i|\tilde{X}_{jk}|.$$

Thus

$$d|\tilde{X}| = \sum_{j=1}^{p}\sum_{k=1}^{p} |\tilde{X}_{jk}|(dx_{jk1} + idx_{jk2})$$

$$= \sum_{j=1}^{p}\sum_{k=1}^{p} |\tilde{X}_{jk}|d\tilde{x}_{jk}$$

$$= \text{tr}\left[(|\tilde{X}_{jk}|)'(d\tilde{X})\right] = |\tilde{X}|\text{tr}\left[\tilde{X}^{-1}(d\tilde{X})\right]$$

observing that $(|\tilde{X}_{jk}|)'$ is the adjoint of $\tilde{X}$ which is $|\tilde{X}|\tilde{X}^{-1}$. When $\tilde{X}$ is a hermitian matrix, X_1 is symmetric and X_2 is skew symmetric. That is, $x_{jk1} = x_{kj1}$ for all j and k, $x_{jk2} = -x_{kj2}$, $j \neq k$ and $x_{jj2} = 0$. Then

$$\frac{\partial |\tilde{X}|}{\partial x_{jk1}} = \begin{cases} |\tilde{X}_{jk}| + |\tilde{X}_{kj}|, & j \neq k \\ |\tilde{X}_{jj}|, & j = k, \end{cases}$$

$$\frac{\partial |\tilde{X}|}{\partial x_{jk2}} = \begin{cases} i|\tilde{X}_{jk}| - i|\tilde{X}_{kj}|, & j \neq k \\ 0, & j = k. \end{cases}$$

Thus

$$d|\tilde{X}| = \sum_{j=1}^{p} |\tilde{X}_{jj}|dx_{jj1} + \sum_{j>k=1}^{p}\left\{|\tilde{X}_{jk}| + |\tilde{X}_{kj}|\right\}dx_{jk1}$$

$$+ i\sum_{j>k=1}^{p}\left\{|\tilde{X}_{jk}| - |\tilde{X}_{kj}|\right\}dx_{jk2}$$

$$= \text{tr}\left[(|\tilde{X}_{jk}|)'(dX_1)\right] + i\text{tr}\left[(|\tilde{X}_{jk}|)'(dX_2)\right]$$

$$= \text{tr}\left[(|\tilde{X}_{jk}|)'(d\tilde{X})\right] = |\tilde{X}|\text{tr}\left[\tilde{X}^{-1}(d\tilde{X})\right].$$

Lemma 3.4 *Let $\tilde{W} = (\tilde{w}_{jk}) = (w_{jk1} + iw_{jk2})$ and $\tilde{U} = (\tilde{u}_{jk}) = (u_{jk1} + iu_{jk2})$ be $p \times p$ matrices of functionally independent complex variables.*

Then, ignoring the sign,

$$\tilde{W} = [\text{tr}(\tilde{U})]I - \tilde{U} \Rightarrow$$
$$d\tilde{W} = (p-1)^2 d\tilde{U} \text{ for a general } \tilde{U}$$

and

$$d\tilde{W} = (p-1)d\tilde{U} \text{ for } \tilde{U} = \tilde{U}^* \text{ or } \tilde{U} = -\tilde{U}^*.$$

Proof $w_{jj1} = \sum_{k\neq j=1}^{p} u_{kk1}$ and $w_{jj2} = \sum_{k\neq j=1}^{p} u_{kk2}$, $w_{jk1} = -u_{jk1}$, $w_{jk2} = -u_{jk2}$ for $j \neq k$.

$$\left(\frac{\partial w_{jkl}, j \neq k}{\partial u_{rkl}, j \neq k} \right) = \text{diag}(-1, ..., -1), \ l = 1, 2.$$

$$\left(\frac{\partial w_{jjl}}{\partial u_{jjl}} \right) = \begin{pmatrix} 0 & 1 & \cdots & 1 \\ 1 & 0 & \cdots & 1 \\ \vdots & \vdots & \ddots & \vdots \\ 1 & 1 & \cdots & 0 \end{pmatrix}, \ l = 1, 2.$$

For the general case the Jacobian matrix is of the form

$$\begin{bmatrix} A & 0 & 0 & 0 \\ 0 & A & 0 & 0 \\ 0 & 0 & -I & 0 \\ 0 & 0 & 0 & -I \end{bmatrix}, \quad A = \begin{bmatrix} 0 & 1 & \cdots & 1 \\ 1 & 0 & \cdots & 1 \\ \vdots & \vdots & \ddots & \vdots \\ 1 & 1 & \cdots & 0 \end{bmatrix}.$$

The determinant in absolute value is $|A|^2 = (p-1)^2$. In the hermitian or skew hermitian case only one A will appear and hence the determinant in absolute value is $(p-1)$. Hence the result.

Theorem 3.15 *Let $\tilde{X}$ be a $p \times p$ nonsingular matrix of functionally independent complex variables. Let $|\tilde{X}|$, $\tilde{X}^{-1}$ and $|\det(\tilde{X})|$ denote the determinant, inverse and absolute value of the determinant respectively. Then, ignoring the sign,*

$$\tilde{Y} = |\tilde{X}|\tilde{X}^{-1} \Rightarrow d\tilde{Y} = (p-1)^2 |\det(\tilde{X}\tilde{X}^*)|^{p(p-2)} d\tilde{X}$$

and

$$\tilde{Z} = \frac{\tilde{X}}{|\tilde{X}|} \Rightarrow d\tilde{Z} = (p-1)^2 |\det(\tilde{X}\tilde{X}^*)|^{-p^2} d\tilde{X}.$$

Proof   Take the differentials to get

$$
\begin{aligned}
(\mathrm{d}\tilde{Y}) &= (\mathrm{d}|\tilde{X}|)\tilde{X}^{-1} + |\tilde{X}|(\mathrm{d}\tilde{X}^{-1}) \\
&= (\mathrm{d}|\tilde{X}|)\tilde{X}^{-1} - |\tilde{X}|\tilde{X}^{-1}(\mathrm{d}\tilde{X})\tilde{X}^{-1} \\
&= |\tilde{X}|\left[\left\{\operatorname{tr}(\mathrm{d}\tilde{V})\right\}I - (\mathrm{d}\tilde{V})\right]\tilde{X}^{-1}
\end{aligned}
$$

where

$$
(\mathrm{d}\tilde{V}) = \tilde{X}^{-1}(\mathrm{d}\tilde{X}) \Rightarrow \mathrm{d}\tilde{V} = |\det(\tilde{X}\tilde{X}^*)|^{-p}\mathrm{d}\tilde{X}.
$$

Let

$$
\begin{aligned}
(\mathrm{d}\tilde{U}) &= (\mathrm{d}\tilde{Y})\tilde{X} = |\tilde{X}|(\mathrm{d}\tilde{W}), \\
(\mathrm{d}\tilde{W}) &= \left\{\operatorname{tr}(\mathrm{d}\tilde{V})\right\}I - (\mathrm{d}\tilde{V}).
\end{aligned}
$$

From Lemma 3.4

$$
\begin{aligned}
\mathrm{d}\tilde{W} &= (p-1)^2\mathrm{d}\tilde{V} = (p-1)^2|\det(\tilde{X}\tilde{X}^*)|^{-p}\mathrm{d}\tilde{X}, \\
\mathrm{d}\tilde{U} &= |\det(\tilde{X}\tilde{X}^*)|^p\mathrm{d}\tilde{Y} = |\det(\tilde{X}\tilde{X}^*)|^{p^2}\mathrm{d}\tilde{W}
\end{aligned}
$$

which means

$$
\mathrm{d}\tilde{Y} = |\det(\tilde{X}\tilde{X}^*)|^{-2p+p^2}\mathrm{d}\tilde{X}.
$$

This establishes the first part. Let

$$
\tilde{Z} = \tilde{Y}^{-1} = \frac{\tilde{X}}{|\tilde{X}|} \Rightarrow
$$
$$
\mathrm{d}\tilde{Z} = |\det(\tilde{Y}\tilde{Y}^*)|^{-2p}\mathrm{d}\tilde{Y}
$$

from Theorem 3.8. That is,

$$
\mathrm{d}\tilde{Z} = |\det(\tilde{Y}\tilde{Y}^*)|^{-2p}(p-1)^2|\det(\tilde{X}\tilde{X}^*)|^{p(p-2)}\mathrm{d}\tilde{X}.
$$

But $|\det(\tilde{Y}\tilde{Y}^*)| = |\det(\tilde{X}\tilde{X}^*)|^{p-1}$. Substituting this the result follows.

Theorem 3.16   *Let $\tilde{X}$ be a $p \times p$ hermitian positive definite matrix of functionally independent complex variables. Then, ignoring the sign,*

$$
\tilde{Y} = |\tilde{X}|\tilde{X}^{-1} \Rightarrow \mathrm{d}\tilde{Y} = (p-1)|\det(\tilde{X}\tilde{X}^*)|^{\frac{1}{2}p(p-2)}\mathrm{d}\tilde{X}
$$

and

$$
\tilde{Z} = \frac{\tilde{X}}{|\tilde{X}|} \Rightarrow \mathrm{d}\tilde{Z} = (p-1)|\det(\tilde{X}\tilde{X}^*)|^{-\frac{p^2}{2}}\mathrm{d}\tilde{X}.
$$

Proof

$$\tilde{Y} = |\tilde{X}|\tilde{X}^{-1} \Rightarrow$$
$$(d\tilde{Y}) = |\tilde{X}| \left\{ \left[\text{tr} \left(\tilde{X}^{-1}(d\tilde{X}) \right) \right] \tilde{X}^{-1} - \tilde{X}^{-1}(d\tilde{X})\tilde{X}^{-1} \right\}.$$

Since $\text{tr}(\cdot)$ is scalar, $\text{tr}(AB) = \text{tr}(BA)$, $[\text{tr}(\cdot)]AB = A[\text{tr}(\cdot)]B$ and since there exists a nonsingular hermitian matrix Q such that $\tilde{X} = Q^2$, this Q is written as $\tilde{X}^{\frac{1}{2}}$, we may write

$$\text{tr} \left(\tilde{X}^{-1}(d\tilde{X}) \right) = \text{tr} \left(\tilde{X}^{-\frac{1}{2}}(d\tilde{X})\tilde{X}^{-\frac{1}{2}} \right)$$

and

$$\left[\text{tr} \left(\tilde{X}^{-1}(d\tilde{X}) \right) \right] \tilde{X}^{-1} = \tilde{X}^{-\frac{1}{2}} \left[\text{tr} \left(\tilde{X}^{-\frac{1}{2}}(d\tilde{X})\tilde{X}^{-\frac{1}{2}} \right) \right] \tilde{X}^{-\frac{1}{2}}.$$

Hence

$$\tilde{X}^{\frac{1}{2}}(d\tilde{Y})\tilde{X}^{\frac{1}{2}} = |\tilde{X}| \left\{ \left[\text{tr}(d\tilde{U}) \right] I - (d\tilde{U}) \right\}$$

where

$$(d\tilde{U}) = \tilde{X}^{-\frac{1}{2}}(d\tilde{X})\tilde{X}^{-\frac{1}{2}}.$$

From Theorem 3.5 and Lemma 3.2 we have

$$d\tilde{U} = |\det(\tilde{X}\tilde{X}^*)|^{-\frac{p}{2}}d\tilde{X}$$

and from Lemma 3.4

$$(d\tilde{V}) = [\text{tr}(d\tilde{U})]I - (d\tilde{U}) \Rightarrow$$
$$d\tilde{V} = (p-1)d\tilde{U} = (p-1)|\det(\tilde{X}\tilde{X}^*)|^{-\frac{p}{2}}d\tilde{X}.$$

Let

$$(d\tilde{W}) = |\tilde{X}|(d\tilde{V}) \Rightarrow$$
$$d\tilde{W} = |\det(\tilde{X}\tilde{X}^*)|^{\frac{p^2}{2}}d\tilde{V} = (p-1)|\det(\tilde{X}\tilde{X}^*)|^{\frac{p(p-1)}{2}}$$

since there are only p^2 variables in the hermitian matrix $(d\tilde{V})$. But

$$(d\tilde{Y}) = \tilde{X}^{-\frac{1}{2}}(d\tilde{W})\tilde{X}^{-\frac{1}{2}} \Rightarrow$$
$$d\tilde{Y} = |\det(\tilde{X}\tilde{X}^*)|^{-\frac{p}{2}}d\tilde{W}$$
$$= (p-1)|\det(\tilde{X}\tilde{X}^*)|^{\frac{p(p-2)}{2}}d\tilde{X}.$$

For proving the second part write

$$\tilde{Z} = \tilde{Y}^{-1} \Rightarrow (d\tilde{Z}) = -\tilde{Y}^{-1}(d\tilde{Y})\tilde{Y}^{-1}$$
$$\Rightarrow d\tilde{Z} = |\det(\tilde{Y}\tilde{Y}^*)|^{-p}d\tilde{Y}$$

ignoring the sign. But

$$|\det(\tilde{Y}\tilde{Y}^*)| = |\det(\tilde{X}\tilde{X}^*)|^{p-1}$$

and

$$d\tilde{Z} = |\det(\tilde{X}\tilde{X}^*)|^{-p(p-1)}d\tilde{Y}.$$

Now substituting for $d\tilde{Y}$ the result follows.

Example 3.11 For the $p \times p$ matrix $\tilde{X} = \tilde{X}^* > 0$ of functionally independent complex variables evaluate the integral

$$A(\alpha) = \int_{\tilde{X} = \tilde{X}^* > 0} |\det(\tilde{X})|^{-\alpha} e^{-\mathrm{tr}(|\tilde{X}|^{-1}\tilde{X})} d\tilde{X}.$$

Solution Let $\tilde{Y} = |\tilde{X}|^{-1}\tilde{X}$. Then from Theorem 3.16, ignoring the sign,

$$d\tilde{Y} = (p-1)|\det(\tilde{X}\tilde{X}^*)|^{-\frac{p^2}{2}}d\tilde{X}$$

and

$$|\tilde{Y}| = |\tilde{X}|^{-(p-1)} \quad \text{or} \quad |\det(\tilde{X}\tilde{X}^*)|^{-(p-1)} = |\det(\tilde{Y}\tilde{Y}^*)|.$$

Take $|\det(\tilde{X}\tilde{X}^*)|^{-1}$ as a specific $(p-1)$-th root and write

$$|\det(\tilde{X}\tilde{X}^*)|^{-1} = |\det(\tilde{Y}\tilde{Y}^*)|^{\frac{1}{p-1}}.$$

Now

$$d\tilde{X} = \frac{|\det(\tilde{X}\tilde{X}^*)|^{\frac{p^2}{2}}}{p-1}d\tilde{Y} = \frac{|\det(\tilde{Y}\tilde{Y}^*)|^{-\frac{p^2}{2(p-1)}}}{p-1}d\tilde{Y}.$$

Then

$$(p-1)A(\alpha) = \int_{\tilde{X}} |\det(\tilde{X})|^{-\alpha} e^{-\mathrm{tr}(|\tilde{X}|^{-1}\tilde{X})} d\tilde{X}$$

$$= \int_{\tilde{X}} |\det(\tilde{X}\tilde{X}^*)|^{-\frac{\alpha}{2}} e^{-\mathrm{tr}(|\tilde{X}|^{-1}\tilde{X})} d\tilde{X}$$

$$= \int_{\tilde{Y} = \tilde{Y}^* > 0} |\det(\tilde{Y})|^{\frac{\alpha-p}{p-1}-p} e^{-\mathrm{tr}(\tilde{Y})} d\tilde{Y}$$

$$= \tilde{\Gamma}_p\left(\frac{\alpha-p}{p-1}\right) \quad \text{for } \mathrm{Re}\left(\frac{\alpha-p}{p-1}\right) > p-1$$

or for $\mathrm{Re}(\alpha) > p^2 - p + 1$, evaluating the integral by using the triangular decomposition in Theorem 3.7.

Theorem 3.17   *Let $\tilde{X} = (\tilde{x}_{jk})$ be a $p \times p$ nonsingular matrix of functionally independent complex variables. Let $\lambda_1, ..., \lambda_p$ be the distinct and nonzero eigenvalues of $\tilde{X}$ such that $\lambda_j + \lambda_k \neq 0$ for all j and k. Then, ignoring the sign,*

$$\tilde{Y} = \tilde{X}^2 \Rightarrow$$

$$d\tilde{Y} = \left\{ \prod_{j=1}^{p} \prod_{k=1}^{p} |\lambda_j + \lambda_k|^2 \right\} d\tilde{X}$$

$$= 2^{2p} |\det(\tilde{X}\tilde{X}^*)| \left\{ \prod_{j>k} |\lambda_j + \lambda_k|^4 \right\} d\tilde{X}$$

for a general $\tilde{X}$ and

$$d\tilde{Y} = 2^p |\det(\tilde{X})| \left\{ \prod_{j>k=1}^{p} |\lambda_j + \lambda_k|^2 \right\} d\tilde{X}$$

for $\tilde{X} = \tilde{X}^*$.

Proof $\tilde{Y} = \tilde{X}^2 = \tilde{X}\tilde{X}$. Taking the differentials one has

$$(d\tilde{\mathbf{Y}}) = \tilde{X}(d\tilde{\mathbf{X}}) + (d\tilde{\mathbf{X}})\tilde{X}. \tag{a}$$

Let $\tilde{X} = X_1 + iX_2$, $\tilde{Y} = Y_1 + iY_2$ where X_1, X_2, Y_1, Y_2 are real, $i = \sqrt{-1}$. Then

$$(d\mathbf{Y_1}) = X_1(d\mathbf{X_1}) - X_2(d\mathbf{X_2}) + (d\mathbf{X_1})X_1 - (d\mathbf{X_2})X_2$$

and

$$(d\mathbf{Y_2}) = X_2(d\mathbf{X_1}) + X_1(d\mathbf{X_2}) + (d\mathbf{X_1})X_2 + (d\mathbf{X_2})X_1.$$

For a general $\tilde{X}$, let $U_1 = \text{vec}(d\mathbf{X_1})$, $U_2 = \text{vec}(d\mathbf{X_2})$, $V_1 = \text{vec}(d\mathbf{Y_1})$ and $V_2 = \text{vec}(d\mathbf{Y_2})$. Then

$$V_1 = AU_1 - BU_2 \text{ and } V_2 = BU_1 + AU_2$$

where

$$A = X_1' \otimes I + I \otimes X_1 \text{ and } B = X_2' \otimes I + I \otimes X_2.$$

Then

$$\frac{\partial(V_1, V_2)}{\partial(U_1, U_2)} = \begin{bmatrix} A & -B \\ B & A \end{bmatrix}.$$

By Lemma 3.1 the determinant is given by

$$\left| \frac{\partial(V_1, V_2)}{\partial(U_1, U_2)} \right| = \left| \begin{matrix} A & -B \\ B & A \end{matrix} \right| = |\det(\tilde{C}\tilde{C}^*)|$$

where $\tilde{C} = A + iB = \tilde{X}' \otimes I + I \otimes \tilde{X}$, $i = \sqrt{-1}$. Following the steps in the real case one has

$$\det(\tilde{C}) = \prod_{j=1}^{p} \prod_{k=1}^{p} (\lambda_j + \lambda_k)$$

$$= 2^p \det(\tilde{X}) \prod_{j>k} (\lambda_j + \lambda_k)^2.$$

Hence

$$|\det(\tilde{C}\tilde{C}^*)| = 2^{2p} |\det(\tilde{X}\tilde{X}^*)| \prod_{j>k} |\lambda_j + \lambda_k|^4.$$

We shall prove the hermitian case through an alternate procedure. When $\tilde{X} = \tilde{X}^*$, let P be a unitary matrix such that

$$P^* \tilde{X} P = \text{diag}(\lambda_1, ..., \lambda_p) = D.$$

Then from equation (a), observing that $D^* = D$ and $Z^* = Z$, one has

$$(d\tilde{W}) = D(d\tilde{Z}) + (d\tilde{Z})D \tag{b}$$

where

$$(d\tilde{W}) = P^*(d\tilde{Y})P, \quad (d\tilde{Z}) = P^*(d\tilde{X})P.$$

Since P is unitary $d\tilde{W} = d\tilde{Y}$ and $d\tilde{Z} = d\tilde{X}$, ignoring the sign. But from (b) one has

$$(d\tilde{W}) = (d\tilde{w}_{jk}) = ((\lambda_j + \lambda_k)d\tilde{z}_{jk}).$$

When $\tilde{X} = \tilde{X}^*$ we have the diagonal elements real and hence, ignoring the sign,

$$\prod_{j=1}^{p} dw_{jj} = \prod_{j=1}^{p} |2\lambda_j| dz_{jj}$$

$$= 2^p |\lambda_1 ... \lambda_p| \prod_{j=1}^{p} dz_{jj}$$

$$= 2^p |\det(\tilde{X})| \prod_{j=1}^{p} dz_{jj}.$$

For $j > k$,

$$\prod_{j>k=1}^{p} d\tilde{w}_{jk} = \left\{ \prod_{j>k=1}^{p} |\lambda_j + \lambda_k|^2 \right\} \left\{ \prod_{j>k=1}^{p} d\tilde{z}_{jk} \right\}.$$

Thus, ignoring the sign,

$$d\tilde{Y} = 2^p |\det(\tilde{X})| \left\{ \prod_{j>k=1}^{p} |\lambda_j + \lambda_k|^2 \right\} d\tilde{X}$$

since $d\tilde{W} = d\tilde{Y}$ and $d\tilde{Z} = d\tilde{X}$.

Remark 3.6 Note that the transformation $\tilde{Y} = \tilde{X}^2$ is not one-to-one. Hence $\tilde{X} = \tilde{Y}^{\frac{1}{2}}$ should be interpreted as a preselected $\tilde{Y}^{\frac{1}{2}}$ such that $\tilde{X}^2 = \tilde{Y}^{\frac{1}{2}} \tilde{Y}^{\frac{1}{2}}$.

Theorem 3.18 *Let $\tilde{X}$ and $\tilde{W}$ be $p \times p$ lower triangular matrices of functionally independent complex variables, $\tilde{W}$ with unit diagonal elements and let $\tilde{D}$ be a diagonal matrix with distinct nonzero diagonal elements $\tilde{\lambda}_1, ..., \tilde{\lambda}_p$. Then*

$$\tilde{X} = \tilde{D}\tilde{W} \Rightarrow d\tilde{X} = \left\{ \prod_{j=1}^{p} |\tilde{\lambda}_j|^{2(j-1)} \right\} d\tilde{D} \, d\tilde{W}$$

and

$$\tilde{X} = \tilde{W}\tilde{D} \Rightarrow d\tilde{X} = \left\{ \prod_{j=1}^{p} |\tilde{\lambda}_j|^{2(p-j)} \right\} d\tilde{D} \, d\tilde{W}.$$

Proof Obviously the transformations are one-to-one. When $\tilde{W}$ is premultiplied by $\tilde{D}$ the j-th row of $\tilde{W}$ is multiplied by $\tilde{\lambda}_j$, $j = 1, ..., p$. Hence $\tilde{X} = \tilde{D}\tilde{W} \Rightarrow \tilde{x}_{jk} = \tilde{\lambda}_j \tilde{w}_{jk}$, $k = 1, ..., j-1$ and $\tilde{x}_{jj} = \tilde{\lambda}_j$, $j = 1, ..., p$. That is,

$$\frac{\partial \tilde{x}_{jk}}{\partial \tilde{w}_{jk}} = |\tilde{\lambda}_j|^2, \ k = 1, ..., j-1 \text{ and } \frac{\partial \tilde{x}_{jj}}{\partial \tilde{\lambda}_j} = 1, \ j = 1, ..., p.$$

Hence the Jacobian is the product of these elements, that is,

$$d\tilde{X} = \left\{ \prod_{j=1}^{p} |\tilde{\lambda}_j|^{2(j-1)} \right\} d\tilde{D} \, d\tilde{W}.$$

For proving the second part note that when $\tilde{W}$ is postmultiplied by $\tilde{D}$ the j-th column of $\tilde{W}$ is multiplied by $\tilde{\lambda}_j$, $j = 1, ..., p$. As before, the diagonal elements give the derivatives unities, $|\tilde{\lambda}_1|^2$ appears $p-1$ times, $|\tilde{\lambda}_2|^2$ appears $p - 2$ times and so on. Hence the result.

When the matrices $\tilde{X}$ and $\tilde{W}$ are upper triangular the roles are reversed and hence we have the following theorem.

Theorem 3.19   Let $\tilde{X}$ and $\tilde{W}$ be $p \times p$ upper triangular matrices of functionally independent complex variables, $\tilde{W}$ with unit diagonal elements. Let $\tilde{D}$ be a diagonal matrix with functionally independent nonzero diagonal elements $\tilde{\lambda}_1, ..., \tilde{\lambda}_p$. Then

$$\tilde{X} = \tilde{D}\tilde{W} \Rightarrow d\tilde{X} = \left\{ \prod_{j=1}^{p} |\tilde{\lambda}_j|^{2(p-j)} \right\} d\tilde{D} \, d\tilde{W}$$

and

$$\tilde{X} = \tilde{W}\tilde{D} \Rightarrow d\tilde{X} = \left\{ \prod_{j=1}^{p} |\tilde{\lambda}_j|^{2(j-1)} \right\} d\tilde{D} \, d\tilde{W}.$$

Remark 3.7   Results in Theorems 3.18 and 3.19 remain the same when the $\tilde{x}_{jj}$, $\tilde{\lambda}_j$, $j = 1, ..., p$ are all real distinct and positive and $\tilde{x}_{jk}$, $\tilde{w}_{jk}$, $j > k$ are all complex.

Example 3.12   Let $\tilde{X} = \tilde{X}^* > 0$ be a $p \times p$ matrix of functionally independent complex random variables having a complex matrix-variate gamma density given by

$$f(\tilde{X}) = \begin{cases} \dfrac{|\det(\tilde{X})|^{\alpha-p} e^{-\text{tr}(\tilde{X})}}{\Gamma_p(\alpha)}, \tilde{X} = \tilde{X}^* > 0, \; \text{Re}(\alpha) > p - 1 \\ \\ 0, \; \text{elsewhere.} \end{cases}$$

Let $\tilde{X} = D\tilde{W}\tilde{W}^* D^*$ where D is a diagonal matrix with real, distinct and positive diagonal elements $\lambda_1, ..., \lambda_p$ and $\tilde{W}$ is a lower triangular matrix with unit diagonal elements. Show that $S_j = \{\tilde{w}_{jk}, k = 1, ..., j-1\}$, $j = 2, ..., p$ are mutually independently distributed.

Solution   Let $\tilde{X} = \tilde{T}\tilde{T}^*$ where $\tilde{T}$ is a lower triangular matrix with real and positive diagonal elements $t_{jj} > 0$, $j = 1, ..., p$. Then from Theorem

3.7
$$d\tilde{X} = 2^p \left\{ \prod_{j=1}^p t_{jj}^{2(p-j)+1} \right\} d\tilde{T}.$$

Note that

$$|\det(\tilde{X})|^{\alpha-p} d\tilde{X} = \left[\prod_{j=1}^p t_{jj}^{2(\alpha-p)} \right] \left[2^p \prod_{j=1}^p t_{jj}^{2(p-j)+1} d\tilde{T} \right]$$

$$= 2^p \left[\prod_{j=1}^p t_{jj}^{2(\alpha-j)+1} \right] d\tilde{T}.$$

Hence the density of $\tilde{T}$, denoted by $g(\tilde{T})$, is given by

$$g(\tilde{T}) d\tilde{T} = \frac{2^p}{\tilde{\Gamma}_p(\alpha)} \left\{ \prod_{j=1}^p t_{jj}^{2(\alpha-j)+1} \right\} e^{-\left\{ \sum_{j \geq k} |\tilde{t}_{jk}|^2 \right\}} d\tilde{T}.$$

Now consider a transformation as in Theorem 3.18 with the λ_j's real and positive. Then

$$\tilde{T} = D\tilde{W} \Rightarrow d\tilde{T} = \left\{ \prod_{j=1}^p \lambda_j^{2(j-1)} \right\} dD \, d\tilde{W}.$$

Note that $\lambda_j = t_{jj}$ and $w_{jj} = 1$.

$$\sum_{j \geq k} |\tilde{t}_{jk}|^2 = \operatorname{tr}(\tilde{T}\tilde{T}^*)$$

$$= \operatorname{tr}(D\tilde{W}\tilde{W}^* D)$$

$$= \sum_{j=1}^p \lambda_j^2 \left(1 + |\tilde{w}_{j1}|^2 + \dots + |\tilde{w}_{j\,j-1}|^2 \right).$$

If the joint density of D and $\tilde{W}$ is denoted by $g_1(D, \tilde{W})$ then

$$g_1(D, \tilde{W}) = \frac{2^p}{\tilde{\Gamma}_p(\alpha)} \left\{ \prod_{j=1}^p \lambda_j^{2\alpha-1} \right\} e^{-\sum_{j=1}^p \lambda_j^2 (1+|\tilde{w}_{j1}|^2+\dots+|\tilde{w}_{j\,j-1}|^2)}.$$

Integrating out the λ_j's one gets the density of $\tilde{W}$, denoted by $h(\tilde{W})$, as

$$h(\tilde{W}) = \frac{[\Gamma(\alpha)]^p}{\tilde{\Gamma}_p(\alpha)} \prod_{j=2}^p \left(1 + |\tilde{w}_{j1}|^2 + \dots + |\tilde{w}_{j\,j-1}|^2 \right)^{-\alpha}.$$

Since this factorizes into functions of S_j, the sets S_j's are mutually independently distributed.

Theorem 3.20   *Let $\tilde{W}$ be a $p \times p$ lower triangular matrix of functionally independent complex variables with diagonal elements unities. Let $D = \mathrm{diag}(\lambda_1, ..., \lambda_p)$ be a diagonal matrix with the λ_j's real, positive and distinct. Let $D^{\frac{1}{2}}$ be such that $D^{\frac{1}{2}}D^{\frac{1}{2}} = D$. Then*

$$\tilde{U} = \tilde{W}D\tilde{W}^* \Rightarrow \mathrm{d}\tilde{U} = \left\{ \prod_{j=1}^{p} \lambda_j^{2(p-j)} \right\} \mathrm{d}D \ \mathrm{d}\tilde{W}$$

and

$$\tilde{V} = D^{\frac{1}{2}}\tilde{W}\tilde{W}^*D^{\frac{1}{2}} \Rightarrow \mathrm{d}\tilde{V} = \left\{ \prod_{j=1}^{p} \lambda_j^{p-1} \right\} \mathrm{d}D \ \mathrm{d}\tilde{W}.$$

Proof   Write

$$\tilde{U} = \tilde{W}D^{\frac{1}{2}}D^{\frac{1}{2}}\tilde{W}^* = \tilde{T}\tilde{T}^*, \ \tilde{T} = \tilde{W}D^{\frac{1}{2}}.$$

From Theorem 3.18, and observing that

$$\mathrm{d}D^{\frac{1}{2}} = 2^{-p} \left\{ \prod_{j=1}^{p} \lambda_j^{-\frac{1}{2}} \right\} \mathrm{d}D$$

since the λ_j's are real and positive, one has

$$\mathrm{d}\tilde{T} = 2^{-p} \left\{ \prod_{j=1}^{p} \lambda_j^{p-j-\frac{1}{2}} \right\} \mathrm{d}D \ \mathrm{d}\tilde{W}.$$

From Theorem 3.7, we have

$$\mathrm{d}\tilde{U} = 2^{p} \left\{ \prod_{j=1}^{p} \left(\lambda_j^{\frac{1}{2}} \right)^{2(p-j)+1} \right\} \mathrm{d}\tilde{T}.$$

Substituting for $\mathrm{d}\tilde{T}$ the result follows. Similarly from Theorems 3.18 and 3.7, the second part follows.

A corresponding result for the upper triangular case can be stated as follows.

Theorem 3.21 Let $\tilde{W}$ be a $p \times p$ upper triangular matrix of functionally independent complex variables with the diagonal elements unities. Let $D = \text{diag}(\lambda_1, ..., \lambda_p)$ with the λ_j's real, positive and distinct. Then

$$\tilde{U} = \tilde{W}D\tilde{W}^* \Rightarrow d\tilde{U} = \left\{ \prod_{j=1}^{p} \lambda_j^{2(j-1)} \right\} dD \ d\tilde{W}$$

and

$$\tilde{V} = D^{\frac{1}{2}}\tilde{W}\tilde{W}^*D^{\frac{1}{2}} \Rightarrow d\tilde{V} = \left\{ \prod_{j=1}^{p} \lambda_j^{p-1} \right\} dD \ d\tilde{W}.$$

Theorem 3.22 Let $\tilde{X} = (\tilde{x}_{jk})$ be a $p \times p$ matrix of functionally independent complex variables with $\tilde{x}_{jj} = 1$, $j = 1, ..., p$ and $\tilde{D} = \text{diag}(\tilde{\lambda}_1, ..., \tilde{\lambda}_p)$ be a diagonal matrix with the $\tilde{\lambda}_j$'s functionally independent and nonzero. Then

$$\tilde{Y} = \tilde{D}\tilde{X} \text{ or } \tilde{X}\tilde{D} \Rightarrow d\tilde{Y} = \left\{ \prod_{j=1}^{p} |\tilde{\lambda}_j|^{2(p-1)} \right\} d\tilde{X}d\tilde{D}$$

and

$$d\tilde{Y} = \left\{ \prod_{j=1}^{p} \lambda_j^{2(p-1)} \right\} d\tilde{X}dD$$

when y_{jj}, λ_j, $j = 1, ..., p$ are real distinct and positive;

$$\tilde{Y} = \tilde{D}\tilde{X}\tilde{D} \Rightarrow d\tilde{Y} = 2^{2p} \left\{ \prod_{j=1}^{p} |\tilde{\lambda}_j|^{2(2p-1)} \right\} d\tilde{X}d\tilde{D}$$

and

$$d\tilde{Y} = 2^p \left\{ \prod_{j=1}^{p} \lambda_j^{4p-3} \right\} d\tilde{X}dD$$

when y_{jj}, λ_j, $j = 1, ..., p$ are real distinct and positive; and

$$\tilde{Y} = D\tilde{X}D^*, \tilde{X} = \tilde{X}^* \Rightarrow d\tilde{Y} = 2^p \left\{ \prod_{j=1}^{p} \lambda_j^{2p-1} \right\} d\tilde{X}dD$$

when y_{jj}, λ_j, $j = 1, ..., p$ are real distinct and positive.

Proof   The method is similar to that used in the previous theorem for the first two cases. We consider the case $\tilde{Y} = \tilde{D}\tilde{X}\tilde{D}$. We have

$$\tilde{y}_{jk} = \tilde{\lambda}_j \tilde{\lambda}_k \tilde{x}_{jk}, \; j \neq k \Rightarrow \frac{\partial \tilde{y}_{jk}}{\partial \tilde{x}_{jk}} = |\tilde{\lambda}_j \tilde{\lambda}_k|^2$$

and

$$\tilde{y}_{jj} = \tilde{\lambda}_j^2 \Rightarrow \frac{\partial \tilde{y}_{jj}}{\partial \tilde{\lambda}_j} = 4|\tilde{\lambda}_j|^2.$$

Since the matrix

$$A = \left(\frac{\partial \tilde{y}_{jj}, j = 1, ..., p}{\partial \tilde{x}_{jk}, j \neq k} \right)$$

is null, we need to consider only the matrices

$$B = \left(\frac{\partial \tilde{y}_{jk}, j \neq k}{\partial \tilde{x}_{jk}, j \neq k} \right) \text{ and } C = \left(\frac{\partial \tilde{y}_{jj}}{\partial \tilde{\lambda}_j}, j = 1, ..., p \right)$$

and the Jacobian is the product of the diagonal elements here. Note that $|\tilde{\lambda}_j|^2$ appears $2(p-1)$ times in the determinant of B and it appears once in $|C|$. Also 4 appears p times. Hence the result.

But when $\tilde{y}_{jk}$, $j \neq k$ are complex and y_{jj}, λ_j, $j = 1, ..., p$ are real then

$$\frac{\partial \tilde{y}_{jk}}{\partial \tilde{x}_{jk}} = (\lambda_j \lambda_k)^2, \; j \neq k \text{ and } \frac{\partial y_{jj}}{\partial \lambda_j} = 2\lambda_j, \; j = 1, ..., p.$$

Then $|B|$ gives λ_j^2 for a total of $2(p-1)$ times and $|C|$ gives only one λ_j. Also 2 appears only p times. Hence the result.

When λ_j, $j = 1, ..., p$ are real then from $\tilde{Y} = D\tilde{X}D^*$, $\tilde{X} = \tilde{X}^*$ note that the y_{jj}'s are real and we have, for $j > k$

$$\tilde{y}_{jk} = \lambda_j \lambda_k \tilde{x}_{jk} \Rightarrow \frac{\partial \tilde{y}_{jk}}{\partial \tilde{x}_{jk}} = (\lambda_j \lambda_k)^2$$

and

$$\frac{\partial y_{jj}}{\partial \lambda_j} = 2\lambda_j.$$

Thus in the Jacobian λ_j^2 appears $p-1$ times from $\frac{\partial \tilde{y}_{jk}}{\partial \tilde{x}_{jk}}$, $j > k$ and hence the result.

Remark 3.8   Note that the transformation $\tilde{Y} = \tilde{D}\tilde{X}\tilde{D}$ is not one-to-one unless $\tilde{\lambda}_{jj}$ is specified in terms of $\tilde{y}_{jj}$ because $\tilde{\lambda}_{jj}^2 = \tilde{y}_{jj}$. When $\tilde{X} = \tilde{X}^*$ and $\tilde{x}_{jj} = 1$, $j = 1,...,p$ there are $2(p(p-1)/2) = p(p-1)$ real variables in $\tilde{X}$ and there are $2p$ real variables in $\tilde{D}$ when the $\tilde{\lambda}_j$'s are complex. Thus there is a total of $p(p+1)$ real variables in $\tilde{X}$ and $\tilde{D}$ combined. But

$$\tilde{Y}^* = (\tilde{D}\tilde{X}\tilde{D}^*)^* = \tilde{D}\tilde{X}\tilde{D}^* = \tilde{Y}$$

when $\tilde{X}^* = \tilde{X}$ and hence $\tilde{Y}$ is hermitian which implies that there are only p^2 real variables in $\tilde{Y}$. Hence when the $\tilde{\lambda}_j$'s are complex the transformation is not one-to-one.

Example 3.13   Let $\tilde{Y}$ be a $p \times p$ complex random matrix having a standard Wishart density

$$f(\tilde{Y}) = \begin{cases} \{|\det(\tilde{Y})|^{n-p}e^{-\mathrm{tr}(\tilde{Y})}/\tilde{\Gamma}_p(n)\}, & \tilde{Y} = \tilde{Y}^* > 0, \ n \geq p \\ 0, & \text{elsewhere.} \end{cases}$$

Evaluate the density of the sample correlation matrix $\tilde{X}$ where $\tilde{Y} = D\tilde{X}D^*$, $D = \mathrm{diag}(\sqrt{y_{11}}, ..., \sqrt{y_{pp}})$, the y_{jj}'s are real and positive.

Solution   Let $D = \mathrm{diag}(\alpha_1, ..., \alpha_p)$, $\alpha_j = \sqrt{y_{jj}}$. Then from Theorem 3.22

$$d\tilde{Y} = 2^p \left\{ \prod_{j=1}^{p} \alpha_j^{2p-1} \right\} d\tilde{X}dD.$$

Thus

$$f(\tilde{Y})d\tilde{Y} = \frac{|\det(D\tilde{X}D^*)|^{n-p}e^{-\mathrm{tr}(D\tilde{X}D^*)}}{\tilde{\Gamma}_p(n)} 2^p \left\{ \prod_{j=1}^{p} \alpha_j^{2p-1} \right\} d\tilde{X}dD.$$

But $\mathrm{tr}(D\tilde{X}D^*) = \mathrm{tr}(D^*D\tilde{X}) = \alpha_1^2 + ... + \alpha_p^2$ since the diagonal elements of $\tilde{X}$ are all unities. Also $|\det(D\tilde{X}D^*)| = \alpha_1^2...\alpha_p^2|\det(\tilde{X})|$. The density of $\tilde{X}$, denoted by $g(\tilde{X})$, is available by integrating out the α_j's. That is,

$$g(\tilde{X}) = \frac{|\det(\tilde{X})|^{n-p}}{\tilde{\Gamma}_p(n)} \left\{ \prod_{j=1}^{p} \int_0^\infty \left[2(\alpha_j^2)^{n-\frac{1}{2}}e^{-\alpha_j^2} \right] d\alpha_j \right\}.$$

But

$$2 \int_0^\infty (u^2)^{n-\frac{1}{2}} e^{-u^2} du = \Gamma(n).$$

Hence the density of $\tilde{X}$ is

$$g(\tilde{X}) = \begin{cases} \{[\Gamma(n)]^p |\det(\tilde{X})|^{n-p} / \tilde{\Gamma}_p(n)\}, & \tilde{X} = \tilde{X}^* > 0 \\ \\ 0, & \text{elsewhere.} \end{cases}$$

Theorem 3.23   *Let $\tilde{X} = (\tilde{x}_{jk})$ and $\tilde{W} = (\tilde{w}_{jk})$ be $p \times p$ nonsingular lower triangular matrices of functionally independent complex variables such that x_{jj}, w_{jj}, $j = 1, ..., p$ are real and positive and $\sum_{k=1}^{j} |\tilde{w}_{jk}|^2 = 1$, $j = 1, ..., p$. Let $D = \text{diag}(\lambda_1, ..., \lambda_p)$ be a diagonal matrix where the λ_j's are real distinct and positive. Then*

$$\tilde{X} = D\tilde{W} \Rightarrow d\tilde{X} = \left\{ \prod_{j=1}^{p} \lambda_j^{2(j-1)} w_{jj}^{-1} \right\} dD d\tilde{W}$$

and

$$\tilde{X} = \tilde{W}D \Rightarrow d\tilde{X} = \left\{ \prod_{j=1}^{p} \lambda_j^{2(p-j)} w_{jj}^{-1} \right\} dD d\tilde{W}.$$

Proof   Let $\tilde{x}_{jk} = x_{jk1} + i x_{jk2}$, $\tilde{w}_{jk} = w_{jk1} + i w_{jk2}$ where $x_{jkm}, w_{jkm}, m = 1, 2$ are real. Then from $\tilde{X} = D\tilde{W}$ we have

$$x_{jk1} = \lambda_j w_{jk1}, \quad x_{jk2} = \lambda_j w_{jk2}, \quad j > k$$

and

$$x_{jj} = \lambda_j w_{jj} = \lambda_j \left[1 - \sum_{k=1}^{j-1} |\tilde{w}_{jk}|^2 \right]^{\frac{1}{2}}.$$

Hence by taking the x-variables in the order x_{jk1}, $j > k$, x_{jk2}, $j > k$, x_{jj}, $j = 1, ..., p$ and the w-variables also in the same order with w_{jj}'s replaced by the λ_j's, the Jacobian matrix is of the form

$$\begin{bmatrix} D_1 & 0 & B_1 \\ 0 & D_2 & B_2 \\ A_1 & A_2 & D_3 \end{bmatrix}$$

where D_1 is a diagonal matrix with λ_j repeated $j-1$ times, $j = 2, ..., p$. $D_1 = D_2$. $D_3 = \text{diag}(1, w_{22}, ..., w_{pp})$. A_1 has the first row null, the j-th row has $-\lambda_j w_{jj}^{-1} w_{jk1}$, $j > k$ corresponding to (x_{jj}, w_{jk1})-th position and other elements zeros. A_2 has the same form with w_{jk1} replaced by w_{jk2}. B_1 has $w_{jk1}, j > k$ corresponding to the (x_{jk1}, λ_j)-th position and others zeros. B_2 has the same form with w_{jk1} replaced by w_{jk2}. When evaluating the determinant, add suitable combinations of the rows of D_1 and D_2 to make A_1 and A_2 null. Then D_3 becomes $\text{diag}(1, w_{22}^{-1}, ..., w_{pp}^{-1})$ with the determinant $\prod_{j=1}^{p} w_{jj}^{-1}$ with $w_{11} = 1$. Take out $\prod_{j=2}^{p} \lambda_j^{2(j-1)}$ from D_1 and D_2 to see the result.

The steps are parallel for the second part and hence omitted. More of this type of Jacobians and those for the upper triangular cases are given as exercises in this chapter.

Remark 3.9   Note that when w_{jj} and λ_j are all real distinct and positive the transformations in Theorem 3.23 are one-to-one.

Exercises

3.3.1   Let $\tilde{X}$ be a $p \times p$ matrix of functionally independent complex variables. Let A and B be $p \times p$ nonsingular matrices of constants. Evaluate the Jacobian of the transformation $\tilde{Y} = A\tilde{X}A^* + B\tilde{X}B^*$.

3.3.2   Let $\tilde{X}, \tilde{W}$ be $p \times p$ nonsingular lower triangular matrices of functionally independent complex variables where $\tilde{W}$ has unit diagonal elements. Let D be a diagonal matrix of real distinct and positive diagonal elements. Evaluate the Jacobians in the following transformations. $\tilde{X} = D^{\frac{1}{2}}\tilde{W}$, $\tilde{Y} = \tilde{W}D^{\frac{1}{2}}$, $\tilde{V} = D^{\frac{1}{2}}\tilde{W}\tilde{W}^*D^{\frac{1}{2}}$, $\tilde{Z} = \tilde{W}^*D\tilde{W}$, $\tilde{T} = D^{\frac{1}{2}}\tilde{W}^*\tilde{W}D^{\frac{1}{2}}$ where $D^{\frac{1}{2}}D^{\frac{1}{2}} = D$.

3.3.3   Redo Exercise 3.3.2 for upper triangular matrices.

3.3.4   Let $\tilde{W} = (\tilde{w}_{jk})$ be a $p \times p$ nonsingular lower triangular matrix of functionally independent complex variables with the w_{jj}'s real and positive. Let $D = \text{diag}(\lambda_1, ..., \lambda_p)$ where the λ_j's are real distinct and positive. Let $\sum_{k=1}^{j} |\tilde{w}_{jk}|^2 = 1$, $j = 1, ..., p$. Evaluate the Jacobians in the following transformations: $\tilde{X} = D^{\frac{1}{2}}\tilde{W}$, $\tilde{Y} = D^{\frac{1}{2}}\tilde{W}\tilde{W}^*D^{\frac{1}{2}}$, $\tilde{Z} = \tilde{W}^*D\tilde{W}$.

3.3.5 Redo Exercise 3.3.4 for upper triangular matrices.

3.3.6 What will be the results in Theorems 3.18 and 3.19 if the diagonal elements of $\tilde{X}$ are real and all the λ_j's are real and distinct? Supply the proofs.

3.3.7 Let $\tilde{X} = \tilde{X}^* > 0$ and $p \times p$. If $\tilde{X} = D\tilde{W}\tilde{W}^*D$ where $\tilde{W}$ is lower triangular with unit diagonal elements and $D = \text{diag}(\lambda_1, ..., \lambda_p)$, $\lambda_1 > ... > \lambda_p > 0$, evaluate $d\tilde{X}$ in terms of dD and $d\tilde{W}$.

3.3.8 Redo Exercise 3.3.7 for the upper triangular case.

Transformations involving eigenvalues and unitary matrices

4.0 Introduction

More complicated transformations involving eigenvalues, unitary matrices and semiunitary matrices will be considered in this chapter.

4.1 Transformations involving skew hermitian matrices

When $\tilde{X}$ is skew hermitian, that is, $\tilde{X} = -\tilde{X}^*$ it is not difficult to show that $I + \tilde{X}$ and $I - \tilde{X}$ are both nonsingular and $\tilde{Z} = 2(I + \tilde{X})^{-1} - I$ is unitary, that is, $\tilde{Z}\tilde{Z}^* = I$. This property will be made use of in the first result that will be discussed here. Also note that

$$2\left(I + \tilde{X}\right)^{-1} - I = \left(I + \tilde{X}\right)^{-1}\left(I - \tilde{X}\right)$$
$$= \left(I - \tilde{X}\right)\left(I + \tilde{X}\right)^{-1}.$$

When $\tilde{X}$ is skew hermitian and of functionally independent complex variables then there are $2[p(p-1)/2]+p = p^2$ real variables in $\tilde{X}$. Let $\tilde{T}$ be a lower triangular matrix of functionally independent complex variables with the diagonal elements being real. Then there are p^2 real variables in $\tilde{T}$ also. Thus combined, there are $2p^2$ real variables in $\tilde{T}$ and $\tilde{X}$. It can be shown that $\tilde{Y} = \tilde{T}\left[2(\tilde{X} + I)^{-1} - I\right]$ can produce a one-to-one transformation when the t_{jj}'s are real and positive.

$$\tilde{Y} = \tilde{T}\tilde{Z}, \ \tilde{Z}\tilde{Z}^* = I, \ \tilde{T} = (\tilde{t}_{jk}), \ \tilde{t}_{jj} = t_{jj} > 0, \ j = 1, ..., p.$$

Then

$$\tilde{Y}\tilde{Y}^* = \tilde{T}\tilde{T}^* \Rightarrow t_{11}^2 = \sum_{k=1}^{p} |\tilde{y}_{1k}|^2.$$

221

Note that when t_{11} is real and positive it is uniquely determined in terms of $\tilde{Y}$. Now consider the first row elements of $\tilde{T}\tilde{T}^*$ that is $t_{11}^2, t_{11}\tilde{t}_{21}, ..., t_{11}\tilde{t}_{p1}$. Hence $\tilde{t}_{21}, ..., \tilde{t}_{p1}$, that is, the first column of $\tilde{T}$ is uniquely determined in terms of $\tilde{Y}$. Now consider the second row of $\tilde{T}\tilde{T}^*$ and so on. Thus $\tilde{T}$ is uniquely determined in terms of $\tilde{Y}$. But $\tilde{Z} = \tilde{T}^{-1}\tilde{Y}$ and hence $\tilde{Z}$, thereby $\tilde{X}$ is uniquely determined in terms of $\tilde{Y}$ with no additional restrictions imposed on the elements of $\tilde{Z}$.

In this chapter the Jacobians will also be written ignoring the sign as in the previous chapters.

Theorem 4.1   *Let* $\tilde{Y}, \tilde{X}$ *and* $\tilde{T} = (\tilde{t}_{jk})$ *be* $p \times p$ *matrices of functionally independent complex variables where* $\tilde{Y}$ *is nonsingular,* $\tilde{X}$ *is skew hermitian and* $\tilde{T}$ *is lower triangular with real and positive diagonal elements. Then, ignoring the sign,*

$$\tilde{Y} = \tilde{T}\left[2(\tilde{X} + I)^{-1} - I\right] = \tilde{T}(I - \tilde{X})(I + \tilde{X})^{-1} \Rightarrow$$

$$d\tilde{Y} = 2^{p^2}\left\{\prod_{j=1}^{p} t_{jj}^{2(p-j)+1}\right\} |\det((I + \tilde{X})(I - \tilde{X}))|^{-p} d\tilde{X} d\tilde{T}.$$

Proof   Taking differentials in $\tilde{Y} = \tilde{T}\left[2(\tilde{X} + I)^{-1} - I\right]$ one has

$$(d\tilde{\mathbf{Y}}) = \tilde{T}\left\{-2(\tilde{X} + I)^{-1}(d\tilde{\mathbf{X}})(\tilde{X} + I)^{-1}\right\}$$
$$+ (d\tilde{\mathbf{T}})\left[2(\tilde{X} + I)^{-1} - I\right].$$

Let

$$\tilde{Z} = 2(\tilde{X} + I)^{-1} - I \Rightarrow (\tilde{X} + I)^{-1} = \frac{1}{2}(I + \tilde{Z})$$

and observe that $\tilde{Z}\tilde{Z}^* = I$. Then

$$\tilde{T}^{-1}(d\tilde{\mathbf{Y}})\tilde{Z}^* = -\frac{1}{2}(\tilde{Z} + I)(d\tilde{\mathbf{X}})(I + \tilde{Z}^*) + \tilde{T}^{-1}(d\tilde{\mathbf{T}}). \qquad (a)$$

Let

$$(d\tilde{\mathbf{W}}) = (\tilde{Z} + I)(d\tilde{\mathbf{X}})(I + \tilde{Z}^*) \Rightarrow$$
$$d\tilde{W} = |\det((I + \tilde{Z})(I + \tilde{Z}^*))|^p d\tilde{X}$$
$$= 2^{2p^2}|\det((I + \tilde{X})(I - \tilde{X}))|^{-p} d\tilde{X}, \qquad (b)$$

from Theorem 3.6, ignoring the sign. Let

$$(d\tilde{U}) = \tilde{T}^{-1}(d\tilde{T}) \Rightarrow d\tilde{U} = \left\{ \prod_{j=1}^{p} t_{jj}^{-(2j-1)} \right\} d\tilde{T} \qquad (c)$$

from Theorem 3.3(d). Let

$$(d\tilde{V}) = \tilde{T}^{-1}(d\tilde{Y})\tilde{Z}^{*} \Rightarrow d\tilde{V} = |\det(\tilde{T}\tilde{T}^{*})|^{-p}d\tilde{Y} \qquad (d)$$

from Theorem 3.2(c) and observing that $\tilde{Z}\tilde{Z}^{*} = I$. Equation ($a$) reduces to

$$(d\tilde{V}) = -\frac{1}{2}(d\tilde{W}) + (d\tilde{U}). \qquad (e)$$

Note that $(d\tilde{W})$ is skew hermitian. Denote

$$\tilde{V} = (\tilde{v}_{jk}) = (v_{jk1}) + i(v_{jk2}),$$
$$\tilde{W} = (\tilde{w}_{jk}) = (w_{jk1}) + i(w_{jk2}), \ \tilde{U} = (u_{jk1}) + i(u_{jk2})$$

where $i = \sqrt{-1}$, $v_{jkm}, w_{jkm}, u_{jkm}$, $m = 1, 2$ are all real. Then from (e)

$$\tilde{V} = -\frac{1}{2}\tilde{W} + \tilde{U} \Rightarrow$$

$$v_{jkm} = \frac{1}{2}w_{jkm} + u_{jkm}, \ j > k, \ m = 1, 2,$$

$$v_{jkm} = -\frac{1}{2}w_{jkm}, \ j < k, \ m = 1, 2,$$

$$v_{jj1} = u_{jj1}, \ v_{jj2} = -\frac{1}{2}w_{jj2}.$$

The matrices of partial derivatives are the following:

$$A_1 = \left(\frac{\partial v_{jk1}}{\partial w_{jk1}}, j > k \right) = \operatorname{diag}\left(\frac{1}{2}, ..., \frac{1}{2} \right)$$

$$= \left(\frac{\partial v_{jk2}}{\partial w_{jk2}}, j > k \right) = A_2$$

with determinant $(\frac{1}{2})^{2(p(p-1)/2)} = 2^{-p(p-1)}$,

$$A_3 = \left(\frac{\partial v_{jk1}}{\partial u_{jk1}}, j > k \right) = I = \left(\frac{\partial v_{jk2}}{\partial u_{jk2}}, j > k \right) = A_4,$$

$$A_5 = \left(\frac{\partial v_{jj1}}{\partial u_{jj1}}, j = 1, ..., p \right) = I_p,$$

$$A_6 = \left(\frac{\partial v_{jj2}}{\partial w_{jj2}}, j = 1, ..., p \right) = \operatorname{diag}\left(-\frac{1}{2}, ..., -\frac{1}{2} \right)$$

with determinant $(-1)^p 2^{-p}$.

The determinants of $A_1, ..., A_6$ contribute towards the Jacobian and the product of the determinants, in absolute value, is

$$2^{-p(p-1)-p} = 2^{-p^2}.$$

Without going through the above procedure one may note from (a) that since $(d\tilde{X})$ has p^2 real variables, multiplication by $\frac{1}{2}$ produces the factor 2^{-p^2} in the Jacobian. From $(b), (c), (d)$ and (e) we have

$$d\tilde{Y} = \left\{ \prod_{j=1}^{p} t_{jj}^{2p} \right\} d\tilde{V} = \left\{ \prod_{j=1}^{p} t_{jj}^{2p} \right\} 2^{-p^2} d\tilde{W} d\tilde{U}$$

$$= \left\{ \prod_{j=1}^{p} t_{jj}^{2p} \right\} 2^{2p^2-p^2} |\det((I + \tilde{X})(I - \tilde{X}))|^{-p} \left\{ \prod_{j=1}^{p} t_{jj}^{-(2j-1)} \right\} d\tilde{X} d\tilde{T}$$

$$= 2^{p^2} \left\{ \prod_{j=1}^{p} t_{jj}^{2(p-j)+1} \right\} |\det((I + \tilde{X})(I - \tilde{X}))|^{-p} d\tilde{X} d\tilde{T}.$$

Example 4.1   Let $\tilde{X}$ be a $p \times p$ skew hermitian matrix of functionally independent complex variables. Then show that

$$\int_{\tilde{X}} |\det((I + \tilde{X})(I - \tilde{X}))|^{-p} d\dot{X} = \frac{\pi^{p^2}}{2^{p(p-1)} \tilde{\Gamma}_p(p)}.$$

Solution   Let $\tilde{Y} = (\tilde{y}_{jk})$ be a $p \times p$ matrix of functionally independent complex variables. Consider the integral

$$\int_{\tilde{Y}} e^{-\text{tr}(\tilde{Y}\tilde{Y}^*)} d\tilde{Y} = \int_{\tilde{Y}} e^{-\sum_{j,k} |\tilde{y}_{jk}|^2} d\tilde{Y}$$

$$= \left\{ \prod_{j,k} \int_{\tilde{y}_{jk}} e^{-|\tilde{y}_{jk}|^2} d\tilde{y}_{jk} \right\} = \pi^{p^2}. \qquad (a)$$

Now consider a transformation $\tilde{Y} = \tilde{T} \left[2(I + \tilde{X})^{-1} - I \right]$ where $\tilde{T} = (\tilde{t}_{jk})$ is lower triangular with t_{jj}'s real and positive and no restrictions on $\tilde{X}$

other than that it is skew hermitian. Then since $\text{tr}(\tilde{Y}\tilde{Y}^*) = \text{tr}(\tilde{T}\tilde{T}^*)$, from Theorem 4.1 and (a) we have

$$\pi^{p^2} = \int_{\tilde{T}, \tilde{X}} e^{-\text{tr}(\tilde{T}\tilde{T}^*)} 2^{p^2} \left\{ \prod_{j=1}^{p} t_{jj}^{2(p-j)+1} \right\}$$

$$\times |\det((I + \tilde{X})(I - \tilde{X}))|^{-p} d\tilde{X} d\tilde{T}.$$

Note that

$$e^{-\text{tr}(\tilde{T}\tilde{T}^*)} = e^{-\sum_{j=1}^{p} t_{jj}^2 - \sum_{j>k} |\tilde{t}_{jk}|^2}.$$

But

$$\int_{\tilde{t}_{jk}} e^{-|\tilde{t}_{jk}|^2} d\tilde{t}_{jk} = \pi \quad \text{and} \quad \int_0^{\infty} t_{jj}^{2(p-j)+1} e^{-t_{jj}^2} dt_{jj} = 2^{-1}\Gamma(p-j+1).$$

Hence

$$\int_{\tilde{T}} e^{-\text{tr}(\tilde{T}\tilde{T}^*)} \left\{ \prod_{j=1}^{p} t_{jj}^{2(p-j)+1} \right\} d\tilde{T} = 2^{-p}\tilde{\Gamma}_p(p).$$

Substituting this the result follows.

Next we consider a representation of a hermitian positive definite matrix $\tilde{Y}$ in terms of a skew hermitian matrix $\tilde{X}$ and a diagonal matrix D such that

$$\tilde{Y} = \left[2(I + \tilde{X})^{-1} - I \right] D \left[2(I + \tilde{X})^{-1} - I \right]^*$$

where $D = \text{diag}(\lambda_1, ..., \lambda_p)$ with the λ_j's real distinct and positive, and the first row elements of $(I + \tilde{X})^{-1}$ real and of specified signs. In this case it can be shown that the transformation is unique. Note that

$$\tilde{Y} = \tilde{Z} D \tilde{Z}^* = \lambda_1 \tilde{z}_1 \tilde{z}_1^* + ... + \lambda_p \tilde{z}_p \tilde{z}_p^* \Rightarrow$$
$$(\tilde{Y} - \lambda_j I)\tilde{z}_j = 0, \quad j = 1, ..., p$$

where $\tilde{z}_1, ..., \tilde{z}_p$ are the columns of $\tilde{Z}$ such that $\tilde{z}_j^* \tilde{z}_j = 1$, $\tilde{z}_j^* \tilde{z}_k = 0, j \neq k$. Since $\lambda_1, ..., \lambda_p$ are the eigenvalues of $\tilde{Y}$, which are assumed to be real distinct and positive, D is uniquely determined in terms of $\tilde{Y}$. Note that $\tilde{z}_j$ is an eigenvector corresponding to λ_j such that $\tilde{z}_j^* \tilde{z}_j = 1, j = 1, ..., p$. Hence $\tilde{z}_j$ is uniquely determined in terms of $\tilde{Y}$ except for a multiple of $-1, -i, i, i = \sqrt{-1}$. If any particular element of $\tilde{z}_j$ is assumed to be real and positive, for example the first element, then $\tilde{z}_j$ is uniquely determined. Thus if the first row elements of $\tilde{Z}$ are real and of specified signs, which is equivalent to saying that the first row elements of $(I + \tilde{X})^{-1}$ are real and of specified signs, then the transformation is unique.

Theorem 4.2   *Let $\tilde{Y}$ and $\tilde{X}$ be $p \times p$ matrices of functionally independent complex variables such that $\tilde{Y}$ is hermitian positive definite, $\tilde{X}$ is skew hermitian and the first row elements of $(I + \tilde{X})^{-1}$ are real and of specified signs. Let $D = \mathrm{diag}(\lambda_1, ..., \lambda_p)$ where the λ_j's are real distinct and positive. Then, ignoring the sign,*

$$\tilde{Y} = \left[2(I + \tilde{X})^{-1} - I\right] D \left[2(I + \tilde{X})^{-1} - I\right]^* \Rightarrow$$

$$d\tilde{Y} = 2^{p(p-1)} \left\{ \prod_{j>k} |\lambda_k - \lambda_j|^2 \right\} |\det((I + \tilde{X})(I - \tilde{X}))|^{-p} d\tilde{X} dD.$$

Proof   Let $\tilde{Z} = 2(I + \tilde{X})^{-1} - I$, $\tilde{X} = -\tilde{X}^*$. Taking the differentials in $\tilde{Y} = \tilde{Z}D\tilde{Z}^*$ we have

$$(d\tilde{Y}) = (d\tilde{Z})D\tilde{Z}^* + \tilde{Z}(dD)\tilde{Z}^* + \tilde{Z}D(d\tilde{Z}^*). \qquad (a)$$

But

$$\begin{aligned}(d\tilde{Z}) &= -2(I + \tilde{X})^{-1}(d\tilde{X})(I + \tilde{X})^{-1} \\ &= -\frac{1}{2}(I + \tilde{Z})(d\tilde{X})(I + \tilde{Z})\end{aligned}$$

and

$$\begin{aligned}(d\tilde{Z}^*) &= 2(I - \tilde{X})^{-1}(d\tilde{X})(I - \tilde{X})^{-1} \\ &= \frac{1}{2}(I + \tilde{Z}^*)(d\tilde{X})(I + \tilde{Z}^*).\end{aligned}$$

From (a) one has

$$\begin{aligned}\tilde{Z}^*(d\tilde{Y})\tilde{Z} &= -\frac{1}{2}(I + \tilde{Z}^*)(d\tilde{X})(I + \tilde{Z})D + (dD) \\ &\quad + \frac{1}{2}D(I + \tilde{Z}^*)(d\tilde{X})(I + \tilde{Z})\end{aligned}$$

observing that $\tilde{Z}^*\tilde{Z} = I$. Let

$$(d\tilde{U}) = \tilde{Z}^*(d\tilde{Y})\tilde{Z} \Rightarrow d\tilde{U} = d\tilde{Y} \text{ since } \tilde{Z}^*\tilde{Z} = I, \qquad (b)$$

$$\begin{aligned}(d\tilde{V}) &= (I + \tilde{Z}^*)(d\tilde{X})(I + \tilde{Z}) \\ &= (I + \tilde{X}^*)^{-1}(4d\tilde{X})(I + \tilde{X})^{-1} \Rightarrow\end{aligned}$$

$$d\tilde{V} = 4^{p^2} |\det((I + \tilde{X})(I - \tilde{X}))|^{-p} d\tilde{X} \qquad (c)$$

if there are p^2 free real variables in $\tilde{X}$. But in our case there are only $p^2 - p$ real variables in $\tilde{X}$ when $\tilde{X}$ is uniquely chosen and hence

$$\mathrm{d}\tilde{V} = 4^{p^2-p}|\det((I+\tilde{X})(I-\tilde{X}))|^{-p}\mathrm{d}\tilde{X}, \qquad (d)$$

and

$$(\mathrm{d}\tilde{U}) = -\frac{1}{2}(\mathrm{d}\tilde{V})D + \frac{1}{2}D(\mathrm{d}\tilde{V}) + (\mathrm{d}D). \qquad (e)$$

From (e) and using the fact that $(\mathrm{d}\tilde{V})$ is skew hermitian and $(\mathrm{d}\tilde{U})$ is hermitian we have

$$du_{jj} = d\lambda_j, \quad du_{jkm} = \pm\frac{1}{2}(\lambda_k - \lambda_j)dv_{jkm}, \ j > k, \ m = 1,2.$$

Thus the determinant of the Jacobian matrix, in absolute value, is

$$\left\{\prod_{j>k}\frac{1}{2}|\lambda_k - \lambda_j|\right\}^2 = 2^{-p(p-1)}\prod_{j>k}|\lambda_k - \lambda_j|^2.$$

That is,

$$\mathrm{d}\tilde{U} = 2^{-p(p-1)}\left\{\prod_{j>k}|\lambda_k - \lambda_j|^2\right\}\mathrm{d}\tilde{V}\mathrm{d}D.$$

Substituting for $\mathrm{d}\tilde{U}$ and $\mathrm{d}\tilde{V}$ from (b) and (c) the result follows.

Example 4.2 For $\tilde{X}$ a $p \times p$ skew hermitian matrix with the first row elements of $(I+\tilde{X})^{-1}$ real and of specified signs show that

$$\int_{\tilde{X}}|\det((I+\tilde{X})(I-\tilde{X}))|^{-p}\mathrm{d}\tilde{X} = \frac{\pi^{\frac{p(p-1)}{2}}(p-1)!(p-2)!...1!}{A}$$

where

$$A = 2^{p(p-1)}\int_D\left\{\prod_{j>k}|\lambda_k - \lambda_j|^2\right\}e^{-\mathrm{tr}(D)}\mathrm{d}D,$$

with $D = \mathrm{diag}(\lambda_1, ..., \lambda_p)$, $\lambda_1 > ... > \lambda_p > 0$.

Solution Consider a $p \times p$ hermitian positive definite matrix $\tilde{Y}$ of functionally independent complex variables. Let

$$B = \int_{\tilde{Y}=\tilde{Y}^*>0}e^{-\mathrm{tr}(\tilde{Y})}\mathrm{d}\tilde{Y} = \int_{\tilde{Y}}|\det(\tilde{Y})|^{p-p}e^{-\mathrm{tr}(\tilde{Y})}\mathrm{d}\tilde{Y}$$

$$= \tilde{\Gamma}_p(p) = \pi^{\frac{p(p-1)}{2}}\Gamma(p)\Gamma(p-1)...\Gamma(1)$$

$$= \pi^{\frac{p(p-1)}{2}}(p-1)!(p-2)!...1!$$

evaluating the integral by using a complex matrix-variate gamma integral. Put

$$\tilde{Y} = \tilde{Z} D \tilde{Z}^*, \quad \tilde{Z} = 2(I + \tilde{X})^{-1} - I$$

as in Theorem 4.2. Then

$$d\tilde{Y} = 2^{p(p-1)} \left\{ \prod_{j>k} |\lambda_k - \lambda_j|^2 \right\}$$
$$\times |\det((I + \tilde{X})(I - \tilde{X}))|^{-p} d\tilde{X} \; dD$$

and

$$B = \int_{\tilde{X}} |\det((I + \tilde{X})(I - \tilde{X}))|^{-p} d\tilde{X}$$
$$\times \int_D 2^{p(p-1)} \left\{ \prod_{j>p} |\lambda_k - \lambda_j|^2 \right\} e^{-\operatorname{tr}(D)} dD.$$

Hence the result.

4.1.1   A direct evaluation of the integral

When a skew hermitian matrix $\tilde{X}$ is used to parametrize a unitary matrix such as $\tilde{Z}$ in Theorem 4.1 can we evaluate the Jacobian by direct integration? This will be examined here. Let

$$A_p = \int_{\tilde{X}} |\det((I + \tilde{X})(I - \tilde{X}))|^{-p} d\tilde{X}.$$

Partition $I + \tilde{X}$ as follows:

$$I + \tilde{X} = \begin{pmatrix} 1 + \tilde{x}_{11} & \tilde{X}_{12} \\ -\tilde{X}_{12}^* & I + \tilde{X}_1 \end{pmatrix}$$

where $\tilde{X}_{12}$ represents the first row of $I + \tilde{X}$ excluding the first element $1 + \tilde{x}_{11}$, and $I + \tilde{X}_1$ is obtained from $I + \tilde{X}$ by deleting the first row and the first column. Note that

$$\det(I + \tilde{X}) = \det(I + \tilde{X}_1) \left[1 + \tilde{x}_{11} + \tilde{X}_{12}(I + \tilde{X}_1)^{-1} \tilde{X}_{12}^* \right].$$

For fixed $(I + \tilde{X}_1)$ let

$$\tilde{U}_{12} = \tilde{X}_{12}(I + \tilde{X}_1)^{-1} \Rightarrow$$
$$d\tilde{U}_{12} = |\det((I + \tilde{X}_1)(I - \tilde{X}_1))|^{-1} d\tilde{X}_{12}$$

and observing that $\tilde{X}_1^* = -\tilde{X}_1$ we have

$$\tilde{X}_{12}(I + \tilde{X}_1)^{-1}\tilde{X}_{12}^* = \tilde{U}_{12}(I - \tilde{X}_1)\tilde{U}_{12}^*.$$

Let Q be a unitary matrix such that

$$Q^*\tilde{X}_1 Q = \text{diag}(i\lambda_1, ..., i\lambda_{p-1})$$

where $i = \sqrt{-1}$ and $\lambda_1, ..., \lambda_{p-1}$ are real. Let

$$\tilde{V}_{12} = \tilde{U}_{12}Q = (\tilde{v}_1, ..., \tilde{v}_{p-1}).$$

Then

$$\tilde{U}_{12}(I - \tilde{X}_1)\tilde{U}_{12}^* = (1 - i\lambda_1)|\tilde{v}_1|^2 + ... + (1 - i\lambda_{p-1})|\tilde{v}_{p-1}|^2$$

and

$$1 + \tilde{x}_{11} + \tilde{X}_{12}(I + \tilde{X}_1)^{-1}\tilde{X}_{12}^* = a - ib$$

where

$$a = 1 + |\tilde{v}_1|^2 + ... + |\tilde{v}_{p-1}|^2$$

and

$$b = (-x_{112} + \lambda_1|\tilde{v}_1|^2 + ... + \lambda_{p-1}|\tilde{v}_{p-1}|^2)$$

observing that $\tilde{x}_{11}$ is purely imaginary, that is, $\tilde{x}_{11} = ix_{112}$ where x_{112} is real. Thus $|(I + \tilde{X})(I + \tilde{X})^*|^{-p}$ yields the factor

$$[(a - ib)(a + ib)]^{-p} = (a^2 + b^2)^{-p}.$$

Then

$$A_p = A_{p-1} \int_{\tilde{V}_{12}} \int_{x_{112}} (a^2 + b^2)^{-p} d\tilde{V}_{12} dx_{112}$$

$$= A_{p-1} \int_{\tilde{V}_{12}} a^{-2p} \int_{x_{112}} \left(1 + \frac{b^2}{a^2}\right)^{-p} d\tilde{V}_{12} dx_{112}.$$

Consider the integral over x_{112}, $-\infty < x_{112} < \infty$. Change x_{112} to b and then to $c = \frac{b}{a}$. Then

$$\int_{x_{112}} \left(1 + \frac{b^2}{a^2}\right)^{-p} dx_{112} = \int_b \left(1 + \frac{b^2}{a^2}\right)^{-p} db$$

$$= a \int_{c=-\infty}^{\infty} (1 + c^2)^{-p} dc$$

$$= 2a \int_0^{\infty} (1 + c^2)^{-p} dc$$

$$= a \frac{\Gamma\left(\frac{1}{2}\right)\Gamma\left(p - \frac{1}{2}\right)}{\Gamma(p)} = k, \text{ say,}$$

by evaluating using a type-2 beta integral after transforming $u = c^2$. Hence

$$A_p = kA_{p-1} \int_{\tilde{V}_{12}} a^{-(2p-1)} d\tilde{V}_{12}$$

$$= kA_{p-1} \int_{-\infty}^{\infty} \cdots \int_{-\infty}^{\infty} \left(1 + |\tilde{v}_1|^2 + \ldots + |\tilde{v}_{p-1}|^2\right)^{-(2p-1)} d\tilde{v}_1 \ldots d\tilde{v}_{p-1}.$$

For evaluating the integral use the polar coordinates. Let $\tilde{v}_j = v_{j1} + iv_{j2}$ where v_{j1} and v_{j2} are real. Let $v_{j1} = r_j \cos\theta_j$, $v_{j2} = r_j \sin\theta_j$, $0 \le r_j < \infty$, $0 \le \theta_j \le 2\pi$. Then denoting the multiple integral by B_{p-1} we have

$$B_{p-1} = \int_{-\infty}^{\infty} \cdots \int_{-\infty}^{\infty} \left(1 + |\tilde{v}_1|^2 + \ldots + |\tilde{v}_{p-1}|^2\right)^{-(2p-1)} d\tilde{v}_1 \ldots d\tilde{v}_{p-1}$$

$$= (2\pi)^{p-1} \int_{r_1=0}^{\infty} \cdots \int_{r_{p-1}=0}^{\infty} r_1 \ldots r_{p-1}$$

$$\times \left(1 + r_1^2 + \ldots + r_{p-1}^2\right)^{-(2p-1)} dr_1 \ldots dr_{p-1}.$$

Evaluating this by a Dirichlet integral we have

$$B_{p-1} = \pi^{p-1} \frac{\Gamma(p)}{\Gamma(2p-1)} \text{ for } p \ge 2.$$

Hence for $p \ge 2$,

$$A_p = A_{p-1} \pi^{p-1} \sqrt{\pi} \frac{\Gamma(p)\Gamma\left(p - \frac{1}{2}\right)}{\Gamma(2p-1)\Gamma(p)}.$$

By using the duplication formula for gamma functions

$$\Gamma(2p-1) = \pi^{-\frac{1}{2}} 2^{2p-2} \Gamma\left(p - \frac{1}{2}\right) \Gamma(p).$$

Hence

$$A_p = A_{p-1} \frac{\pi^p}{2^{2p-2}\Gamma(p)}.$$

Repeating this process we have

$$A_p = \frac{\pi^p}{2^{2p-2}\Gamma(p)} \frac{\pi^{p-1}}{2^{2(p-1)-2}\Gamma(p-1)} \cdots \frac{\pi}{2^{2-2}\Gamma(1)}$$

$$= \frac{\pi^{p(p+1)/2}}{2^{p(p-1)}\Gamma(p)\Gamma(p-1)\ldots\Gamma(1)} = \frac{\pi^{p^2}}{2^{p(p-1)}\tilde{\Gamma}_p(p)}.$$

This is what we obtained in Example 4.1.

Exercises

4.1.1 Let $\tilde{X}$ be a $p \times p$ hermitian matrix of functionally independent complex variables. Show that

$$\int_{\tilde{X}} e^{-\text{tr}(\tilde{X}\tilde{X}^*)} d\tilde{X} = 2^{-\frac{p(p-1)}{2}} \pi^{\frac{p^2}{2}}.$$

4.1.2 By using Exercise 4.1.1 or otherwise show that

$$\int_{\infty > \lambda_1 > \ldots > \lambda_p > -\infty} \left\{ \prod_{j>k} |\lambda_k - \lambda_j|^2 \right\} \exp \left\{ -\sum_{j=1}^{p} \lambda_j^2 \right\} d\lambda_1 \ldots d\lambda_p$$

$$= 2^{-\frac{p(p-1)}{2}} \pi^{\frac{p}{2}} \left[\prod_{j=1}^{p} j! \right].$$

4.1.3 Show that

$$\Gamma_p \left(\frac{p+1}{2} \right) \Gamma_p \left(\frac{p}{2} \right) = 2^{-p(p-1)/2} \pi^{p/2} \tilde{\Gamma}_p(p).$$

4.1.4 Show that

$$\int_{\infty > y_1 > \ldots > y_p > -\infty} \left\{ \prod_{j<k} |y_j - y_k| \right\} e^{-\frac{1}{2} \sum_{j=1}^{p} y_j^2} dy_1 \ldots dy_p$$

$$= 2^{p/2} \left\{ \prod_{j=1}^{p} \Gamma \left(\frac{j}{2} \right) \right\}.$$

4.1.5 Show that

$$\frac{1}{(2\pi)^{p/2}} \int_{-\infty}^{\infty} \cdots \int_{-\infty}^{\infty} \left\{ \prod_{j>r} (x_j - x_r)^{2k} \right\} e^{-\frac{1}{2}(x_1^2 + \ldots + x_p^2)} dx_1 \ldots dx_p$$

$$= \prod_{j=1}^{p} \frac{\Gamma(1+jk)}{\Gamma(1+k)} \quad \text{for } \text{Re}(k) > -p^{-1}.$$

4.2   Transformations involving unitary matrices

When dealing with unitary matrices a basic property to be noted is the following:

$$\tilde{U}\tilde{U}^* = I \Rightarrow (d\tilde{U})\tilde{U}^* + \tilde{U}(d\tilde{U}^*) = 0$$
$$\Rightarrow \tilde{U}^*(d\tilde{U}) = -(d\tilde{U}^*)\tilde{U}.$$

But

$$\left[\tilde{U}^*(d\tilde{U})\right]^* = (d\tilde{U}^*)\tilde{U}$$

which means that $\tilde{U}^*(d\tilde{U})$ is a skew hermitian matrix. The wedge product of $\tilde{U}^*(d\tilde{U})$, namely, $\wedge \left[\tilde{U}^*(d\tilde{U})\right]$ enters into the picture when evaluating the Jacobians involving unitary transformations. Hence this will be denoted by $d\tilde{G}$ for convenience. Starting from $\tilde{U}^*\tilde{U} = I$ one has $(d\tilde{U})\tilde{U}^*$.

Notation 4.1   When $\tilde{U}$ is a $p \times p$ unitary matrix of functionally independent complex variables, $\tilde{U}^*$ its conjugate transpose and $(d\tilde{U})$ the matrix of differentials then the wedge product in $(d\tilde{G}) = (d\tilde{U})\tilde{U}^*$ will be denoted by $d\tilde{G}$. That is, ignoring the sign,

$$d\tilde{G} = \wedge \left[(d\tilde{U})\tilde{U}^*\right] = \wedge \left[\tilde{U}(d\tilde{U}^*)\right].$$

If the diagonal elements or the elements in one row of this unitary matrix $\tilde{U}$ are assumed to be real then the skew hermitian matrix $(d\tilde{U})\tilde{U}^*$ will be denoted by $(d\tilde{G}_1)$ and its wedge product by

$$d\tilde{G}_1 = \wedge \left[(d\tilde{U})\tilde{U}^*\right].$$

Theorem 4.3   *Let $\tilde{T}$ be a $p \times p$ lower triangular matrix and $\tilde{U}$ a $p \times p$ unitary matrix of functionally independent complex variables. Let $\tilde{X} = \tilde{T}\tilde{U}$. Then (1) for all the diagonal elements $t_{jj}, j = 1, ..., p$ of $\tilde{T}$ being real and positive, other elements in $\tilde{T}$ complex and all elements in $\tilde{U}$ complex*

$$\tilde{X} = \tilde{T}\tilde{U} \Rightarrow d\tilde{X} = \left\{\prod_{j=1}^{p} t_{jj}^{2(p-j)+1}\right\} d\tilde{T}d\tilde{G}$$

with $(\mathrm{d}\tilde{G}) = (\mathrm{d}\tilde{U})\tilde{U}^*$ and, (2) for all elements in $\tilde{T}$ being complex, all the diagonal elements in $\tilde{U}$ real and all other elements in $\tilde{U}$ complex

$$\tilde{X} = \tilde{T}\tilde{U} \Rightarrow \mathrm{d}\tilde{X} = \left\{ \prod_{j=1}^{p} |\tilde{t}_{jj}|^{2(p-j)} \right\} \mathrm{d}\tilde{T}\mathrm{d}\tilde{G}_1,$$

with $(\mathrm{d}\tilde{G}_1) = (\mathrm{d}\tilde{U})\tilde{U}^*$.

Proof   Taking differentials in $\tilde{X} = \tilde{T}\tilde{U}$ one has

$$(\mathrm{d}\tilde{X}) = (\mathrm{d}\tilde{T})\tilde{U} + \tilde{T}(\mathrm{d}\tilde{U}).$$

Postmultiplying by $\tilde{U}^*$ and observing that $\tilde{U}\tilde{U}^* = I$ we have

$$(\mathrm{d}\tilde{X})\tilde{U}^* = (\mathrm{d}\tilde{T}) + \tilde{T}(\mathrm{d}\tilde{U})\tilde{U}^*. \tag{a}$$

Case 1   Let the diagonal elements in $\tilde{T}$ be real and positive and all other elements in $\tilde{T}$ and $\tilde{U}$ be complex. Let

$$(\mathrm{d}\tilde{V}) = (\mathrm{d}\tilde{X})\tilde{U}^* \Rightarrow \mathrm{d}\tilde{V} = \mathrm{d}\tilde{X} \tag{b}$$

ignoring the sign, since $\tilde{U}$ is unitary. Let $(\mathrm{d}\tilde{G}) = (\mathrm{d}\tilde{U})\tilde{U}^*$ and its wedge product be $\mathrm{d}\tilde{G}$. Then

$$(\mathrm{d}\tilde{V}) = (\mathrm{d}\tilde{T}) + \tilde{T}(\mathrm{d}\tilde{G}) \tag{c}$$

where $(\mathrm{d}\tilde{G})$ is skew hermitian. Write

$$\tilde{V} = (\tilde{v}_{jk}), \ \tilde{v}_{jk} = v_{jk1} + iv_{jk2}, \ i = \sqrt{-1}$$
$$\tilde{T} = (\tilde{t}_{jk}), \ \tilde{t}_{jk} = t_{jk1} + it_{jk2}, \ j > k, \ t_{jj1} = t_{jj} > 0, t_{jj2} = 0$$
$$(\mathrm{d}\tilde{G}) = (\mathrm{d}\tilde{g}_{jk}), \ \mathrm{d}\tilde{g}_{jk} = \mathrm{d}g_{jk1} + i\mathrm{d}g_{jk2}, \mathrm{d}g_{jk1} = -\mathrm{d}g_{kj1}, \mathrm{d}g_{jk2} = \mathrm{d}g_{kj2}.$$

From (c)

$$\mathrm{d}\tilde{v}_{jk} = \mathrm{d}\tilde{t}_{jk} + (\tilde{t}_{j1}\mathrm{d}\tilde{g}_{1k} + \ldots + \tilde{t}_{jj}\mathrm{d}\tilde{g}_{jk}), \ j \geq k$$
$$= (\tilde{t}_{j1}\mathrm{d}\tilde{g}_{1k} + \ldots + \tilde{t}_{jj}\mathrm{d}\tilde{g}_{jk}), \ j < k.$$

The new variables are $\mathrm{d}v_{jkm} = \hat{v}_{jkm}$, $\mathrm{d}t_{jkm} = \hat{t}_{jkm}$, $j \geq k$, $\mathrm{d}g_{jkm} = \hat{g}_{jkm}$, $j \geq k$, $m = 1, 2$. The matrices of partial derivatives are easily seen to be the following:

$$\left(\frac{\partial \hat{v}_{jj1}}{\partial \hat{t}_{jj}} \right) = I, \ \left(\frac{\partial \hat{v}_{jkm}, j > k}{\partial \hat{t}_{jkm}, j > k} \right) = I, \ m = 1, 2,$$

$$\left(\frac{\partial \hat{v}_{kj1}, j > k}{\partial \hat{g}_{kj1}, j > k} \right) = A, \ \left(\frac{\partial \hat{v}_{kj2}, j > k}{\partial \hat{g}_{kj2}, j > k} \right) = B$$

where A and B are triangular matrices with t_{jj} repeated $p - j$ times,

$$\left(\frac{\partial \hat{v}_{jj2}}{\partial \hat{g}_{jj2}}\right) = \text{diag}(t_{11}, ..., t_{pp}).$$

By using the above identity matrices one can wipe out other submatrices in the same rows and columns and using the triangular blocks one can wipe out other blocks below it when evaluating the determinant of the Jacobian matrix and finally the determinant in absolute value reduces to the form

$$|A|\,|B|\,t_{11}...t_{pp} = \prod_{j=1}^{p} t_{jj}^{2(p-j)+1}.$$

Hence the result.

Case 2   Let the diagonal elements of $\tilde{U}$ be real and all other elements in $\tilde{U}$ and $\tilde{T}$ complex. Starting from equation (c), observing that (d$\tilde{G}$) is (d$\tilde{G}_1$) in this case with the wedge product d$\tilde{G}_1$, and taking the variables d$\tilde{v}_{jk}$'s in the order dv_{jk1}, $j \geq k$, dv_{jk1}, $j < k$, dv_{jk2}, $j \geq k$, dv_{jk2}, $j < k$ and the other variables in the order dt_{jk1}, $j \geq k$, dg_{jk1}, $j > k$, dt_{jk2}, $j \geq k$, dg_{jk2}, $j > k$ we have the following configuration in the Jacobian matrix:

$$\begin{bmatrix} I & * & 0 & * \\ 0 & -A_1 & 0 & A_2 \\ 0 & * & I & * \\ 0 & -A_2 & 0 & -A_1 \end{bmatrix}$$

where the matrices marked by $*$ can be made null by operating with the first and third column submatrices when taking the determinant. Thus they can be taken as null matrices, and A_1 and A_2 are triangular matrices with respectively t_{jj1} and t_{jj2} repeated $p - j$ times in the diagonal. The Jacobian matrix can be reduced to the form

$$\begin{bmatrix} A & B \\ -B & A \end{bmatrix}, \quad A = \begin{bmatrix} I & 0 \\ 0 & -A_1 \end{bmatrix}, \quad B = \begin{bmatrix} 0 & 0 \\ 0 & A_2 \end{bmatrix}.$$

Then from Lemma 3.1 the determinant is given by

$$\begin{vmatrix} A & B \\ -B & A \end{vmatrix} = |\det((A+iB)(A+iB)^*)|, \quad i = \sqrt{-1}$$

$$= |\det((-A_1+iA_2)(-A_1+iA_2)^*)|$$

$$= \left\{ \prod_{j=1}^{p} |\tilde{t}_{jj}|^{2(p-j)} \right\}$$

since $-A_1 + iA_2$ is triangular with the diagonal elements $-t_{jj1} + it_{jj2}$ repeated $p - j$ times, giving $t_{jj1}^2 + t_{jj2}^2 = |t_{jj}|^2$ repeated $p - j$ times in the final determinant and hence the result.

By using Theorem 4.3 one can obtain expressions for the integral over $\tilde{U}$ of $d\tilde{G}$ and $d\tilde{G}_1$. These will be stated as corollaries here and the proofs will be given after stating both the corollaries.

Corollary 4.3.1   Let $\tilde{U}_1$ be a $p \times p$ unitary matrix with the diagonal elements real, $\tilde{U}_1\tilde{U}_1^* = I$. Let $(d\tilde{G}_1) = (d\tilde{U}_1)\tilde{U}_1^*$. Let the full unitary group of such $p \times p$ matrices $\tilde{U}_1$ be denoted by $\tilde{O}_{1(p)}$. Then

$$\int_{\tilde{O}_{1(p)}} d\tilde{G}_1 = \frac{\pi^{p(p-1)}}{\tilde{\Gamma}_p(p)}.$$

Corollary 4.3.2   Let $\tilde{U}$ be a $p \times p$ unitary matrix with all elements complex, $\tilde{U}\tilde{U}^* = I$. Let $(d\tilde{G}) = (d\tilde{U})\tilde{U}^*$. Let the full unitary group of such $p \times p$ matrices $\tilde{U}$ be denoted by $\tilde{O}_{(p)}$. Then

$$\int_{\tilde{O}_{(p)}} d\tilde{G} = \frac{2^p \pi^{p^2}}{\tilde{\Gamma}_p(p)}.$$

Proofs   Let $\tilde{X}$ be a $p \times p$ matrix of functionally independent complex variables. Let

$$B = \int_{\tilde{X}} e^{-\operatorname{tr}(\tilde{X}\tilde{X}^*)} d\tilde{X} = \int_{\tilde{X}} e^{-\sum_{j,k} |\tilde{x}_{jk}|^2} d\tilde{X} = \pi^{p^2}$$

since

$$\int_{\tilde{x}_{jk}} e^{-|\tilde{x}_{jk}|^2} d\tilde{x}_{jk} = \int_{-\infty}^{\infty} \int_{-\infty}^{\infty} e^{-(x_{jk1}^2 + x_{jk2}^2)} dx_{jk1} dx_{jk2} = \pi.$$

Consider the transformation used in Theorem 4.3 with t_{jj}'s real and positive. Then

$$\operatorname{tr}(\tilde{X}\tilde{X}^*) = \operatorname{tr}(\tilde{T}\tilde{T}^*)$$

and let

$$B = \int_{\tilde{X}} e^{-\mathrm{tr}(\tilde{X}\tilde{X}^*)} d\tilde{X}$$

$$= \int_{\tilde{T}} \int_{\tilde{U}} \left\{ \prod_{j=1}^{p} t_{jj}^{2(p-j)+1} \right\} e^{-\sum_{j \geq k} |\tilde{t}_{jk}|^2} d\tilde{T} \, d\tilde{G}.$$

But

$$\int_0^\infty t_{jj}^{2(p-j)+1} e^{-t_{jj}^2} dt_{jj} = 2^{-1} \Gamma(p-j+1) \quad \text{for } p-j+1 > 0$$

and for $j > k$

$$\int_{\tilde{t}_{jk}} e^{-|\tilde{t}_{jk}|^2} d\tilde{t}_{jk} = \pi.$$

Then the integral over $\tilde{T}$ gives

$$2^{-p} \pi^{\frac{p(p-1)}{2}} \prod_{j=1}^{p} \Gamma(p-j+1) = 2^{-p} \tilde{\Gamma}_p(p).$$

Hence

$$\int_{\tilde{U}} d\tilde{G} = \frac{2^p \pi^{p^2}}{\tilde{\Gamma}_p(p)}.$$

Now consider the transformation used in Theorem 4.3 with all the elements in $\tilde{T}$ complex and the diagonal elements of $\tilde{U}$ real. Then

$$B = \int_{\tilde{T}} \int_{\tilde{U}_1} \left\{ \prod_{j=1}^{p} |\tilde{t}_{jj}|^{2(p-j)} \right\} e^{-\sum_{j \geq k} |\tilde{t}_{jk}|^2} d\tilde{T} \, d\tilde{G}_1.$$

Note that

$$\prod_{j>k} \left\{ \int_{\tilde{t}_{jk}} e^{-|\tilde{t}_{jk}|^2} d\tilde{t}_{jk} \right\} = \pi^{\frac{p(p-1)}{2}}.$$

Let $\tilde{t}_{jj} = \tilde{t} = t_1 + it_2, i = \sqrt{-1}$. Put $t_1 = r\cos\theta$ and $t_2 = r\sin\theta$. Let the integral over $\tilde{t}_{jj}$ be denoted by a_j. Then

$$a_j = \int_{\tilde{t}} |\tilde{t}|^{2(p-j)} e^{-|\tilde{t}|^2} d\tilde{t}$$

$$= \int_{-\infty}^{\infty} \int_{-\infty}^{\infty} (t_1^2 + t_2^2)^{(p-j)} e^{-(t_1^2 + t_2^2)} dt_1 dt_2$$

$$= 4 \int_0^\infty \int_0^\infty (t_1^2 + t_2^2)^{(p-j)} e^{-(t_1^2 + t_2^2)} dt_1 dt_2$$

$$= 4 \int_{\theta=0}^{\pi/2} \int_{r=0}^{\infty} (r^2)^{(p-j)} e^{-r^2} (r) dr \, d\theta$$

$$= \pi\Gamma(p-j+1) \text{ for } p-j+1 > 0.$$

Then

$$B = \pi^p \tilde{\Gamma}_p(p) \int_{\tilde{U}_1} d\tilde{G}_1 = \pi^{p^2} \Rightarrow \int_{\tilde{U}_1} d\tilde{G}_1 = \frac{\pi^{p(p-1)}}{\tilde{\Gamma}_p(p)}$$

which establishes the result.

Example 4.3   Evaluate the integral

$$A(\alpha) = \int_{\tilde{X}} |\det(\tilde{X}\tilde{X}^*)|^\alpha e^{-\text{tr}(\tilde{X}\tilde{X}^*)} d\tilde{X}$$

where $\tilde{X}$ is a $p \times p$ matrix of functionally independent complex variables.

Solution   Put $\tilde{X} = \tilde{T}\tilde{U}$, $\tilde{U}\tilde{U}^* = I$, $\tilde{T}$ is lower triangular with real distinct and positive diagonal elements. Then

$$|\det(\tilde{X}\tilde{X}^*)|^\alpha = \prod_{j=1}^p t_{jj}^{2\alpha},$$

$$\text{tr}(\tilde{X}\tilde{X}^*) = \text{tr}(\tilde{T}\tilde{T}^*) = \sum_{j \geq k} |\tilde{t}_{jk}|^2$$

$$= \sum_{j=1}^p t_{jj}^2 + \sum_{j>k} |\tilde{t}_{jk}|^2,$$

$$d\tilde{X} = \left[\prod_{j=1}^p t_{jj}^{2(p-j)+1}\right] d\tilde{T} d\tilde{G},$$

$$\int_{\tilde{U}} d\tilde{G} = \frac{2^p \pi^{p^2}}{\tilde{\Gamma}_p(p)}.$$

Thus

$$A(\alpha) = \int_{\tilde{T},\tilde{U}} \left[\prod_{j=1}^p t_{jj}^{2\alpha+2(p-j)+1} e^{-t_{jj}^2}\right] e^{-\sum_{j>k} |\tilde{t}_{jk}|^2} d\tilde{G}$$

$$= \left[\prod_{j=1}^p \int_0^\infty t_{jj}^{2\alpha+2(p-j)+1} e^{-t_{jj}^2} dt_{jj}\right] \left[\prod_{j>k} \int_{-\infty}^\infty e^{-|\tilde{t}_{jk}|^2} d\tilde{t}_{jk}\right] \int_{\tilde{U}} d\tilde{G}$$

$$= \left[2^{-p} \prod_{j=1}^p \Gamma(\alpha+p-j+1)\right] \left[\pi^{p(p-1)/2}\right] \left[\frac{2^p \pi^{p^2}}{\tilde{\Gamma}_p(p)}\right]$$

for $\mathrm{Re}(\alpha + p) > p - 1$ or $\mathrm{Re}(\alpha) > -1$. That is,

$$A(\alpha) = \pi^{p^2} \frac{\tilde{\Gamma}_p(\alpha + p)}{\tilde{\Gamma}_p(p)} \text{ for } \mathrm{Re}(\alpha) > -1.$$

Theorem 4.4   *Let $\tilde{X}$ be a $p \times p$ hermitian matrix of functionally indepen-*
dent complex variables with real distinct eigenvalues $\lambda_1 > \lambda_2 > ... > \lambda_p$. Let
$\tilde{U}$ be a $p \times p$ unitary matrix with real diagonal elements such that $\tilde{U}\tilde{U}^* = I$
and let $\tilde{X} = \tilde{U} D \tilde{U}^*$, $D = \mathrm{diag}(\lambda_1, ..., \lambda_p)$. Let $(\mathrm{d}\tilde{G}_1) = \tilde{U}(\mathrm{d}\tilde{U})$ where $(\mathrm{d}\tilde{U})$
is the matrix of differentials in $\tilde{U}$ and $\mathrm{d}\tilde{G}_1$ the wedge product in $(\mathrm{d}\tilde{G}_1)$.
Then

$$\tilde{X} = \tilde{U} D \tilde{U}^* \Rightarrow \mathrm{d}\tilde{X} = \left\{ \prod_{j>k} |\lambda_k - \lambda_j|^2 \right\} \mathrm{d}D \; \mathrm{d}\tilde{G}_1.$$

Proof   Take the differentials in $\tilde{X} = \tilde{U} D \tilde{U}^*$ to get

$$(\mathrm{d}\tilde{X}) = (\mathrm{d}\tilde{U})D\tilde{U}^* + \tilde{U}(\mathrm{d}D)\tilde{U}^* + \tilde{U}D(\mathrm{d}\tilde{U}^*).$$

Premultiply by $\tilde{U}^*$, postmultiply by $\tilde{U}$ and observe that $(\mathrm{d}\tilde{G}_1)$ is skew
hermitian. Then one has

$$(\mathrm{d}\tilde{W}) = (\mathrm{d}\tilde{G}_1)D + (\mathrm{d}D) - D(\mathrm{d}\tilde{G}_1)$$

where $(\mathrm{d}W) = \tilde{U}^*(\mathrm{d}X)\tilde{U}$ with $\mathrm{d}\tilde{W} = \mathrm{d}\tilde{X}$. Using the same steps as in the
proof of Theorem 4.2 we have

$$\mathrm{d}\tilde{W} = \left\{ \prod_{j>k} |\lambda_k - \lambda_j|^2 \right\} \mathrm{d}D \; \mathrm{d}\tilde{G}_1.$$

Example 4.4   Let $D = \mathrm{diag}(\lambda_1, ..., \lambda_p)$ where the λ_j's are real distinct
and positive or let $\lambda_1 > ... > \lambda_p > 0$. Show that

(i)
$$\int_D \left\{ \prod_{j>k} |\lambda_k - \lambda_j|^2 \right\} \mathrm{e}^{-\mathrm{tr}(D)} \mathrm{d}D = \frac{[\tilde{\Gamma}_p(p)]^2}{\pi^{p(p-1)}}$$

and

(ii) $\quad \int_D \left\{ \prod_{j>k} |\lambda_k - \lambda_j|^2 \right\} \left\{ \prod_{j=1}^p |\lambda_j|^{\alpha-p} \right\} e^{-\text{tr}(D)} dD = \dfrac{\tilde{\Gamma}_p(\alpha)\tilde{\Gamma}_p(p)}{\pi^{p(p-1)}}.$

Solution   Let $\tilde{Y}$ be a $p \times p$ hermitian positive definite matrix, $\tilde{U}$ a unitary matrix with real diagonal elements such that

$$\tilde{U}^* \tilde{Y} \tilde{U} = D = \text{diag}(\lambda_1, ..., \lambda_p).$$

From the matrix-variate gamma integral

$$\tilde{\Gamma}_p(\alpha) = \int_{\tilde{Y}=\tilde{Y}^*>0} |\det(\tilde{Y})|^{\alpha-p} e^{-\text{tr}(\tilde{Y})} d\tilde{Y} \text{ for } \text{Re}(\alpha) > p-1.$$

Hence

$$\tilde{\Gamma}_p(p) = \int_{\tilde{Y}=\tilde{Y}^*>0} e^{-\text{tr}(\tilde{Y})} d\tilde{Y}.$$

Let

$$\tilde{Y} = \tilde{U} D \tilde{U}^* \Rightarrow$$
$$d\tilde{Y} = \left\{ \prod_{j>k} |\lambda_k - \lambda_j|^2 \right\} dD \, d\tilde{G}_1, \ (d\tilde{G}_1) = \tilde{U}^*(d\tilde{U}).$$

Note that

$$\text{tr}(\tilde{Y}) = \text{tr}(\tilde{U} D \tilde{U}^*) = \text{tr}(D\tilde{U}^*\tilde{U}) = \text{tr}(D)$$

and

$$|\det(\tilde{Y})|^{\alpha-p} = \left(\prod_{j=1}^p |\lambda_j| \right)^{\alpha-p}.$$

Then

$$\tilde{\Gamma}_p(p) = \int_D \left\{ \prod_{j>k} |\lambda_k - \lambda_j|^2 \right\} e^{-\text{tr}(D)} dD \int_{\tilde{U}} d\tilde{G}_1.$$

But from Corollary 4.3.1

$$\int_{\tilde{O}_{1(p)}} d\tilde{G}_1 = \frac{\pi^{p(p-1)}}{\tilde{\Gamma}_p(p)}.$$

Substituting this, results (i) and (ii) follow.

4.2.1   Semiunitary transformations

The next theorem is for a rectangular matrix and involves semiunitary transformations. This result can be looked upon as a generalization of Theorem 4.3 or Theorem 4.3 may be taken as a particular case of the next theorem.

Theorem 4.5   *Let $\tilde{X}$ be a $p \times n$, $n \geq p$, matrix of functionally independent complex variables. Let $\tilde{X} = \tilde{T}\tilde{U}$ where $\tilde{T}$ is a $p \times p$ nonsingular lower triangular matrix and $\tilde{U}$ is a $p \times n$ semiunitary matrix such that $\tilde{U}\tilde{U}^* = I$. Let $\tilde{Q} = \left(\begin{smallmatrix} \tilde{U} \\ \tilde{V} \end{smallmatrix}\right)$ so that $\tilde{Q}\tilde{Q}^* = I$ where $\tilde{V}$ is $(n-p) \times n$. Let the j-th row of $\tilde{Q}$ be denoted by $\tilde{q}_j$ and the j-th row of $(d\tilde{U})$ by $d\tilde{u}_j$. Then, ignoring the sign,*

$$\tilde{X} = \tilde{T}\tilde{U} \Rightarrow$$

$$d\tilde{X} = \left\{ \prod_{j=1}^{p} |\tilde{t}_{jj}|^{2(n-j)} \right\} d\tilde{T} \; h(\tilde{U}_1)$$

where

$$h(\tilde{U}_1) = \wedge_{j=1}^{p} \wedge_{k=j+1}^{n} (d\tilde{u}_j)\tilde{q}_k^*$$

when all the elements in $\tilde{T}$ are complex, all the leading diagonal elements in $\tilde{U}$ are real and all other elements in $\tilde{U}$ are complex, and

$$\tilde{X} = \tilde{T}\tilde{U} \Rightarrow$$

$$d\tilde{X} = \left\{ \prod_{j=1}^{p} t_{jj}^{2(n-j)+1} \right\} d\tilde{T} \; h(\tilde{U}_2),$$

where

$$h(\tilde{U}_2) = \wedge_{j=1}^{p} \wedge_{k=j+1}^{n} (d\tilde{u}_j)\tilde{q}_k^*$$

when all the diagonal elements of $\tilde{T}$ are real and positive, all other elements in $\tilde{T}$ are complex and all elements in $\tilde{U}$ are complex.

Proof   Take the differentials in $\tilde{X} = \tilde{T}\tilde{U}$ to get

$$\tilde{T}^{-1}(d\tilde{X}) = \tilde{T}^{-1}(d\tilde{T})\tilde{U} + (d\tilde{U}). \tag{a}$$

Take a set of another $(n-p)$ orthonormal vectors or consider the $(n-p) \times n$ matrix $\tilde{V}$ such that $\tilde{Q}$, a matrix formed by augmenting $\tilde{U}$ with $\tilde{V}$, is an $n \times n$ unitary matrix. That is,

$$\tilde{Q} = \begin{pmatrix} \tilde{U} \\ \tilde{V} \end{pmatrix}, \quad \tilde{Q}\tilde{Q}^* = I.$$

Postmultiply (a) with $\tilde{Q}^*$ to get

$$
\begin{aligned}
\tilde{T}^{-1}(d\tilde{X})\tilde{Q}^* &= \tilde{T}^{-1}(d\tilde{T})\tilde{U}\tilde{Q}^* + (d\tilde{U})\tilde{Q}^* \\
&= \tilde{T}^{-1}(d\tilde{T})[I,0] + [(d\tilde{U})\tilde{U}^*,(d\tilde{U})\tilde{V}^*].
\end{aligned} \tag{b}
$$

Case 1 Let all the elements in $\tilde{T}$ be complex and the diagonal elements in $\tilde{U}$ be real and other elements in $\tilde{U}$ complex. Let

$$
\begin{aligned}
(d\tilde{W}_1) &= \tilde{T}^{-1}(d\tilde{X})\tilde{Q}^*, \quad (d\tilde{Z}_1) = \tilde{T}^{-1}(d\tilde{T}) \\
(d\tilde{G}_1) &= (d\tilde{U})\tilde{U}^*, \quad (d\tilde{H}_1) = (d\tilde{U})\tilde{V}^*.
\end{aligned} \tag{c}
$$

Then

$$
d\tilde{W}_1 = |\det(\tilde{T}\tilde{T}^*)|^{-n} d\tilde{X} = \left\{ \prod_{j=1}^{p} |\tilde{t}_{jj}|^{-2n} \right\} d\tilde{X}
$$

by Theorem 3.2(c) of Chapter 3 and

$$
d\tilde{Z}_1 = \left\{ \prod_{j=1}^{p} |\tilde{t}_{jj}|^{-2j} \right\} d\tilde{T}
$$

by Theorem 3.3 of Chapter 3. Note that all the free variables are already accounted for in the above two wedge products along with $(d\tilde{G}_1)$ and $(d\tilde{H}_1)$. Hence

$$
d\tilde{X} = \left\{ \prod_{j=1}^{p} |\tilde{t}_{jj}|^{2(n-j)} \right\} d\tilde{T}\, h(\tilde{U}_1).
$$

Case 2 Let all the elements of $\tilde{U}$ be complex, all the diagonal elements in $\tilde{T}$ be real and positive, and all other elements in $\tilde{T}$ be complex. Then in this case denote the matrices of differentials in (c) by $(d\tilde{W})$, $(d\tilde{Z})$, $(d\tilde{G})$ and $(d\tilde{H})$ respectively. Note that $d\tilde{W} = d\tilde{W}_1$. From Theorem 4.3 we note that

$$
d\tilde{Z} = \left\{ \prod_{j=1}^{p} t_{jj}^{-(2j-1)} \right\} d\tilde{T}.
$$

Since all the independent variables are accounted for in the wedge products $d\tilde{W}$, $d\tilde{Z}$, $d\tilde{G}$ and $d\tilde{H}$ the result follows.

Proceeding as in the corollaries to Theorem 4.3 we can evaluate integrals over unitary matrices here also. These will be stated next.

Corollary 4.5.1   *Let $\tilde{U}_1 = (\tilde{u}_{jk})$ be a $p \times n$, $n \geq p$ semiunitary matrix with u_{jj}, $j = 1, ..., p$ real and all other elements complex such that $\tilde{U}_1 \tilde{U}_1^* = I$. Let $h(\tilde{U}_1)$ be the wedge product as stated in Theroem 4.5. Let $\tilde{V}_{1(p,n)}$ be the manifold of all such $\tilde{U}_1$, $\tilde{U}_1 \in \tilde{V}_{1(p,n)}$. Then*

$$\int_{\tilde{V}_{1(p,n)}} h(\tilde{U}_1) = \frac{\pi^{p(n-1)}}{\bar{\Gamma}_p(n)}.$$

Corollary 4.5.2   *Let $\tilde{U}$ be a $p \times n$, $n \geq p$ semiunitary matrix with all elements complex, $\tilde{U}\tilde{U}^* = I$. Let $h(\tilde{U}_2)$ be the wedge product as stated in Theorem 4.5. Let $\tilde{V}_{(p,n)}$ be the manifold of such $\tilde{U}$, $\tilde{U} \in \tilde{V}_{(p,n)}$. Then*

$$\int_{\tilde{V}_{(p,n)}} h(\tilde{U}_2) = \frac{2^p \pi^{np}}{\bar{\Gamma}_p(n)}.$$

Corollary 4.5.3   *Let $\tilde{X}$ be a $p \times n$, $n \geq p$ matrix as defined in Theorem 4.5 and let $\tilde{S} = \tilde{X}\tilde{X}^*$. Then*

$$\mathrm{d}\tilde{X} = 2^{-p}|\det(\tilde{S})|^{n-p}\mathrm{d}\tilde{S} \; h(\tilde{U}_2)$$

where $h(\tilde{U}_2)$ is the same as the one appearing in Theorem 4.5.

This result follows by replacing $\mathrm{d}\tilde{T}$ of Theorem 4.5 by $\mathrm{d}\tilde{S}$ with the help of the triangular decomposition in Theorem 3.7 of Chapter 3.

Transformations involving unitary and semiunitary matrices and decompositions involving skew hermitian matrices are used in statistical distribution theory in connection with the distributions of random eigenvalues and test statistics which are functions of such eigenvalues. Some sample references are the following: Carter, Khatri and Srivastava (1976), Fujikoshi (1971), Giri (1965, 1972), Goodman (1957), Gupta (1971, 1973, 1976), Hayakawa (1972a), Hirakawa (1975), Kabe (1966a, 1968), Khatri (1964, 1965, 1966, 1969), Lee, Krishnaiah and Chang (1977), Li, Pillai and Chang (1970), Nagarsenker and Das (1975), Pillai and Hsu (1979), Pillai and Jouris (1971), Pillai and Li (1970), Pillai and Young (1971), Singh (1982), Sugiyama (1972), and Waikar, Chang and Krishnaiah (1972).

Exercises

4.2.1 Let $\tilde{Z}$ be a $p \times n$ matrix of functionally independent complex random variables, B a $q \times n$ matrix of complex constants of rank $q \leq n$ and A an $n \times n$ hermitian positive definite matrix of complex constants. Let $f(\cdot, \cdot)$ be a real valued positive function for which the following integral exists. Then for $n \geq p + q$ show that

$$\int_{\tilde{Z}A^{-1}\tilde{Z}^*=\tilde{U},\ B\tilde{Z}^*=\tilde{V}^*} f\left(\tilde{Z}A^{-1}\tilde{Z}^*, B\tilde{Z}^*\right) \mathrm{d}\tilde{Z} = \frac{\pi^{p(n-q)}}{\tilde{\Gamma}_p(n-q)}|\det(A)|^p$$

$$\times |\det(BAB^*)|^{-p} f(\tilde{U}, \tilde{V}^*)|\det(\tilde{U} - \tilde{V}(BAB^*)^{-1}\tilde{V}^*)|^{n-q-p}.$$

4.2.2 Let $\tilde{Y}$ and $\tilde{X}$ be $p \times p$ hermitian positive semidefinite matrices of functionally independent complex variables of rank $r \leq p$ each such that $\tilde{Y} = A\tilde{X}A^*$ where A is a nonsingular matrix of constants. Let

$$Y = \begin{pmatrix} \tilde{Y}_{11} & \tilde{Y}_{12} \\ \tilde{Y}_{12}^* & \tilde{Y}_{22} \end{pmatrix}, \quad X = \begin{pmatrix} \tilde{X}_{11} & \tilde{X}_{12} \\ \tilde{X}_{12}^* & \tilde{X}_{22} \end{pmatrix}$$

where $\tilde{Y}_{11}$ and $\tilde{X}_{11}$ are $r \times r$ nonsingular and hermitian positive definite, and $\tilde{Y}_{22}$ and $\tilde{X}_{22}$ are $(p-r) \times (p-r)$. Let A^* be partitioned as $A^* = (A_1^*, A_2^*)$ where A_1 is $r \times p$. Then show that, ignoring the sign,

$$\mathrm{d}\tilde{Y} = \frac{|\det(A_1\tilde{X}A_1^*)|^{p-r}|\det(AA^*)|^r}{|\tilde{X}_{11}|^{p-r}}\mathrm{d}\tilde{X}.$$

4.2.3 Let $\tilde{X}$ be a $p \times n$, $n \geq p$ matrix of complex random variables having the density

$$f(\tilde{X})\mathrm{d}\tilde{X} = \frac{e^{-\mathrm{tr}\left(A^{-1}\tilde{X}\tilde{X}^*\right)}}{\pi^{np}|\det(A)|^n}\mathrm{d}\tilde{X}, \quad A = A^* > 0$$

where A is a constant hermitian positive definite matrix. Then show by using Corollary 4.5.3 or otherwise that the density of $\tilde{S} = \tilde{X}\tilde{X}^*$ is given by

$$g(\tilde{S})\mathrm{d}\tilde{S} = \frac{e^{-\mathrm{tr}\left(A^{-1}\tilde{S}\right)}|\det(\tilde{S})|^{n-p}}{|\det(A)|^n\tilde{\Gamma}_p(n)}\mathrm{d}\tilde{S}, \quad \tilde{S} = \tilde{S}^* > 0.$$

4.2.4   If Z, A, B are all real in Exercise 4.2.1 then show that the integral evaluates to the following:

$$
\pi^{\frac{1}{2}p(n-q)-\frac{1}{4}p(p-1)} \left\{ \prod_{j=1}^{p} \Gamma \left(\frac{n-q-j+1}{2} \right) \right\}^{-1} |A|^{\frac{1}{2}p}
$$
$$
\times |BAB'|^{-\frac{1}{2}p} f(S, V') |U - V(BAB')V'|^{\frac{1}{2}(n-p-q-1)}.
$$

4.2.5   When the matrices are real in Exercise 4.2.2 show that

$$
dY = \frac{|A_1 X A_1'|^{\frac{1}{2}(p-r)} |A|^{r+1}}{|X_{11}|^{\frac{1}{2}(p-r)}} dX.
$$

4.2.6   Show that

$$
\int_0^1 \cdots \int_0^1 \left\{ \prod_{j>r} (x_j - x_r)^{2k} \right\} \left\{ \prod_{j=1}^{p} x_j^{\alpha-1}(1-x_j)^{\beta-1} dx_j \right\}
$$
$$
= \prod_{j=1}^{p} \frac{\Gamma(1+jk)\Gamma(\alpha+(j-1)k)\Gamma(\beta+(j-1)k)}{\Gamma(1+k)\Gamma(\alpha+\beta+(p+j-2)k)}
$$

for $\mathrm{Re}(\alpha) > 0$, $\mathrm{Re}(\beta) > 0$, $\mathrm{Re}(k) > -\min \left\{ n^{-1}, \mathrm{Re}\left(\frac{\alpha}{p-1} \right), \mathrm{Re}\left(\frac{\beta}{p-1} \right) \right\}$.

4.3 Transformations involving singular values

In the previous sections we have been dealing with hermitian and skew hermitian matrices. If we have an arbitrary square matrix of complex variables to be diagonalized then one has to go through what is known as a singular value decomposition.

Definition 4.1 Singular values in the complex case Let $\tilde{X}$ be an $n \times p$, $n \geq p$ matrix of rank r and of functionally independent complex variables. Let $\tilde{U}$ be $n \times r$ and $\tilde{V}$ be $p \times r$ semiunitary matrices such that $\tilde{U}^*\tilde{U} = I_r$ and $\tilde{V}^*\tilde{V} = I_r$. Then there exist a diagonal matrix $\tilde{D} = \text{diag}(\tilde{\mu}_1, ..., \tilde{\mu}_r)$ and two matrices $\tilde{U}$ and $\tilde{V}$ such that $\tilde{X} = \tilde{U}\tilde{D}\tilde{V}^*$. These $\tilde{\mu}_1, ..., \tilde{\mu}_r$ are called the singular values of $\tilde{X}$.

The following results are immediate consequences of the definition itself:

$$\tilde{X}\tilde{X}^* = (\tilde{U}\tilde{D}\tilde{V}^*)(\tilde{V}\tilde{D}^*\tilde{U}^*) = \tilde{U}\tilde{D}\tilde{D}^*\tilde{U}^*$$

and

$$\tilde{X}^*\tilde{X} = \tilde{V}\tilde{D}\tilde{D}^*\tilde{V}^*.$$

Note that

$$\tilde{D}\tilde{D}^* = \text{diag}\left(|\tilde{\mu}_1|^2, ..., |\tilde{\mu}_r|^2\right)$$

and thus $|\tilde{\mu}_1|^2, ..., |\tilde{\mu}_r|^2$ are the eigenvalues of $\tilde{X}\tilde{X}^*$ as well as those of $\tilde{X}^*\tilde{X}$.

Consider the simple case of $r = p$, that is, the matrix $\tilde{X}$ is of full rank p. Then there are $2np$ real variables in $\tilde{X}$. There are $2np - p^2$ real variables in $\tilde{U}$ and p^2 real variables in $\tilde{V}$. But there are $2p$ real variables in $\tilde{D}$ if the singular values are of functionally independent complex variables. Thus there are $2np + 2p$ variables in $\tilde{U}\tilde{D}\tilde{V}^*$. Hence we can impose $2p$ conditions on the elements of $\tilde{U}$ and $\tilde{V}$. Let us assume that the leading diagonal elements of $\tilde{V}$ and $\tilde{U}$ are real.

4.3.1 Square matrices of full rank

Let $\tilde{X}, \tilde{U}, \tilde{V}, \tilde{D}$ be $p \times p$ nonsingular matrices of functionally independent complex variables such that $\tilde{U}$ and $\tilde{V}$ are unitary and $\tilde{D}$ is diagonal with diagonal elements $\tilde{\mu}_1, ..., \tilde{\mu}_p$. Let

$$\tilde{X} = \tilde{U}\tilde{D}\tilde{V}^* = \tilde{\mu}_1\tilde{u}_1\tilde{v}_1^* + ... + \tilde{\mu}_p\tilde{u}_p\tilde{v}_p^* \qquad (i)$$

where $\tilde{u}_1, ..., \tilde{u}_p$ and $\tilde{v}_1, ..., \tilde{v}_p$ are the columns of $\tilde{U}$ and $\tilde{V}$ respectively. Then

$$\tilde{X}\tilde{X}^* = \tilde{U}\tilde{D}\tilde{D}^*\tilde{U}^*$$

which means that $|\tilde{\mu}_1|^2, ..., |\tilde{\mu}_p|^2$ are the eigenvalues of $\tilde{X}\tilde{X}^*$ with the corresponding eigenvectors $\tilde{u}_1, ..., \tilde{u}_p$. Thus if $\tilde{\mu}_1, ..., \tilde{\mu}_p$ are functionally independent then $\tilde{u}_1, ..., \tilde{u}_p$ are uniquely determined, up to a constant multiple of $\pm 1, \pm i$, in terms of the elements of $\tilde{X}$. Similarly since

$$\tilde{X}^*\tilde{X} = \tilde{V}\tilde{D}^*\tilde{D}\tilde{V}^*$$

the vectors $\tilde{v}_1, ..., \tilde{v}_p$ are uniquely determined, up to a constant multiple, in terms of $\tilde{X}$. Then from (i), $\tilde{\mu}_1, ..., \tilde{\mu}_p$ are uniquely determined in terms of $\tilde{X}$ since

$$\tilde{u}_j^*\tilde{X}\tilde{v}_j = \tilde{\mu}_j.$$

Note also that there are 4^p choices for each of $\tilde{U}$ and $\tilde{V}$ alone but combined there are only 4^p choices of $\tilde{U}$ and $\tilde{V}$. Hence we have a unique representation $\tilde{X} = \tilde{U}\tilde{D}\tilde{V}^*$ in the following two situations: (i) The elements of any particular row or of the diagonal of $\tilde{U}$ as well as those of $\tilde{V}$ are real, these elements either in $\tilde{U}$ or in $\tilde{V}$ are of specified signs and the elements of the diagonal matrix $\tilde{D}$ are functionally independent complex variables; (ii) The elements in any row or of the diagonal of $\tilde{U}$ or those of $\tilde{V}$ are real and of specified signs and $\tilde{\mu}_j$, $j = 1, ..., p$ are real, distinct and nonzero.

Theorem 4.6   *Let $\tilde{X}$ be a $p \times p$ nonsingular matrix of functionally independent complex variables. Let $\tilde{U}$ and $\tilde{V}$ be $p \times p$ unitary matrices with real elements for any of its rows or diagonal and those elements either in $\tilde{U}$ or in $\tilde{V}$ be positive, and $\tilde{D} = \text{diag}(\tilde{\mu}_1, ..., \tilde{\mu}_p)$ be a diagonal matrix of functionally independent complex variables. Then*

$$\tilde{X} = \tilde{U}\tilde{D}\tilde{V}^* \Rightarrow$$

$$d\tilde{X} = \left\{ \prod_{j<k} \left| |\tilde{\mu}_j|^2 - |\tilde{\mu}_k|^2 \right| \right\} d\tilde{D}d\tilde{G}_1 d\tilde{H}_1$$

where

$$(d\tilde{G}_1) = \tilde{U}^*(d\tilde{U}) \text{ and } (d\tilde{H}_1) = (d\tilde{V}^*)\tilde{V}.$$

Proof   Taking the differentials in $\tilde{X} = \tilde{U}\tilde{D}\tilde{V}^*$ one has

$$(d\tilde{X}) = (d\tilde{U})\tilde{D}\tilde{V}^* + \tilde{U}(d\tilde{D})\tilde{V}^* + \tilde{U}\tilde{D}(d\tilde{V}^*). \tag{a}$$

Premultiplying by $\tilde{U}^*$ and postmultiplying by $\tilde{V}$ one has

$$\tilde{U}^*(d\tilde{X})\tilde{V} = (d\tilde{G}_1)\tilde{D} + (d\tilde{D}) + \tilde{D}(d\tilde{H}_1). \tag{b}$$

Let

$$(d\tilde{W}) = \tilde{U}^*(d\tilde{X})\tilde{V} \Rightarrow d\tilde{W} = d\tilde{X}, \tag{c}$$

ignoring the sign, since $\tilde{U}$ and $\tilde{V}$ are unitary. Let

$$(d\tilde{G}_1) = (d\tilde{g}_{jk}), \ (d\tilde{H}_1) = (d\tilde{h}_{jk}), \ (d\tilde{W}) = (d\tilde{w}_{jk}), \ (d\tilde{D}) = (d\tilde{\mu}_j),$$

$$\tilde{g}_{jk} = g_{jk1} + ig_{jk2}, \ \tilde{h}_{jk} = h_{jk1} + ih_{jk2},$$

$$\tilde{w}_{jk} = w_{jk1} + iw_{jk2}, \ \tilde{\mu}_j = \mu_{j1} + i\mu_{j2}.$$

Let $\hat{g}_{jkm} = dg_{jkm}$, $\hat{h}_{jkm} = dh_{jkm}$, $\hat{w}_{jkm} = dw_{jkm}$, $\hat{\mu}_{jm} = d\mu_{jm}$, $m = 1, 2$.
Take the matrix of partial derivatives by taking the $\hat{w}$-variables in the order

$$(\hat{w}_{121}, \hat{w}_{211}), (\hat{w}_{131}, \hat{w}_{311}), ..., (\hat{w}_{122}, \hat{w}_{212}), (\hat{w}_{132}, \hat{w}_{312}), ...$$

and the $\hat{g}$- and $\hat{h}$-variables in the order

$$(\hat{g}_{121}, \hat{h}_{121}), (\hat{g}_{131}, \hat{h}_{131}), ..., (\hat{g}_{122}, \hat{h}_{122}), (\hat{g}_{132}, \hat{h}_{132}), ...$$

Note that the elements in $(d\tilde{D})$ appears only with the diagonal elements of $(d\tilde{W})$ and since the corresponding matrices of partial derivatives are identity matrices we need to consider only the nondiagonal elements of $(d\tilde{W})$. Since p restrictions are put on the diagonal elements of $\tilde{U}$ and $\tilde{V}$ we need to consider only the upper triangular elements of $(d\tilde{G}_1)$ and $(d\tilde{H}_1)$. Note also that $(d\tilde{G}_1)$ and $(d\tilde{H}_1)$ are skew hermitian matrices. The matrix of partial derivatives will give several blocks of the following type:

$$\frac{\partial(\hat{w}_{121}, \hat{w}_{211})}{\partial(\hat{g}_{121}, \hat{h}_{121})} = \begin{bmatrix} -\mu_{21} & \mu_{11} \\ \mu_{11} & -\mu_{21} \end{bmatrix}, \ \frac{\partial(\hat{w}_{121}, \hat{w}_{211})}{\partial(\hat{g}_{122}, \hat{h}_{122})} = \begin{bmatrix} -\mu_{22} & -\mu_{12} \\ -\mu_{12} & -\mu_{22} \end{bmatrix}$$

and other blocks in this set will be null matrices;

$$\frac{\partial(\hat{w}_{131}, \hat{w}_{311})}{\partial(\hat{g}_{131}, \hat{h}_{131})} = \begin{bmatrix} -\mu_{31} & \mu_{11} \\ \mu_{11} & -\mu_{31} \end{bmatrix}, \ \frac{\partial(\hat{w}_{131}, \hat{w}_{311})}{\partial(\hat{g}_{132}, \hat{h}_{132})} = \begin{bmatrix} -\mu_{32} & -\mu_{12} \\ -\mu_{12} & -\mu_{32} \end{bmatrix}$$

and other blocks in this set are null matrices, and so on. The above form is available by taking the variables in the form $-\hat{g}_{jk1}$, $j < k$, $\hat{g}_{jk1}$, $j > k$ and $\hat{h}_{jk1}$, $j < k$, $-\hat{h}_{jk1}$, $j > k$ to keep the skew hermitian format. Multiply the columns corresponding to $\hat{g}_{211}, \hat{g}_{311}, \hat{g}_{321}, ...$ and the rows corresponding to $\hat{w}_{211}, \hat{w}_{311}, ...$ by -1. Then the final configuration of the Jacobian matrix corresponding to $\hat{g}$- and $\hat{h}$-variables will be of the form

$$\begin{bmatrix} A & B \\ -B & A \end{bmatrix} \text{ with determinant } |\det(A + iB)|$$

see also Lemma 3.1. But A and B are diagonal block matrices with the diagonal blocks of $A + iB$ of the form

$$
\begin{bmatrix} \mu_{21} - i\mu_{22} & \mu_{11} - i\mu_{12} \\ \mu_{11} + i\mu_{12} & \mu_{21} + i\mu_{22} \end{bmatrix}, \quad
\begin{bmatrix} \mu_{31} - i\mu_{32} & \mu_{11} - i\mu_{12} \\ \mu_{11} + i\mu_{12} & \mu_{31} + i\mu_{32} \end{bmatrix}, \cdots
$$

with determinants

$$
\left(|\bar{\mu}_2|^2 - |\bar{\mu}_1|^2 \right), \left(|\bar{\mu}_3|^2 - |\bar{\mu}_1|^2 \right), \ldots.
$$

Hence the final determinant of the Jacobian matrix is of the form

$$
\prod_{j>k} \left| \left[|\bar{\mu}_j|^2 - |\bar{\mu}_k|^2 \right] \right|
$$

and the result follows.

Remark 4.1   If it is the second case, namely the diagonal elements of $\tilde{U}$ are real and of specified sign, $\tilde{V}$ is arbitrary, that is, $\tilde{V} \in \tilde{O}_{(p)}$, $\tilde{U} \in \tilde{O}_{1(p)}$, see also Corollaries 4.3.1 and 4.3.2, and the μ_j's real, distinct and nonzero then, replace $|\bar{\mu}_j|^2$ by μ_j^2, $j = 1, ..., p$ and $d\tilde{H}_1$ by dH in Theorem 4.6.

Remark 4.2   When the singular values of $\tilde{X}$ in Theorem 4.6 are real we take $\tilde{U}$ with p elements real and $\tilde{V}$ over the full unitary group. There are 2^p choices still since the p real elements could be positive or negative with the corresponding sign change in $\tilde{V}$. Hence for a unique transformation $\int_{\tilde{O}_{(p)}} d\tilde{H}$ must be divided by 2^p.

Example 4.5   Evaluate $A = \int_{0 < \tilde{X} < I} |\det(\tilde{X})|^\alpha d\tilde{X}$, for $p = 2$ when $\tilde{X}$ is of functionally independent complex variables with the singular values $\bar{\mu}_1$ and $\bar{\mu}_2$ such that $0 < |\bar{\mu}_1|^2 < |\bar{\mu}_2|^2 < 1$ if complex, and $0 < \mu_1^2 < \mu_2^2 < 1$ if real.

Solution   Case 1: Let the singular values be real. Then writing

$$
\tilde{X} = \tilde{U} D \tilde{V}^*, \quad \tilde{U} \in \tilde{O}_{1(p)}, \quad \tilde{V} \in \tilde{O}_{(p)},
$$
$$
D = \text{diag}(\mu_1, \mu_2), \quad 0 < \mu_1 < \mu_2 < 1,
$$
$$
A = \int_{0 < \mu_1^2 < \mu_2^2 < 1} |\mu_1 \mu_2|^\alpha (\mu_2^2 - \mu_1^2) d\mu_1 d\mu_2 \int_{\tilde{U}} d\tilde{G}_1 \int_{\tilde{V}} d\tilde{G}.
$$

From Corollaries 4.3.1 and 4.3.2 one has, see also Remark 4.2,

$$\frac{1}{2^p} \int_{\tilde{U}} d\tilde{G}_1 \int_{\tilde{V}} d\tilde{G} = \frac{1}{2^p} \frac{\pi^{p(p-1)}}{\tilde{\Gamma}_p(p)} \frac{2^p \pi^{p^2}}{\tilde{\Gamma}_p(p)} \text{ for } p = 2$$

$$= \pi^4 \text{ for } p = 2, \text{ and}$$

$$\int \int (\mu_1\mu_2)^\alpha (\mu_2^2 - \mu_1^2) d\mu_1 d\mu_2 = \int_{\mu_2^2=0}^{1} [\int_{\mu_1^2=0}^{\mu_2^2} |\mu_1\mu_2|^\alpha (\mu_2^2 - \mu_1^2) d\mu_1] d\mu_2$$

$$= 4 \int_{\mu_2=0}^{1} [\int_{\mu_1=0}^{\mu_2} (\mu_1\mu_2)^\alpha (\mu_2^2 - \mu_1^2) d\mu_1] d\mu_2$$

$$= \int_{0<u_1<u_2<1} (u_1 u_2)^{\frac{(\alpha-1)}{2}} (u_2 - u_1) du_1 du_2$$

$$= \frac{4}{(\alpha+1)(\alpha+2)(\alpha+3)}.$$

Hence, in the real case,

$$A = \frac{4\pi^4}{(\alpha+1)(\alpha+2)(\alpha+3)}.$$

Case 2: Let $\tilde{X} = \tilde{U}\tilde{D}\tilde{V}^*, \tilde{U} \in \tilde{O}_{1(p)}, \tilde{V} \in \tilde{O}_{1(p)}, D = \text{diag}(\tilde{\mu}_1, \tilde{\mu}_2)$. Note that

$$|\det(\tilde{X})|^\alpha = |\tilde{\mu}_1 \tilde{\mu}_2|^\alpha = [|\tilde{\mu}_1|^2 |\tilde{\mu}_2|^2]^{\alpha/2}.$$

Hence

$$A = \int_{0<|\tilde{\mu}_1|^2<|\tilde{\mu}_2|^2<1} [|\tilde{\mu}_1|^2|\tilde{\mu}_2|^2]^{\alpha/2} [|\tilde{\mu}_2|^2 - |\tilde{\mu}_1|^2] d\tilde{\mu}_1 d\tilde{\mu}_2 \int_{\tilde{U}} d\tilde{G}_1 \int_{\tilde{V}} d\tilde{H}_1.$$

But from Corollary 4.3.1

$$\int_{\tilde{U}} d\tilde{G}_1 \int_{\tilde{V}} d\tilde{H}_1 = \left[\frac{\pi^{p(p-1)}}{\Gamma_p(p)}\right]^2 = \pi^2 \text{ for } p = 2. \qquad (a)$$

Let $\tilde{\mu}_j = \mu_{j1} + i\mu_{j2}, \mu_{j1} = r_j \cos\theta_j, \mu_{j2} = r_j \sin\theta_j$. Since $|\tilde{\mu}_j|^2$ is even in μ_{j1} and μ_{j2} we have

$$\int_{\tilde{D}} [|\tilde{\mu}_1|^2|\tilde{\mu}_2|^2]^{\alpha/2} [|\tilde{\mu}_2|^2 - |\tilde{\mu}_1|^2] d\tilde{\mu}_1 d\tilde{\mu}_2$$

$$= 4^2 \int_{\theta_1=0}^{\pi/2} d\theta_1 \int_{\theta_2=0}^{\pi/2} d\theta_2$$

$$\times \int_{0<r_1<r_2<1} (r_1^2)^{\alpha/2}(r_2^2)^{\alpha/2}[r_2^2 - r_1^2]dr_1dr_2$$

$$= 4\pi^2 \int_{0<r_1<r_2<1} (r_1^2 r_2^2)^{\alpha/2}[r_2^2 - r_1^2]dr_1dr_2$$

$$= \pi^2 \int_{0<u_1<u_2<1} (u_1 u_2)^{\frac{(\alpha-1)}{2}}[u_2 - u_1]du_1du_2$$

$$= \frac{4\pi^2}{(\alpha+1)(\alpha+2)(\alpha+3)}. \tag{b}$$

From (a) and (b) one has, in the complex case,

$$A = \frac{4\pi^4}{(\alpha+1)(\alpha+2)(\alpha+3)}.$$

4.3.2 Rectangular matrices of full rank

Now we consider extending Theorem 4.6 to the case of rectangular matrices of full rank. Let $\tilde{X}$ be $n \times p$, $n \geq p$ and of rank p. Let $\tilde{U}$ be $n \times p$ and semiunitary, that is, $\tilde{U}^*\tilde{U} = I_p$, and $\tilde{V}$ be $p \times p$ and unitary. Then we have the following result which can be considered to be a generalization of Theorem 4.6.

Theorem 4.7   *Let $\tilde{X}$ be $n \times p$, $n \geq p$ of rank p and of functionally independent complex variables. Let $\tilde{U}$ be $n \times p$ and semiunitary, and $\tilde{V}$ be $p \times p$ and unitary where the elements in any particular row or the leading diagonal elements of $\tilde{U}$ and $\tilde{V}$ are real and these elements either in $\tilde{U}$ or in $\tilde{V}$ being of specified signs. Let $\tilde{D} = \mathrm{diag}(\tilde{\mu}_1, ..., \tilde{\mu}_p)$ where $\tilde{\mu}_1, ..., \tilde{\mu}_p$ are functionally independent complex variables. Then the unique representation*

$$\tilde{X} = \tilde{U}\tilde{D}\tilde{V}^* \Rightarrow$$

$$d\tilde{X} = [|\tilde{\mu}_1|...|\tilde{\mu}_p|]^{2(n-p)} \left\{ \prod_{j>k} ||\tilde{\mu}_j|^2 - |\tilde{\mu}_k|^2| \right\} d\tilde{D}d\tilde{G}_1 d\tilde{H}_1$$

where

$$(d\tilde{G}_1) = \tilde{U}^*(d\tilde{U}) \text{ and } (d\tilde{H}_1) = (d\tilde{V}^*)\tilde{V}$$

and the wedge products $d\tilde{G}_1$ and $d\tilde{H}_1$ are as given in Theorem 4.5 and Theorem 4.3 respectively.

Proof   Proceed as in the real case of Theorem 2.25 and use the steps in the proof of Theorem 4.6 to obtain the result.

Example 4.6   Let $\tilde{P} = \tilde{P}^* > 0$, $\tilde{Q} = \tilde{Q}^* > 0$ where $\tilde{P}$ is $p \times p$ and $\tilde{Q}$ is $n \times n$. Let $\tilde{P}$ and $\tilde{Q}$ be constant matrices and $\tilde{X}$ an $n \times p$, $n \geq p$, matrix of functionally independent complex variables. Let $\tilde{X} = \tilde{U}\tilde{D}\tilde{V}^*$, $\tilde{D} = \text{diag}(\tilde{\mu}_1, ..., \tilde{\mu}_p)$, $\tilde{U}^*\tilde{U} = I_p$, $\tilde{V}^*\tilde{V} = I_p$, $\tilde{U}$ be $n \times p$ and $\tilde{V}$ be $p \times p$, and the $\tilde{\mu}_j$'s be functionally independent and nonzeros, the diagonal elements of $\tilde{U}$ and $\tilde{V}$ being real and those of $\tilde{U}$ be of specified signs. Let

$$A = \int_{\tilde{X}} e^{-\text{tr}[\tilde{P}^{-1}\tilde{X}^*\tilde{Q}^{-1}\tilde{X}]} d\tilde{X}.$$

Show that

(i) $A = \pi^{np}|\det(\tilde{P})|^n|\det(\tilde{Q})|^p$

and

(ii) $\int_{\tilde{D}} | \,|\tilde{\mu}_1|...|\tilde{\mu}_p|\, |^{2(n-p)} \left\{ \prod_{j>k} |\,|\tilde{\mu}_j|^2 - |\tilde{\mu}_k|^2\,| \right\} e^{-\text{tr}(\tilde{D}^*\tilde{D})} d\tilde{D}$

$$= \frac{\tilde{\Gamma}_p(p)\tilde{\Gamma}_p(n)}{\pi^{p(p-2)}}.$$

Solution   Since $\tilde{P}$ and $\tilde{Q}$ are hermitian positive definite write $\tilde{P}^{-1} = \tilde{P}_1\tilde{P}_1^*$ and $\tilde{Q}^{-1} = \tilde{Q}_1\tilde{Q}_1^*$ and apply Theorem 3.2. Then

$$A = |\det(\tilde{P})|^n|\det(\tilde{Q})|^p \int_{\tilde{Y}} e^{-\text{tr}(\tilde{Y}^*\tilde{Y})} d\tilde{Y}. \tag{a}$$

Direct integration over $\tilde{Y}$ yields

$$\int_{\tilde{Y}} e^{-\text{tr}(\tilde{Y}^*\tilde{Y})} d\tilde{Y} = \pi^{np}. \tag{b}$$

From (a) and (b) result (i) follows. Now start with (b) and apply Theorem 4.7. Then

$$\pi^{np} = \int_{\tilde{D}} \left\{ \prod_{j=1}^{p}(|\tilde{\mu}_j|)^{2(n-p)} \right\} \left\{ \prod_{j>k} |\,|\tilde{\mu}_j|^2 - |\tilde{\mu}_k|^2\,| \right\} e^{-\text{tr}(\tilde{D}^*\tilde{D})} d\tilde{D}$$

$$\times \int_{\tilde{U}} d\tilde{G}_1 \int_{\tilde{V}} d\tilde{H}_1 \tag{c}$$

where $d\tilde{G}_1$ and $d\tilde{H}_1$ are given in Theorem 4.3. From Corollary 4.3.1

$$\int_{\tilde{V}} d\tilde{H}_1 = \frac{\pi^{p(p-1)}}{\tilde{\Gamma}_p(p)} \qquad (d)$$

and from Corollary 4.5.2

$$\int_{\tilde{U}} d\tilde{G}_1 = \frac{\pi^{p(n-1)}}{\tilde{\Gamma}_p(n)}. \qquad (e)$$

Then from (c), (d) and (e) the result in (ii) follows.

Exercises

4.3.1   Extend Theorems 2.21 and 2.22 to the complex case.

4.3.2   Extend Theorem 2.24 to the complex case.

4.3.3   For $u_1 > u_2 > ... > u_p > 0$ and for $\text{Re}(\alpha) > p - 1$ show that

$$\int_{u_1 > ... > u_p > 0} \left\{ \prod_{j>k} |u_k^2 - u_j^2|^2 \right\} \left\{ \prod_{j=1}^{p} |u_j|^{2\alpha - 2p + 1} \right\} e^{-\sum_{j=1}^{p} u_j^2} du_1 ... du_p$$

$$= 2^{-p} \frac{\tilde{\Gamma}_p(\alpha)\tilde{\Gamma}_p(p)}{\pi^{p(p-1)}}.$$

4.3.4   For $m \geq p + 2$, and denoting the following integral by I, show that

$$I = \frac{1}{n!} \int_{\lambda_1 > 0} ... \int_{\lambda_p > 0} \lambda_1^{-1} e^{-\frac{1}{2} \sum_{j=1}^{p} \lambda_j}$$

$$\times \left\{ \prod_{j=1}^{p} \lambda_j^{\frac{m}{2} - \frac{p+1}{2}} \right\} \left\{ \prod_{j>k} |\lambda_j - \lambda_k| \right\} \prod_{j=1}^{p} d\lambda_j$$

$$= n^{-1} \int_{\lambda_1 > ... > \lambda_p > 0} e^{-\frac{1}{2} \sum_{j=1}^{p} \lambda_j} \left\{ \prod_{j=1}^{p} \lambda_j^{\frac{m}{2} - \frac{p+3}{2}} \right\}$$

$$\times [\lambda_2 ... \lambda_p + \lambda_1 \lambda_3 ... \lambda_p + ... + \lambda_1 ... \lambda_{p-1}]$$

$$\times \left\{ \prod_{j>k} |\lambda_j - \lambda_k| \right\} \prod_{j=1}^{p} d\lambda_j.$$

CHAPTER 5

Some special functions
of matrix argument

5.0 Introduction

In extending the theory of scalar functions of scalar variables to those of
matrix variables we encounter several problems. If the matrix is rectangular then powers, inverses, and so on are not usually defined. If one tries
to define these by using concepts such as g-inverse still one may not have
a unique entity. Even if we confine ourselves to real square nonsingular
matrices there are still difficulties. If A is a real nonsingular matrix still its
square root, namely $A^{\frac{1}{2}}$, need not be uniquely defined. Many such problems can be avoided if we confine ourselves to symmetric positive definite
matrices when real and hermitian positive definite matrices when complex.
Hence we will consider the development of the theory of scalar functions
of matrix argument when the matrices involved are either real symmetric
positive definite or hermitian positive definite. In this chapter we will deal
with only the real case and in the next chapter we will consider the complex
case.

Consider a $p \times p$ symmetric positive definite matrix $X = (x_{jk})$ with
the x_{jk}'s being either real scalar random variables or real scalar mathematical variables. Since X is assumed to be symmetric there are $p(p+1)/2$
functionally independent scalar variables in X. Consider a scalar function
of X. For example , $f_1(X) = |X|$, the determinant of X, $f_2(X) = e^{-\text{tr}(X)}$,
$f_3(X) = |X|^2 + 5$ are all scalar functions of the matrix X. All the matrices
appearing in this chapter are $p \times p$ real symmetric positive definite unless
stated otherwise. That is, all the eigenvalues are strictly positive unless
stated otherwise.

5.1 Elementary functions of matrix argument

First we will consider exponential, gamma and beta functions and then in
the next section we will deal with general hypergeometric functions. The
basic ideas in the various approaches in defining a scalar function of matrix

argument will be introduced in this section and the details of the development will be given in the next section. Since most of the applications of the theory are found in statistical distribution theory some of the basic matrix-variate statistical distributions will also be introduced in this section.

5.1.1   Matrix-variate real gamma density

The *real matrix-variate gamma*, namely,

$$\Gamma_p(\alpha) = \pi^{p(p-1)/4}\Gamma(\alpha)\Gamma\left(\alpha - \frac{1}{2}\right)\Gamma(\alpha-1)...\Gamma\left(\alpha - \frac{p-1}{2}\right), \quad (5.1.1)$$

for $\mathrm{Re}(\alpha) > \frac{p-1}{2}$ was introduced in Example 1.24 of Chapter 1, and has the integral representation

$$\Gamma_p(\alpha) = \int_{A>0} |A|^{\alpha - \frac{p+1}{2}} e^{-\mathrm{tr}(A)} \mathrm{d}A, \quad \mathrm{Re}(\alpha) > \frac{p-1}{2}.$$

From the above integral representation one can define a matrix-variate gamma density as follows. Let $X = X' > 0$ be a real matrix random variable and $B = B' > 0$ be a constant matrix. Then

$$f(X) = \frac{|X|^{\alpha - \frac{p+1}{2}} e^{-\mathrm{tr}(BX)}}{|B|^{-\alpha}\Gamma_p(\alpha)}, \quad X = X' > 0, \ B = B' > 0, \quad (5.1.2)$$

for $\mathrm{Re}(\alpha) > \frac{p-1}{2}$ and $f(X) = 0$ elsewhere, is known as the *matrix-variate real gamma density*. From (5.1.2) we can have the following identity:

$$|B|^{-\alpha} \equiv \frac{1}{\Gamma_p(\alpha)} \int_{X>0} |X|^{\alpha - \frac{p+1}{2}} e^{-\mathrm{tr}(BX)} \mathrm{d}X \quad (5.1.3)$$

for $\mathrm{Re}(\alpha) > \frac{p-1}{2}$, and X and B are $p \times p$ real symmetric positive definite matrices.

5.1.2   Laplace transform

If x is a real positive scalar variable and if $f(x)$ is a function of x then the *Laplace transform of $f(x)$*, denoted by $L_f(t)$, whenever it exists, is given by the following:

Definition 5.1   Laplace transform: scalar variable case

$$L_f(t) = \int_0^\infty e^{-tx} f(x) dx \qquad (5.1.4)$$

where t is an arbitrary parameter. If $f(x)$ is a density associated with the scalar random variable $x > 0$ then $L_f(-t)$ is called the *moment generating function* of x. If $f(x_1, ..., x_k)$ is a function of k real scalar variables $0 < x_j < \infty$, $j = 1, ..., k$ then the Laplace transform of f is defined as follows:

Definition 5.2   Laplace transform: multivariable case

$$L_f(t_1, ..., t_k) = \int_0^\infty ... \int_0^\infty e^{-(t_1 x_1 + ... + t_k x_k)} f(x_1, ..., x_k) dx_1 ... dx_k \quad (5.1.5)$$

where $t_1, ..., t_k$ are arbitrary constants.

Our aim here is to extend this idea to cover the case of matrix variables. Let $f(X)$ be a scalar function of a real symmetric positive definite $p \times p$ matrix X. Consider a $p \times p$ matrix T of parameters such that the diagonal elements of T are t_{jj}, $j = 1, ..., p$ and the nondiagonal elements are $\frac{1}{2} t_{jk}$, $j, k = 1, ..., p$, $j \neq k$, subject to the condition that $t_{jk} = t_{kj}$ for all j and k. If such matrices X and T are considered then note that

$$\text{tr}(TX) = \text{tr}(XT) = \sum_{j \geq k} t_{jk} x_{jk}. \qquad (5.1.6)$$

Then by using the definition of Laplace transform for the multivariable case (5.1.5) we can define the Laplace transform for the matrix variable case as follows:

Definition 5.3   Laplace transform: matrix variable case

$$L_f(T) = \int_{X>0} e^{-\text{tr}(TX)} f(X) dX \qquad (5.1.7)$$

where $f(X)$ is a scalar function of the real symmetric positive definite matrix X, and T is defined in (5.1.6). The integral in (5.1.7) may not exist. When it exists (5.1.7) gives the *Laplace transform* of the function $f(X)$. If $f(X)$ is a density function then $L_f(-T)$ is called the *moment generating function* of $f(X)$ and it is denoted by $E(e^{\text{tr}(TX)})$ where E denotes the *expected value*.

Here $T = (\eta_{jk}t_{jk}), t_{jk} = t_{kj}$, for all j and k, $\eta_{jk} = 1, j = k$ and $\eta_{jk} = \frac{1}{2}, j \neq k$. The class of such $p \times p$ matrices is going to appear very often in our discussion and hence we will denote this class by S_p^*.

Notation 5.1 $S_p^* =$ the class of $p \times p$ matrices of the type T described above, $T \in S_p^*$.

Remark 5.1   In (5.1.5) all the scalar variables x_j's are positive whereas in (5.1.7) all the x_{jk}'s need not be positive. The individual x_{jk}'s could take negative values provided X remains symmetric positive definite.

Remark 5.2   If the parameter matrix T in (5.1.7) is taken as $T = (t_{jk})$ with $t_{jk} = t_{kj}$ then (5.1.7) remains as the Laplace transform of $f(X)$ where the variables are taken as x_{jj}, $j = 1, ..., p$ and $2x_{jk}$, $j > k$.

Theorem 5.1   *If X is a $p \times p$ real symmetric positive definite matrix having the matrix-variate gamma density*

$$g(X) = \frac{|X|^{\alpha - \frac{p+1}{2}} e^{-\mathrm{tr}(BX)}}{|B|^{-\alpha}\Gamma_p(\alpha)}, \quad X > 0, \ B > 0, \ \mathrm{Re}(\alpha) > \frac{p-1}{2} \qquad (5.1.8)$$

then the Laplace transform of $g(X)$ is $|I + TB^{-1}|^{-\alpha}$ and the moment generating function is $|I - TB^{-1}|^{-\alpha}$ where B is assumed to be symmetric positive definite, T is a matrix as defined in (5.1.6), that is, $T \in S_p^*$, and $I \pm TB^{-1} > 0$.

Proof   Taking the Laplace transform as the expected value of $e^{-\mathrm{tr}(TX)}$ one has

$$E\left(e^{-\mathrm{tr}(TX)}\right) = \int_{X>0} \frac{|A|^{\alpha - \frac{p+1}{2}} e^{-\mathrm{tr}((T+B)X)}}{|B|^{-\alpha}\Gamma_p(\alpha)} dX. \qquad (5.1.9)$$

Since T is symmetric with arbitrary elements and B is symmetric positive definite with constant elements, without loss of generality we may assume that $T + B$ can be written as AA' where A is a $p \times p$ nonsingular matrix. But

$$\mathrm{tr}((T+B)X) = \mathrm{tr}(AA'X) = \mathrm{tr}(A'XA).$$

Then $U = A'XA \Rightarrow dU = |A|^{p+1}dX$ by Theorem 1.20 of Chapter 1. Thus

$$E\left[e^{-\mathrm{tr}(TX)}\right] = \frac{|T+B|^{-\alpha}}{|B|^{-\alpha}\Gamma_p(\alpha)} \int_{U>0} |U|^{\alpha - \frac{p+1}{2}} e^{-\mathrm{tr}(U)} dU$$

$$= \frac{|T+B|^{-\alpha}}{|B|^{-\alpha}} = |I + TB^{-1}|^{-\alpha}.$$

Is it possible that a determinant of the type $|I + TB^{-1}|^{-\alpha}$ uniquely determines a scalar function $f(X)$ of the matrix argument X if it is known that this determinant is the Laplace transform of some function $f(X)$? One can extend the results associated with the Laplace and inverse Laplace transforms in the multivariable case and establish the results by which $f(X)$ is uniquely determined.

There are three different developments available in the literature for the theory of functions of matrix argument. One approach is through the matrix-variate Laplace and inverse Laplace transforms, see for example, Bochner (1944,1951) and Herz (1955). Another is through series representations with the help of zonal polynomials, see for example, James (1961, 1964, 1968, 1976) and Constantine (1963). The third approach due to this author, Mathai (1978,1980, 1993), is through a generalized matrix transform. The basic ideas in all these approaches will be described here briefly.

5.1.3   Functions of matrix argument through Laplace transform

If $f(X)$ is a scalar function of a $p \times p$ real symmetric positive definite matrix X and if $h(T)$ is the Laplace transform of f then the uniqueness of f through $h(T)$ can be established if f and h satisfy some conditions. Let $Z = X + iY$, $i = \sqrt{-1}$, where X and Y are real symmetric matrices belonging to the class S_p^* of Notation 5.1, and X is symmetric positive definite. Let $f(Z)$ be a complex analytic function of the $p \times p$ matrix Z. Let $f(Z)$ be symmetric in the sense that $f(Z) = f(Q^*ZQ), QQ^* = I$ for all unitary (or orthonormal when real) $p \times p$ matrices Q, where Q^* denotes the conjugate transpose of Q. Then f is a function of the p eigenvalues of Z. Also f is a complex analytic function of the p elementary symmetric functions $s_1 = \lambda_1 + \ldots + \lambda_p, s_2 = \sum_{jk} \lambda_j \lambda_k, \ldots, s_p = \lambda_1 \lambda_2 \ldots \lambda_p$ where $\lambda_1, \ldots, \lambda_p$ are the eigenvalues of Z. Let for A real symmetric positive definite

$$h(Z) = \int_{A=A'>0} e^{-\text{tr}(AZ)} f(A) dA. \qquad (5.1.10)$$

If the integral is absolutely convergent in some generalized right half plane $\text{Re}(Z) = X > X_0$ it represents a complex analytic function $h(Z)$. If

$$\int |h(X + iY)| dY < \infty$$

and

$$\lim_{|X| \to \infty} \int |h(X + iY)| dY = 0$$

for some $X = X' > X_0 = X_0' > 0$, $X \in S_p^*, Y \in S_p^*$, then Cauchy (French mathematician A.L. Cauchy (1789–1857)) inversion formula can be applied to write the inverse Laplace transform, that is,

$$\frac{1}{(2\pi i)^{p(p+1)/2}} \int_{\mathrm{Re}(Z)=X>X_0} e^{\mathrm{tr}(AZ)} h(Z) \mathrm{d}Z = \begin{cases} f(A), & \text{for } A > 0 \\ 0, & \text{elsewhere.} \end{cases} \quad (5.1.11)$$

Remark 5.3 If Z in equation (5.1.11) is taken as $Z = X + iY$ with $X = (x_{jk})$, $x_{jk} = x_{kj}$, $Y = (y_{jk})$, $y_{jk} = y_{kj}$, that is, the nondiagonal elements are not multiplied by $\frac{1}{2}$, then the left side of (5.1.11) is to be multiplied by $2^{p(p-1)/2}$. That is, for $Z = X + iY$, $X = X' > 0$, $Y = Y'$

$$\frac{2^{p(p-1)/2}}{(2\pi i)^{p(p+1)/2}} \int_{\mathrm{Re}(Z)=X>X_0} e^{\mathrm{tr}(AZ)} h(Z) \mathrm{d}Z = \begin{cases} f(A), & \text{for } A > 0 \\ 0, & \text{elsewhere.} \end{cases}$$

Suppose that $f(A)$ in (5.1.10) is $|A|^{\alpha - \frac{p+1}{2}}$ for $A = A' > 0$ and $\mathrm{Re}(\alpha) > \frac{p-1}{2}$. Then (5.1.10) yields ·

$$h(Z) = \Gamma_p(\alpha) |Z|^{-\alpha}.$$

Now from (5.1.11) we have the following particular case, which will be used quite often later on.

$$\frac{1}{(2\pi i)^{\frac{p(p+1)}{2}}} \int_{\mathrm{Re}(Z)=X>X_0} |Z|^{-\alpha} e^{\mathrm{tr}(UZ)} \mathrm{d}Z = \frac{|U|^{\alpha - \frac{p+1}{2}}}{\Gamma_p(\alpha)} \quad (5.1.12)$$

for $\mathrm{Re}(\alpha) > \frac{p-1}{2}$, $U = U' > 0$ and $Z = X + iY$, $X \in S_p^*$, $Y \in S_p^*$, $X = X' > 0$. For $Z = X + iY$, $X = X' > 0$, $Y = Y'$ where the nondiagonal elements of X and Y are not multiplied by $\frac{1}{2}$ then

$$\frac{2^{\frac{p(p-1)}{2}}}{(2\pi i)^{\frac{p(p+1)}{2}}} \int_{\mathrm{Re}(Z)=X>X_0} |Z|^{-\alpha} e^{\mathrm{tr}(UZ)} \mathrm{d}Z = \frac{|U|^{\alpha - \frac{p+1}{2}}}{\Gamma_p(\alpha)}$$

for $\mathrm{Re}(\alpha) > \frac{p-1}{2}$, $U = U' > 0$.

With the help of the above results one can develop the theory of special functions of matrix argument. It is assumed that all the special functions appearing in this chapter are symmetric in the sense $f(A) = f(Q^* A Q), QQ^* = I$. A convolution theorem for the Laplace transform can be stated as follows:

Theorem 5.2   *Let f_1 and f_2 be scalar functions of $p \times p$ real symmetric positive definite matrices. Let g_1 and g_2 be their Laplace transforms respectively. Let*

$$f_3(X) = \int_0^X f_1(X - S) f_2(S) dS \qquad (5.1.13)$$

where the integration is done over all $S = S' > 0$ and $X - S > 0$, $X = X'$. Then the Laplace transform of f_3 is $g_1 g_2$.

5.1.4   Matrix-variate real beta density

As a direct application of Theorem 5.2 one can develop a matrix-variate beta, analogous to the one in the scalar case.

Definition 5.4   Matrix-variate type-1 real beta function

It is defined by the integral

$$\int_0^I |S|^{\alpha - \frac{p+1}{2}} |I - S|^{\beta - \frac{p+1}{2}} dS = B_p^{\cdot}(\alpha, \beta) \qquad (5.1.14)$$

for $\mathrm{Re}(\alpha) > \frac{p-1}{2}$, $\mathrm{Re}(\beta) > \frac{p-1}{2}$ where $B_p(\alpha, \beta)$ is called the *matrix-variate real beta function*. Note that the integration is done over $S = S' > 0, I - S > 0$.

With the help of (5.1.14) we can define a type-1 matrix-variate real beta density as follows:

Definition 5.5   Matrix-variate type-1 real beta density

$$f(X) = \frac{|X|^{\alpha - \frac{p+1}{2}} |I - X|^{\beta - \frac{p+1}{2}}}{B_p(\alpha, \beta)}, \quad 0 < X < I, \qquad (5.1.15)$$

for $\mathrm{Re}(\alpha) > \frac{p-1}{2}$, $\mathrm{Re}(\beta) > \frac{p-1}{2}$ and $f(X) = 0$ elsewhere, where the notation $0 < X < I$ means that $X = X' > 0, I - X > 0$. With the help of Theorem 5.2 we will show that $B_p(\alpha, \beta) = \Gamma_p(\alpha) \Gamma_p(\beta) / \Gamma_p(\alpha + \beta)$. This will be given as an example and the steps will be given in detail so that these steps can be used to establish many other results later on.

Example 5.1   Show that

$$B_p(\alpha, \beta) = \Gamma_p(\alpha) \Gamma_p(\beta) / \Gamma_p(\alpha + \beta)$$

for $\mathrm{Re}(\alpha) > \frac{p-1}{2}$, $\mathrm{Re}(\beta) > \frac{p-1}{2}$.

Solution   Let X be a $p \times p$ real symmetric positive definite matrix and let $X = CC'$ where C is nonsingular. Consider a transformation of the type $U = CSC'$ which gives $dU = |C|^{p+1}dS$. For $0 < S < I$ we have $0 < U < X$ and then from (5.1.14) we have

$$B_p(\alpha, \beta)|X|^{\alpha+\beta-\frac{p+1}{2}} = \int_0^X |U|^{\alpha-\frac{p+1}{2}}|X - U|^{\beta-\frac{p+1}{2}}dU. \qquad (5.1.16)$$

Taking the Laplace transform on both sides of (5.1.16) we have the

$$\text{left side} = B_p(\alpha, \beta) \int_{X>0} |X|^{\alpha+\beta-\frac{p+1}{2}}e^{-\text{tr}(TX)}dX$$

and the

$$\text{right side} = \int_{X>0} e^{-\text{tr}(TX)} \left\{ \int_0^X |U|^{\alpha-\frac{p+1}{2}}|X - U|^{\beta-\frac{p+1}{2}}dU \right\} dX$$

$$= |T|^{-\alpha}\Gamma_p(\alpha)|T|^{-\beta}\Gamma(\beta) \qquad (5.1.17)$$

for $\text{Re}(\alpha) > \frac{p-1}{2}$, $\text{Re}(\beta) > \frac{p-1}{2}$ which follows from Theorem 5.2. The integral on the left side of (5.1.17) gives $|T|^{-(\alpha+\beta)}\Gamma_p(\alpha+\beta)B_p(\alpha, \beta)$ which establishes the result.

For the sake of illustration we will establish this result by another method also. From the definition of a matrix-variate gamma one has

$$\Gamma_p(\alpha)\Gamma_p(\beta) = \int_{X_1>0} |X_1|^{\alpha-\frac{p+1}{2}}e^{-\text{tr}(X_1)}dX_1$$

$$\times \int_{X_2>0} |X_2|^{\beta-\frac{p+1}{2}}e^{-\text{tr}(X_2)}dX_2$$

$$= \int_{X_1>0}\int_{X_2>0} |X_1|^{\alpha-\frac{p+1}{2}}|X_2|^{\beta-\frac{p+1}{2}}$$

$$\times e^{-\text{tr}(X_1+X_2)}dX_1 dX_2. \qquad (a)$$

Change X_2 to U for fixed X_1 by the transformation

$$X_2 = X_1^{\frac{1}{2}}UX_1^{\frac{1}{2}} \Rightarrow dX_2 = |X_1|^{\frac{p+1}{2}}dU.$$

But

$$|X_2| = |X_1^{\frac{1}{2}}UX_1^{\frac{1}{2}}| = |X_1||U|$$

and

$$\operatorname{tr}(X_1 + X_2) = \operatorname{tr}\left(X_1 + X_1^{\frac{1}{2}} U X_1^{\frac{1}{2}}\right) = \operatorname{tr}\left[X_1^{\frac{1}{2}}(I + U)X_1^{\frac{1}{2}}\right]$$
$$= \operatorname{tr}\left[(I + U)X_1\right] = \operatorname{tr}\left[(I + U)^{\frac{1}{2}} X_1 (I + U)^{\frac{1}{2}}\right].$$

The integral on the right of (a) is then

$$\Gamma_p(\alpha)\Gamma_p(\beta) = \int_{X_1 > 0} \int_{U > 0} |X_1|^{\alpha + \beta - \frac{p+1}{2}} |U|^{\beta - \frac{p+1}{2}}$$
$$\times e^{-\operatorname{tr}\left[(I + U)^{\frac{1}{2}} X_1 (I + U)^{\frac{1}{2}}\right]} dX_1 dU. \qquad (b)$$

Integrating out X_1 in (b) by using a gamma integral one has

$$\Gamma_p(\alpha)\Gamma_p(\beta) = \Gamma_p(\alpha + \beta) \int_{U > 0} |U|^{\beta - \frac{p+1}{2}} |I + U|^{-(\alpha + \beta)} dU. \qquad (c)$$

That is,

$$\int_{U > 0} |U|^{\beta - \frac{p+1}{2}} |I + U|^{-(\alpha + \beta)} dU = \frac{\Gamma_p(\alpha)\Gamma_p(\beta)}{\Gamma_p(\alpha + \beta)}$$
$$= B_p(\alpha, \beta). \qquad (d)$$

Put

$$V = (I + U)^{-1} \Rightarrow V^{-1} = I + U$$
$$\Rightarrow U = V^{-1} - I$$
$$\Rightarrow |U| = |V|^{-1} |I - V|$$

and $dU = |V|^{-(p+1)} dV$ (see Theorem 1.27 of Chapter 1). The integral on the left side now becomes

$$\int_{U > 0} |U|^{\beta - \frac{p+1}{2}} |I + U|^{-(\alpha + \beta)} dU = \int_0^I |V|^{\alpha - \frac{p+1}{2}} |I - V|^{\beta - \frac{p+1}{2}} dV. \qquad (e)$$

This estblishes the result. From (d) and (e) and from the symmetry of $B_p(\alpha, \beta)$ one has the following:

$$\int_0^I |V|^{\alpha - \frac{p+1}{2}} |I - V|^{\beta - \frac{p+1}{2}} dV = \int_0^I |X|^{\beta - \frac{p+1}{2}} |I - X|^{\alpha - \frac{p+1}{2}} dX$$
$$= \int_{U > 0} |U|^{\beta - \frac{p+1}{2}} |I + U|^{-(\alpha + \beta)} dU$$
$$= \int_{W > 0} |W|^{\alpha - \frac{p+1}{2}} |I + W|^{-(\alpha + \beta)} dW$$
$$= B_p(\alpha, \beta) = B_p(\beta, \alpha) \qquad (5.1.18)$$

for $\mathrm{Re}(\alpha, \beta) > \frac{p-1}{2}$. The notation $\mathrm{Re}(\cdot, \cdot) > \frac{p-1}{2}$ means that each quantity is such that $\mathrm{Re}(\cdot) > \frac{p-1}{2}$. From this we can define a real type-2 beta density as follows:

Definition 5.6   Matrix-variate type-2 real beta density

$$f(X) = \frac{|X|^{\alpha - \frac{p+1}{2}}|I + X|^{-(\alpha+\beta)}}{B_p(\alpha, \beta)} \tag{5.1.19}$$

for $X = X' > 0$, $\mathrm{Re}(\alpha, \beta) > (p-1)/2$ and $f(X) = 0$ elsewhere.

Example 5.2   Let the $p \times p$ real random matrices X_j, $j = 1, 2$ be independently gamma distributed with the densities

$$f_j(X_j) = \frac{|X_j|^{\alpha_j - \frac{p+1}{2}} e^{-\mathrm{tr}(BX_j)}}{|B|^{-\alpha_j} \Gamma_p(\alpha_j)}$$

for $X_j = X_j' > 0$, $\mathrm{Re}(\alpha_j) > (p-1)/2$ and $f_j(X_j) = 0$ elsewhere, $j = 1, 2$, where $B = B' > 0$ is a constant matrix. Show that

(i) $U = X_1 + X_2 \sim$ gamma $(\alpha_1 + \alpha_2, B)$;

(ii) $V = (X_1 + X_2)^{-\frac{1}{2}} X_1 (X_1 + X_2)^{-\frac{1}{2}} \sim$ type-1 beta (α_1, α_2);

(iii) U and V are independently distributed

where $\sim$ means *distributed as* and the parameters are given in the brackets.

Solution   Since X_1 and X_2 are independently distributed the joint density of X_1 and X_2, denoted by $f(X_1, X_2)$, is the product of the marginal densities. That is,

$$\begin{aligned}
f(X_1, X_2) &= f_1(X_1)f_2(X_2) \\
&= \frac{|X_1|^{\alpha_1 - \frac{p+1}{2}}|X_2|^{\alpha_2 - \frac{p+1}{2}} e^{-\mathrm{tr}(B(X_1+X_2))}}{|B|^{-(\alpha_1+\alpha_2)}\Gamma_p(\alpha_1)\Gamma_p(\alpha_2)}
\end{aligned}$$

for $X_j = X_j' > 0$, $j = 1, 2$, $B = B' > 0$. We will make a few transformations and the nonzero part of the resulting densities will be denoted by $f_j(\cdot, \cdot)$, $j = 1, 2, \ldots$ and the marginal densities by $g_j(\cdot)$, $j = 1, 2, \ldots$. Let

$U = X_1 + X_2$ for fixed X_2 which gives $dU = dX_1$, $X_2 = U - X_1$ and $0 < X_1 < U$ (that is, $U = U' > 0$, $X_1 = X_1' > 0$, $U - X_1 > 0$). Then

$$f_1(X_1, U) = \frac{|X_1|^{\alpha_1 - \frac{p+1}{2}} |U - X_1|^{\alpha_2 - \frac{p+1}{2}} e^{-tr(BU)}}{|B|^{-(\alpha_1 + \alpha_2)} \Gamma_p(\alpha_1) \Gamma_p(\alpha_2)}$$

for $U > 0$, $0 < X_1 < U$. Also

$$|U - X_1| = |U| \, |I - U^{-1} X_1| = |U| |I - X_1 U^{-1}|$$
$$= |U| |I - U^{-\frac{1}{2}} X_1 U^{-\frac{1}{2}}|$$

where $U^{\frac{1}{2}}$ is the symmetric square root of $U = U' > 0$. Change X_1 to $V = U^{-\frac{1}{2}} X_1 U^{-\frac{1}{2}}$ for fixed U. Then $X_1 = U^{\frac{1}{2}} V U^{\frac{1}{2}}$ and $dV = |U|^{-\frac{p+1}{2}} dX_1$. Then the joint density of U and V becomes

$$f_2(V, U) = \frac{|V|^{\alpha_1 - \frac{p+1}{2}} |I - V|^{\alpha_2 - \frac{p+1}{2}}}{|B|^{-(\alpha_1 + \alpha_2)} \Gamma_p(\alpha_1) \Gamma_p(\alpha_2)}$$
$$\times |U|^{\alpha_1 + \alpha_2 - \frac{p+1}{2}} e^{-tr(BU)}$$

for $0 < V < I, U > 0$. Observe that

$$0 < X_1 < U \Rightarrow 0 < U^{-\frac{1}{2}} X_1 U^{-\frac{1}{2}} < I, \; 0 < V < I.$$

The joint density $f_2(V, U)$ of U and V is a product of two functions of U and V each. Hence U and V are independently distributed. This establishes (iii). Note that the function $|U|^{\alpha_1 + \alpha_2 - \frac{p+1}{2}} e^{-tr(BU)}$ has the normalizing constant

$$C_1 = \left[|B|^{-(\alpha_1 + \alpha_2)} \Gamma_p(\alpha_1 + \alpha_2) \right]^{-1}$$

to make it a density (see the real matrix-variate gamma density in Section 5.1.1). Now multiply and divide the right side of $f_2(V, U)$ by $\Gamma_p(\alpha_1 + \alpha_2)$ to get the marginal densities $g_1(U)$ and $g_2(V)$ of U and V respectively, that is,

$$f_2(V, U) = g_1(U) g_2(V)$$

where

$$g_1(U) = \frac{|U|^{\alpha_1 + \alpha_2 - \frac{p+1}{2}} e^{-tr(BU)}}{|B|^{-(\alpha_1 + \alpha_2)} \Gamma_p(\alpha_1 + \alpha_2)}$$

for $U = U' > 0$, $Re(\alpha_1 + \alpha_2) > (p-1)/2$, $B = B' > 0$ and $g_1(U) = 0$ elsewhere and

$$g_2(V) = \frac{\Gamma_p(\alpha_1 + \alpha_2)}{\Gamma_p(\alpha_1) \Gamma_p(\alpha_2)}$$
$$\times |V|^{\alpha_1 - \frac{p+1}{2}} |I - V|^{\alpha_2 - \frac{p+1}{2}}$$

for $0 < V < I$, $\text{Re}(\alpha_j) > (p-1)/2$, $j = 1,2$ and $g_2(V) = 0$ elsewhere. Thus g_1 and g_2 establish (i) and (ii). Note that

$$V \sim \text{type-1 beta } (\alpha_1, \alpha_2) \Rightarrow$$
$$I - V \sim \text{type-1 beta } (\alpha_2, \alpha_1).$$

Remark 5.4   A particular case of the matrix-variate gamma density given in Section 5.1.1 is the Wishart density which is one of the fundamental densities in multivariate statistical analysis associated with the multivariate Gaussian distribution. (The Gaussian distribution is named after the German mathematician C.F. Gauss (1777–1855)).

Example 5.3   Denoting the Laplace transform of a scalar function $f(X)$ of a $p \times p$ real symmetric positive definite matrix X, with respect to the parameter matrix $T \in S_p^*$, by $h(T) = L_T(f)$, prove the following basic results:

(i) $$L_T\left[|X|^{-\frac{p+1}{2}}\Gamma_p\left(\frac{p+1}{2}\right)f(X)\right] = \int_{U>T} h(U)dU;$$

(ii) $$L_T\left[\left\{\Gamma_p\left(\frac{p+1}{2}\right)|X|^{-\frac{p+1}{2}}\right\}^n f(X)\right]$$

$$= \int_{W_1>T}\int_{W_2>W_1}\cdots\int_{W_n>W_{n-1}} h(W_n)dW_1...dW_n$$

where $W_j = W_j' > 0$, $W_j - T > 0$, $W_{j+1} - W_j > 0$, $j = 1,...,n-1$; and

(iii) $$L_T(1) = |T|^{-\frac{p+1}{2}}\Gamma_p\left(\frac{p+1}{2}\right).$$

Solution   Note that since $X = X' > 0$ we have

$$|X|^{-\frac{p+1}{2}} = \int_{U>0} \frac{e^{-\text{tr}(UX)}}{\Gamma_p\left(\frac{p+1}{2}\right)}dU, \ U = U' > 0$$

from (5.1.3). Let

$$M = \int_{X>0} |X|^{-\frac{p+1}{2}}e^{-\text{tr}(TX)}f(X)dX$$

$$
= \int_{X>0} e^{-\text{tr}(TX)} f(X) \left[\int_{V>0} \frac{e^{-\text{tr}(VX)}}{\Gamma_p\left(\frac{p+1}{2}\right)} dV \right] dX
$$

$$
= \frac{1}{\Gamma_p\left(\frac{p+1}{2}\right)} \int_{V>0} \int_{X>0} e^{-\text{tr}((T+V)X)} f(X) dX dV
$$

$$
= \frac{1}{\Gamma_p\left(\frac{p+1}{2}\right)} \int_{V>0} h(T+V) dV
$$

$$
= \frac{1}{\Gamma_p\left(\frac{p+1}{2}\right)} \int_{U>T} h(U) dU,
$$

by transforming $T + V = U (\Rightarrow dV = dU$ since T is a constant). This establishes (i). By repeated application of $(i), (ii)$ follows. Note that (iii) is a particular case.

Example 5.4   Establish the following:

(i) $L_T(|X|^n) = |T|^{-n-\frac{p+1}{2}} \Gamma_p \left(n + \frac{p+1}{2} \right), n > -1;$

For $X > B$, $B = B' > 0$, $\nu > -1$,

(ii) $L_T(|X - B|^\nu) = |T|^{-(\nu + \frac{p+1}{2})} e^{-\text{tr}(TB)} \Gamma_p \left(\nu + \frac{p+1}{2} \right).$

Solution   Note that (i) is a particular case of (ii) and hence consider (ii).

$$
L_T(|X - B|^\nu) = \int_{X>B} |X - B|^\nu e^{-\text{tr}(TX)} dX
$$

$$
= e^{-\text{tr}(BT)} \int_{Y>0} |Y|^\nu e^{-\text{tr}(TY)} dY
$$

for $Y = X - B$. Since $\text{tr}(TY) = \text{tr}(T^{\frac{1}{2}} Y T^{\frac{1}{2}})$ transform $U = T^{\frac{1}{2}} Y T^{\frac{1}{2}}$ and integrate out by using a gamma integral to get the result. Note that $\nu = \nu + \frac{p+1}{2} - \frac{p+1}{2}$.

5.1.5   Functions of matrix argument through zonal polynomial

Another approach for the development of functions of matrix argument is through *zonal polynomials*. Zonal polynomials associated with a matrix Z are certain symmetric functions in the eigenvalues of Z. A detailed description of various types of polynomials and the development of the theory is available from Mathai, Provost and Hayakawa (1995). Consider a $p \times p$ real symmetric positive definite matrix X. Let V_k be the vector space of homogeneous polynomials $g(X)$ of degree k in the $p(p+1)/2$ different elements of the $p \times p$ symmetric matrix X. Consider a congruent transformation $X \rightarrow LXL'$ by a nonsingular $p \times p$ matrix L. A subspace $V_s \subset V_k$ is called invariant if $LV_s \subset V_s$ for all nonsingular matrices L. If V_s has no proper invariant subspace it is called an irreducible invariant subspace. It can be shown that V_k decomposes into a direct sum of irreducible invariant subspaces V_K corresponding to each partition $K = (k_1, k_2, ..., k_p)$, $k_1 + k_2 + ... + k_p = k$, into not more than p parts. Each subspace contains a unique one dimensional subspace invariant under the orthogonal group of linear transformations. These subspaces are generated by the zonal polynomials $U_K(X)$ which when normalized in a certain fashion give the zonal polynomials $C_K(X)$. Explicit forms of these polynomials are available for small values of k. For large values of k it will be extremely difficult to compute these polynomials. For handling elementary special functions of matrix argument we need a few properties of these zonal polynomials. These will be listed here without proofs since the proofs are too involved. These properties will be sufficient to establish the results that we will discuss later on. One basic result which is an immediate consequence of the definition itself is that when X is a 1×1 matrix, namely, a scalar quantity x,

$$C_K(X) = x^k. \qquad (5.1.20)$$

Hence one can look upon $C_K(X)$ as a generalization of x^k. The exponential function has the following expansion.

$$e^{\mathrm{tr}(X)} = \sum_{k=0}^{\infty} \frac{1}{k!}(\mathrm{tr}(X))^k = \sum_{k=0}^{\infty} \sum_K \frac{C_K(X)}{k!}. \qquad (5.1.21)$$

The binomial expansion is the following. For $I - X > 0$, that is, $X = X' > 0$ and all the eigenvalues of X between 0 and 1,

$$|I - X|^{-\alpha} = \sum_{k=0}^{\infty} \sum_K \frac{(\alpha)_K}{k!} C_K(X) \qquad (5.1.22)$$

where

$$(\alpha)_K = \prod_{j=1}^{p} \left(\alpha - \frac{j-1}{2} \right)_{k_j}$$

with $K = (k_1, ..., k_p)$, $k_1 + ... + k_p = k$.

$$\int_{O_{(p)}} C_K(H'XHT)\mathrm{d}H = C_K(X)C_K(T)/C_K(I) \qquad (5.1.23)$$

where I is the identity matrix, the integral is over the orthogonal group of $p \times p$ matrices and dH is the *invariant Haar measure*. (This measure is named after the Hungarian mathematician A. Haar (1885–1933)).

$$\int_{X=X'>0} e^{-\text{tr}(ZX)} |X|^{\alpha - \frac{p+1}{2}} C_K(XT) dX = |Z|^{-\alpha} C_K(TZ^{-1}) \Gamma_p(\alpha, K)$$

(5.1.24)

for $\text{Re}(\alpha) > (p-1)/2$ where

$$\Gamma_p(\alpha, K) = \pi^{p(p-1)/4} \prod_{j=1}^{p} \Gamma\left(\alpha + k_j - \frac{j-1}{2}\right).$$

An inverse Laplace transform for the zonal polynomials can be stated as follows:

$$\frac{1}{(2\pi i)^{\frac{1}{2}p(p+1)}} \int_{\text{Re}(Z)=X>X_0} e^{\text{tr}(SZ)} |Z|^{-\alpha} C_K(Z) dZ$$

$$= \frac{1}{\Gamma_p(\alpha, K)} |S|^{\alpha - \frac{1}{2}(p+1)} C_K(S)$$

(5.1.25)

for $Z = X + iY$, $X \in S_p^*$, $Y \in S_p^*$, $X = X' > 0$. But if X and Y are simply symmetric with X positive definite, that is, if the nondiagonal elements of X and Y are not multiplied by $\frac{1}{2}$ then the left side of (5.1.25) is to be multiplied by $2^{p(p-1)/2}$.

A type-1 beta integral can be given as follows:

$$\int_0^I |X|^{\alpha - \frac{p+1}{2}} |I - X|^{\beta - \frac{p+1}{2}} C_K(TX) dX = \frac{\Gamma_p(\alpha, K) \Gamma_p(\beta)}{\Gamma_p(\alpha + \beta, K)} C_K(T). \quad (5.1.26)$$

By making a change of variable and taking a particular case one can obtain the following integral:

$$\int_{0<S<A} |S|^{\alpha - \frac{1}{2}(p+1)} C_K(ZS) dS = \frac{\Gamma_p(\alpha, K) \Gamma_p\left(\frac{p+1}{2}\right)}{\Gamma_p\left(\alpha + \frac{p+1}{2}, K\right)} |A|^{\alpha} C_K(ZA).$$

(5.1.27)

The definition of a general hypergeometric function in terms of zonal polynomial will be given in Section 5.2.

5.1.6 The Weyl fractional integral and fractional derivative

Let $S = (s_{jk}) = S' > 0$ be a real $p \times p$ symmetric positive definite matrix and $f(S)$ be a rapidly decreasing scalar function of S. By this we mean that $f(S)$ and all its partial derivatives decrease faster than any polynomial in the elements of S. Let $T = (\eta_{jk} t_{jk})$ with $t_{jk} = t_{kj}$, $\eta_{jk} = \frac{1}{2}$, $j \neq k$ and $\eta_{jj} = 1$ for all $j, k = 1, ..., p$, that is, $T \in S_p^*$. Let

$$\hat{D} = \hat{D}_T = (-1)^p \left| \frac{\partial}{\partial \hat{T}} \right|, \quad D = D_S = (-1)^p \left| \frac{\partial}{\partial S} \right| \qquad (5.1.28)$$

where $\left| \frac{\partial}{\partial \hat{T}} \right|$ denotes the determinant of the matrix of partial differential operators where the (j, k)-th element in $\frac{\partial}{\partial \hat{T}}$ is given by $\eta_{jk} \frac{\partial}{\partial t_{jk}}$. Then $\hat{D}$ operating on a scalar function $f(T)$ of the matrix T, that is $\hat{D} f(T)$, means to evaluate $\frac{\partial f}{\partial \hat{T}}$, take its determinant and multiply by $(-1)^p$. The (j, k)-th element in $\frac{\partial}{\partial S}$ is simply $\frac{\partial}{\partial s_{jk}}$.

Definition 5.7   Weyl integral   The Weyl fractional integral of order α of $f(S)$, denoted by $G(S, \alpha)$, is defined as

$$G(S, \alpha) = W^{-\alpha} f(S)$$

$$= \frac{1}{\Gamma_p(\alpha)} \int_{U > S} |U - S|^{\alpha - \frac{p+1}{2}} f(U) dU, \qquad (5.1.29)$$

for $S = S' > 0$, $U = U' > 0$, $U - S > 0$ and $\text{Re}(\alpha) > \frac{p-1}{2}$, whenever the integral converges absolutely. When $-\alpha$ is replaced by α then $W^\alpha G(S, \alpha)$ will be called Weyl fractional derivative operating on $G(S, \alpha)$. (The Weyl integral is named after the German mathematician H. Weyl (1885-1955)).

Lemma 5.1   For $\text{Re}(\alpha) > \frac{p-1}{2}$, $\text{Re}(\beta) > \frac{p-1}{2}$

$$W^{-\alpha} W^{-\beta} = W^{-(\alpha+\beta)} = W^{-\beta} W^{-\alpha} \qquad (i)$$

$$\hat{D}^n W^{-\alpha} = W^{-\alpha} \hat{D}^n = W^{-(\alpha-n)}, \quad n = 0, 1, ... \qquad (ii)$$

$$\hat{D}^n W^{-\alpha} = \hat{D}^{n+q} W^{-(\alpha+q)}, \quad n = 0, 1, ..., \quad q = 0, 1, ... \qquad (iii)$$

and

$$W^m = \hat{D}^m, \quad m = 0, 1, ... \qquad (iv)$$

When working in the class of matrices of the type T in (5.1.28), $T \in S_p^*$, then D in $(ii), (iii), (iv)$ is $\hat{D}$. When D is treated as an inverse operator of $W^{-\alpha}$ then we will be using the operator in the class S_p^* and thus D is replaced by $\hat{D}$. We will give an outline of the proof for (i) and verify the remaining in an example later on. Note that by definition

$$W^{-\alpha}f(S) = \frac{1}{\Gamma_p(\alpha)} \int_{U>S} |U - S|^{\alpha - \frac{p+1}{2}} f(U) dU$$

$$= \frac{1}{\Gamma_p(\alpha)} \int_{V>0} |V|^{\alpha - \frac{p+1}{2}} f(V + S) dV$$

$$= G(S, \alpha).$$

Then

$$W^{-\beta} \left[W^{-\alpha} f(S) \right] = W^{-\beta} G(S, \alpha)$$

$$= \frac{1}{\Gamma_p(\beta)} \int_{V>0} |V|^{\beta - \frac{p+1}{2}} G(S + V, \alpha) dV$$

$$= \frac{1}{\Gamma_p(\beta)} \int_{V>0} |V|^{\beta - \frac{p+1}{2}}$$

$$\times \left[\frac{1}{\Gamma_p(\alpha)} \int_{U>0} |U|^{\alpha - \frac{p+1}{2}} f(U + V + S) dU \right] dV.$$

Change $X = U + V \Rightarrow V = X - U$ and $0 < U < X$. Take X out of $|X - U|$ and change $X^{\frac{1}{2}} Y X^{\frac{1}{2}} = U \Rightarrow dU = |X|^{\frac{p+1}{2}} dY$. Then one has

$$W^{-\beta} \left[W^{-\alpha} f(S) \right] = \frac{1}{\Gamma_p(\alpha) \Gamma_p(\beta)} \int_{0<Y<I} |Y|^{\alpha - \frac{p+1}{2}} |I - Y|^{\beta - \frac{p+1}{2}}$$

$$\times \int_{X>0} |X|^{\alpha + \beta - \frac{p+1}{2}} f(X + S) dX$$

$$= \frac{1}{\Gamma_p(\alpha + \beta)} \int_{X>0} |X|^{\alpha + \beta - \frac{p+1}{2}} f(X + S) dX$$

by integrating out Y using a type-1 beta integral. Now the X-integral on the right is $W^{-(\alpha+\beta)} f(S)$. This establishes (i).

If $G(S, \alpha) = W^{-\alpha} f(S)$ then naturally one may wonder whether the inverse is uniquely determined, that is, given $G(S, \alpha)$ can we find $f(S)$ uniquely ? The answer is in the affirmative. It can be shown, see the details in Richards (1984) and the references therein, that $f(S)$ is uniquely determined through $G(S, \alpha)$.

Theorem 5.3   *When $G(S, \alpha)$ defined in (5.1.29) and $f(S)$ are rapidly decreasing then*

$$f(S) = W^{\alpha} G(S, \alpha) \tag{5.1.30}$$

$$= \frac{1}{\Gamma_p(m - \alpha)} \int_{U>S} \left(\hat{D}^m G(U, \alpha) \right) |U - S|^{m - \alpha - \frac{p+1}{2}} dU$$

where m is any nonnegative integer for which $\hat{D}^m G(U, \alpha)$ is absolutely integrable on $U = U' > 0$ and $\mathrm{Re}(m - \alpha) > \frac{p-1}{2}$.

Example 5.5   Verify (ii) of Lemma 5.1 and Theorem 5.3 for (a) $f(S) = e^{-\mathrm{tr}(S)}$ and (b) $f(S) = |S|^{-\beta}$.

Solution   Let

$$f(S) = e^{-\mathrm{tr}(S)} \Rightarrow$$
$$W^{-\alpha} f(S) = \frac{1}{\Gamma_p(\alpha)} \int_{U>S} |U - S|^{\alpha - \frac{p+1}{2}} e^{-\mathrm{tr}(U)} dU, \ \mathrm{Re}(\alpha) > \frac{p-1}{2}$$
$$= \frac{1}{\Gamma_p(\alpha)} \int_{V>0} |V|^{\alpha - \frac{p+1}{2}} e^{-\mathrm{tr}(V+S)} dV$$
$$= e^{-\mathrm{tr}(S)}.$$

Hence

$$\hat{D}^n \left[W^{-\alpha} f(S) \right] = \hat{D}^n e^{-\mathrm{tr}(S)} = e^{-\mathrm{tr}(S)}.$$

Thus

$$\hat{D}^n W^{-\alpha} = W^{-\alpha} \hat{D}^n = W^{-(\alpha-n)}$$

for $\mathrm{Re}(\alpha - n) > \frac{p-1}{2}$, $n = 0, 1, \ldots$. Then

$$G(S, \alpha) = e^{-\mathrm{tr}(S)} = \hat{D}^m G(S, \alpha).$$

Now let us evaluate the right side of (5.1.30).

$$W^{\alpha} G(S, \alpha) = \frac{1}{\Gamma_p(m - \alpha)} \int_{U>S} \left(\hat{D}^m G(U, \alpha) \right) |U - S|^{m - \alpha - \frac{p+1}{2}} dU$$
$$= \frac{e^{-\mathrm{tr}(S)}}{\Gamma_p(m - \alpha)} \int_{V>0} |V|^{m - \alpha - \frac{p+1}{2}} e^{-\mathrm{tr}(V)} dV$$
$$= e^{-\mathrm{tr}(S)},$$

for $\mathrm{Re}(m - \alpha) > \frac{p-1}{2}$. This verifies (a). Now consider $f(S) = |S|^{-\beta}$. The steps are parallel. We get

$$W^{-\alpha} f(S) = |S|^{-(\beta-\alpha)} \frac{\Gamma_p(\beta - \alpha)}{\Gamma_p(\beta)} = G(S, \alpha)$$

for $\mathrm{Re}(\beta - \alpha) > \frac{p-1}{2}$, $\mathrm{Re}(\beta) > \frac{p-1}{2}$. Then

$$\hat{D}^m \left[W^{-\alpha} f(S) \right] = \frac{\Gamma_p(\beta - \alpha)}{\Gamma_p(\beta)} \hat{D}^m |S|^{-(\beta-\alpha)}.$$

Now replace $|S|^{-(\beta-\alpha)}$ by an equivalent integral representation. That is, for $\mathrm{Re}(\beta - \alpha) > \frac{p-1}{2}$

$$
\begin{aligned}
\hat{D}^m\left[W^{-\alpha}f(S)\right] &= \frac{\Gamma_p(\beta-\alpha)}{\Gamma_p(\beta)}\frac{\hat{D}_S^m}{\Gamma_p(\beta-\alpha)}\int_{U>0}|U|^{\beta-\alpha-\frac{p+1}{2}}e^{-\mathrm{tr}(US)}dU \\
&= \frac{1}{\Gamma_p(\beta)}\int_{U>0}|U|^{\beta-\alpha+m-\frac{p+1}{2}}e^{-\mathrm{tr}(US)}dU \\
&= |S|^{-(\beta-\alpha+m)}\frac{\Gamma_p(\beta-\alpha+m)}{\Gamma_p(\beta)}, \quad \mathrm{Re}(\beta-\alpha+m) > \frac{p-1}{2} \\
&= \hat{D}^m G(S,\alpha),
\end{aligned}
$$

by observing that $\mathrm{tr}(US) = \sum_j u_{jj}s_{jj} + 2\sum_{j\geq k}s_{jk}u_{jk}$ and

$$
\hat{D}_S\, e^{-\mathrm{tr}(US)} = |U|e^{-\mathrm{tr}(US)}
$$

when U and S belong simply to the class of symmetric matrices. Now the right side of (5.1.30) gives

$$
\begin{aligned}
W^\alpha G(S,\alpha) &= \frac{\Gamma_p(\beta-\alpha+m)}{\Gamma_p(m-\alpha)\Gamma_p(\beta)}\int_{U>S}|U|^{-(\beta-\alpha+m)}|U-S|^{m-\alpha-\frac{p+1}{2}}dU \\
&= \frac{\Gamma_p(\beta-\alpha+m)}{\Gamma_p(\beta)\Gamma_p(m-\alpha)}\int_{V>0}|V|^{m-\alpha-\frac{p+1}{2}}|V+S|^{-(\beta-\alpha+m)}dV \\
&= |S|^{-\beta} = f(S)
\end{aligned}
$$

evaluating by using a type-2 beta integral. This verifies Theorem 5.3 for (b).

5.1.7   The M-transform

Note that $G(S,\alpha)$ in (5.1.29) still has $p(p+1)/2$ scalar variables in S and an additional parameter α. Thus intuitively one can expect a formula such as the one in (5.1.30) to recover $f(S)$ from $G(S,\alpha)$. Now consider $G(0,\alpha)$, that is,

$$
G(0,\alpha) = \frac{1}{\Gamma_p(\alpha)}\int_{U>0}|U|^{\alpha-\frac{p+1}{2}}f(U)dU. \tag{5.1.31}
$$

Is it possible to recover $f(S)$, a function of $p(p+1)/2$ real scalar variables, uniquely from $G(0,\alpha)$, a function of one variable α? Obviously the answer is in the negative. But when $f(U)$ is a symmetric function in the sense $f(AB) = f(BA)$ for all A and B for which AB and BA are defined and when $G(0, \alpha)$ exists it is seen that the class of functions $f(U)$ satisfying the integral equation (5.1.31) enjoy properties analogous to the ones enjoyed by the various types of special functions in the real scalar case. Motivated by this fact the generalized matrix transform or M-transform is defined as follows:

Definition 5.8   The M-transform   Consider the class of $p \times p$ real symmetric definite matrices and the null matrix 0. If S is a member of this class it is either strictly positive definite or strictly negative definite or null. Let α be a complex variable such that $\text{Re}(\alpha) > \frac{p-1}{2}$. Let $f(S)$ be a symmetric function in the sense described above. Then the M-transform of $f(S)$, denoted by $M_\alpha(f)$, is defined as

$$M_\alpha(f) = \int_{U=U'>0} |U|^{\alpha - \frac{p+1}{2}} f(U) dU \qquad (5.1.32)$$

whenever $M_\alpha(f)$ exists.

Note that (5.1.32) is (5.1.29) where S goes to a null matrix or $M_\alpha(f) = W^\alpha f(0)$. We will show later that the unique hypergeometric function defined through a matrix-variate Laplace transform is a member of the class of hypergeometric functions defined through (5.1.32). As examples of symmetric functions we may also observe that

$$e^{\pm\text{tr}(AB)} = e^{\pm\text{tr}(BA)},$$
$$|I \pm AB|^\alpha = |I \pm BA|^\alpha$$

for nonsingular $p \times p$ matrices A and B and thus, $|X|^\alpha, \text{tr}(X), e^{\pm\text{tr}(X)}$ and $|I \pm X|^\alpha$ are all symmetric functions in the sense described above. A convolution property for the M-transform can be stated as follows.

Theorem 5.4   Let $f_1(U)$ and $f_2(U)$ be two symmetric scalar functions of the $p \times p$ real symmetric positive definite matrix U with M-transforms $M_\rho(f_1) = g_1(\rho)$ and $M_\rho(f_2) = g_2(\rho)$. Let

$$f_3(S) = \int_{U>0} |U|^\beta f_1 \left(U^{\frac{1}{2}} S U^{\frac{1}{2}} \right) f_2(U) dU$$

then the M-transform of f_3, that is $M_\rho(f_3)$, is given by

$$M_\rho(f_3) = g_1(\rho) g_2 \left(\beta - \rho + \frac{p+1}{2} \right)$$

when f_1 and f_2 allow interchange of integrals and when the M-trasforms exist.

Example 5.6 As an illustration of the use of M-transform, establish result (ii) of Example 5.4 by using M-transform in the sense that both sides have the same M-transforms.

Solution The result established in Example 5.4 is the following:

$$L_T\left(|X - B|^\nu\right) = \Gamma_p\left(\nu + \frac{p+1}{2}\right)|T|^{-\left(\nu + \frac{p+1}{2}\right)}e^{-\mathrm{tr}(TB)}.$$

We will re-establish this by showing that the M-transforms on both sides are one and the same. The

$$\text{left side} = \int_{X>B}|X - B|^\nu e^{-\mathrm{tr}(TX)}dX$$
$$= e^{-\mathrm{tr}(BT)}\int_{Y>0}|Y|^\nu e^{-\mathrm{tr}(TY)}dY.$$

The M-transform of the left side with respect to the parameter ρ is

$$\int_{T>0}|T|^{\rho - \frac{p+1}{2}}\left\{e^{-\mathrm{tr}(BT)}\int_{Y>0}|Y|^\nu e^{-\mathrm{tr}(TY)}dY\right\}dT.$$

Writing $\nu = \nu + \frac{p+1}{2} - \frac{p+1}{2}$ and transforming $U = T^{\frac{1}{2}}YT^{\frac{1}{2}}$ the Y-integral gives $\Gamma_p\left(\nu + \frac{p+1}{2}\right)|T|^{-\left(\nu + \frac{p+1}{2}\right)}$. Hence

$$M_\rho(\text{left side}) = \Gamma_p\left(\nu + \frac{p+1}{2}\right)\int_{T>0}|T|^{\rho - \nu - \frac{p+1}{2} - \frac{p+1}{2}}e^{-\mathrm{tr}(BT)}dT$$
$$= \Gamma_p\left(\nu + \frac{p+1}{2}\right)\Gamma_p\left(\rho - \nu - \frac{p+1}{2}\right)|B|^{-\rho + \nu + \frac{p+1}{2}}$$

for $\mathrm{Re}(\nu) > -1$, $\mathrm{Re}(\rho - \nu) > p$ and

$$M_\rho(\text{right side}) = \Gamma_p\left(\nu + \frac{p+1}{2}\right)\int_{T>0}|T|^{\rho - \frac{p+1}{2} - \nu - \frac{p+1}{2}}e^{-\mathrm{tr}(TB)}dT$$
$$= \Gamma_p\left(\nu + \frac{p+1}{2}\right)\Gamma_p\left(\rho - \nu - \frac{p+1}{2}\right)|B|^{-\rho + \nu + \frac{p+1}{2}}.$$

This establishes the result.

5.1.8   Matrix-variate Dirichlet integrals and Dirichlet distributions

The Dirichlet integrals and Dirichlet densities in the scalar case can be considered to be extensions of the univariate type-1 and type-2 beta integrals and beta densities. These are discussed in the literature, see for example Exton (1976), Gupta and Richards (1987) and Jack (1964-65). Our concern here is to give the matrix-variate analogues for $p \times p$ real symmetric positive definite matrices so that the scalar cases will be available from there when $p = 1$.

Definition 5.9   Type-1 Dirichlet integral   Let $X_1, ..., X_k$ be $p \times p$ real symmetric positive definite matrices. Let

$$f_1(X_1, ..., X_k) = \left\{ \prod_{j=1}^{k} |X_j|^{\alpha_j - \frac{p+1}{2}} \right\} |I - X_1 - ... - X_k|^{\alpha_{k+1} - \frac{p+1}{2}} \quad (5.1.33)$$

for $\mathrm{Re}(\alpha_j) > \frac{p-1}{2}$, $j = 1, ..., k+1$ where the X_j's are defined over the region $\Omega = \{(X_1, ..., X_k) | 0 < X_j < I, \ 0 < X_1 + ... + X_k < I\}$ and $f_1 = 0$ elsewhere. Then the integral of f_1 over Ω is known as the type-1 Dirichlet integral.

Theorem 5.5   *If Ω is as defined above then*

$$D_1 = \int_\Omega f_1(X_1, ..., X_k) dX_1...dX_k$$

$$= \int_\Omega \left\{ \prod_{j=1}^{k} |X_j|^{\alpha_j - \frac{p+1}{2}} \right\} |I - X_1 - ... - X_k|^{\alpha_{k+1} - \frac{p+1}{2}} dX_1...dX_k$$

$$= \frac{\Gamma_p(\alpha_1)...\Gamma_p(\alpha_{k+1})}{\Gamma_p(\alpha_1 + \alpha_2 + ... + \alpha_{k+1})}. \quad (5.1.34)$$

Proof   Consider the following transformation. Let

$$X_1 = Y_1$$
$$X_2 = (I - Y_1)^{\frac{1}{2}} Y_2 (I - Y_1)^{\frac{1}{2}}$$
$$X_3 = (I - Y_1)^{\frac{1}{2}} (I - Y_2)^{\frac{1}{2}} Y_3 (I - Y_2)^{\frac{1}{2}} (I - Y_1)^{\frac{1}{2}} \quad (5.1.35)$$
$$\vdots$$
$$X_k = (I - Y_1)^{\frac{1}{2}} (I - Y_2)^{\frac{1}{2}}...(I - Y_{k-1})^{\frac{1}{2}} Y_k (I - Y_{k-1})^{\frac{1}{2}}...(I - Y_1)^{\frac{1}{2}}.$$

Since $0 < X_j < I$, $0 < X_1 + ... + X_k < I$ we have $0 < Y_j < I$, $0 < I - Y_j < I$, $j = 1, ..., k$. Let $(I - Y_j)^{1/2}$ denote the unique symmetric square root of $(I - Y_j)$. Here X_j is free of $Y_{j+1}, ..., Y_k$, $j = 1, ..., k-1$ and hence the Jacobian matrix is a triangular block matrix, see also Theorem 2.29 of Chapter 2. Then

$$dX_1...dX_k = |I - Y_1|^{(k-1)(\frac{p+1}{2})}|I - Y_2|^{(k-2)(\frac{p+1}{2})}...|I - Y_{k-1}|^{\frac{p+1}{2}}dY_1...dY_k$$

and

$$|I - X_1 - ... - X_k| = |I - Y_1||I - Y_2|...|I - Y_k|.$$

This is achieved by successively removing from the left and right $|I - Y_1|^{\frac{1}{2}}, |I - Y_2|^{\frac{1}{2}}, ...$ Hence

$$D_1 = \int_0^I |Y_1|^{\alpha_1 - \frac{p+1}{2}}|I - Y_1|^{\alpha_2 + ... + \alpha_{k+1} - \frac{p+1}{2}}dY_1 \int_0^I |Y_2|^{\alpha_2 - \frac{p+1}{2}}$$

$$\times |I - Y_2|^{\alpha_3 + ... + \alpha_{k+1} - \frac{p+1}{2}}dY_2 ... \int_0^I |Y_k|^{\alpha_k - \frac{p+1}{2}}|I - Y_k|^{\alpha_{k+1} - \frac{p+1}{2}}dY_k$$

$$= \frac{\Gamma_p(\alpha_1)\Gamma_p(\alpha_2 + ... + \alpha_{k+1})}{\Gamma_p(\alpha_1 + ... + \alpha_{k+1})} \frac{\Gamma_p(\alpha_2)\Gamma_p(\alpha_3 + ... + \alpha_{k+1})}{\Gamma_p(\alpha_2 + ... + \alpha_{k+1})} ...$$

$$\times \frac{\Gamma_p(\alpha_k)\Gamma_p(\alpha_{k+1})}{\Gamma_p(\alpha_k + \alpha_{k+1})} = \frac{\Gamma_p(\alpha_1)...\Gamma_p(\alpha_{k+1})}{\Gamma_p(\alpha_1 + \alpha_2 + ... + \alpha_{k+1})} \qquad (5.1.36)$$

for $\mathrm{Re}(\alpha_j) > \frac{p-1}{2}$, $j = 1, ..., k+1$. Note that a density can be created by normalizing f_1.

Definition 5.10   The density f_1/D_1 is known as *the matrix-variate type-1 Dirichlet density in the real case.*

As corollaries to Theorem 5.5 we can state the following results.

Corollary 5.5.1   *Let $X_1, ..., X_k$, have a type-1 matrix-variate Dirichlet density as in Definition 5.10. Let $Y_1 = X_1$, $Y_2 = X_1 + X_2, ..., Y_k = X_1 + ... + X_k$. Then the joint density of $Y_1, ..., Y_k$ is given by*

$$g(Y_1, ..., Y_k) = C_1|Y_1|^{\alpha_1 - \frac{p+1}{2}}|Y_2 - Y_1|^{\alpha_2 - \frac{p+1}{2}}$$

$$\times ...|Y_k - Y_{k-1}|^{\alpha_k - \frac{p+1}{2}}|I - Y_k|^{\alpha_{k+1} - \frac{p+1}{2}},$$

for $0 < Y_1 < ... < Y_k < I$, and $g(Y_1, ..., Y_k) = 0$ elsewhere, where $C_1 = 1/D_1$.

This result corresponds to the joint density of order statistics in the scalar case.

Corollary 5.5.2   *If the matrix random variables $X_1, ..., X_k$ are jointly type-1 Dirichlet distributed as in Definition 5.10 then any subset of the matrix variables $X_1, ..., X_k$ are also type-1 Dirichlet distributed.*

Proof   In D_1 integrate out X_1. Note that $0 < X_1 < I - X_2 - ... - X_k$. Also

$$|I - X_1 - ... - X_k| = |(I - X_2 - ... - X_k) - X_1|$$
$$= |I - X_2 - ... - X_k|$$
$$\times |I - UX_1U|,$$
$$U = (I - X_2 - ... - X_k)^{-\frac{1}{2}}.$$

Let

$$Y = (I - X_2 - ... - X_k)^{-\frac{1}{2}} X_1 (I - X_2 - ... - X_k)^{-\frac{1}{2}} \Rightarrow$$
$$dY = |I - X_2 - ... - X_k|^{-\frac{p+1}{2}} dX_1.$$

The integral over X_1 gives

$$\int_{0 < X_1 < I - X_2 - ... - X_k} |X_1|^{\alpha_1 - \frac{p+1}{2}} |I - X_1 - ... - X_k|^{\alpha_{k+1} - \frac{p+1}{2}} dX_1$$

$$= |I - X_2 - ... - X_k|^{\alpha_1 + \alpha_{k+1} - \frac{p+1}{2}} \int_{0 < Y < I} |Y|^{\alpha_1 - \frac{p+1}{2}} |I - Y|^{\alpha_{k+1} - \frac{p+1}{2}} dY$$

$$= \frac{\Gamma_p(\alpha_1)\Gamma_p(\alpha_{k+1})}{\Gamma_p(\alpha_1 + \alpha_{k+1})} |I - X_2 - ... - X_k|^{\alpha_1 + \alpha_{k+1} - \frac{p+1}{2}}.$$

Substituting this back we see that the joint density of $X_2, ..., X_k$ is again a type-1 Dirichlet with the last parameter equal to $\alpha_1 + \alpha_{k+1}$. Any X_j could have been integrated out instead of X_1 and the same steps could have been continued any number of times and hence the result.

Corollary 5.5.3   *If $X_1, ..., X_k$ have a real type-1 matrix-variate Dirichlet density as in Definition 5.10 then $U = X_1 + ... + X_k$ and $V = I - U$ are both real type-1 beta distributed.*

Proof   The method of transformation of variables as described above or generalized Laplace transform method or other methods can be used to establish these results. We will establish these by using the M-transform. Let $g(V)$ be the density of V. Since V is a function of $X_1, ..., X_k$ the integration can be done over the joint density of $X_1, ..., X_k$. That is,

$$
\begin{aligned}
M_\rho\left[g(V)\right] &= \int_V |V|^{\rho - \frac{p+1}{2}} g(V) dV \\
&= \int_{X_1} \cdots \int_{X_k} \left[f_1(X_1, ..., X_k)/D_1\right] \\
&\quad \times |I - X_1 - ... - X_k|^{\rho - \frac{p+1}{2}} dX_1 ... dX_k
\end{aligned}
\tag{a}
$$

where $[f_1(X_1, ..., X_k)/D_1]$ is the type-1 Dirichlet density. The only change in the parameters in (a) is that α_{k+1} is replaced by $\alpha_{k+1} + \rho - \frac{p+1}{2}$. The integral is then available from the normalizing constant D_1. That is,

$$
M_\rho\left[g(V)\right] = \frac{\Gamma_p\left(\rho - \frac{p+1}{2} + \alpha_{k+1}\right) \Gamma_p\left(\alpha_1 + ... + \alpha_k\right)}{\Gamma_p\left(\alpha_{k+1}\right) \Gamma_p\left(\alpha_1 + ... + \alpha_{k+1} + \rho - \frac{p+1}{2}\right)}.
$$

Comparing this with the $(\rho - \frac{p+1}{2})$-th moment of $|V|$ coming from a matrix-variate type-1 beta density, one has one function for $g(V)$ as the type-1 beta density with parameters $(\alpha_{k+1}, \alpha_1 + ... + \alpha_k)$. From the uniqueness of the density of V the result follows. Hence $U = I - V$ is type-1 beta with parameters $(\alpha_1 + ... + \alpha_k, \alpha_{k+1})$.

Definition 5.11   Real matrix-variate type-2 Dirichlet function, integral and density

Let

$$
f_2(X_1, ..., X_k) = \left\{ \prod_{j=1}^{k} |X_j|^{\alpha_j - \frac{p+1}{2}} \right\} |I + X_1 + ... + X_k|^{-(\alpha_1 + ... + \alpha_{k+1})}
$$

$$
\tag{5.1.37}
$$

for $\operatorname{Re}(\alpha_j) > \frac{p-1}{2}$, $j = 1, ..., k+1$, $X_j = X_j' > 0$, $j = 1, ..., k$ and $f_2(X_1, ..., X_k) = 0$ elsewhere. Then f_2 is known as a *real matrix-variate type-2 Dirichlet function* and

$$
D_2 = \int_{X_1 > 0} \cdots \int_{X_k > 0} f_2(X_1, ..., X_k) dX_1 ... dX_k
$$

is called a *real matrix-variate Type-2 Dirichlet integral*. It is easy to show that D_2 gives the same gamma product in (5.1.34) which is also given by D_1. Also f_2/D_2 is called a *real matrix-variate type-2 Dirichlet density*.

We can establish the following results on type-2 Dirichlet functions.

Theorem 5.6   *Let $X_1, ..., X_k$ have the type-2 Dirichlet density f_2/D_2 and let $X_0 = X_1 + ... + X_k$. Then $Y_j = (I + X_0)^{-\frac{1}{2}} X_j (I + X_0)^{-\frac{1}{2}}$, $j = 1, ..., k$ are jointly distributed as type-1 Dirichlet.*

Proof   The Jacobian of the transformation $(X_1, ..., X_k)$ to $(Y_1, ..., Y_k)$ can be seen to be $|I + X_0|^{-(k+1)(\frac{p+1}{2})}$, see also Theorem 2.31 of Chapter 2. Substituting in f_2/D_2 one has the following:

$$f_2(X_1, ..., X_k)dX_1...dX_k = C|Y_1|^{\alpha_1 - \frac{p+1}{2}}...|Y_k|^{\alpha_k - \frac{p+1}{2}}$$
$$\times |I + X_0|^{-\alpha_{k+1} + \frac{p+1}{2}} dY_1...dY_k$$

where C^{-1} is the gamma product in (5.1.34). Observe that

$$I - Y_1 - ... - Y_k = I - (I + X_0)^{-\frac{1}{2}} X_0 (I + X_0)^{-\frac{1}{2}}$$
$$= (I + X_0)^{-\frac{1}{2}}[(I + X_0) - X_0](I + X_0)^{-\frac{1}{2}}$$
$$= (I + X_0)^{-1}.$$

Substituting this the result is established.

Corollary 5.6.1   *Let $Y_1, ..., Y_k$ be type-1 Dirichlet distributed and with the density f_1/D_1. Let $Y_0 = Y_1 + ... + Y_k$ and*

$$X_j = (I - Y_0)^{-\frac{1}{2}} Y_j (I - Y_0)^{-\frac{1}{2}}, \quad j = 1, ..., k.$$

Then $X_1, ..., X_k$ have a real matrix-variate type-2 Dirichlet distribution f_2/D_2.

This result follows trivially by computing the Jacobian with the help of Theorem 2.30 of Chapter 2. A generalization of matrix-variate Dirichlet family of distributions is the *Liouville family of distributions*.

5.1.9   Real matrix-variate Liouville function and Liouville distribution

Let $X_1, ..., X_k$ be $p \times p$ real symmetric positive definite matrices. Let

$$f_3(X_1, ..., X_k) = g\left(\sum_{j=1}^{k} X_j\right) \prod_{j=1}^{k} |X_j|^{\alpha_j - \frac{p+1}{2}} \qquad (5.1.38)$$

for $\mathrm{Re}(\alpha_j) > \frac{p-1}{2}$, $j = 1, ..., k$ where g is an arbitrary positive continuous scalar function supported on the cone of positive definite $p \times p$ matrices. g is assumed to satisfy the integrability condition,

$$\int_{U>0} |U|^{\alpha - \frac{p+1}{2}} g(U)\mathrm{d}U < \infty, \quad \alpha = \alpha_1 + ... + \alpha_k.$$

Let

$$D_3 = \int_{X_1>0} ... \int_{X_k>0} f_3(X_1, ..., X_k)\mathrm{d}X_1...\mathrm{d}X_k. \qquad (5.1.39)$$

(The Liouville function and the Liouville integral are named after the French mathematician J. Liouville (1809–1882)). Then f_3/D_3 is known as the *real matrix-variate Liouville density*. Properties corresponding to the Dirichlet family of distributions f_1/D_1 and f_2/D_2 can also be worked out for the Liouville family, see for example Gupta and Richards (1987). There are other such families of general structures known as spherically symmetric and elliptically contoured distributions. For details on these see, for example, Anderson and Girshick (1944), Fang and Anderson (1990), Fang, Kotz and Ng (1990), and Mathai, Provost and Hayakawa (1995).

5.1.10   Some applications

Functions of matrix argument are widely used in almost all branches of mathematical statistics. A particular case of the matrix-variate gamma density of (5.1.2) is the Wishart density which is one of the fundamental distributions in multivariate statistical analysis. Several test statistics for testing hypotheses on the parameters of multivariate Gaussian models are scalar functions of matrix-variate gamma, matrix-variate beta and other matrix-variate random variables. The matrix-variate type-2 beta variable is associated with generalized analysis of variance problems, design of experiments, regression problems, model building and so on. Dirichlet densities are associated with ordered variables, spacings and so on. Various such applications may be seen from books on multivariate statistical analysis, for example, Anderson (1971), Srivastava and Khatri (1979) and Muirhead (1982).

Functions of matrix argument also arise in a variety of problems in mathematical physics and statistical mechanics. Some of the topics and applications are available from Mehta (1967) and Girko (1990). Quadratic and bilinear forms in vector and matrix variable cases occur in many applied areas. Some details of the theory and applications can be seen from Mathai and Provost (1992) and Mathai, Provost and Hayakawa (1995). A large number of topics, where matrix-variate functions of elliptically contoured classes are applied, are available from Fang and Anderson (1990).

Exercises

5.1.1   If the $p \times p$ real random matrix X has a type-1 beta density with the parameters α_1 and α_2 show the following:

$$(i) \quad U = (I - X)^{-\frac{1}{2}} X (I - X)^{-\frac{1}{2}}$$
$$\sim \text{ type-2 beta } (\alpha_1, \alpha_2);$$
$$(ii) \quad V = X^{-1} - I$$
$$\sim \text{ type-2 beta } (\alpha_2, \alpha_1).$$

5.1.2   If the $p \times p$ real random matrix X has a type-2 beta density with the parameters (α_1, α_2), that is,

$$f(X) = c|X|^{\alpha_1 - \frac{p+1}{2}} |I + X|^{-(\alpha_1 + \alpha_2)}$$

for $X = X' > 0$, $\text{Re}(\alpha_j) > (p-1)/2$, $j = 1, 2$ and $f(X) = 0$ elsewhere, where

$$c = \frac{\Gamma_p(\alpha_1 + \alpha_2)}{\Gamma_p(\alpha_1)\Gamma_p(\alpha_2)},$$

show the following:

$$(i) \quad U = X^{-1} \sim \text{ type-2 beta } (\alpha_2, \alpha_1);$$
$$(ii) \quad V = (I + X)^{-1} \sim \text{ type-1 beta } (\alpha_2, \alpha_1);$$
$$(iii) \quad W = (I + X)^{-\frac{1}{2}} X (I + X)^{-\frac{1}{2}} \sim \text{ type-1 beta } (\alpha_1, \alpha_2).$$

5.1.3   Let the Laplace transform of a scalar function $f(X)$ of the $p \times p$ real symmetric positive definite matrix X be denoted by $g(T) = L_T(f)$ where T is the parameter matrix given in Definition 5.3. Then establish the following:

(i)
$$g(T + A) = L_T(e^{-\text{tr}(AX)} f(X));$$

(ii)
$$\Delta^n g(T) = L_T(|X|^n f(X)), \Delta = (-1)^p \left| \frac{\partial}{\partial T} \right|,$$

where $|\frac{\partial}{\partial T}|$ denotes the determinant of the matrix of the operator $\frac{\partial}{\partial T}$. Note that $T = (\eta_{jk}t_{jk})$ where $\eta_{jk} = \frac{1}{2}$, $j \neq k$ and $\eta_{jk} = 1$, $j = k$ with $t_{jk} = t_{kj}$ but the (j,k)-th element in the operator $\frac{\partial}{\partial T}$ is taken as $\frac{\partial}{\partial t_{jk}}$.

5.1.4   Establish the result in Example 5.1 by using M-transform.

5.1.5   Let the real symmetric positive definite $p \times p$ matrix random variables $X_0, X_1, ..., X_k$ be independently distributed, and have the matrix-variate gamma densities

$$f_j(X_j; \alpha_j, B) = \frac{|X_j|^{\alpha_j - \frac{p+1}{2}} e^{-\text{tr}(BX_j)}}{|B|^{-\alpha_j}\Gamma_p(\alpha_j)}, \quad X_j > 0, \ B > 0, \ \text{Re}(\alpha_j) > \frac{p-1}{2}$$

and $f(X_j; \alpha_j, B) = 0$ elsewhere, $j = 0, 1, ..., k$. Let $X = X_0 + X_1 + ... + X_k$ and $U_j = X^{-1/2}X_j X^{-1/2}$, $j = 1, ..., k$ where $X^{1/2}$ is the symmetric square root of X. Then show that the joint distribution of $U_1, ..., U_k$ is a real matrix-variate type-1 Dirichlet.

5.1.6   Let X_j, $j = 0, 1, ...k$ be distributed as in Exercise 5.1.5 above. Then show that $Y_j = X_0^{-\frac{1}{2}}X_j X_0^{-\frac{1}{2}}$, $j = 1, ..., k$ have a real matrix-variate type-2 Dirichlet distribution.

5.1.7   Let $X_1, ..., X_k$ be real $p \times p$ matrix-variate random variables having a real matrix-variate Liouville distribution f_3/D_3 given by (5.1.38) and (5.1.39). Show that every subset of matrices from the set $\{X_1, ..., X_k\}$ also have a distribution belonging to the same Liouville family of distributions.

5.1.8   Let $\hat{D}$ be the differential operator defined in (5.1.28) excluding $(-1)^p$. By using (5.1.12) show that

$$\hat{D}^m \left[\frac{|U|^{\alpha - \frac{p+1}{2}}}{\Gamma_p(\alpha)} \right] = \frac{|U|^{(\alpha - m) - \frac{p+1}{2}}}{\Gamma_p(\alpha - m)}$$

for $U = U' > 0$, $\text{Re}(\alpha) > \frac{p-1}{2}$, $\text{Re}(\alpha - m) > \frac{p-1}{2}$, $m = 0, 1, ...$

5.1.9   By using Exercise 5.1.8 or otherwise prove $(ii), (iii)$ and (iv) of Lemma 5.1 on Weyl fractional integrals and fractional derivatives.

5.1.10   Prove Theorem 5.3.

5.2   Hypergeometric functions of matrix argument

A series representation of the hypergeometric function for the scalar variable case is the following where the standard notation $_rF_s$ is used:

$$
_rF_s = {_rF_s}(a_1, ..., a_r; b_1, ..., b_s; x)
$$

$$
= \sum_{k=0}^{\infty} \frac{(a_1)_k...(a_r)_k}{(b_1)_k...(b_s)_k} \frac{x^k}{k!}
\tag{5.2.1}
$$

where $a_1, ..., a_r$ are called the upper parameters and $b_1, ..., b_s$ the lower parameters where, for example, the notation $(a)_m$ means

$$
(a)_m = a(a+1)...(a+m-1), \quad (a)_0 = 1, \quad a \neq 0.
$$

The series on the right converges for all x when $s \geq r$, it converges for $|x| < 1$ when $r = s+1$ and diverges for $r > s+1$. Convergence conditions for $x = 1$ or $x = -1$ can be worked out separately. Note that when $r = s = 0$ we have the exponential series, that is,

$$
_0F_0(\; ; \; ; x) = e^x, \quad \text{and} \quad {_1F_0}(\alpha; \; ; x) = (1-x)^{-\alpha}
$$

for $|x| < 1$.

Another definition, for the scalar case, in terms of a Mellin–Barnes type representation is the following: (English mathematician E.W. Barnes (1874–1953), Finnish mathematician R.H. Mellin (1854–1933)):

$$
_rF_s(a_1, ..., a_r; b_1, ..., b_s; -x) = \frac{C}{2\pi i} \int_L \frac{\left\{ \prod_{j=1}^r \Gamma(a_j - \rho) \right\} \Gamma(\rho) z^{-\rho}}{\prod_{j=1}^s \Gamma(b_j - \rho)} d\rho
\tag{5.2.2}
$$

where

$$
C = \left\{ \prod_{j=1}^s \Gamma(b_j) \right\} \Big/ \left\{ \prod_{j=1}^r \Gamma(a_j) \right\}
$$

and L is a suitable contour. A sufficient set of conditions for the existence of (5.2.2) is that $\mathrm{Re}(a_j) > \mathrm{Re}(\rho) > 0$, $j = 1, ..., r$, $\mathrm{Re}(b_j) > \mathrm{Re}(\rho) > 0$, $j = 1, ..., s$, $s \geq r$ or $r = s+1$ and $|x| < 1$. With the help of (5.2.2) or by other means one can establish the following results:

$$
_{r+1}F_s(a_1, ..., a_r, c; b_1, ..., b_s; -x^{-1})
$$

$$
= \frac{x^c}{\Gamma(c)} \int_{t=0}^{\infty} e^{-tx} {_rF_s}(a_1, ..., a_r; b_1, ..., b_s; -t) t^{c-1} dt
\tag{5.2.3}
$$

and

$$_rF_{s+1}(a_1, ..., a_r; b_1, ..., b_s, c; -t)t^{c-1}$$

$$= \frac{\Gamma(c)}{2\pi i} \int_L e^{tz} {}_rF_s(a_1, ..., a_r; b_1, ..., b_s; -z^{-1})z^{-c}dz \qquad (5.2.4)$$

where L is a suitable contour, $\text{Re}(c) > 0$, $\text{Re}(z) > 0$, $t > 0$, and other conditions mentioned earlier for the existence of $_rF_s$ are also assumed.

5.2.1 Hypergeometric functions of matrix argument through Laplace transform

Hypergeometric functions of matrix argument can be defined by using matrix-variate Laplace transform as that function satisfying the following pair of integral equations corresponding to the equations in (5.2.3) and (5.2.4), see for example, Herz (1955):

$$_{r+1}F_s(a_1, ..., a_r, c; b_1, ..., b_s; -\Lambda^{-1})|\Lambda|^{-c}$$

$$= \frac{1}{\Gamma_p(c)} \int_{U=U'>0} e^{-\text{tr}(\Lambda U)} {}_rF_s(a_1, ..., a_r; b_1, ..., b_s; -U)|U|^{c-\frac{p+1}{2}}dU \qquad (5.2.5)$$

and

$$_rF_{s+1}(a_1, ..., a_r; b_1, ..., b_s, c; -\Lambda)|\Lambda|^{c-\frac{p+1}{2}}$$

$$= \frac{\Gamma_p(c)}{(2\pi i)^{\frac{p(p+1)}{2}}} \int_{\text{Re}(Z)=X>X_0} e^{\text{tr}(\Lambda Z)} {}_rF_s(a_1, ..., a_r; b_1, ..., b_s; -Z^{-1})|Z|^{-c}dZ$$

$$(5.2.6)$$

where $Z = X + iY$, $X \in S_p^*$, $Y \in S_p^*$, $X > 0$, see also Notation 5.1.

Remark 5.5 If the nondiagonal elements of X and Y are not weighted with $\frac{1}{2}$, that is, if $X = X' > 0$, $Y = Y'$, $X = (x_{jk})$, $Y = (y_{jk})$ then the right side of (5.2.6) is to be multiplied by $2^{p(p-1)/2}$.

But equations (5.2.5) and (5.2.6) do not give the hypergeometric functions explicitly except for special cases. Taking

$$_0F_0(\; ; \; -\Lambda) = e^{-\text{tr}(\Lambda)}$$

and using (5.2.5) one has $_1F_0$ explicitly. But higher functions in this category are not available from (5.2.5) and (5.2.6). But these equations enable us to study the properties of a hypergeometric function defined through this pair of equations. The conditions to be satisfied are that $\Lambda = \Lambda' > 0$, $\text{Re}(c) > \frac{p-1}{2}$, $s \geq r$ or $r = s + 1$ and $||\Lambda|| < 1$ where

$\|\Lambda\|$ denotes a norm of Λ. The parameters $b_1, ..., b_s$ are such that none of $b_j - \frac{k-1}{2}$, $j = 1, ..., s$, $k = 1, ..., p$ is a negative integer or zero. If any of the $b_j - \frac{k-1}{2}$, $j = 1, ..., s$, $k = 1, ..., p$ is a negative integer or zero then there should be an a_j, $j = 1, ..., r$ such that $\left(a_j - \frac{k-1}{2}\right)_m = 0$ first before any $\left(b_l - \frac{k-1}{2}\right)_m = 0$ for $j = 1, ..., r$, $k = 1, ..., p$, $l = 1, ..., s$. All the matrices appearing in the hypergeometric function defined through (5.2.5) and (5.2.6) are assumed to be $p \times p$ symmetric positive definite when real and hermitian positive definite when complex. Also the hypergeometric function itself is assumed to be symmetric in the sense of Definition 5.8.

Note that starting with a $_0F_0(\ ; \ ; -X)$, which is taken as $e^{-\text{tr}(X)}$, one can bring in a new upper parameter by using (5.2.5) and a lower parameter by (5.2.6). Thus systematically one can build up the hypergeometric function $_rF_s$ for all r and s. We will illustrate this in the following example.

Example 5.7   Show that the binomial function of matrix argument

$$|I + V|^{-c} = {}_1F_0\left(c; \ ; -V^{-1}\right)|V|^{-c}$$

$$= \frac{1}{\Gamma_p(c)} \int_{\Lambda > 0} e^{-\text{tr}(\Lambda V)} e^{-\text{tr}(\Lambda)} |\Lambda|^{c - \frac{p+1}{2}} \, d\Lambda,$$

and the *Bessel function of matrix argument*

$$_0F_1(\ ; c; -\Lambda) = |\Lambda|^{-\left(c - \frac{p+1}{2}\right)} \frac{\Gamma_p(c)}{(2\pi i)^{\frac{p(p+1)}{2}}}$$

$$\times \int_{\text{Re}(Z) = X > X_0} e^{\text{tr}(\Lambda Z)} e^{-\text{tr}(Z^{-1})} |Z|^{-c} dZ \qquad (5.2.7)$$

$$= \frac{\Gamma_p(c)}{(2\pi i)^{\frac{p(p+1)}{2}}} \int_{\text{Re}(Z) = X > X_0} e^{\text{tr}(Z)}$$

$$\times \ e^{-\text{tr}(Z^{-1}\Lambda)} |Z|^{-c} dZ \qquad (5.2.8)$$

where $Z = X + iY$, $X \in S_p^*$, $Y \in S_p^*$, $X = X' > 0$, see also Notation 5.1.

Solution   Taking $_0F_0(\ ; \ ; -V^{-1}) = e^{-\text{tr}(V^{-1})}$, from (5.2.6) we have (5.2.7). Take $r = s = 0$ on the right side of (5.2.5), replace $_0F_0(\ ; \ ; -\Lambda)$ by $e^{-\text{tr}(\Lambda)}$, write

$$\text{tr}((I + V)\Lambda)) = \text{tr}(U'\Lambda U)$$

where $I + V = UU'$. Then since Λ is real $W = U'\Lambda U \Rightarrow d\Lambda = |U|^{-\frac{p+1}{2}} dW$. Now from (5.2.5), with the help of the gamma integral, one has $|I + V|^{-c}$. But

$$|I + V|^{-c} = |V|^{-c} |I + V^{-1}|^{-c}$$

$$= |V|^{-c} {}_1F_0\left(c; \ ; -V^{-1}\right).$$

Replacing $\Lambda^{\frac{1}{2}} Z \Lambda^{\frac{1}{2}}$ by some other matrix in (5.2.7) the result in (5.2.8) follows.

Now we look at a method of increasing one paramter each to the sets of numerator and denominator parameters in a hypergeometric function. This can be done by convoluting with a type-1 beta function. This can be seen from the next example where the convolution is with a $_0F_0$. But if a general $_rF_s$ is used then one ends up with a $_{r+1}F_{s+1}$

Example 5.8   Show that the *confluent hypergeometric function of matrix argument*

$$
\begin{aligned}
_1F_1(a; c; -\Lambda) &= \frac{\Gamma_p(c)}{\Gamma_p(a)\Gamma_p(c-a)} \int_0^I |U|^{a-\frac{p+1}{2}} |I-U|^{c-a-\frac{p+1}{2}} \\
&\quad \times {}_0F_0(\ ;\ ; -(\Lambda U)) dU \qquad\qquad (5.2.9)
\end{aligned}
$$

$$
\begin{aligned}
&= \frac{\Gamma_p(c)}{(2\pi i)^{\frac{p(p+1)}{2}}} \int_{Re(Z)=X>X_0} e^{tr(Z)} \\
&\quad \times |Z|^{-c} |I + \Lambda Z^{-1}|^{-a} dZ \qquad\qquad (5.2.10)
\end{aligned}
$$

$$
= e^{-tr(\Lambda)} {}_1F_1(c-a; c; \Lambda) \qquad\qquad (5.2.11)
$$

$$
\begin{aligned}
&= \frac{\Gamma_p(c)}{\Gamma_p(a)\Gamma_p(c-a)} |\Lambda|^{-(c-\frac{p+1}{2})} \int_{0<V<\Lambda} e^{-tr(V)} \\
&\quad \times |V|^{a-\frac{p+1}{2}} |\Lambda - V|^{c-a-\frac{p+1}{2}} dV \qquad\qquad (5.2.12)
\end{aligned}
$$

where Λ, U, V are $p \times p$ symmetric positive definite matrices, $Z = X + iY$ with $X \in S_p^*$, $Y \in S_p^*$, and $Re(c, a, c-a) > \frac{p-1}{2}$ which means that the real part of each is greater than $\frac{p-1}{2}$.

Solution   Note that

$$
_0F_0(\ ;\ ; -\Lambda U) = e^{-tr(\Lambda U)}
$$

and

$$
tr(\Lambda U) = tr(\Lambda^{\frac{1}{2}} U \Lambda^{\frac{1}{2}}) = tr(V),
$$

where $V = \Lambda^{\frac{1}{2}} U \Lambda^{\frac{1}{2}}$ with $(\cdot)^{\frac{1}{2}}$ denoting the symmetric square root of $(\cdot)$ for convenience. We could have used the representation $\Lambda = WW'$ where W is nonsingular. Make the transformation U to V; then

$$
dU = |\Lambda|^{-\frac{p+1}{2}} dV,
$$

$$
|I - U| = |I - \Lambda^{-\frac{1}{2}} V \Lambda^{-\frac{1}{2}}| = |\Lambda|^{-1} |\Lambda - V|
$$

and $0 < U < I \Rightarrow 0 < V < \Lambda$. Under this substitution (5.2.9) changes to (5.2.12) and the result to be proved is then

$$|\Lambda|^{c-\frac{p+1}{2}}{}_1F_1(a; c; -\Lambda) = \frac{\Gamma_p(c)}{\Gamma_p(a)\Gamma_p(c-a)} \int_0^\Lambda |V|^{a-\frac{p+1}{2}}$$
$$\times |\Lambda - V|^{c-a-\frac{p+1}{2}} e^{-\text{tr}(V)} dV. \qquad (a)$$

This can be proved by showing that both sides have the same Laplace transforms. Taking the parameter matrix as T the Laplace transform of the left side of (a) is available from (5.2.5) for $r = 1, s = 1$. That is,

$$L_T(\text{left side}) = \Gamma_p(c)|T|^{-c}{}_2F_1\left(a, c : c; -T^{-1}\right)$$

for $\text{Re}(c) > \frac{p-1}{2}$. But

$$\begin{aligned}{}_2F_1\left(a, c; c; -T^{-1}\right) &= {}_1F_0\left(a; \;; -T^{-1}\right) \\ &= |I + T^{-1}|^{-a} \\ &= |I + T|^{-a}|T|^{a}. \end{aligned}$$

The Laplace transform of the right side is available from the convolution result in Theorem 5.2 with

$$f_1(V) = |V|^{c-a-\frac{p+1}{2}}, \quad f_2(V) = |V|^{a-\frac{p+1}{2}} e^{-\text{tr}(V)}.$$

Then

$$L_T(f_1) = \Gamma_p(c-a)|T|^{-(c-a)}, \quad \text{Re}(c-a) > \frac{p-1}{2}$$

and

$$L_T(f_2) = \Gamma_p(a)|I + T|^{-a}, \quad \text{Re}(a) > \frac{p-1}{2}.$$

Thus

$$\frac{\Gamma_p(c)}{\Gamma_p(a)\Gamma_p(c-a)} L_T(f_1)L_T(f_2) = \Gamma_p(c)|T|^{-(c-a)}|I + T|^{-a}.$$

This establishes (5.2.9). Note that (5.2.10) follows from (5.2.6). In (5.2.9) change U to $I - U$, take out $e^{-\text{tr}(\Lambda)}$ and evaluate the integral by using (5.2.9) to obtain (5.2.11). This formula (5.2.11) is also known as *Kummer's formula for the confluent hypergeometric function* ${}_1F_1$. (This formula and many other related results are due to the German mathematician E.E. Kummer (1810–1893)). From (a) above the result in (5.2.12) follows.

We will also look at some limiting properties of hypergeometric functions. This will be established with the help of the following lemma.

Lemma 5.2   *For ϵ scalar and A a $p \times p$ matrix with p finite*

$$\lim_{\epsilon \to 0} |I + \epsilon A|^{-\frac{1}{\epsilon}} = e^{-\mathrm{tr}(A)}.$$

Proof   Let $\lambda_1, ..., \lambda_p$ be the eigenvalues of A. Then the determinant

$$|I + \epsilon A| = \prod_{j=1}^{p} (1 + \epsilon \lambda_j).$$

But

$$\lim_{\epsilon \to 0} (1 + \epsilon \lambda_j)^{-\frac{1}{\epsilon}} = e^{-\lambda_j}$$

and then

$$\lim_{\epsilon \to 0} |I + \epsilon A|^{-\frac{1}{\epsilon}} = e^{-\lambda_1 - ... - \lambda_p} = e^{-\mathrm{tr}(A)}$$

since $\mathrm{tr}(A) = \lambda_1 + ... + \lambda_p$.

Example 5.9   Show that

$$\lim_{a \to \infty} {}_1F_1\left(a; c; -\frac{Z}{a}\right) = \lim_{\epsilon \to 0} {}_1F_1\left(\frac{1}{\epsilon}; c; -\epsilon Z\right)$$
$$= {}_0F_1(\ ; c; -Z). \qquad (5.2.13)$$

Solution   Consider the representation in (5.2.10) for the confluent hypergeometric function ${}_1F_1$. Change Λ to $\epsilon \Lambda$ and a to $1/\epsilon$ and then take the limit for $\epsilon \to 0$ with the help of Lemma 5.2 to see that

$$\lim_{\epsilon \to 0} \left|I + \epsilon \Lambda Z^{-1}\right|^{-1/\epsilon} = e^{-\mathrm{tr}(\Lambda Z^{-1})}.$$

Now check with the integral representation for a ${}_0F_1$ in (5.2.7) to see the result.

Example 5.10   Incomplete gamma integral   For $0 < X < B$, $B = B' > 0$ and $n > -1$, show that

$$L_T\left(|X|^n\right) = |B|^{n + \frac{p+1}{2}} \frac{\Gamma_p\left(\frac{p+1}{2}\right) \Gamma_p\left(n + \frac{p+1}{2}\right)}{\Gamma_p(n + p + 1)}$$
$$\times {}_1F_1\left(n + \frac{p+1}{2}; n + p + 1; -B^{\frac{1}{2}} T B^{\frac{1}{2}}\right)$$

where X and B are $p \times p$ symmetric positive definite matrices.

Solution   Transform $X = B^{\frac{1}{2}} Y B^{\frac{1}{2}}$ to obtain

$$L_T(|X|^n) = \int_0^B |X|^n e^{-\operatorname{tr}(TX)} dX$$

$$= |B|^{n+\frac{p+1}{2}} \int_0^I |Y|^n e^{-\operatorname{tr}((B^{\frac{1}{2}} T B^{\frac{1}{2}})Y)} dY.$$

Now compare with the integral in Example 5.8 to see the result.

Note that this integral is the generalization of the incomplete gamma function to the matrix-variate case.

With the help of the pair of equations in (5.2.5) and (5.2.6) and the basic results given above one can generalize almost all results on hypergeometric functions in the scalar case, except the ones involving certain fractional powers and multiplication formula for gamma functions, to those of the matrix-variate case.

5.2.2   Hypergeometric functions of matrix argument through zonal polynomial

Another approach in the development of hypergeometric functions of matrix argument is through *zonal polynomials*. Zonal polynomials and functions defined through them are discussed in Section 5.1.5. One can give a series representation of a hypergeometric function through zonal polynomials as follows, see for example Constantine (1963).

Definition 5.12

$$_rF_s(a_1, ..., a_r; b_1, ..., b_s; X) = \sum_{k=0}^{\infty} \sum_K \frac{(a_1)_K ... (a_r)_K}{(b_1)_K ... (b_s)_K} \frac{C_K(X)}{k!} \qquad (5.2.14)$$

where $K = (k_1, k_2, ..., k_p)$, $k_1 \geq k_2 \geq ... \geq k_p \geq 0$, $k_1 + k_2 + ... + k_p = k$, $\sum_K$ denotes the summation over all partitions of the non-negative integer k into not more than p parts, $s \geq r$, or $r = s + 1$ and $\|X\| < 1$, X is a $p \times p$ real symmetric positive definite matrix with $\|X\|$ denoting a norm of X, $C_K(X)$ is the zonal polynomial and $(a)_K$ is defined in (5.1.22). The representation in (5.2.14) gives a multiple series representation for the hypergeometric function once the zonal polynomials are evaluated. Even though the computations of zonal polynomials for higher values of k are difficult this definition is the only one which will give an explicit computable form for a hypergeometric function. The definition through Laplace transform and the one through M-transform, to be discussed later, do not give explicit forms except for the particular cases $_0F_0$ and $_1F_0$, but are more suitable for establishing theoretical results.

Example 5.11   Establish the following results for the *Gauss' hypergeometric function* $_2F_1$.

$$_2F_1(\alpha, \beta; \gamma; X) = \frac{\Gamma_p(\gamma)}{\Gamma_p(\alpha)\Gamma_p(\gamma - \alpha)} \int_0^I |\Lambda|^{\alpha - \frac{p+1}{2}}$$

$$\times |I - \Lambda|^{\gamma - \alpha - \frac{p+1}{2}} |I - \Lambda X|^{-\beta} d\Lambda \qquad (5.2.15)$$

for $\text{Re}(\alpha, \gamma - \alpha) > \frac{p-1}{2}$;

$$= |I - X|^{-\beta} {}_2F_1\left(\gamma - \alpha, \beta; \gamma; -X(I - X)^{-1}\right) (5.2.16)$$

$$= |I - X|^{-\alpha} {}_2F_1\left(\alpha, \gamma - \beta; \gamma; -X(I - X)^{-1}\right) (5.2.17)$$

$$= |I - X|^{\gamma - \alpha - \beta} {}_2F_1(\gamma - \alpha, \gamma - \beta; \gamma; X); \qquad (5.2.18)$$

$$_2F_1(\alpha, \beta; \gamma; I) = \frac{\Gamma_p(\gamma)\Gamma_p(\gamma - \alpha - \beta)}{\Gamma_p(\gamma - \alpha)\Gamma_p(\gamma - \beta)} \qquad (5.2.19)$$

$$= \sum_{k=0}^{\infty} \sum_K \frac{(\alpha)_K(\beta)_K}{(\gamma)_K} \frac{C_K(I)}{k!} \qquad (5.2.20)$$

and

$$\lim_{\beta \to \infty} {}_2F_1\left(\alpha, \beta; \gamma; \frac{X}{\beta}\right) = {}_1F_1(\alpha; \gamma; X). \qquad (5.2.21)$$

Solution   (5.2.15) can be established by using the convolution property of the Laplace transform or Theorem 5.2. This can also be established by expanding $|I - \Lambda X|^{-\beta}$ in terms of zonal polynomials, interchanging the sum and integrals and integrating out Λ by using a type-1 matrix-variate beta integral. Note also that

$$|I - \Lambda X|^{-\beta} = |I - X + X - \Lambda X|^{-\beta}$$
$$= |I - X|^{-\beta} |I + (I - X)^{-1} X (I - \Lambda)|^{-\beta}. \qquad (a)$$

In (5.2.15) make the change as in (a) above, take out $|I - X|^{-\beta}$, change the variable Λ to $I - \Lambda$ and interpret in the light of (5.2.15) to see (5.2.16). Note that $(I - X)^{-1}X = \left(X^{-1} - I\right)^{-1}$ which is real symmetric positive definite when $X = X' > 0$ and thus the argument of the new $_2F_1$ in (5.2.16) remains symmetric positive definite. The same procedure applies for establishing

(5.2.17) after using the fact that the upper parameters α and β can be interchanged. But (5.2.18) needs the following series of transformations.

$$U = I - \Lambda \Rightarrow 0 < U < I, \ d\Lambda = dU, \text{ ignoring the sign;}$$
$$V = U^{-1} - I \Rightarrow V > 0, \ dU = |I + V|^{-(p+1)}dV;$$
$$W = (I - X)^{\frac{1}{2}}V(I - X)^{\frac{1}{2}} \Rightarrow W > 0, \ dV = |I - X|^{-\frac{p+1}{2}}dW;$$
$$T = I + W \Rightarrow T > I, \ dW = dT;$$
$$Y = T^{-1} \Rightarrow 0 < Y < I, \ dT = |Y|^{-(p+1)}dY.$$

Under these transformations $|I - X|^{\gamma-\alpha-\beta}$ comes out and the remaining is a $_2F_1(\gamma-\alpha, \gamma-\beta; \gamma; X)$ by identifying with (5.2.15). From Definition 5.12 note that

$$|I - \Lambda X|^{-\beta} = \sum_{k=0}^{\infty} \sum_K (\beta)_K \frac{C_K(\Lambda X)}{k!}$$

for $\|X\| < 1$ since $\|\Lambda\|$ is already < 1. Now interchange the sum and the integral in (5.2.15), which is valid since it is a type-1 beta function, integrate out by using (5.1.26) and interpret the resulting series with the help of Definition 5.12 to see that it is a $_2F_1$. Now replace X by an identity matrix I and evaluate the integral by using a type-1 beta integral to see that the $_2F_1$ when the matrix argument is an identity matrix reduces to a gamma product as in (5.2.19). Then compare with the series expansion given by Definition 5.12 to see (5.2.20). Now apply Lemma 5.2 on $|I - \Lambda X|^{-\beta}$ after replacing X by X/β and then interpret by using (5.2.9) to see (5.2.21).

By using the definition itself one can establish the following results. In all these, expand the hypergeometric function appearing inside the integral, interchange the integral and the sum, the steps are valid since they are type-1 beta integrals, integrate out by using (5.1.26) and use (5.2.19), (5.2.20) and the Definition 5.12 to see the results.

For $\text{Re}(\rho, \beta - \rho, \gamma - \alpha - \rho) > \frac{p-1}{2}$,

$$\int_0^I |X|^{\rho-\frac{p+1}{2}}|I - X|^{\beta-\rho-\frac{p+1}{2}} \, _2F_1(\alpha, \beta; \gamma; X)dX$$

$$= \frac{\Gamma_p(\gamma)\Gamma_p(\rho)\Gamma_p(\beta - \rho)\Gamma_p(\gamma - \alpha - \rho)}{\Gamma_p(\beta)\Gamma_p(\gamma - \alpha)\Gamma_p(\gamma - \rho)}. \tag{5.2.22}$$

For $\text{Re}(\rho, \sigma, \gamma + \sigma - \alpha - \beta) > \frac{p-1}{2}$,

$$\int_0^I |X|^{\rho-\frac{p+1}{2}}|I - X|^{\sigma-\frac{p+1}{2}} \, _2F_1(\alpha, \beta; \gamma; X)dX$$

$$= \frac{\Gamma_p(\rho)\Gamma_p(\sigma)}{\Gamma_p(\rho+\sigma)} {}_3F_2(\alpha,\beta,\rho;\gamma,\rho+\sigma;I). \tag{5.2.23}$$

5.2.3 Bessel function of matrix argument

From (5.2.8) replacing c by $c + \frac{p+1}{2}$ one has

$$\frac{{}_0F_1\left(\;;c+\frac{p+1}{2};-\Lambda\right)}{\Gamma_p\left(c+\frac{p+1}{2}\right)} = \frac{1}{(2\pi i)^{\frac{p(p+1)}{2}}} \int_{\mathrm{Re}(Z)=X>X_0} e^{\mathrm{tr}(Z)}$$
$$\times\, e^{-\mathrm{tr}(Z^{-1}\Lambda)}|Z|^{-c-\frac{p+1}{2}} dZ$$

for $\Lambda = \Lambda' > 0$, $Z = X + iY$, $X \in S_p^*$, $Y \in S_p^*$, $X > 0$. The quantity on the left is defined as the Bessel function of matrix argument $A_c(\Lambda)$.

Definition 5.13 **Bessel function of matrix argument** (named after the German mathematician F.W. Bessel (1784–1846)). It is denoted by $A_\gamma(S)$ and defined for all $S = S' > 0$ as follows:

$$A_\gamma(S) = \frac{1}{\Gamma_p\left(\gamma+\frac{p+1}{2}\right)} {}_0F_1\left(\;;\gamma+\frac{p+1}{2};-S\right). \tag{5.2.24}$$

An immediate consequence of (5.2.24) and (5.2.5) is the following result which could be interpreted as a Laplace transform by suitably selecting Λ and S.

$$\int_{S=S'>0} e^{-\mathrm{tr}(\Lambda S)}|S|^{\delta-\frac{p+1}{2}} A_\gamma(S) dS$$
$$= \frac{1}{\Gamma_p\left(\gamma+\frac{p+1}{2}\right)} \int_{S>0} e^{-\mathrm{tr}(\Lambda S)}|S|^{\delta-\frac{p+1}{2}} {}_0F_1\left(\;;\gamma+\frac{p+1}{2};-S\right) dS$$
$$= \frac{\Gamma_p(\delta)}{\Gamma_p\left(\gamma+\frac{p+1}{2}\right)} |\Lambda|^{-\delta} {}_1F_1\left(\delta;\gamma+\frac{p+1}{2};-\Lambda^{-1}\right). \tag{5.2.25}$$

Other elementary special functions and orthogonal polynomials can be defined in terms of zonal polynomials. Consider the univeriate Laguerre polynomial $L_k^a(x)$. Let $S = S' > 0$ be a real $p \times p$ matrix. If x in $L_k^a(x)$ is replaced by $\mathrm{tr}(S)$ and written as a sum

$$L_k^{p(\gamma+\frac{p+1}{2})-1}(\mathrm{tr}(S)) = \sum_K L_K^\gamma(S) \tag{5.2.26}$$

then one has a method of extending Laguerre polynomials to the matrix-variate case for the partition K of k into not more than p parts, that is, $L_K^\gamma(S)$ for $K = (k_1, ..., k_p)$, $k_1 \geq k_2 \geq ... \geq k_p$, $k_1 + ... + k_p = k$.

Definition 5.14   Laguerre polynomial of matrix argument for the partition K (This polynomial is named after the French mathematician E.N. Laguerre (1834–1886)).   For the partition given above, $S = S' > 0$ and $\mathrm{Re}(\gamma) > -1$

$$L_K^\gamma(S) = e^{\mathrm{tr}(S)} \int_{\Lambda=\Lambda'>0} A_\gamma(\Lambda S) e^{-\mathrm{tr}(\Lambda)} |\Lambda|^\gamma C_K(\Lambda) d\Lambda$$

$$= \Gamma_p\left(\gamma + \frac{p+1}{2}, K\right) \frac{1}{(2\pi i)^{\frac{p(p+1)}{2}}}$$

$$\times \int_{\mathrm{Re}(Z)>X_0} e^{\mathrm{tr}(Z)} |Z|^{-\gamma - \frac{p+1}{2}} C_K\left(I - SZ^{-1}\right) dZ \quad (5.2.27)$$

where $C_K(\cdot)$ are the zonal polynomials, $\Gamma_p(\cdot, K)$ is defined in (5.1.24) and $Z = X + iY$, $X \in S_p^*$, $Y \in S_p^*$, $X > 0$. If the nondiagonal elements of X and Y are not weighted by $\frac{1}{2}$ then multiply the right side of (5.2.27) by $2^{p(p-1)/2}$. An orthogonal property for Laguerre polynomials can be stated as follows. Let K be a partition of the nonnegative integer k and M a partition of m, $m \neq k$, into not more than p parts each or let K and M be different partitions of the same integer k. Then $L_K^\gamma(S)$ and $L_M^\gamma(S)$ are orthogonal on $S = S' > 0$ with respect to the weight function $|S|^\gamma e^{-\mathrm{tr}(S)}$ and the result is the following:

$$\int_{S=S'>0} e^{-\mathrm{tr}(S)} |S|^\gamma L_K^\gamma(S) L_M^\gamma(S) dS$$

$$= \begin{cases} 0, \ K \neq M \\ k! C_K(I_p) \Gamma_p\left(\gamma + \frac{p+1}{2}, K\right), \ K = M, \ m = k. \end{cases} \quad (5.2.28)$$

Other orthogonal polynomials of scalar variables can be extended to the matrix-variate cases with the help of zonal polynomials. Some discussion of this aspect can be seen from Mathai, Provost and Hayakawa (1995) and the references therein.

Definition 5.15   Laguerre function of matrix argument   For the real $p \times p$ matrix $X = X' > 0$, the Laguerre function of matrix argument is defined as follows:

$$L_\nu^{(\gamma)}(X) = \frac{\Gamma_p\left(\gamma + \nu + \frac{p+1}{2}\right)}{\Gamma_p\left(\gamma + \frac{p+1}{2}\right)} {}_1F_1\left(-\nu; \gamma + \frac{p+1}{2}; X\right) \quad (5.2.29)$$

for $\text{Re}(\gamma) > -1$, $\text{Re}(\gamma + \nu) > -1$, $\text{Re}(\nu) < \frac{1-p}{2}$. It is easy to note that

$$L_{\nu}^{(\gamma)}(X) = \frac{e^{\text{tr}(X)}|X|^{-\gamma}}{\Gamma_p(-\nu)}$$

$$\times \int_{0<\Lambda<X} e^{-\text{tr}(\Lambda)}|\Lambda|^{\gamma+\nu}|X - \Lambda|^{-\nu-\frac{p+1}{2}} d\Lambda \quad (5.2.30)$$

where the integration is done over real $\Lambda = \Lambda' > 0$ and $X - \Lambda > 0$. This result can be established as follows. In (5.2.12) put $a = \gamma + \nu + \frac{p+1}{2}$ and $c = \gamma + \frac{p+1}{2}$. Then one gets a $_1F_1$ with the argument $-X$. Now apply (5.2.11) to write a $_1F_1$ with the argument X which then coincides with (5.2.29) above. Some more results are given as problems at the end of this section.

Definition 5.16   Jacobi polynomial of matrix argument   For a real $p \times p$ matrix $X = X' > 0$, such that $0 < X < I$ and for $\text{Re}(\gamma) > -1$ the Jacobi polynomial $P_{\nu}^{(\gamma,\delta)}(X)$ is defined as follows:

$$P_{\nu}^{(\gamma,\delta)}(X) = \frac{\Gamma_p\left(\gamma+\nu+\frac{p+1}{2}\right)}{\Gamma_p\left(\gamma+\frac{p+1}{2}\right)}$$

$$\times \, _2F_1\left(-\nu, \gamma+\delta+\nu+\frac{p+1}{2}; \gamma+\frac{p+1}{2}; X\right). \quad (5.2.31)$$

It is easy to note that

$$P_{\nu}^{(\gamma,\delta)}(X) = \frac{\Gamma_p\left(\gamma+\nu+\frac{p+1}{2}\right)}{\Gamma_p\left(\gamma+\frac{p+1}{2}\right)}|I - X|^{-\delta}$$

$$\times \, _2F_1\left(\gamma+\nu+\frac{p+1}{2}, -\delta-\nu; \gamma+\frac{p+1}{2}; X\right). \quad (5.2.32)$$

This follows from (5.2.18).

Definition 5.17   Gegenbauer function of matrix argument (This function is named after the German mathematician L. Gegenbauer (1849–1903)). Let T be a non-symmetric nonsingular $p \times p$ matrix. For $\text{Re}(\delta) > \frac{p}{2} - 1$, Gegenbauer functions are defined by the following formulae:

$$C_{2\nu}^{(\delta)}(T) = (-1)^{p\nu} \frac{\Gamma_p\left(\delta+\frac{1}{2}\right)}{\Gamma_p\left(\delta+\nu+\frac{1}{2}\right)} P_{\nu}^{(-\frac{1}{2},\delta-\frac{p}{2})}(T'T) \quad (5.2.33)$$

and

$$C_{2\nu+1}^{(\delta)}(T) = (-1)^{p\nu}\frac{\Gamma_p\left(\delta+\frac{1}{2}\right)}{\Gamma_p\left(\delta+\nu+\frac{1}{2}\right)}|T|P_\nu^{\left(\frac{1}{2},\delta-\frac{p}{2}\right)}(T'T). \quad (5.2.34)$$

When $\delta = \frac{p}{2}$ we have the *Legendre function*. (The Legendre function is named after the French mathematician A.M. Legendre (1752–1833)).

5.2.4   Hypergeometric function of matrix argument through M-transform

We will denote the class of hypergeometric functions defined through M-transform by $_rF^*{}_s(a_1, ..., a_r; b_1, ..., b_s; -X)$ to distinguish from the unique function defined through matrix-variate Laplace and inverse Laplace transforms, or abbreviated by $_rF_s^*$ where X is a $p \times p$ real symmetric positive definite matrix and $_rF_s^*$ itself is assumed to be a symmetric function in the sense $_rF_s^* = {_rF_s^*}(Q'XQ), QQ' = I$ for all orthogonal matrices Q.

Definition 5.18   A hypergeometric function is defined as that class of functions satisfying the following integral equation:

$$\int_{X=X'>0} |X|^{\rho-\frac{p+1}{2}} {_rF_s^*}(a_1, ..., a_r; b_1, ..., b_s; -X)\mathrm{d}X$$

$$= \frac{\left\{\prod_{j=1}^s \Gamma_p(b_j)\right\}}{\left\{\prod_{j=1}^r \Gamma_p(a_j)\right\}} \frac{\left\{\prod_{j=1}^r \Gamma_p(a_j - \rho)\right\} \Gamma_p(\rho)}{\left\{\prod_{j=1}^s \Gamma_p(b_j - \rho)\right\}} \quad (5.2.35)$$

whenever the gamma products are defined. It can be shown that at least one function $_rF_s^*$ defined through the integral equation (5.2.35) agrees with the $_rF_s$ defined through the generalized Laplace transform ((5.2.5) and (5.2.6)) and through zonal polynomials (5.2.14). This will be illustrated in the following example by showing that the M-transforms on both sides of the equations are one and the same.

Example 5.12   Show that one of the functions satisfying (5.2.35) also satisfies the equations (5.2.5) and (5.2.6).

Solution   Consider the right side of (5.2.5). Take $f_1(\Lambda) = e^{-\mathrm{tr}(\Lambda)}$ and $f_2(\Lambda) = {_rF_s}(a_1, ..., a_r; b_1, ..., b_s; -\Lambda)$, and $\beta = c - \frac{p+1}{2}$. Then take the M-transform with the help of Theorem 5.4 on the convolution property of M-transforms and apply Definition 5.18. Take the M-transform of the left side of (5.2.5) by making a transformation $V = \Lambda^{-1}$. Then we see

that the M-transform on both sides are one and the same. To establish
(5.2.6) consider the following transformation. On the right side of (5.2.6)
put $U = \Lambda^{\frac{1}{2}} Z \Lambda^{\frac{1}{2}} \Rightarrow dU = |\Lambda|^{\frac{p+1}{2}} dZ$. Remove $|\Lambda|^{c-\frac{p+1}{2}}$ from both sides of
(5.2.6). Now we have to show that

$$
\begin{aligned}
{}_r F_{s+1} &= {}_r F_{s+1}(a_1, ..., a_r; b_1, ..., b_s, c; -\Lambda) \\
&= \frac{\Gamma_p(c)}{(2\pi i)^{\frac{p(p+1)}{2}}} \int_{\mathrm{Re}(U)>\Lambda^{\frac{1}{2}} X_0 \Lambda^{\frac{1}{2}}} e^{\mathrm{tr}(U)} \\
&\quad \times {}_r F_s(a_1, ..., a_r; b_1, ..., b_s; -U^{-1}\Lambda)|U|^{-c} dU.
\end{aligned}
$$

Consider the M-transform of the right side. Assume interchange of inte-
grals. Make a change of variable $S = U^{-\frac{1}{2}} \Lambda U^{-\frac{1}{2}}$. Assuming integrability
over V, and in the light of (5.2.35) one has

$$
|U|^{-c} \int_{\Lambda>0} |\Lambda|^{\rho-\frac{p+1}{2}} {}_r F_s(a_1, ..., a_r; b_1, ..., b_s; -U^{-1}\Lambda) d\Lambda
$$
$$
= |U|^{-(c-\rho)} \frac{\left\{\prod_{j=1}^s \Gamma_p(b_j)\right\} \left\{\prod_{j=1}^r \Gamma_p(a_j - \rho)\right\} \Gamma_p(\rho)}{\left\{\prod_{j=1}^r \Gamma_p(a_j)\right\} \left\{\prod_{j=1}^s \Gamma_p(b_j - \rho)\right\}}.
$$

Now the U-integral is simply the inverse Laplace representation of $\{\Gamma_p(c - \rho)\}^{-1}$, see (5.1.12). Thus the M-transforms on both sides of (5.2.6) are one
and the same, which establishes the result.

Thus if ${}_r F_s^*(Z)$ is symmetric in the sense described in Definition 5.8,
analytic in some generalized right half plane $\mathrm{Re}(Z) = X > X_0$ and rapidly
decreasing so that Cauchy inversion formula can be applied then ${}_r F_s^*$ seems
to be unique. In any case the class of functions defined through the M-
transform contains the unique hypergeometric function defined through the
Laplace and inverse Laplace integral equations (5.2.5) and (5.2.6).

Note that all the results established through Laplace transform and
zonal polynomial, that is from (5.2.7) to (5.2.32), can also be established
with the help of the M-transform in Definition 5.18. As an illustration we
will do one more example here.

Example 5.13   Show that

$$
\begin{aligned}
f(Y) &= \int_0^I |X|^{a-\frac{p+1}{2}} |I - X|^{b-\frac{p+1}{2}} {}_2F_1(\alpha, \beta; \gamma; -YX) dX \\
&= \frac{\Gamma_p(a)\Gamma_p(b)}{\Gamma_p(a+b)} {}_3F_2(\alpha, \beta, a; \gamma, a+b; -Y)
\end{aligned} \tag{5.2.36}
$$

for $\text{Re}(a, b) > \frac{p-1}{2}$, where it is assumed that $f(Y)$ is such that

$$f(XY) = f(YX) = f\left(X^{\frac{1}{2}}YX^{\frac{1}{2}}\right) = f\left(Y^{\frac{1}{2}}XY^{\frac{1}{2}}\right).$$

Solution   Take the M-transform of the integral by taking

$$f_1(X) = {}_2F_1(\alpha, \beta; \gamma; -X)$$

and

$$f_2(X) = |X|^{a-\frac{p+1}{2}}|I - X|^{b-\frac{p+1}{2}}$$

and applying Theorem 5.4. Then from (5.2.35) one has

$$g_1(\rho) = \frac{\Gamma_p(\gamma)}{\Gamma_p(\alpha)\Gamma_p(\beta)} \frac{\Gamma_p(\alpha - \rho)\Gamma_p(\beta - \rho)\Gamma_p(\rho)}{\Gamma_p(\gamma - \rho)}$$

and from a type-1 beta integral

$$g_2(\rho) = \frac{\Gamma\left(a + \rho - \frac{p+1}{2}\right)\Gamma_p(b)}{\Gamma_p\left(a + b + \rho - \frac{p+1}{2}\right)}.$$

Then

$$\begin{aligned}
g_1(\rho)g_2\left(-\rho + \frac{p+1}{2}\right) &= \frac{\Gamma_p(\gamma)\Gamma_p(b)}{\Gamma_p(\alpha)\Gamma_p(\beta)} \\
&\quad \times \frac{\Gamma_p(\alpha - \rho)\Gamma_p(\beta - \rho)\Gamma_p(a - \rho)\Gamma_p(\rho)}{\Gamma_p(\gamma - \rho)\Gamma_p(a + b - \rho)} \\
&= \frac{\Gamma_p(a)\Gamma_p(b)}{\Gamma_p(a + b)} \\
&\quad \times M_\rho\left({}_3F_2(\alpha, \beta, a; \gamma, a + b; -X)\right)
\end{aligned}$$

which establishes the result.

The result in (5.2.36) can also be established by the Laplace transform technique or through zonal polynomial for the unique ${}_2F_1$ and hence we have kept the notation ${}_2F_1$ instead of ${}_2F_1^*$. Note also that if the ${}_2F_1$ in the integral in (5.2.36) is replaced by a general ${}_rF_s$ the integral leads to a ${}_{r+1}F_{s+1}$ which can be proved by the M-transform technique as in the above example or by using zonal polynomials or Laplace transform.

5.2.5 Whittaker function of matrix argument

When $f(X)$ is a scalar function of a symmetric positive definite matrix it is evident that

$$\int_A^B f(X)dX \neq \int_0^B f(X)dX - \int_0^A f(X)dX$$

for $p > 1$. As an example note that

$$\int_A^B dX = \int_{A<X<B} dX = \int_{0<Y<B-A} dY$$

$$= |B - A|^{\frac{p+1}{2}} \int_{0<Z<I} dZ$$

by changing $Z = (B - A)^{-\frac{1}{2}} Y (B - A)^{-\frac{1}{2}} \Rightarrow dY = |B - A|^{\frac{p+1}{2}} dZ$. But

$$\int_{0<Z<I} dZ = \frac{\Gamma_p \left(\frac{p+1}{2} \right) \Gamma_p \left(\frac{p+1}{2} \right)}{\Gamma_p(p + 1)}$$

by evaluating it using a type-1 beta integral, see for example (5.1.18). Thus

$$\int_0^B dX = |B|^{\frac{p+1}{2}} \int_{0<Z<I} dZ$$

and

$$\int_0^A dX = |A|^{\frac{p+1}{2}} \int_{0<Z<I} dZ.$$

But

$$|B - A|^{\frac{p+1}{2}} \neq |B|^{\frac{p+1}{2}} - |A|^{\frac{p+1}{2}} \text{ for } p > 1.$$

In the scalar variable case one can obtain incomplete gamma functions from the relation

$$\int_{0<x<a} x^{\alpha-1} e^{-x} dx = \Gamma(\alpha) - \int_{x>a} x^{\alpha-1} e^{-x} dx, \quad \text{Re}(\alpha) > 0.$$

But in the matrix-variate case, for $p > 1$,

$$\int_{0<X<A} |X|^{\alpha - \frac{p+1}{2}} e^{-\text{tr}(X)} dX \neq \Gamma_p(\alpha) - \int_{X>A} |X|^{\alpha - \frac{p+1}{2}} e^{-\text{tr}(X)} dX.$$

Thus one has to work out the two integrals separately to get analogues of incomplete gamma functions.

Whittaker function in the scalar case is associated with an incomplete gamma integral. (Whittaker functions are named after the English mathematician E. Whittaker (1873–1956)). Motivated by this fact we will define a Whittaker function of matrix argument through an incomplete matrix-variate gamma integral. Observe that

$$\int_{X>A} |X|^{\alpha-\frac{p+1}{2}} e^{-\mathrm{tr}(X)} dX = e^{-\mathrm{tr}(A)} \int_{Y>0} |Y+A|^{\alpha-\frac{p+1}{2}} e^{-\mathrm{tr}(Y)} dY$$

$$= |A|^{\alpha} e^{-\mathrm{tr}(A)} \int_{Z>0} |I+Z|^{\alpha-\frac{p+1}{2}} e^{-\mathrm{tr}(AZ)} dZ$$

by making the transformations $Y = X - A$, $Z = A^{-\frac{1}{2}} Y A^{-\frac{1}{2}}$. We will define a function $M(\alpha, \beta; A)$ as follows:

$$M(\alpha, \beta; A) = \int_{X=X'>0} |X|^{\alpha-\frac{p+1}{2}} |I+X|^{\beta-\frac{p+1}{2}} e^{-\mathrm{tr}(AX)} dX \qquad (5.2.37)$$

for $A = A' > 0$, $\mathrm{Re}(\alpha) > \frac{p-1}{2}$ and no restriction on β. Then we have the following properties from the definition itself.

$$M\left(\alpha, \frac{p+1}{2}; A\right) = |A|^{-\alpha} \Gamma_p(\alpha), \quad \mathrm{Re}(\alpha) > \frac{p-1}{2}. \qquad (5.2.38)$$

$$\int_{X>B} |X|^{\nu} e^{-\mathrm{tr}(TX)} dX = |B|^{\nu+\frac{p+1}{2}} e^{-\mathrm{tr}(BT)}$$
$$\times M\left(\frac{p+1}{2}, \nu+\frac{p+1}{2}; B^{\frac{1}{2}} T B^{\frac{1}{2}}\right). \qquad (5.2.39)$$

$$\int_{X>0} |X+A|^{\nu} e^{-\mathrm{tr}(TX)} dX = |A|^{\nu+\frac{p+1}{2}}$$
$$\times M\left(\frac{p+1}{2}, \nu+\frac{p+1}{2}; A^{\frac{1}{2}} T A^{\frac{1}{2}}\right). \qquad (5.2.40)$$

But for $p = 1$, that is , in the scalar variable case the function $M(\cdot, \cdot; \cdot)$ is associated with a Whittaker function. Hence we define a Whittaker function $W_{.,.}(\cdot)$ in terms of $M(\cdot, \cdot; \cdot)$ as follows.

$$M(\mu, \nu; A) = |A|^{-\frac{\mu+\nu}{2}} \Gamma_p(\mu) e^{\frac{1}{2}\mathrm{tr}(A)} W_{\frac{1}{2}(\nu-\mu), \frac{1}{2}(\nu+\mu-(p+1)/2)}(A) \qquad (5.2.41)$$

for $A = A' > 0$, $\mathrm{Re}(\mu) > \frac{p-1}{2}$ and no restriction on ν. In terms of the integral representation we have

$$\int_{Z>0} |Z|^{\mu-\frac{p+1}{2}} |I+Z|^{\nu-\frac{p+1}{2}} e^{-\mathrm{tr}(AZ)} dZ$$

$$= |A|^{-\frac{\mu+\nu}{2}} \Gamma_p(\mu) e^{\frac{1}{2}\mathrm{tr}(A)}$$
$$\times W_{\frac{1}{2}(\nu-\mu), \frac{1}{2}(\nu+\mu-(p+1)/2)}(A). \qquad (5.2.42)$$

We may write $\alpha = \frac{1}{2}(\nu - \mu)$ and $\beta = \frac{1}{2}(\nu + \mu - \frac{p+1}{2})$ to get an integral representation for $W_{\alpha,\beta}(A)$ from (5.2.42). By using this definition one can establish a number of results on Whittaker function which will generalize the corresponding univariate results to the matrix-variate case. A few of these will be stated here as theorems and some more are given in the exercises and many more such results can be established.

Theorem 5.7   For $B = B' > 0$, $U = U' > 0$, $M = M' > 0$, $\mathrm{Re}(q) > \frac{p-1}{4}$,

$$\int_{X>U} |X + B|^{2\alpha - \frac{p+1}{2}} |X - U|^{2q - \frac{p+1}{2}} e^{-\mathrm{tr}(MX)} dX$$

$$= |U + B|^{\alpha + q - \frac{p+1}{2}} |M|^{-(\alpha + q)} e^{\frac{1}{2}\mathrm{tr}[(B-U)M]} \Gamma_p(2q)$$

$$\times W_{(\alpha - q),(\alpha + q - (p+1)/4)}(S), \quad S = (U + B)^{\frac{1}{2}} M (U + B)^{\frac{1}{2}}.$$

Proof.   Consider the left side and make the following transformations.

$$Y = X - U, \ Z = (B + U)^{-\frac{1}{2}} Y (B + U)^{-\frac{1}{2}} \Rightarrow dZ = |B + U|^{-\frac{p+1}{2}} dY.$$

Then we have the

$$\text{left side} = e^{-\mathrm{tr}(MU)} |B + U|^{2\alpha + 2q - \frac{p+1}{2}}$$

$$\times \int_{Z>0} |Z|^{2q - \frac{p+1}{2}} |I + Z|^{2\alpha - \frac{p+1}{2}} e^{-\mathrm{tr}\left[M(B+U)^{\frac{1}{2}} Z (B+U)^{\frac{1}{2}} \right]} dZ$$

for $\mathrm{Re}(2q) > \frac{p-1}{2}$. Now writing the integral by using (5.2.42) the result follows by observing that $\frac{\mu + \nu}{2} = \alpha + q$, $\frac{\nu - \mu}{2} = \alpha - q$, $\frac{1}{2}(\mu + \nu - (p+1)/2) = \alpha + q - (p+1)/4$.

Theorem 5.8   For $A = A' > 0$, $B = B' > 0$, $\mathrm{Re}(\nu) > \frac{p-1}{2}$, $\mathrm{Re}(1 - \mu - \nu) > 0$,

$$\int_{X>0} |X|^{\nu - \frac{p+1}{2}} |X + A|^{\mu - \frac{p+1}{2}} e^{-\mathrm{tr}(BX^{-1})} dX$$

$$= |B|^{\frac{1}{2}(\nu - \frac{p+1}{2})} |A|^{\mu + \frac{1}{2}(\nu - \frac{p+1}{2})} \Gamma_p\left(\frac{p+1}{2} - \mu - \nu \right)$$

$$\times e^{\frac{1}{2}\mathrm{tr}(BA^{-1})} W_{(\mu + \frac{1}{2}(\nu - (p+1)/2)),(-\frac{1}{2}\nu)}\left(A^{-\frac{1}{2}} B A^{-\frac{1}{2}} \right).$$

Proof. Consider the left side and make the transformation $X^{-1} = Y \Rightarrow$
$dX = |Y|^{-(p+1)}dY$, $Y = Y' > 0$. Then the

$$\text{left side} = \int_{Y>0} |Y|^{-\nu-\frac{p+1}{2}} |Y^{-1} + A|^{\mu-\frac{p+1}{2}} e^{-\text{tr}(BY)}dY$$

$$= |A|^{\nu+\mu-\frac{p+1}{2}} \int_{Z>0} |Z|^{\frac{p+1}{2}-\nu-\mu-\frac{p+1}{2}}$$

$$\times |I + Z|^{\mu-\frac{p+1}{2}} e^{-\text{tr}\left[A^{-\frac{1}{2}}BA^{-\frac{1}{2}}Z\right]}dZ.$$

Now compare the integral with (5.2.42) to establish the result. The condition needed is that $\text{Re}\left(\frac{p+1}{2} - \nu - \mu\right) > \frac{p-1}{2}$.

Theorem 5.9 *For $A = A' > 0$,*

$$\int_{X>0} |I + X|^{-\alpha}e^{-\text{tr}(AX)}dX = |A|^{\frac{\alpha}{2}-\frac{p+1}{2}}\Gamma_p\left(\frac{p+1}{2}\right)e^{\frac{1}{2}\text{tr}(A)}$$

$$\times W_{-\frac{\alpha}{2},\frac{1}{2}(-\alpha+(p+1)/2)}(A).$$

This follows from (5.2.42) by putting $\nu = -\alpha + \frac{p+1}{2}$ and $\mu = \frac{p+1}{2}$.

Theorem 5.10 *For $U = U' > 0$,*

$$\int_{X>U} |X|^{-\alpha}e^{-\text{tr}(X)}dX = |U|^{-\frac{\alpha}{2}}e^{-\frac{1}{2}\text{tr}(U)}\Gamma_p\left(\frac{p+1}{2}\right)$$

$$\times W_{-\frac{\alpha}{2},\frac{1}{2}(-\alpha+(p+1)/2)}(U).$$

This follows from (5.2.42) after making the substitutions $Y = X - U$, $Z = U^{-\frac{1}{2}}YU^{-\frac{1}{2}}$.

Theorems 5.7–5.10 generalize the following univariate results on the Whittaker function given in Gradshteyn and Ryshik (1980): page 320 formula 3.384(3) or P.320(3.384(3)), P.340(3.471(7)), P.318(3.382(3)), P.318 (3.381(6)) respectively.

Remark 5.6 There are many results available in the literature covering various properties of hypergeometric functions in the scalar variable case. A large number of these results, excluding possibly the ones involving various powers, especially fractional powers, of the scalar argument and multiplication formula for gamma functions, can be extended to the matrix-variate case by using the definitions through M-transform, zonal polynomial and matrix-variate Laplace transform. The basic properties discussed so far in Sections 5.1 and 5.2 are sufficient to achieve this and hence further discussion of the topic is omitted in order to save space.

Now we will define generalized hypergeometric functions belonging to the class of Meijer's G-function and Fox's H-function for the matrix-variate case. These will generalize the corresponding theory for the scalar variable case.

Remark 5.7   When going from a univariate function to the multivariate case there is no unique way of defining the multivariate analogue unless several additional restrictions are imposed. For example there is no unique function which can be called the multivariate analogue of a gamma function or gamma density.

5.2.6   Meijer's G- and Fox's H-functions of matrix argument through the M-transform

The theory and applications of Meijer's G- and Fox's H-functions of scalar variables are available in the literature, see for example Mathai and Saxena (1973, 1978) and Mathai (1993). (English mathematician C. Fox (1897–1977), Dutch mathematician C.S. Meijer (1904–1974)). Since there is no unique way of going from the scalar to the vector or matrix cases one cannot expect to find a unique G- or H-function of matrix argument. In what follows we will define a class of functions satisfying certain integral equations which will be called the G- and H-functions of matrix argument. It is easy to note that results analogous to the ones in the scalar case can be established by using these definitions. Since the G-function is a particular case of an H-function we will define the H-function first.

Definition 5.19   Fox's H-function of matrix argument

Let $H(S)$ denote Fox's H-function of the matrix argument S where S is $p \times p$ real symmetric positive definite and let $H(S)$ be a symmetric function in the sense $H(S) = H(Q'SQ), QQ' = I$ for all orthogonal matrices Q. We will use notations analogous to the ones in the scalar case.

$$H(S) = H_{r,s}^{m,n} \left[S \middle| \begin{array}{c} (a_1, \alpha_1), ..., (a_r, \alpha_r) \\ (b_1, \beta_1), ..., (b_s, \beta_s) \end{array} \right]$$

and $H(S)$ will be defined by the following integral equation:

$$\int_{S>0} |S|^{\rho - \frac{p+1}{2}} H(S) dS = \frac{\left\{\prod_{j=1}^{m} \Gamma_p (b_j + \beta_j \rho)\right\}}{\left\{\prod_{j=m+1}^{s} \Gamma_p \left(\frac{p+1}{2} - b_j - \beta_j \rho\right)\right\}}$$

$$\times \frac{\left\{\prod_{j=1}^{n} \Gamma_p \left(\frac{p+1}{2} - a_j - \alpha_j \rho\right)\right\}}{\left\{\prod_{j=n+1}^{r} \Gamma_p (a_j + \alpha_j \rho)\right\}} \qquad (5.2.43)$$

where all the gammas are assumed to exist, α_j, $j = 1, ..., r$ and β_j, $j = 1, ..., s$ are real and positive, the poles of $\Gamma_p(b_j + \beta_j \rho)$, $j = 1, ..., m$ and those of $\Gamma_p(\frac{p+1}{2} - a_j - \alpha_j \rho)$, $j = 1, ..., n$ are assumed to be separated from each other, and other corresponding conditions for the scalar case are assumed to remain the same.

Definition 5.20   Meijer's G-function of matrix argument

In Definition 5.19 put $\alpha_1 = \alpha_2 = ... = \alpha_r = 1$, $\beta_1 = \beta_2 = ... = \beta_s = 1$. Then the H-function reduces to the G-function. The notation as in the scalar case will be used.

$$G(S) = G_{r,s}^{m,n} \left[S \middle| \begin{matrix} a_1, ..., a_r \\ b_1, ..., b_s \end{matrix} \right]$$

and

$$\int_{S>0} |S|^{\rho - \frac{p+1}{2}} G(S) dS = \frac{\left\{\prod_{j=1}^{m} \Gamma_p (b_j + \rho)\right\}}{\left\{\prod_{j=m+1}^{s} \Gamma_p \left(\frac{p+1}{2} - b_j - \rho\right)\right\}}$$

$$\times \frac{\left\{\prod_{j=1}^{n} \Gamma_p \left(\frac{p+1}{2} - a_j - \rho\right)\right\}}{\left\{\prod_{j=n+1}^{r} \Gamma_p (a_j + \rho)\right\}}. \qquad (5.2.44)$$

Note that almost all results, except series expansions, involving G- and H-functions of scalar argument can be extended to the corresponding matrix-variate cases by using Definitions 5.18, 5.19 and 5.20.

Example 5.14   Show that for $\text{Re}(\alpha) > \frac{p-1}{2}$, $S = S' > 0$

$$G_{1,1}^{1,1} \left(S \middle|_0^{\frac{p+1}{2} - \alpha} \right) = \Gamma_p(\alpha) |I + S|^{-\alpha}.$$

Solution Take the M-transform of the right side with respect to the parameter ρ to obtain the following:

$$M_\rho(\text{right side}) = \Gamma_p(\alpha) \int_{S>0} |S|^{\rho-\frac{p+1}{2}} |I+S|^{-\alpha} dS$$

$$= \Gamma_p(\alpha) \frac{\Gamma_p(\rho)\Gamma_p(\alpha-\rho)}{\Gamma_p(\alpha)}$$

$$= \Gamma_p(\rho)\Gamma_p(\alpha-\rho)$$

for $\text{Re}(\rho) > \frac{p-1}{2}$, $\text{Re}(\alpha-\rho) > \frac{p-1}{2}$. Now interpret this as the M-transform of a G-function to see the result.

Some more results of this type are given as exercises at the end of this section. A large number of results on G- and H-functions of scalar variables can be extended to the matrix-variate case by using the Definitions 5.19 and 5.20.

5.2.7 Some applications

Hypergeometric functions of matrix argument occur in a wide variety of statistical distribution problems. Some of these are available from Constantine (1963), James (1961), Mathai (1981, 1993), Mathai and Saxena (1978), and Sitgreaves (1952). A noncentral Wishart density can be written in terms of a Bessel function of matrix argument. Incomplete matrix-variate gamma and beta integrals lead to the hypergeometric functions of the type $_1F_1$ and $_2F_1$. Certain incomplete matrix-variate gamma integral, input-output, growth-decay, storage-consumption type models, certain covariance structures, certain quadratic and bilinear forms and so on are also associated with Whittaker function of matrix argument, see for example, Mathai and Pederzoli (1996, 1996b)) and Mathai (1993a). Various types of orthogonal polynomials of matrix argument and some of their applications can be seen from Mathai, Provost and Hayakawa (1995). A survey of the applications of functions of matrix argument in econometric theory and other applications can be found in Chikuse (1992, 1992a) and Chikuse and Davis (1986). Some of the applications in mathematical physics and statistical mechanics may be seen from Mehta (1967). Meijer's G and Fox's H-function of scalar variables are widely applied in many areas, see for example, Mathai and Saxena (1973, 1978) and Mathai and Haubold (1988). Many such problems have the matrix-variate analogues where the G and H-function of matrix argument are applicable. These are not yet explored in the literature.

Exercises

5.2.1   If the Bessel function $J_\nu(X)$ for $X = X' > 0$ is defined as

$$J_\nu\left(2X^{\frac{1}{2}}\right) = \frac{|X|^{\frac{\nu}{2}}}{\Gamma_p\left(\frac{p+1}{2} + \nu\right)} {}_0F_1\left(\; ; \frac{p+1}{2} + \nu; -X\right)$$

and the G-function as in Section 5.2.6 then show that

$$G_{0,2}^{1,0}\left(X|_{\alpha,\beta}\right) = |X|^{\frac{\alpha+\beta}{2}} J_{\alpha-\beta}\left(2X^{\frac{1}{2}}\right).$$

5.2.2   Show that

(i) $$G_{0,1}^{1,0}(X|a) = |X|^a e^{-\text{tr}(X)};$$

(ii) $$G_{1,1}^{1,0}\left(X|_\beta^\alpha\right) = \frac{1}{\Gamma_p(\alpha - \beta)}|X|^\beta |I - X|^{(\alpha-\beta)-\frac{p+1}{2}}$$

for $0 < X = X' < I$, $\text{Re}(\alpha - \beta) > \frac{p-1}{2}$;

(iii) $$G_{1,1}^{1,1}\left(AX|_\beta^{\frac{p+1}{2}-\alpha+\beta}\right) = \Gamma_p(\alpha)|A|^\beta |X|^\beta |I + AX|^{-\alpha}$$

for $A = A' > 0$, $X = X' > 0$, $\text{Re}(\beta) > \frac{p-1}{2}$, $\text{Re}(\alpha - \beta) > \frac{p-1}{2}$;

(iv) $$G_{1,2}^{1,1}\left(X|_{b,c}^a\right) = \frac{\Gamma_p\left(\frac{p+1}{2} - a + b\right)}{\Gamma_p\left(\frac{p+1}{2} + b - c\right)}|X|^b$$

$$\times {}_1F_1\left(\frac{p+1}{2} - a + b; \frac{p+1}{2} + b - c; -X\right)$$

for $X = X' > 0$, $\text{Re}(b - a + 1) > 0$, $\text{Re}(b - c + 1) > 0$.

5.2.3   Let the real symmetric $p \times p$ matrix S have a matrix-variate gamma density as given in (5.1.2) with $\alpha = n/2$, $n \geq p$ and $B = \frac{1}{2}V^{-1}$ for $V = V' > 0$. In this case S is said to have a Wishart density with parameter matrix V and degrees of freedom n. Let

$$S = \begin{bmatrix} S_{11} & S_{12} \\ S_{21} & S_{22} \end{bmatrix}, \; V = \begin{bmatrix} V_{11} & V_{12} \\ V_{21} & V_{22} \end{bmatrix}, \; V^{-1} = \begin{bmatrix} V^{11} & V^{12} \\ V^{21} & V^{22} \end{bmatrix}$$

where S_{11} and V_{11} are $r \times r$, S_{22} and V_{22} are $s \times s$, $r + s = p$. Let

$$U = S_{11}^{-\frac{1}{2}} S_{12} S_{22}^{-1} S_{21} S_{11}^{-\frac{1}{2}}$$

where $S_{11}^{\frac{1}{2}}$ denotes the symmetric square root of the symmetric positive definite matrix S_{11}. This U is known as the sample canonical correlation matrix when the sample comes from a multivariate normal population. Show by using the Laplace transform technique as well as by using M-transforms that the density of U is given by the following:

$$g(U) = \Gamma_r\left(\frac{n}{2}\right) |I - P|^{\frac{n}{2}} \left[\Gamma_r\left(\frac{s}{2}\right) \Gamma_r\left(\frac{n}{2} - \frac{s}{2}\right)\right]^{-1}$$
$$\times \; |U|^{\frac{s}{2} - \frac{r+1}{2}} |I - U|^{\frac{n}{2} - \frac{r+s+1}{2}}$$
$$\times \; {}_2F_1\left(\frac{n}{2}, \frac{n}{2}; \frac{s}{2}; P^{\frac{1}{2}} U P^{\frac{1}{2}}\right)$$

for $0 < U < I$, $0 < P < I$ and $g(U) = 0$ elsewhere. (For more details see Mathai (1981)).

5.2.4 For the Bessel function given in Definition 5.13 show that

$$A_{\gamma+\delta}(\Lambda) = \frac{1}{\Gamma_p(\delta)} \int_0^I A_\gamma(\Lambda S) |S|^\gamma |I - S|^{\delta - \frac{p+1}{2}} dS$$

for $\text{Re}(\gamma) > -1$, $\text{Re}(\delta) > \frac{p-1}{2}$. and

$$A_{\gamma+\delta}(\Lambda) |\Lambda|^{\gamma+\delta} = \frac{1}{\Gamma_p(\delta)} \int_0^\Lambda A_\gamma(S) |S|^\gamma |\Lambda - S|^{\delta - \frac{p+1}{2}} dS.$$

5.2.5 For the Bessel and Laguerre functions given in Definitions 5.13 and 5.15 show that

$$\int_{\Lambda > 0} e^{-\text{tr}(\Lambda T)} A_\gamma(\Lambda S) |\Lambda|^{\gamma+\nu} d\Lambda$$
$$= e^{-\text{tr}(ST^{-1})} L_\nu^{(\gamma)}\left(ST^{-1}\right) |T|^{-\gamma-\nu-\frac{p+1}{2}}.$$

5.2.6 For the Whittaker function defined in Section 5.2.5 show that for $\text{Re}(\beta - \alpha) > \frac{p-3}{4}$, $A = A' > 0$,

$$\int_{S>0} |S|^{\beta - \alpha - \frac{p+1}{4}} |I + S|^{\alpha + \beta - \frac{p+1}{4}} e^{-\text{tr}(AS)} dS$$
$$= |A|^{-\beta - \frac{p+1}{4}} \Gamma_p\left(\beta - \alpha + \frac{p+1}{4}\right) e^{\frac{1}{2}\text{tr}(A)} W_{\alpha,\beta}(A).$$

5.2.7   Show that

$$W_{\alpha,(\alpha+(p+1)/4)}(A) = \frac{|A|^{\alpha+\frac{p+1}{4}}}{\Gamma_p\left(\frac{p+1}{2}\right)} e^{-\frac{1}{2}\text{tr}(A)}$$

$$\times \int_{S>0} |I+S|^{2\alpha} e^{-\text{tr}(AS)} dS.$$

5.2.8   Show that for $A = A' > 0$, $\text{Re}(\pm\beta - \alpha) > \frac{p-3}{4}$, the M-transform of the Whittaker function is given by the following.

$$M_\rho\left[e^{\frac{1}{2}\text{tr}(A)}W_{\alpha,\beta}(A)\right] = \frac{1}{\Gamma_p\left(\beta-\alpha+\frac{p+1}{4}\right)} \int_{A>0} |A|^{\beta+\rho-\frac{p+1}{4}}$$

$$\times \int_{S>0} |S|^{\beta-\alpha-\frac{p+1}{4}} |I+S|^{\alpha+\beta-\frac{p+1}{4}} e^{-\text{tr}(AS)} dS$$

$$= \frac{\Gamma_p\left(\beta+\rho+\frac{p+1}{4}\right)\Gamma_p(-\alpha-\rho)\Gamma_p\left(-\beta+\rho+\frac{p+1}{4}\right)}{\Gamma_p\left(\beta-\alpha+\frac{p+1}{4}\right)\Gamma_p\left(-\alpha-\beta+\frac{p+1}{4}\right)}$$

where ρ is such that the poles of the gammas in $\Gamma_p(-\alpha-\rho)$ are separated from those of the gammas in $\Gamma_p\left(\beta+\rho+\frac{p+1}{4}\right)$ and $\Gamma_p\left(\rho-\beta+\frac{p+1}{4}\right)$.

5.2.9   Show that for $\text{Re}(\nu \pm \beta) > \frac{p-3}{4}$, $\text{Re}(\beta - \alpha) > \frac{p-3}{4}$, $X = X' > 0$

$$\int_{X>0} |X|^{\nu-\frac{p+1}{4}} e^{-\frac{1}{2}\text{tr}(X)} W_{\alpha,\beta}(X) dX$$

$$= \frac{\Gamma_p\left(\frac{p+1}{4}+\nu+\beta\right)\Gamma_p\left(\frac{p+1}{4}+\nu-\beta\right)}{\Gamma_p\left(\frac{p+1}{2}+\nu-\alpha\right)}.$$

5.2.10   Show that for $A = A' > 0$, $X = X' > 0$, $\text{Re}(\beta-\alpha) > \frac{p-3}{4}$, $\text{Re}(\beta-\alpha-\mu) > \frac{p-3}{4}$, $\text{Re}(\mu) > \frac{p-1}{2}$,

$$\int_{X>I} |X-I|^{\mu-\frac{p+1}{4}} |X|^{-\beta-\frac{p+1}{4}} e^{\frac{1}{2}\text{tr}(AX)} W_{\alpha,\beta}(AX) dX$$

$$= |A|^{-\frac{\mu}{2}}\Gamma_p(\mu) e^{\frac{1}{2}\text{tr}(A)} W_{\alpha+\frac{\mu}{2},\beta-\frac{\mu}{2}}(A)$$

$$\times \frac{\Gamma_p\left(\beta-\alpha-\mu+\frac{p+1}{4}\right)}{\Gamma_p\left(\beta-\alpha+\frac{p+1}{4}\right)}.$$

For more results on Whittaker function of matrix argument see Mathai and Pederzoli (1995,1996,1996a, 1996b).

5.3   Appell's and Humbert's functions

Appell's functions F_1, F_2, F_3, F_4 and Humbert's functions of real scalar variables will be generalized to these functions with real matrix argument in this section. All the matrices considered are real symmetric positive definite and $p \times p$ unless specified otherwise. Several results on these functions including limiting properties and integral properties will be established. Many of these correspond to the results in the scalar variable cases. The case $p = 1$ will be the scalar case in the results to follow. The M-transform will be used to define these functions.

Appell's functions and Humbert's functions of scalar variables appear in many fields such as statistical distribution theory, heat flow, astrophysics and related areas. Some of these may be seen from Exton (1976) and Srivastava and Karlsson (1985). Contributions to the theory in the scalar variables case have been made by many people, see for example Buschman (1965), Exton (1976), Slater (1966), and Srivastava and Karlsson (1985). In multivariate statistical analysis in the area of non-null distributions of test statistics the corresponding functions of matrix arguments are needed. The need for such functions in econometric problems may be seen from Chikuse and Davis (1986), and for more on the theoretical aspects and problems in physics see Haubold and Mathai (1994), and Mathai (1993c). The theory discussed in this section will find immediate applications in many disciplines. These functions are named after the French mathematicians P.E. Appell (1855–1930) and M.G. Humbert (1859–1921).

All the functions appearing in this section are assumed to be symmetric functions in the sense

$$f(X_1Y_1, ..., X_nY_n) = f(Y_1X_1, ..., Y_nX_n). \qquad (5.3.1)$$

A definition of a function of several matrix arguments in terms of zonal polynomials will involve invariant polynomials, and the multiple series representations will become quite involved, see for example, Mathai, Provost and Hayakawa (1995). Hence we will define these funtions through the M-transform introduced in Section 5.1.7. The M-transform in the many matrix variables case will be defined with respect to the parameters $\rho_1, ..., \rho_n$ as follows:

$$
\begin{aligned}
M(f) &= M_{\rho_1, ..., \rho_n}[f(X_1, ..., X_n)] \\
&= \int_{X_1>0} \cdots \int_{X_n>0} |X_1|^{\rho_1 - \frac{p+1}{2}} \cdots |X_n|^{\rho_n - \frac{p+1}{2}} \\
&\quad \times f(X_1, ..., X_n)\mathrm{d}X_1...\mathrm{d}X_n
\end{aligned}
\qquad (5.3.2)
$$

whenever $M(f)$ exists. The symmetry condition in (5.3.1) will have the following effect. For any scalar function $g(X)$ of the real symmetric positive definite matrix X,

$$g(X) = g(QQ'X) = g(Q'XQ)$$
$$= g\left(\text{diag}(\lambda_1, ..., \lambda_p)\right), \ QQ' = Q'Q = I \qquad (5.3.3)$$

where $\lambda_1, ..., \lambda_p$ are the eigenvalues of X. Thus effectively g is a function of the p eigenvalues of X and $\lambda_j > 0$, $j = 1, ..., p$.

In order to establish some limiting properties as well as integral properties we will require Dirichlet functions of matrix arguments. These are discussed in Section 5.1.8.

5.3.1   Appell's functions of matrix arguments

These will be defined in terms of their M-transforms as those functions having the following M-transforms with respect to the parameters ρ_1, ρ_2. In the following definitions and theorems the notation $\text{Re}(*, **, ...) > \alpha$ will mean that $\text{Re}(*) > \alpha$, $\text{Re}(**) > \alpha, ...$.

Definition 5.21: $F_1 = F_1(a, b, b'; c; -X_1, -X_2)$.

$$M(F_1) = \int_{X_1>0} \int_{X_2>0} |X_1|^{\rho_1 - \frac{p+1}{2}} |X_2|^{\rho_2 - \frac{p+1}{2}}$$
$$\times \ F_1(a, b, b'; c; -X_1, -X_2) \mathrm{d}X_1 \mathrm{d}X_2$$
$$= \frac{\Gamma_p(c)}{\Gamma_p(a)\Gamma_p(b)\Gamma_p(b')} \frac{\Gamma_p(\rho_1)\Gamma_p(\rho_2)}{\Gamma_p(c - \rho_1 - \rho_2)}$$
$$\times \ \Gamma_p(a - \rho_1 - \rho_2)\Gamma_p(b - \rho_1)\Gamma_p(b' - \rho_2) \qquad (5.3.4)$$

for $\text{Re}(\rho_1, \rho_2, a - \rho_1 - \rho_2, b - \rho_1, b' - \rho_2, c - \rho_1 - \rho_2) > \frac{p-1}{2}$.

Definition 5.22: $F_2 = F_2(a, b, b'; c, c'; -X_1, -X_2)$.

$$M(F_2) = \int_{X_1>0} \int_{X_2>0} |X_1|^{\rho_1 - \frac{p+1}{2}} |X_2|^{\rho_2 - \frac{p+1}{2}}$$
$$\times \ F_2(a, b, b'; c, c'; -X_1, -X_2) \mathrm{d}X_1 \mathrm{d}X_2$$
$$= \frac{\Gamma_p(c)\Gamma_p(c')}{\Gamma_p(a)\Gamma_p(b)\Gamma_p(b')} \frac{\Gamma_p(\rho_1)\Gamma_p(\rho_2)}{\Gamma_p(c - \rho_1)\Gamma_p(c' - \rho_2)}$$
$$\times \ \Gamma_p(b - \rho_1)\Gamma_p(b' - \rho_2)\Gamma_p(a - \rho_1 - \rho_2) \qquad (5.3.5)$$

for $\mathrm{Re}(\rho_1, \rho_2, a - \rho_1 - \rho_2, b - \rho_1, b' - \rho_2, c - \rho_1, c' - \rho_2) > \frac{p-1}{2}$.

Definition 5.23: $F_3 = F_3(a, a', b, b'; c; -X_1, -X_2)$.

$$M(F_3) = \frac{\Gamma_p(c)}{\Gamma_p(a)\Gamma_p(a')\Gamma_p(b)\Gamma_p(b')} \frac{\Gamma_p(\rho_1)\Gamma_p(\rho_2)}{\Gamma_p(c - \rho_1 - \rho_2)}$$
$$\times \Gamma_p(a - \rho_1)\Gamma_p(a' - \rho_2)\Gamma_p(b - \rho_1)\Gamma_p(b' - \rho_2) \qquad (5.3.6)$$

for $\mathrm{Re}(a - \rho_1, a' - \rho_2, b - \rho_1, b' - \rho_2, \rho_1, \rho_2, c - \rho_1 - \rho_2) > \frac{p-1}{2}$.

Definition 5.24: $F_4 = F_4(a, b; c, c'; -X_1, -X_2)$.

$$M(F_4) = \frac{\Gamma_p(c)\Gamma_p(c')}{\Gamma_p(a)\Gamma_p(b)} \frac{\Gamma_p(a - \rho_1 - \rho_2)}{\Gamma_p(c - \rho_1)\Gamma_p(c' - \rho_2)}$$
$$\times \Gamma_p(b - \rho_1 - \rho_2)\Gamma_p(\rho_1)\Gamma_p(\rho_2) \qquad (5.3.7)$$

for $\mathrm{Re}(\rho_1, \rho_2, a - \rho_1 - \rho_2, b - \rho_1 - \rho_2, c - \rho_1, c' - \rho_2) > \frac{p-1}{2}$.

Several properties will be established by using these definitions in (5.3.4)–(5.3.7). One result on a hypergeometric function $_rF_s$, when the argument is a sum of several matrices, will be established first. This is needed later on.

Theorem 5.11   *The M-transform of* $_rF_s(a_1, ..., a_r; b_1, ..., b_s; -(X_1 + ... + X_n))$ *is given by*

$$M(_rF_s) = \int_{X_1 > 0} \cdots \int_{X_n > 0} |X_1|^{\rho_1 - \frac{p+1}{2}} \cdots |X_n|^{\rho_n - \frac{p+1}{2}}$$
$$\times\, _rF_s(a_1, ..., a_r; b_1, ..., b_s; -(X_1 + ... + X_n))dX_1...dX_n$$
$$= \frac{\left\{\prod_{j=1}^{s} \Gamma_p(b_j)\right\}}{\left\{\prod_{j=1}^{r} \Gamma_p(a_j)\right\}} \left\{\prod_{j=1}^{n} \Gamma_p(\rho_j)\right\}$$
$$\times \frac{\left\{\prod_{j=1}^{r} \Gamma_p(a_j - \rho_1 - ... - \rho_n)\right\}}{\left\{\prod_{j=1}^{s} \Gamma_p(b_j - \rho_1 - ... - \rho_n)\right\}} \qquad (5.3.8)$$

for $\mathrm{Re}(\rho_j, a_m - \rho_1 - ... - \rho_n, b_k - \rho_1 - ... - \rho_n) > \frac{p-1}{2}$, $j = 1, ..., n$, $m = 1, ..., r$, $k = 1, ..., s$.

Proof   Consider the following transformations.

$$Y = X_1 + ... + X_n \Rightarrow X_1 = Y - X_2 - ... - X_n$$

$$\Rightarrow dX_1 = dY, \text{ for fixed } X_2, ..., X_n,$$
$$X_1 = Y^{\frac{1}{2}}[I - Y^{-\frac{1}{2}}X_2Y^{-\frac{1}{2}} - ... - Y^{-\frac{1}{2}}X_nY^{-\frac{1}{2}}]Y^{\frac{1}{2}}$$

where $Y^{\frac{1}{2}}$ is the unique symmetric square root of $Y = Y' > 0$. Put

$$V_j = Y^{-\frac{1}{2}}X_jY^{-\frac{1}{2}} \Rightarrow dV_j = |Y|^{-\frac{p+1}{2}}dX_j$$

and

$$X_j = Y^{\frac{1}{2}}V_jY^{\frac{1}{2}}, \ j = 2, ..., n$$

for fixed Y, which give

$$X_1 = Y^{\frac{1}{2}}[I - V_2 - ... - V_n]Y^{\frac{1}{2}},$$

see also Theorem 1.20 of Chapter 1. Then

$$M(_rF_s) = \int_{Y>0} |Y|^{\rho_1 + ... + \rho_n - \frac{p+1}{2}}$$
$$\times \ _rF_s(a_1, ..., a_r; b_1, ..., b_s; -Y)dY$$
$$\times \int ... \int |I - V_2 - ... - V_n|^{\rho_1 - \frac{p+1}{2}}$$
$$\times |V_2|^{\rho_2 - \frac{p+1}{2}} ... |V_n|^{\rho_n - \frac{p+1}{2}}dV_2...dV_n.$$

Note that

$$I - V_2 - ... - V_n = (I - V_3 - ... - V_n)^{\frac{1}{2}}$$
$$\times \left[I - (I - V_3 - ... - V_n)^{-\frac{1}{2}}V_2(I - V_3 - ... - V_n)^{-\frac{1}{2}}\right]$$
$$\times (I - V_3 - ... - V_n)^{\frac{1}{2}}.$$

Put

$$U_2 = (I - V_3 - ... - V_n)^{-\frac{1}{2}}V_2(I - V_3 - ... - V_n)^{-\frac{1}{2}}$$
$$\Rightarrow dU_2 = |I - V_3 - ... - V_n|^{-\frac{p+1}{2}}dV_2$$

for fixed $V_3, ..., V_n$ and $0 < U_2 < I$. Now integrate out U_2 by using a type-1 beta integral to get

$$\frac{\Gamma_p(\rho_1)\Gamma_p(\rho_2)}{\Gamma_p(\rho_1 + \rho_2)}|I - V_3 - ... - V_n|^{\rho_1 + \rho_2 - \frac{p+1}{2}}.$$

Integrate out $V_3, ..., V_n$ to get the final form as $\left\{\prod_{j=1}^n \Gamma_p(\rho_j)\right\}/\Gamma_p(\rho_1 + ... + \rho_n)$, see also Theorem 5.5. Now integrate out Y by using the M-transform of an $_rF_s$ in Section 5.2.4. Note that $\Gamma_p(\rho_1 + ... + \rho_n)$ is cancelled and the result follows.

Theorem 5.12   *For* $\operatorname{Re}(b, b', c - b - b') > \frac{p-1}{2}$,

$$F_1 = F_1(a, b, b'; c; -X_1, -X_2)$$

$$= \frac{\Gamma_p(c)}{\Gamma_p(b)\Gamma_p(b')\Gamma_p(c - b - b')}$$

$$\times \int \int |U_1|^{b - \frac{p+1}{2}} |U_2|^{b' - \frac{p+1}{2}} |I - U_1 - U_2|^{c - b - b' - \frac{p+1}{2}}$$

$$\times \left| I + U_1^{\frac{1}{2}} X_1 U_1^{\frac{1}{2}} + U_2^{\frac{1}{2}} X_2 U_2^{\frac{1}{2}} \right|^{-a} dU_1 dU_2$$

where the integration is over $0 < U_j < I$, $j = 1, 2$, $U_1 + U_2 < I$.

Proof   Take the M-transform of the right side with respect to X_1 and X_2 and with the parameters ρ_1 and ρ_2. Then

$$\int_{X_1 > 0} \int_{X_2 > 0} |X_1|^{\rho_1 - \frac{p+1}{2}} |X_2|^{\rho_2 - \frac{p+1}{2}} |I + U_1^{\frac{1}{2}} X_1 U_1^{\frac{1}{2}} + U_2^{\frac{1}{2}} X_2 U_2^{\frac{1}{2}}|^{-a} dX_1 dX_2$$

$$= |U_1|^{-\rho_1} |U_2|^{-\rho_2} \frac{\Gamma_p(\rho_1)\Gamma_p(\rho_2)\Gamma_p(a - \rho_1 - \rho_2)}{\Gamma_p(a)}$$

for $\operatorname{Re}(\rho_1, \rho_2, a - \rho_1 - \rho_2) > \frac{p-1}{2}$. This is evaluated by using a type-2 Dirichlet integral D_2 of Section 5.1.8 after making the transformation $Y_j = U_j^{\frac{1}{2}} X_j U_j^{\frac{1}{2}}$ for fixed U_j. Integrating out the U_j's by using a type-1 Dirichlet integral D_1 of Section 5.1.8, the M-transform of the right side reduces to that of an F_1 and hence the result.

Note that if $-X_1$ and $-X_2$ are replaced by X_1 and X_2, that is, for $F_1(a, b, b'; c; X_1, X_2)$ the same integral representation also holds good. But in this case, in order for the matrix $I - U_1^{\frac{1}{2}} X_1 U_1^{\frac{1}{2}} - U_2^{\frac{1}{2}} X_2 U_2^{\frac{1}{2}}$ to remain positive definite, we need the condition

$$\|U_1^{\frac{1}{2}} X_1 U_1^{\frac{1}{2}} + U_2^{\frac{1}{2}} X_2 U_2^{\frac{1}{2}}\| \leq \|U_1^{\frac{1}{2}} X_1 U_1^{\frac{1}{2}}\| + \|U_2^{\frac{1}{2}} X_2 U_2^{\frac{1}{2}}\|$$
$$\leq \|X_1\| + \|X_2\| < 1$$

since $\|U_1\| \leq 1$, $\|U_2\| \leq 1$ where $\|(\cdot)\|$ denotes a norm of the matrix $(\cdot)$. Thus we need the condition $\|X_1\| + \|X_2\| < 1$ in this case.

The proofs are similar to the one above in the next few theorems and hence these will be stated without proofs.

Theorem 5.13   *For* $\mathrm{Re}(a, c - a) > \frac{p-1}{2}$,

$$F_1 = F_1(a, b, b'; c; -X_1, -X_2)$$

$$= \frac{\Gamma_p(c)}{\Gamma_p(a)\Gamma_p(c-a)} \int_0^I |U|^{a-\frac{p+1}{2}} |I - U|^{c-a-\frac{p+1}{2}}$$

$$\times \left| I + U^{\frac{1}{2}} X_1 U^{\frac{1}{2}} \right|^{-b} \left| I + U^{\frac{1}{2}} X_2 U^{\frac{1}{2}} \right|^{-b'} dU.$$

Note that for $F_1(a, b, b'; c; X_1, X_2)$ the same theorem also holds good provided $\|X_1\| < 1$ and $\|X_2\| < 1$.

Theorem 5.14   *For* $\mathrm{Re}(b, b', c - b, c' - b') > \frac{p-1}{2}$,

$$F_2 = F_2(a, b, b'; c, c'; -X_1, -X_2)$$

$$= \frac{\Gamma_p(c)\Gamma_p(c')}{\Gamma_p(b)\Gamma_p(b')\Gamma_p(c-b)\Gamma_p(c'-b')} \int_0^I \int_0^I |U_1|^{b-\frac{p+1}{2}}$$

$$\times |U_2|^{b'-\frac{p+1}{2}} |I - U_1|^{c-b-\frac{p+1}{2}} |I - U_2|^{c'-b'-\frac{p+1}{2}}$$

$$\times \left| I + U_1^{\frac{1}{2}} X_1 U_1^{\frac{1}{2}} + U_2^{\frac{1}{2}} X_2 U_2^{\frac{1}{2}} \right|^{-a} dU_1 dU_2.$$

The same theorem holds good for $F_2(a, b, b'; c, c'; X_1, X_2)$ also provided $\|X_1\| + \|X_2\| < 1$.

Theorem 5.15   *For* $\mathrm{Re}(b, b', c - b - b') > \frac{p-1}{2}$,

$$F_3 = F_3(a, a', b, b'; c; -X_1, -X_2)$$

$$= \frac{\Gamma_p(c)}{\Gamma_p(b)\Gamma_p(b')\Gamma_p(c-b-b')} \int \int |U_1|^{b-\frac{p+1}{2}}$$

$$\times |U_2|^{b'-\frac{p+1}{2}} |I - U_1 - U_2|^{c-b-b'-\frac{p+1}{2}}$$

$$\times \left| I + U_1^{\frac{1}{2}} X_1 U_1^{\frac{1}{2}} \right|^{-a} \left| I + U_2^{\frac{1}{2}} X_2 U_2^{\frac{1}{2}} \right|^{-a'} dU_1 dU_2$$

where the integration is over $0 < U_j < I$, $j = 1, 2$, $U_1 + U_2 < I$.

The same theorem is valid for $F_3(a, a', b, b'; c; X_1, X_2)$ also for $\|X_1\| < 1$ and $\|X_2\| < 1$.

Theorem 5.16   *For* $\operatorname{Re}(a, b) > \frac{p-1}{2}$,

$$F_4 = F_4(a, b; c, c'; -X_1, -X_2)$$

$$= \frac{1}{\Gamma_p(a)\Gamma_p(b)} \int_{T_1>0} \int_{T_2>0} e^{-\operatorname{tr}(T_1+T_2)} |T_1|^{a-\frac{p+1}{2}}$$

$$\times |T_2|^{b-\frac{p+1}{2}} {}_0F_1\left(\ ; c; -T_1^{\frac{1}{2}}T_2^{\frac{1}{2}}X_1T_1^{\frac{1}{2}}T_2^{\frac{1}{2}}\right)$$

$$\times {}_0F_1\left(\ ; c'; -T_1^{\frac{1}{2}}T_2^{\frac{1}{2}}X_2T_2^{\frac{1}{2}}T_1^{\frac{1}{2}}\right) dT_1 dT_2.$$

Proof   Take the M-transform of the right side with respect to X_1 and X_2 and after making the transformation

$$Y_j = T_1^{\frac{1}{2}}T_2^{\frac{1}{2}}X_jT_2^{\frac{1}{2}}T_1^{\frac{1}{2}} \Rightarrow dY_j = |T_1T_2|^{\frac{p+1}{2}}dX_j \text{ for fixed } T_1, T_2.$$

Then we get

$$|T_1|^{-\rho_1-\rho_2}|T_2|^{-\rho_1-\rho_2}\frac{\Gamma_p(\rho_1)\Gamma_p(\rho_2)\Gamma_p(c)\Gamma_p(c')}{\Gamma_p(c-\rho_1)\Gamma_p(c'-\rho_2)}.$$

Integrating out T_1 and T_2 by using gamma integrals one gets $\Gamma_p(a - \rho_1 - \rho_2)\Gamma_p(b - \rho_1 - \rho_2)$ and the right side is then the M-transform of an F_4. This establishes the result.

5.3.2   Humbert's functions of matrix arguments

Again M-transform will be used to define Humbert's functions, as those functions having the following M-transforms with respect to the parameters ρ_1, ρ_2.

Definition 5.25: $\Phi_1 = \Phi_1(a, b; c; -X_1, -X_2)$.

$$M(\Phi_1) = \frac{\Gamma_p(c)}{\Gamma_p(a)\Gamma_p(b)} \frac{\Gamma_p(a - \rho_1 - \rho_2)}{\Gamma_p(c - \rho_1 - \rho_2)}$$
$$\times \Gamma_p(b - \rho_1)\Gamma_p(\rho_1)\Gamma_p(\rho_2)$$

for $\operatorname{Re}(\rho_1, \rho_2, b - \rho_1, a - \rho_1 - \rho_2, c - \rho_1 - \rho_2) > \frac{p-1}{2}$.

Definition 5.26: $\Phi_2 = \Phi_2(b, b'; c; -X_1, -X_2)$.

$$M(\Phi_2) = \frac{\Gamma_p(c)}{\Gamma_p(b)\Gamma_p(b')} \frac{\Gamma_p(b - \rho_1)}{\Gamma_p(c - \rho_1 - \rho_2)}$$
$$\times \Gamma_p(b' - \rho_2)\Gamma_p(\rho_1)\Gamma_p(\rho_2)$$

for $\mathrm{Re}(\rho_1, \rho_2, b - \rho_1, b' - \rho_2, c - \rho_1 - \rho_2) > \frac{p-1}{2}$.

Definition 5.27: $\Phi_3 = \Phi_3(b; c; -X_1, -X_2)$.

$$M(\Phi_3) = \frac{\Gamma_p(c)}{\Gamma_p(b)} \frac{\Gamma_p(b - \rho_1)\Gamma_p(\rho_1)\Gamma_p(\rho_2)}{\Gamma_p(c - \rho_1 - \rho_2)}$$

for $\mathrm{Re}(\rho_1, \rho_2, b - \rho_1, c - \rho_1 - \rho_2) > \frac{p-1}{2}$.

Definition 5.28: $\Psi_1 = \Psi_1(a, b; c, c'; -X_1, -X_2)$.

$$M(\Psi_1) = \frac{\Gamma_p(c)\Gamma_p(c')}{\Gamma_p(a)\Gamma_p(b)} \frac{\Gamma_p(a - \rho_1 - \rho_2)}{\Gamma_p(c - \rho_1)\Gamma_p(c' - \rho_2)}$$
$$\times \Gamma_p(b - \rho_1)\Gamma_p(\rho_1)\Gamma_p(\rho_2)$$

for $\mathrm{Re}(\rho_1, \rho_2, b - \rho_1, c - \rho_1, c' - \rho_2, a - \rho_1 - \rho_2) > \frac{p-1}{2}$.

Definition 5.29: $\Psi_2 = \Psi_2(a; c, c'; -X_1, -X_2)$.

$$M(\Psi_2) = \frac{\Gamma_p(c)\Gamma_p(c')}{\Gamma_p(a)} \frac{\Gamma_p(a - \rho_1 - \rho_2)\Gamma_p(\rho_1)\Gamma_p(\rho_2)}{\Gamma_p(c - \rho_1)\Gamma_p(c' - \rho_2)}$$

for $\mathrm{Re}(\rho_1, \rho_2, c - \rho_1, c' - \rho_2, a - \rho_1 - \rho_2) > \frac{p-1}{2}$.

Definition 5.30: $\Xi_1 = \Xi_1(a, a', b; c; -X_1, -X_2)$.

$$M(\Xi_1) = \frac{\Gamma_p(c)}{\Gamma_p(a)\Gamma_p(a')\Gamma_p(b)} \frac{\Gamma_p(a - \rho_1)\Gamma_p(a' - \rho_2)}{\Gamma_p(c - \rho_1 - \rho_2)}$$
$$\times \Gamma_p(b - \rho_1)\Gamma_p(\rho_1)\Gamma_p(\rho_2)$$

for $\mathrm{Re}(\rho_1, \rho_2, a - \rho_1, a' - \rho_2, b - \rho_1, c - \rho_1 - \rho_2) > \frac{p-1}{2}$.

Definition 5.31: $\Xi_2 = \Xi_2(a, b; c; -X_1, -X_2)$.

$$M(\Xi_2) = \frac{\Gamma_p(c)}{\Gamma_p(a)\Gamma_p(b)} \frac{\Gamma_p(a - \rho_1)}{\Gamma_p(c - \rho_1 - \rho_2)}$$
$$\times \Gamma_p(b - \rho_1)\Gamma_p(\rho_1)\Gamma_p(\rho_2)$$

for $\mathrm{Re}(\rho_1, \rho_2, a - \rho_1, b - \rho_1, c - \rho_1 - \rho_2) > \frac{p-1}{2}$.

A few results will be established on Humbert's functions. The first few are integral representations and the remaining ones are some limiting properties. For $p = 1$ one gets the corresponding results for the scalar case.

Theorem 5.17   *For* $\text{Re}(a, c - a) > \frac{p-1}{2}$,

$$\Phi_1 = \Phi_1(a, b; c; -X_1, -X_2)$$

$$= \frac{\Gamma_p(c)}{\Gamma_p(a)\Gamma_p(c - a)} \int_0^I |U|^{a - \frac{p+1}{2}} |I - U|^{c - a - \frac{p+1}{2}}$$

$$\times \left|I + U^{\frac{1}{2}} X_1 U^{\frac{1}{2}}\right|^{-b} e^{-\text{tr}(U X_2)} dU.$$

Proof   Take the M-transform of the right side with respect to X_1 and X_2. Then

$$\int_{X_1 > 0} |X_1|^{\rho_1 - \frac{p+1}{2}} \left|I + U^{\frac{1}{2}} X_1 U^{\frac{1}{2}}\right|^{-b} dX_1 = \frac{\Gamma_p(\rho_1)\Gamma_p(b - \rho_1)}{\Gamma_p(b)} |U|^{-\rho_1}$$

for $\text{Re}(\rho_1, b - \rho_1) > \frac{p-1}{2}$, evaluated by using D_2 of Section 5.1.8 after making the transformation $Y_1 = U^{\frac{1}{2}} X_1 U^{\frac{1}{2}}$ and

$$\int_{X_2 > 0} |X_2|^{\rho_2 - \frac{p+1}{2}} \exp\left\{-\text{tr}\left(U^{\frac{1}{2}} X_2 U^{\frac{1}{2}}\right)\right\} dX_2 = |U|^{-\rho_2} \Gamma_p(\rho_2)$$

for $\text{Re}(\rho_2) > \frac{p-1}{2}$, by using a gamma integral. Now integrate out U by using a type-1 beta integral or D_1 of Section 5.1.8 for $n = 1$ to see that the M-transform of the right side reduces to that of a Φ_1 and hence the result.

Note that from Theorem 5.11 one can have the following result.

Theorem 5.18   *For* $\text{Re}(a, c - a) > \frac{p-1}{2}$,

$$\frac{\Gamma_p(c)}{\Gamma_p(a)\Gamma_p(c - a)} \int_0^I |U|^{a - \frac{p+1}{2}}$$

$$\times |I - U|^{c - a - \frac{p+1}{2}} \exp\left\{-\text{tr}\left(U^{\frac{1}{2}}(X_1 + X_2)U^{\frac{1}{2}}\right)\right\} dU$$

$$= {}_1F_1(a; c; -(X_1 + X_2)).$$

Proof   Take the M-transform of the right side with respect to X_1 and X_2. Integrate out U by using a type-1 beta integral. Note that this M-transform is a particular case of the M-transform in Theorem 5.11 with $r = s = 1$ and hence the result. Also see (5.2.9).

In the next two theorems the proofs are similar and hence these will be stated without proofs.

Theorem 5.19   *For* $\text{Re}(b, b', c - b - b') > \frac{p-1}{2}$,

$$\Phi_2 = \Phi_2(b, b'; c; -X_1, -X_2)$$

$$= \frac{\Gamma_p(c)}{\Gamma_p(b)\Gamma_p(b')\Gamma_p(c-b-b')} \int \int |U_1|^{b-\frac{p+1}{2}} |U_2|^{b'-\frac{p+1}{2}}$$

$$\times |I - U_1 - U_2|^{c-b-b'-\frac{p+1}{2}}$$

$$\times \exp\left\{ -\text{tr}\left(U_1^{\frac{1}{2}} X_1 U_1^{\frac{1}{2}} + U_2^{\frac{1}{2}} X_2 U_2^{\frac{1}{2}} \right) \right\} dU_1 dU_2$$

where the integration is over $0 < U_j < I$, $j = 1, 2$, $U_1 + U_2 < I$.

Theorem 5.20   *For* $\text{Re}(d, d', c - d - d', b, b', c - b - b') > \frac{p-1}{2}$,

$$\Phi_2 = \Phi_2(b, b'; c; -X_1, -X_2)$$

$$= \frac{\Gamma_p(c)}{\Gamma_p(d)\Gamma_p(d')\Gamma_p(c-d-d')} \int \int |U_1|^{d-\frac{p+1}{2}} |U_2|^{d'-\frac{p+1}{2}}$$

$$\times |I - U_1 - U_2|^{c-d-d'-\frac{p+1}{2}} {}_1F_1\left(b; d; -U_1^{\frac{1}{2}} X_1 U_1^{\frac{1}{2}} \right)$$

$$\times {}_1F_1\left(b'; d'; -U_2^{\frac{1}{2}} X_2 U_2^{\frac{1}{2}} \right) dU_1 dU_2$$

where the integration is over $0 < U_j < I$, $j = 1, 2$, $U_1 + U_2 < I$.

Theorem 5.21   *For* $\text{Re}(b, b', c - b - b') > \frac{p-1}{2}$,

$$\Phi_2 = \Phi_2(b, b'; c; -X_1, -X_2)$$

$$= \frac{\Gamma_p(c) 2^{\frac{p(p-1)}{2}}}{(2\pi i)^{\frac{p(p+1)}{2}}} \int e^{\text{tr}(T)} |T|^{-c}$$

$$\times |I + X_1 T^{-1}|^{-b} |I + X_2 T^{-1}|^{-b'} dT$$

where $T = T_1 + iT_2$, $T_1 \in S_p^*$, $T_2 \in S_p^*$, *see also Notation 5.1, with* $T_1 > 0$ *and it is assumed that* $T^{-1} = VV'$ *with* $V'X_jV > 0$, $j = 1, 2$.

Proof   Take the M-transform of the right side with respect to X_1 and X_2 to get

$$|T|^{\rho_1 + \rho_2} \frac{\Gamma_p(\rho_1)\Gamma_p(b - \rho_1)}{\Gamma_p(b)} \frac{\Gamma_p(\rho_2)\Gamma_p(b' - \rho_2)}{\Gamma_p(b')}$$

by making the transformation $Y_j = V'X_jV$ for fixed V and evaluating by using type-2 beta integrals. Note that

$$|I + X_j T^{-1}| = |I + X_j VV'| = |I + V'X_j V|$$

for $|V| \neq 0$. Integrating out T by using the inverse Laplace representation of the gamma function, see (5.1.12), the result follows.

The following theorem can be proved by taking the M-transform and interpreting the result.

Theorem 5.22   *For* $\text{Re}(b, b', c - b - b') > \frac{p-1}{2}$,

$$\Xi_1 = \Xi_1(a, b', b; c; -X_1, -X_2)$$

$$= \frac{\Gamma_p(c)}{\Gamma_p(b)\Gamma_p(b')\Gamma_p(c - b - b')} \int \int |U_1|^{b - \frac{p+1}{2}} |U_2|^{b' - \frac{p+1}{2}}$$

$$\times |I - U_1 - U_2|^{c - b - b' - \frac{p+1}{2}} \left| I + U_1^{\frac{1}{2}} X_1 U_1^{\frac{1}{2}} \right|^{-a} e^{-\text{tr}(U_2 X_2)} dU_1 dU_2$$

where the integration is over $0 < U_j < I$, $j = 1, 2$, $U_1 + U_2 < I$.

If the roles of (X_1, U_1) and (X_2, U_2) are interchanged then one gets another result of the same type having two gammas with ρ_2 and one with ρ_1 in the numerator. A similar result is available if X_1 and X_2 are interchanged.

5.3.3   Humbert's functions as limiting forms

A number of results giving Humbert's functions as limiting forms of Appell's functions will be established. We need two results, one is Lemma 5.2 given in Section 5.2 and other will be given here as a lemma.

Lemma 5.3   *For* $|z| \to \infty$ *and b bounded*

$$\lim_{z \to \infty} \left\{ z^{-pb} \frac{\Gamma_p(z - b)}{\Gamma_p(z)} \right\} = 1.$$

Proof   From the ratio of the gamma functions the factor $\pi^{\frac{p(p-1)}{4}}$ is cancelled and then

$$\frac{\Gamma_p(z - b)}{\Gamma_p(z)} = \prod_{j=1}^{p} \left\{ \frac{\Gamma(z - b - \frac{j-1}{2})}{\Gamma(z - \frac{j-1}{2})} \right\}.$$

Now apply the asymptotic formula for gamma functions, namely, for $|z| \to \infty$ and a bounded

$$\Gamma(z + a) \to (2\pi)^{\frac{1}{2}} z^{z + a - \frac{1}{2}} e^{-z}.$$

Apply this to each gamma in the ratio by taking a as $-b - \frac{i-1}{2}$ and $-\frac{i-1}{2}$ respectively to establish the result.

Theorem 5.23

$$\lim_{\epsilon \to 0} F_1 \left(a, b, \frac{1}{\epsilon}; c; -X_1, -\epsilon X_2 \right) = \Phi_1(a, b; c; -X_1, -X_2).$$

Proof   Take the integral representation of F_1 from Theorem 5.13. Now

$$\lim_{\epsilon \to 0} |I + \epsilon U^{\frac{1}{2}} X_2 U^{\frac{1}{2}}|^{-1/\epsilon} = e^{-\text{tr}\left(U^{\frac{1}{2}} X_2 U^{\frac{1}{2}} \right)}$$

by Lemma 5.2. Then the right side of Theorem 5.13 reduces to the integral representation of Φ_1 of Theorem 5.17 which establishes the result. Since the integral is of the beta type the limits can be taken inside the integral.

Theorem 5.24

$$\lim_{\epsilon \to 0} F_1 \left(\frac{1}{\epsilon}, b, b'; c; -\epsilon X_1, -\epsilon X_2 \right) = \Phi_2(b, b'; c; -X_1, -X_2).$$

Proof   Take the integral representation of F_1 in Theorem 5.12. Then

$$\lim_{\epsilon \to 0} \left| I + \epsilon U_1^{\frac{1}{2}} X_1 U_1^{\frac{1}{2}} + \epsilon U_2^{\frac{1}{2}} X_2 U_2^{\frac{1}{2}} \right|^{-\frac{1}{\epsilon}} = e^{-\text{tr}\left(U_1^{\frac{1}{2}} X_1 U_1^{\frac{1}{2}} + U_2^{\frac{1}{2}} X_2 U_2^{\frac{1}{2}} \right)}$$

by Lemma 5.2. From the integral representation of Φ_2 in Theorem 5.19 the result follows.

Theorem 5.25

$$\lim_{\epsilon \to 0} F_3 \left(a, \frac{1}{\epsilon}, b, b'; c; -X_1, -\epsilon X_2 \right) = \Xi_1(a, b', b; c; -X_1, -X_2).$$

Proof   Take the integral representation of F_3 from Theorem 5.15 and apply steps similar to the ones in the proof of Theorem 5.23 to see the result. In a similar fashion we can prove the following theorems.

Theorem 5.26

$$\lim_{\epsilon \to 0} F_3\left(\frac{1}{\epsilon}, a', b, b'; c; -\epsilon X_1, -X_2\right) = \hat{\Xi}_1(a', b, b'; c; -X_1, -X_2)$$

where $\hat{\Xi}_1$ has the same form as in Ξ_1 with ρ_1 and ρ_2 interchanged.

Theorem 5.27

$$\lim_{\epsilon \to 0} F_1\left(a, \frac{1}{\epsilon}, b'; c; -\epsilon X_1, -X_2\right) = \hat{\Phi}_1(a, b'; c; -X_1, -X_2)$$

and $\hat{\Phi}_1$ has the same form of Φ_1 with ρ_1 and ρ_2 interchanged.

Theorem 5.28

$$\lim_{\epsilon \to 0} F_1\left(a, \frac{1}{\epsilon}, \frac{1}{\epsilon}; c; -\epsilon X_1, -\epsilon X_2\right) = {}_1F_1(a; c; -(X_1 + X_2)).$$

After taking the limits with the help of Lemma 5.2 use Theorem 5.11 to establish the result.

Theorem 5.29

$$\lim_{\epsilon \to 0} F_2\left(a, b; \frac{1}{\epsilon}; c, c'; -X_1, -\epsilon X_2\right) = \Psi_1(a, b; c, c'; -X_1, -X_2).$$

Proof Use Definition 5.22 and Lemma 5.3 on the gamma ratio $\Gamma_p(b' - \rho_2)/\Gamma_p(b')$ for $\lim_{\epsilon \to 0} = \lim_{n \to \infty}$, $n = \frac{1}{\epsilon}$, to obtain $n^{-p\rho_2}$. Make a change of the variable $\epsilon X_2 = Y_2 \Rightarrow dY_2 = \epsilon^{\frac{p(p+1)}{2}} dX_2$ and $|Y_2| = |X_2|\epsilon^p$. Then $\epsilon^{-p\rho_2}$ comes out and cancels with $n^{-p\rho_2}$. The remaining is the M-transform of Ψ_1 and hence the result.

Similar steps can be used to prove the following results.

Theorem 5.30

$$\lim_{\epsilon \to 0} F_2\left(a, \frac{1}{\epsilon}, \frac{1}{\epsilon}; c, c'; -\epsilon X_1, -\epsilon X_2\right) = \Psi_2(a; c, c'; -X_1, -X_2).$$

$$\lim_{\epsilon \to 0} F_3 \left(a, a', b, \frac{1}{\epsilon}; c; -X_1, -\epsilon X_2 \right) = \Xi_1(a, a', b; c; -X_1, -X_2).$$

$$\lim_{\epsilon \to 0} F_3 \left(a, \frac{1}{\epsilon}, b, \frac{1}{\epsilon}; c; -X_1, -\epsilon^2 X_2 \right) = \Xi_2(a, b; c; -X_1, -X_2).$$

$$\lim_{\epsilon \to 0} \eta_1 \left(\frac{1}{\epsilon}, b, b'; c; -\epsilon X_1, -X_2 \right) = \Phi_2(b', b; c; -X_1, -X_2).$$

This last one also follows from the integral representation of Ξ_1 in Theorem 5.22 and Lemma 5.2.

Some of the above results and the various results stated in the following theorem do not seem to have been recorded in the literature for the scalar cases either.

Theorem 5.31

$$\lim_{\epsilon \to 0} \Phi_2 \left(b, \frac{1}{\epsilon}; c; -X_1, -\epsilon X_2 \right) = \Phi_3(b; c; -X_1, -X_2).$$

$$\lim_{\epsilon \to 0} \Psi_1 \left(a, \frac{1}{\epsilon}; c, c'; -\epsilon X_1, -X_2 \right) = \Psi_2(a; c, c'; -X_1, -X_2).$$

$$\lim_{\epsilon \to 0} \Xi_1 \left(a, \frac{1}{\epsilon}, b; c; -X_1, -\epsilon X_2 \right) = \Xi_2(a, b; c; -X_1, -X_2).$$

$$\lim_{\epsilon \to 0} \Xi_1 \left(a, a', \frac{1}{\epsilon}; c; -\epsilon X_1, -X_2 \right) = \Phi_2(a, a'; c; -X_1, -X_2).$$

$$\lim_{\epsilon \to 0} \Xi_2 \left(a, \frac{1}{\epsilon}; c; -\epsilon X_1, -X_2 \right) = \Phi_3(a; c; -X_1, -X_2).$$

$$\lim_{\epsilon \to 0} \Xi_2 \left(\frac{1}{\epsilon}, b; c; -\epsilon X_1, -X_2 \right) = \Phi_3(b; c; -X_1, -X_2).$$

For $U = X_1 + X_2$,

$$_1F_1(a; c; -U) = \lim_{\epsilon \to 0} \Phi_1 \left(a, \frac{1}{\epsilon}; c; -\epsilon X_1, -X_2 \right).$$

$$_0F_1(; c; -U) = \lim_{\epsilon \to 0} \Phi_2 \left(\frac{1}{\epsilon}, \frac{1}{\epsilon}; c; -\epsilon X_1, -\epsilon X_2 \right)$$

$$= \lim_{\epsilon \to 0} \Phi_3 \left(\frac{1}{\epsilon}; c; -\epsilon X_1, -X_2 \right)$$

$$= \lim_{\epsilon \to 0} \Xi_1 \left(\frac{1}{\epsilon}, \frac{1}{\epsilon}, \frac{1}{\epsilon}; c; -\epsilon^2 X_1, -\epsilon X_2 \right)$$

$$= \lim_{\epsilon \to 0} \Xi_2 \left(\frac{1}{\epsilon}, \frac{1}{\epsilon}; c; -\epsilon^2 X_1, -X_2 \right).$$

Note that one could have used some convenient integral representations given in Theorems 5.12–5.22 to define Appell's and Humbert's functions. Then these functions are uniquely defined. Further, the definitions given in this section can be obtained from those integrals by taking the M-transforms. The limiting properties given in Theorems 5.23–5.31 could also be obtained from such integral representations. But for establishing various types of results it is already noted that the definitions through M-transforms are the most convenient ones. It is also interesting to see that the whole families of functions defined through these M-transforms enjoy the properties listed as theorems in this section.

Exercises

5.3.1   Taking the integral in Theorem 5.12 as a definition for F_1, derive Theorems 5.13, 5.23, 5.24, 5.27, 5.28 and the result in (5.3.4).

5.3.2   Taking the integral in Theorem 5.14 as a definition for F_2, derive the result in (5.3.5), Theorem 5.29 and the first result in Theorem 5.30.

5.3.3   Taking the integral in Theorem 5.15 as a definition for F_3, derive the result in (5.3.6), Theorems 5.25, 5.26 and the second and third results in Theorem 5.30.

5.3.4   Taking the integral in Theorem 5.16 as a definition for F_4, derive the result in (5.3.7).

5.3.5   Taking the integral in Theorem 5.17 as a definition for Φ_1, derive the M-transform in Definition 5.25 and the seventh result in Theorem 5.31.

5.3.6   Taking the integral in Theorem 5.19 as a definition for Φ_2 derive the M-transform in Definition 5.26, Theorems 5.20, 5.21 and the first and eighth results in Theorem 5.31.

5.3.7   Taking the integral in Theorem 5.22 as a definition for Ξ_1 derive the M-transform in Definition 5.30, and the third, fourth and tenth results in Theorem 5.31.

5.4   Kampé de Fériet's functions of matrix argument

A class of functions which will be called the Kampé de Fériet's function $F_{s:m:n}^{r:q:k}(\cdot)$ of matrix arguments will be defined here when the argument matrices are real symmetric positive definite and several results on this class of functions including limiting properties and integral properties will be established. Some of these correspond to the results in the scalar variable cases. As in the previous section, M-transform will be used to define this function.

Kampé de Fériet's functions of scalar variables appear in many applied areas such as statistical distribution theory, astrophysics and other related areas. Some of these topics may be seen from Exton (1976) and Srivastava and Karlsson (1985). Contributions to the theoretical development of these functions in the scalar variable case have been made by several people, see for example Buschman (1965), Exton (1976), Saxena (1970), Slater (1966), and Srivastava and Karlsson (1985). In the nonnull distribution problems in multivariate statistical analysis the corresponding functions of matrix arguments are needed. Functions of several matrix arguments are also needed in several econometric problems, see for example Chikuse and Davis (1986). Thus the theory developed in this section will find immediate applications in many disciplines. These functions are named after the Belgian mathematician J.M. Kampé de Fériet (1893–1982).

As in the previous section all matrices appearing in this section are symmetric positive definite when real and hermitian positive definite when complex. Other notations remain the same as in the previous section. All the functions appearing in this section are assumed to be symmetric functions in the sense of (5.3.1) of Section 5.3.

5.4.1   Definition

Kampé de Fériet's functions will be defined in terms of the M-transform as that class of functions having the following M-transform with respect to the parameters ρ_1, ρ_2.

Definition 5.32: $F_{s:m:n}^{r:q:k} = F_{s:m:n}^{r:q:k}\left[-\binom{X_1}{X_2} \middle| \begin{array}{l} (a_r) : (b_q) : (c_k) \\ (\alpha_s) : (\beta_m) : (\gamma_n) \end{array} \right].$

$$M(F) = \int_{X_1>0} \int_{X_2>0} |X_1|^{\rho_1 - \frac{p+1}{2}} |X_2|^{\rho_2 - \frac{p+1}{2}}$$

$$\times F_{s:m:n}^{r:q:k} \left[-\binom{X_1}{X_2} \middle| \begin{matrix} (a_r) : (b_q) : (c_k) \\ (\alpha_s) : (\beta_m) : (\gamma_n) \end{matrix} \right] dX_1 dX_2$$

$$= C \frac{\left\{ \prod_{j=1}^{r} \Gamma_p(a_j - \rho_1 - \rho_2) \right\} \left\{ \prod_{j=1}^{q} \Gamma_p(b_j - \rho_1) \right\}}{\left\{ \prod_{j=1}^{s} \Gamma_p(\alpha_j - \rho_1 - \rho_2) \right\} \left\{ \prod_{j=1}^{m} \Gamma_p(\beta_j - \rho_1) \right\}}$$

$$\times \frac{\left\{ \prod_{j=1}^{k} \Gamma_p(c_j - \rho_2) \right\}}{\left\{ \prod_{j=1}^{n} \Gamma_p(\gamma_j - \rho_2) \right\}} \Gamma_p(\rho_1) \Gamma_p(\rho_2) \qquad (5.4.1)$$

for $\text{Re}(\rho_1, \rho_2, a_j - \rho_1 - \rho_2, \ j = 1, ..., r, \ \alpha_j - \rho_1 - \rho_2, \ j = 1, ..., s, \ b_j - \rho_1, \ j = 1, ..., q, \ \beta_j - \rho_1, \ j = 1, ..., m, \ c_j - \rho_2, \ j = 1, ..., k, \ \gamma_j - \rho_2, \ j = 1, ..., n) > \frac{p-1}{2}$
where

$$C = \frac{\left\{ \prod_{j=1}^{s} \Gamma_p(\alpha_j) \right\} \left\{ \prod_{j=1}^{m} \Gamma_p(\beta_j) \right\} \left\{ \prod_{j=1}^{n} \Gamma_p(\gamma_j) \right\}}{\left\{ \prod_{j=1}^{r} \Gamma_p(a_j) \right\} \left\{ \prod_{j=1}^{q} \Gamma_p(b_j) \right\} \left\{ \prod_{j=1}^{k} \Gamma_p(c_j) \right\}}.$$

In writing the parameters we have used the simplified notation. Here, for example, (a_r) stands for the sequence of parameters $a_1, ..., a_r$. Some elementary properties follow from this definition itself. Note that in Definition 5.32 we have used the general form considered in Srivastava and Karlsson (1985). The case $q = k$ and $m = n$, and $p = 1$ is often known in the literature as the Kampé de Fériet's function in the scalar case. Moreover

$$F_{1:0:0}^{1:1:1} = F_1; \quad F_{0:1:1}^{1:1:1} = F_2; \quad F_{1:0:0}^{0:2:2} = F_3; \quad F_{0:1:1}^{2:0:0} = F_4 \qquad (5.4.2)$$

where F_1, F_2, F_3, F_4 are Appell's functions defined in Section 5.3.

$$F_{0:m:n}^{0:q:k} = {}_qF_m^*(-X_1)_k F_n^*(-X_2) \qquad (5.4.3)$$

where ${}_qF_m^*$ and ${}_kF_n^*$ are hypergeometric functions of the matrix arguments X_1 and X_2 respectively, see also Section 5.2.4.

$$F_{s:0:0}^{r:0:0} = {}_rF_s^*(a_1, ..., a_r; \alpha_1, ..., \alpha_s; -(X_1 + X_2)). \qquad (5.4.4)$$

This result follows from Theorem 5.11.

5.4.2 Some integral representations

A few results on the integral representations of the Kampé de Fériet's functions will be given here.

Theorem 5.32   *For* $\operatorname{Re}(a_j, \alpha_j - a_j, b, c) > \frac{p-1}{2}$, $j = 1, ..., r$,

$$F = F_{r:0:0}^{r:1:1}\left[-\binom{X_1}{X_2}\Bigg|\begin{matrix}(a_r) : b : c\\(\alpha_r) : :\end{matrix}\right]$$

$$= \left\{\prod_{j=1}^{r}\frac{\Gamma_p(\alpha_j)}{\Gamma_p(a_j)\Gamma_p(\alpha_j - a_j)}\right\}$$

$$\times \int_0^I\int_0^I |U_1|^{a_1-\frac{p+1}{2}}...|U_r|^{a_r-\frac{p+1}{2}}|I - U_1|^{\alpha_1-a_1-\frac{p+1}{2}}...$$

$$\times |I - U_r|^{\alpha_r-a_r-\frac{p+1}{2}}\left|I + U_1^{\frac12}...U_r^{\frac12}X_1 U_r^{\frac12}...U_1^{\frac12}\right|^{-b}$$

$$\times \left|I + U_1^{\frac12}...U_r^{\frac12}X_2 U_r^{\frac12}...U_1^{\frac12}\right|^{-c}dU_1...dU_r.$$

Proof   Take the M-transform of the right side with respect to X_1 and X_2 and with the parameters ρ_1 and ρ_2. The integrals are

$$I_{X_1} = \int_{X_1>0} |X_1|^{\rho_1-\frac{p+1}{2}}\left|I + U_1^{\frac12}...U_r^{\frac12}X_1 U_r^{\frac12}...U_1^{\frac12}\right|^{-b}dX_1$$

$$= |U_1|^{-\rho_1}...|U_r|^{-\rho_1}\frac{\Gamma_p(\rho_1)\Gamma_p(b - \rho_1)}{\Gamma_p(b)}$$

for $\operatorname{Re}(\rho_1, b - \rho_1) > \frac{p-1}{2}$ by making the transformation

$$Y_j = U_1^{\frac12}...U_r^{\frac12}X_j U_r^{\frac12}...U_1^{\frac12} \Rightarrow$$

$$dY_j = |U_1...U_r|^{\frac{p+1}{2}}dX_j; \quad |X_j| = |U_1...U_r|^{-1}|Y_j|$$

and a similar integral for X_2. Integrate out the U_j's by using type-1 beta integrals. Then one gets

$$\int_0^I |U_j|^{a_j-\rho_1-\rho_2-\frac{p+1}{2}}|I - U_j|^{\alpha_j-a_j-\frac{p+1}{2}}$$

$$= \frac{\Gamma_p(a_j - \rho_1 - \rho_2)\Gamma_p(\alpha_j - a_j)}{\Gamma_p(\alpha_j - \rho_1 - \rho_2)}$$

for $\operatorname{Re}(a_j - \rho_1 - \rho_2, \alpha_j - a_j) > \frac{p-1}{2}$, $j = 1, ..., r$. Now take the product to obtain the result.

Remark 5.8   The result in Theorem 5.32 can be generalized in many directions. (i) Delete some of the U_j's from either the determinant containing X_1 or the one containing X_2. Then the resulting F will be different.

(ii) Replace $\left|I + U_1^{\frac{1}{2}}...U_r^{\frac{1}{2}} X_j U_r^{\frac{1}{2}}...U_1^{\frac{1}{2}}\right|$, $j = 1, 2$ by two different general hypergeometric functions. (iii) Combine (i) and (ii) to get new results.

Theorem 5.33 For $\operatorname{Re}(a, b_j, \beta_j - b_j, \; j = 1, ..., q; \; c_j, \gamma_j - c_j, \; j = 1, ..., k) > \frac{p-1}{2}$

$$F = F_{0:q:k}^{1:q:k}\left[-\begin{pmatrix}X_1\\X_2\end{pmatrix}\begin{vmatrix}a:(b_q):(c_k)\\:(\beta_q):(\gamma_k)\end{vmatrix}\right]$$

$$= \left\{\prod_{j=1}^{q} \frac{\Gamma_p(\beta_j)}{\Gamma_p(b_j)\Gamma_p(\beta_j - b_j)}\right\}\left\{\prod_{j=1}^{k} \frac{\Gamma_p(\gamma_j)}{\Gamma_p(c_j)\Gamma_p(\gamma_j - c_j)}\right\}$$

$$\times \int_0^I ... \int_0^I \left\{\prod_{j=1}^{q} |U_j|^{b_j - \frac{p+1}{2}}|I - U_j|^{\beta_j - b_j - \frac{p+1}{2}}\right\}$$

$$\times \left\{\prod_{j=1}^{k} |V_j|^{c_j - \frac{p+1}{2}}|I - V_j|^{\gamma_j - c_j - \frac{p+1}{2}}\right\}$$

$$\times \left|I + U_1^{\frac{1}{2}}...U_q^{\frac{1}{2}} X_1 U_q^{\frac{1}{2}}...U_1^{\frac{1}{2}} + V_1^{\frac{1}{2}}...V_k^{\frac{1}{2}} X_2 V_k^{\frac{1}{2}}...V_1^{\frac{1}{2}}\right|^{-a}$$

$$\times dU_1...dU_q dV_1...dV_k.$$

The steps for proving this result are the same as in the previous theorem. The case for $q = k$ in the scalar case is available in Exton (1976).

Remark 5.9   This theorem can also be generalized in many directions. Suppose that $|I + ...|^{-a}$ is replaced by a $_rF_s^*$. Then following through the same procedure and using the M-transform of a $_rF_s^*$ we have the following theorem:

Theorem 5.34   For $\operatorname{Re}(a_j, \; j = 1, ..., r, \; \alpha_j \neq -\frac{\lambda}{2}, \; \lambda = 0, 1, ..., \; j = 1, ..., s, \; s \geq r, \; b_j, \; \beta_j - b_j, \; j = 1, ..., q, \; c_j, \; \gamma_j - c_j, \; j = 1, ..., k) > \frac{p-1}{2}$

$$F = F_{s:q:k}^{r:q:k}\left[-\begin{pmatrix}X_1\\X_2\end{pmatrix}\begin{vmatrix}(a_r):(b_q):(c_k)\\(\alpha_s):(\beta_q):(\gamma_k)\end{vmatrix}\right]$$

$$= C_1 \int_0^I ... \int_0^I \left\{\prod_{j=1}^{q} |U_j|^{b_j - \frac{p+1}{2}}|I - U_j|^{\beta_j - b_j - \frac{p+1}{2}}\right\}$$

$$\times \left\{\prod_{j=1}^{k} |V_j|^{c_j - \frac{p+1}{2}}|I - V_j|^{\gamma_j - c_j - \frac{p+1}{2}}\right\}$$

$$\times \ _rF_s^*(a_1, ..., a_r; \alpha_1, ..., \alpha_s; -Z)dU_1...dU_q dV_1...dV_k$$

where

$$Z = U_1^{\frac{1}{2}}...U_q^{\frac{1}{2}} X_1 U_q^{\frac{1}{2}}...U_1^{\frac{1}{2}} + V_1^{\frac{1}{2}}...V_k^{\frac{1}{2}} X_2 V_k^{\frac{1}{2}}...V_1^{\frac{1}{2}}$$

and C_1 is the same constant part appearing in Theorem 5.33.

Theorem 5.35   For $\operatorname{Re}(b_j, \ c_j, \ \beta_j - b_j - c_j, \ j = 1, ..., q, \ b_{q+1}, c_{q+1}) > \frac{p-1}{2}$

$$F = F_{q:0:0}^{0:q+1:q+1} \left[-\binom{X_1}{X_2} \bigg| \begin{matrix} : (b_{q+1}) : (c_{q+1}) \\ (\beta_q) : : \end{matrix} \right]$$

$$= \left\{ \prod_{j=1}^{q} \frac{\Gamma_p(\beta_j)}{\Gamma_p(b_j)\Gamma_p(c_j)\Gamma_p(\beta_j - b_j - c_j)} \right\}$$

$$\times \int ... \int \left\{ \prod_{j=1}^{q} |U_j|^{b_j - \frac{p+1}{2}} |V_j|^{c_j - \frac{p+1}{2}} |I - U_j - V_j|^{\beta_j - b_j - c_j - \frac{p+1}{2}} \right\}$$

$$\times \left| I + U_1^{\frac{1}{2}}...U_q^{\frac{1}{2}} X_1 U_q^{\frac{1}{2}}...U_1^{\frac{1}{2}} \right|^{-b_{q+1}}$$

$$\times \left| I + V_1^{\frac{1}{2}}...V_q^{\frac{1}{2}} X_2 V_q^{\frac{1}{2}}...V_1^{\frac{1}{2}} \right|^{-c_{q+1}} dU_1...dU_q dV_1...dV_q$$

where the integration is over $0 < U_j < I, \ 0 < V_j < I, \ U_j + V_j < I, \ j = 1, ..., q$.

Proof   Evaluate the M-transform of the right side. Then integrate out the U_j's and V_j's by using q type-1 Dirichlet integrals to get the result.

Remark 5.10   This result can be generalized in many directions. Replace the determinant containing X_1 and X_2 by two different hypergeometric functions. This will generate a set of new results.

Theorem 5.36   For $\operatorname{Re}(a, \ \alpha - a, \ \beta_j, \ j = 1, ..., m, \ b_j, \ j = 1, ..., q, c) > \frac{p-1}{2}$,

$$F = F_{1:m:0}^{1:q:1} \left[-\binom{X_1}{X_2} \bigg| \begin{matrix} a : (b_q) : c \\ \alpha : (\beta_m) : \end{matrix} \right]$$

$$= \frac{\Gamma_p(\alpha)}{\Gamma_p(a)\Gamma_p(\alpha - a)} \int_{0<U<I} |U|^{a - \frac{p+1}{2}} |I - U|^{\alpha - a - \frac{p+1}{2}}$$

$$\times \ _qF_m \left(b_1, ..., b_q; \beta_1, ..., \beta_m; -U^{\frac{1}{2}} X_1 U^{\frac{1}{2}} \right)$$

$$\times \left| I + U^{\frac{1}{2}} X_2 U^{\frac{1}{2}} \right|^{-c} dU.$$

Proof Take the M-transform of the right side with respect to X_1 and X_2 and with parameters ρ_1 and ρ_2.

$$\int_{X_1} |X_1|^{\rho_1 - \frac{p+1}{2}} {}_q F_s \left(b_1, ..., b_q; \beta_1, ..., \beta_m; -U^{\frac{1}{2}} X_1 U^{\frac{1}{2}} \right) dX_1$$

$$= \frac{\left\{ \prod_{j=1}^m \Gamma_p(\beta_j) \right\}}{\left\{ \prod_{j=1}^q \Gamma_p(b_j) \right\}} |U|^{-\rho_1} \frac{\left\{ \prod_{j=1}^q \Gamma_p(b_j - \rho_1) \right\}}{\left\{ \prod_{j=1}^m \Gamma_p(\beta_j - \rho_1) \right\}} \Gamma_p(\rho_1) \qquad (a)$$

for $\mathrm{Re}(\rho_1, b_j - \rho_1, j = 1, ..., q, \beta_j - \rho_1, j = 1, ..., m) > \frac{p-1}{2}$ and

$$\int_{X_2} |X_2|^{\rho_2 - \frac{p+1}{2}} \left| I + U^{\frac{1}{2}} X_2 U^{\frac{1}{2}} \right|^{-c} dX_2$$

$$= |U|^{-\rho_2} \frac{\Gamma_p(\rho_2) \Gamma_p(c - \rho_2)}{\Gamma_p(c)}, \qquad (b)$$

for $\mathrm{Re}(\rho_2, c - \rho_2) > \frac{p-1}{2}$. Integrate out U.

$$\int_U |U|^{a - \rho_1 - \rho_2 - \frac{p+1}{2}} |I - U|^{\alpha - a - \frac{p+1}{2}} dU$$

$$= \frac{\Gamma_p(a - \rho_1 - \rho_2) \Gamma_p(\alpha - a)}{\Gamma_p(\alpha - \rho_1 - \rho_2)}, \qquad (c)$$

for $\mathrm{Re}(\alpha - a, a - \rho_1 - \rho_2) > \frac{p-1}{2}$. From (a), (b) and (c) the M-transform of the right side is the M-transform of an

$$F_{1:m:0}^{1:q:1} \text{ multiplied by } \Gamma_p(\alpha - a) \frac{\Gamma_p(a)}{\Gamma_p(\alpha)}.$$

Hence the result.

Remark 5.11 Note that the structure in this theorem can be used to give integral representation for several types of F's. For example if $|I + U X_2|^{-c}$ is replaced by

$$_k F_n^* \left(c_1, ..., c_k; \gamma_1, ..., \gamma_n; -U^{\frac{1}{2}} X_2 U^{\frac{1}{2}} \right)$$

then a representation for an $F_{1:m:n}^{1:q:k}$ is obtained. One can introduce a type-1 Dirichlet integral for the U to get new results.

5.4.3   Some limiting forms

For establishing these we will make use of the Lemmas 5.2 and 5.3. For establishing limiting forms of the Kampé de Fériet's functions either we have to assume the validity of interchange of limits and M-transforms or one can use a convenient integral representation where the interchange of limits and integrals can be justified.

Theorem 5.37

$$\lim_{\epsilon \to 0} F_{s:m:n}^{r:q:k} \left[-\begin{pmatrix} \epsilon^{q'} X_1 \\ \epsilon^{k'} X_2 \end{pmatrix} \middle| \begin{matrix} (a_r) : (\delta) : \\ (\alpha_s) : (\beta_m) : (\gamma_n) \end{matrix} \right]$$

$$= F_{s:m:n}^{r:q-q':k-k'} \left[-\begin{pmatrix} X_1 \\ X_2 \end{pmatrix} \middle| \begin{matrix} (a_r) : (b_{q-q'}) : (a_{k-k'}) \\ (\alpha_s) : (\beta_m) : (\gamma_n) \end{matrix} \right]$$

where (δ) denotes the sequence of parameters

$$(\delta) = (b_{q-q'}), \frac{1}{\epsilon}, ..., \frac{1}{\epsilon} : (c_{k-k'}), \frac{1}{\epsilon}, ..., \frac{1}{\epsilon}.$$

Proof   Make the transformation

$$Y_1 = \epsilon^{q'} X_1 \Rightarrow dY_1 = \epsilon^{\frac{q'p(p+1)}{2}} dX_1$$

and

$$Y_2 = \epsilon^{k'} X_2.$$

Then take the M-transform of the left side and apply Lemma 5.3 to see that the corresponding gammas disappear and the result follows. By the same technique we can prove the following result.

Theorem 5.38

$$\lim_{\epsilon \to 0} F_{s:0:0}^{r:q:k} \left[-\begin{pmatrix} \epsilon^q X_1 \\ \epsilon^k X_2 \end{pmatrix} \middle| \begin{matrix} (a_r) : \frac{1}{\epsilon}, ..., \frac{1}{\epsilon} : \frac{1}{\epsilon}, ..., \frac{1}{\epsilon} \\ (\alpha_s) : : \end{matrix} \right]$$

$$= {}_r F_s^* (a_1, ..., a_r; \alpha_1, ..., \alpha_s : -(X_1 + X_2)).$$

Remark 5.12   Note that q' of the denominator gammas from the set (β_q) can be wiped out by changing X_1 to $X_1/\epsilon^{q'}$ and similarly k' of the denominator gammas from the set (γ_k) can be eliminated by changing X_2 to $X_2/\epsilon^{k'}$, replacing the corresponding parameters by $1/\epsilon$ and then taking the limits when $\epsilon \to 0$. This can be seen from the definition in terms of M-transform. If any of the integral representations is used to establish this result we may need some manipulations of matrices. This will be illustrated here from Theorem 5.33. Suppose that β_q is to be eliminated. Change X_1 to X_1/ϵ and β_q to $1/\epsilon$. In the right side expression in Theorem 5.33 transfer the ϵ to U_q in the last factor and change U_q/ϵ to W_q

$$\Rightarrow dU_q = \epsilon^{\frac{p(p+1)}{2}} dW_q.$$

Then

$$|U_q|^{b_q - \frac{p+1}{2}} |I - U_q|^{\frac{1}{\epsilon} - b_q - \frac{p+1}{2}} dU_q = \epsilon^{\frac{p(p+1)}{2}} |\epsilon W_q|^{b_q - \frac{p+1}{2}}$$
$$\times |I - \epsilon W_q|^{\frac{1}{\epsilon} - b_q - \frac{p+1}{2}} dW_q.$$

From Lemma 5.3, when $\beta_q = 1/\epsilon$,

$$\frac{\Gamma_p(\beta_q)}{\Gamma_p(\beta_q - b_q)} \to \epsilon^{-pb_q} \text{ as } \epsilon \to 0.$$

Thus from Lemma 5.2 one has

$$\frac{\Gamma_p(\beta_q)}{\Gamma_p(\beta_q - b_q)} |U_q|^{b_q - \frac{p+1}{2}} |I - U_q|^{\frac{1}{\epsilon} - b_q - \frac{p+1}{2}}$$

$$\to |W_q|^{b_q - \frac{p+1}{2}} e^{-\text{tr}(W_q)}$$

as $\epsilon \to 0$ since

$$|I - \epsilon W_q|^{-b_q - \frac{p+1}{2}} \to 1$$

and

$$|I - \epsilon W_q|^{-1/\epsilon} \to e^{-\text{tr}(W_q)}.$$

One can interchange the limits and integrals since the integrals over the U_j's are type-1 beta integrals . Now proceed with the proof of Theorem 5.33 to see that W_q produces only a $\Gamma_p(b_q - \rho_1)$ but U_j for $j = 1, ..., q - 1$ produce $\Gamma_p(b_j - \rho_1)\Gamma_p(\beta_j - b_j)/\Gamma_p(\beta_j - \rho_1)$. Hence the result.

For eliminating equal numbers of gammas from the sets (a_r) and (α_s) replace the corresponding a_j's and α_j's by $1/\epsilon$ and take the limit when $\epsilon \to 0$. If r' of the gammas from the set (a_r) are to be eliminated then replace the corresponding a_j's by $1/\epsilon$, X_1 and X_2 by $\epsilon^{r'} X_1$ and $\epsilon^{r'} X_2$ respectively and then take the limit when $\epsilon \to 0$. Similarly if s' of the gammas from the set (α_s) are to be eliminated then replace the corresponding α_j's by $1/\epsilon$, X_1 and X_2 by $X_1/\epsilon^{s'}$ and $X_2/\epsilon^{s'}$ respectively and then take the limit when $\epsilon \to 0$. Thus all sorts of limiting forms can be established by using the ideas mentioned above.

5.4.4   Some reduction formulae

When $X_1 = X_2 = X$ a number of reduction formulae can be written giving Kampé de Fériet's functions in terms of hypergeometric functions. A large number of such results for the scalar case are given in Srivastava and Karlsson (1985). In the scalar case one can start with some expansions and rearrangements to come up with these reduction formulae. But in the matrix case such a technique is not suitable even if a definition is given in terms of series involving generalized zonal polynomials. Also note that the multiplication formula for $\Gamma(mz)$, $m = 2, 3, \ldots$ does not hold for $\Gamma_p(mz)$, $m = 3, 4, \ldots$ for a general p. For $p = 2$, $m = 2$ one can get a formula but the structure is different from that for $p = 1$. Thus a number of results in Srivastava and Karlsson (1985) may not have their matrix analogues. A few simple ones are given here.

Theorem 5.39

$$F^{r:1:1}_{r:0:0}\left[-\begin{pmatrix}X\\X\end{pmatrix}\middle|\begin{matrix}(a_r):b:c\\(\alpha_r)::\end{matrix}\right]$$

$$= {}_{r+1}F^*_r(a_1, \ldots, a_r, b+c; \alpha_1, \ldots, \alpha_r; -X).$$

Proof   In Theorem 5.32 put $X_1 = X_2 = X$ and take the M-transform of the right side with respect to X and with the parameter ρ. Note that when $X_1 = X_2 = X$,

$$|I + V|^{-b}|I + V|^{-c} = |I + V|^{-(b+c)}$$

where

$$V = U_1^{\frac{1}{2}}\ldots U_r^{\frac{1}{2}} X U_r^{\frac{1}{2}}\ldots U_1^{\frac{1}{2}}.$$

Now proceed as in the proof of Theorem 5.32 to see the result.

As a corollary we may note that when $r = 0$ one has

$$|I + AX|^{-(b+c)} = {}_1F^*_0(\ ; b+c; -AX)$$

$$= F^{0:1:1}_{0:0:0}\left[-\begin{pmatrix}AX\\AX\end{pmatrix}\middle|\begin{matrix}:b:c\\::\end{matrix}\right], \quad A = A' > 0.$$

Theorem 5.40
$$F_{r:0:0}^{r:0:0}\left[-\begin{pmatrix}X\\X\end{pmatrix}\Big|\begin{matrix}(a_r):: \\ (\alpha_r):: \end{matrix}\right]$$
$$= {}_rF_r^*(a_1,...,a_r;\alpha_1,...,\alpha_r;-2X).$$

This follows from Theorem 5.32 by replacing both $|I+V_j|^{-b}$ and $|I+V_j|^{-c}$ by $e^{-\mathrm{tr}(V_j)}$, $j=1,2$ where

$$V_j = U_1^{\frac{1}{2}}...U_r^{\frac{1}{2}}X_jU_r^{\frac{1}{2}}...U_1^{\frac{1}{2}}$$

and then looking at the M-transform of the right side for $X_1 = X_2 = X$.

Remark 5.13 The ideas in this section can be used to define functions corresponding to $F_{s:m:n}^{r:q:k}$ for many matrix variables $X_1,...,X_t$ as well as to extend the generalized triple series given in Srivastava and Karlsson (1985) to matrix cases. Properties corresponding to the ones in this section can be established for such functions also.

5.4.5   An application in astrophysics problems

A stellar model considered by Haubold and Mathai (1994) is that the density distribution in the sun is of the form

$$\rho(r) = \rho_0(1 - (r/R)^\delta)^\gamma$$

for $\delta > 0$ and $\gamma > 0$ where R is the radius of the sun and r an arbitrary distance from the center of the sun. The mass of the sun in this case will be the following:

$$M(r) = 4\pi \int_0^r t^2 \rho(t)\,dt$$
$$= 4\pi\rho_0 \int_0^r t^2 \left[1 - (t/R)^\delta\right]^\gamma dt$$
$$= \frac{4\pi\rho_0}{3} R^3 \left(\frac{r}{R}\right)^3 {}_2F_1\left(-\gamma, \frac{3}{\delta}; \frac{3}{\delta}+1; \left(\frac{r}{R}\right)^\delta\right), \qquad (5.4.5)$$

when γ is assumed to be a positive integer, where ${}_2F_1$ is a Gauss' hypergeometric function of scalar argument. The pressure at radius r will then be the following:

$$P(r) = P_0 - G \int_0^r \frac{M(t)\rho(t)}{t^2}\,dt$$

$$= P_0 - \frac{4\pi G}{\delta^2} \rho_0^2 R^2 \sum_{m=0}^{\gamma} \frac{(-\gamma)_m (r/R)^{m\delta+2}}{\left(\frac{3}{\delta}+m\right)\left(\frac{2}{\delta}+m\right)}$$

$$\times \; {}_2F_1\left(-\gamma, \frac{2}{\delta}+m; \frac{2}{\delta}+m+1; \left(\frac{r}{R}\right)^\delta\right) \qquad (5.4.6)$$

when γ is assumed to be a positive integer, where G is the gravitational constant. This $P(r)$ can be written as a Kampé de Fériet's function of scalar argument as follows:

$$P(r) = P_0 - \frac{2}{3}\pi G \rho_0^2 r^2 F_{1:2:0}^{1:3:1}\left[\left.\begin{pmatrix}(r/R)^\delta\\(r/R)^\delta\end{pmatrix}\right| \begin{matrix} 2/\delta : -\gamma, 3/\delta, 2/\delta : -\gamma \\ 2/\delta+1 : 3/\delta+1, 2/\delta+1 : \end{matrix}\right]$$

and then assuming that the pressure at the surface is zero, that is, $P(R) = 0$ one has

$$P_0 = \frac{2}{3}\pi G \rho_0^2 R^2 F_{1:2:0}^{1:3:1}\left[\left.\begin{pmatrix}1\\1\end{pmatrix}\right| \begin{matrix} 2/\delta : -\gamma, 3/\delta, 2/\delta : -\gamma \\ 2/\delta+1 : 3/\delta+1, 2/\delta+1 : \end{matrix}\right].$$

More on the theory and the details of such applications of Kampé de Fériet's functions can be seen from Haubold and Mathai (1994), Mathai and Haubold (1988), Mathai and Pederzoli (1993) and Pederzoli (1995).

Exercises

5.4.1   Starting with an integral representation for a Kampé de Fériet's function, establish (5.4.3).

5.4.2   Establish the results in (5.4.2) by using the definitions of F_1, F_2, F_3 and F_4 through M-transform as well as the definition through integrals given in Exercises 5.3.1–5.3.4.

5.4.3   Starting with an integral representation for a Kampé de Fériet's function, establish (5.4.4)

5.4.4   Evaluate the Humbert's function Φ_1 as a limiting form of a Kampé de Fériet's function.

5.4.5   Evaluate the Humbert's function Ξ_1 as a limiting form of a Kampé de Fériet's function.

5.5   Lauricella functions of matrix arguments

Lauricella functions f_A, f_B, f_C and f_D of several matrix arguments are defined in this section when the matrices are real symmetric positive definite, and several results on these functions are established which correspond to the results in the scalar variable cases. It will be shown that some analogues do not exist when the arguments are matrices. As in the previous sections, M-transform is used to define these functions. These functions are named after the Italian mathematician G. Lauricella (1867–1913).

Lauricella functions of scalar variables are widely applied in topics such as statistical distribution theory, genetics, differential equations, heat flow and astrophysics problems, see Exton (1976) and Mathai and Saxena (1987) for some examples. Some other multivariable special functions in the scalar case, in the category of the basic of q-hypergeometric functions, can be found in Floreanini, Lapointe and Vinet (1994). Lauricella functions of matrix arguments arise when dealing with nonnull distributions of test statistics, see for example Mathai (1989), Mathai (1993, 1993b) and Mathai and Rathie (1980). Applications of Lauricella functions to some econometric problems are available from Chikuse and Davis (1986). Functions of several matrices through invariant polynomials can be found in Davis (1979, 1981).

As in the previous sections, we will assume that all the functions appearing here are symmetric in the sense described in (5.3.1) and (5.3.3) of Section 5.3. Also the M-transform will be taken with respect to the paramters $\rho_1, ..., \rho_n$.

5.5.1   Lauricella function f_A of matrix arguments

Definition 5.33   A function

$$f_A = f_A(a, b_1, ..., b_n; c_1, ..., c_n; -X_1, ..., -X_n)$$

satisfying the following integral equation will be called Lauricella function f_A of the matrix arguments $X_1, ..., X_n$ with upper parameters $a, b_1, ..., b_n$ and lower parameters $c_1, ..., c_n$.

$$M(f_A) = \int_{X_1>0} ... \int_{X_n>0} |X_1|^{\rho_1 - \frac{p+1}{2}} ... |X_n|^{\rho_n - \frac{p+1}{2}}$$
$$\times f_A(a, b_1, ..., b_n; c_1, ..., c_n; -X_1, ..., -X_n) dX_1 ... dX_n$$
$$= \frac{\left\{\prod_{j=1}^n \Gamma_p(c_j)\right\} \left\{\prod_{j=1}^n \Gamma_p(b_j - \rho_j)\right\}}{\Gamma_p(a) \left\{\prod_{j=1}^n \Gamma_p(b_j)\right\}}$$

$$\times \; \frac{\Gamma_p(a - \rho_1 - \ldots - \rho_n)\left\{\prod_{j=1}^n \Gamma_p(\rho_j)\right\}}{\left\{\prod_{j=1}^n \Gamma_p(c_j - \rho_j)\right\}} \tag{5.5.1}$$

for $\operatorname{Re}(b_j - \rho_j, \; c_j - \rho_j, \; \rho_j, \; a - \rho_1 - \ldots - \rho_n) > \frac{p-1}{2}$, $j = 1, \ldots, n$. Several results can be established by using this definition. Some of these on integral representations of f_A will be given here. Corresponding results are available in the literature for the scalar variable cases.

Theorem 5.41

$$f_A = f_A(a, b_1, \ldots, b_n; c_1, \ldots, c_n; -X_1, \ldots, -X_n)$$

$$= \frac{\left\{\prod_{j=1}^n \Gamma_p(c_j)\right\}}{\left\{\prod_{j=1}^n \Gamma_p(b_j)\Gamma_p(c_j - b_j)\right\}} \int_0^I \ldots \int_0^I |U_1|^{b_1 - \frac{p+1}{2}}$$

$$\times \ldots |U_n|^{b_n - \frac{p+1}{2}} |I - U_1|^{c_1 - b_1 - \frac{p+1}{2}} \ldots |I - U_n|^{c_n - b_n - \frac{p+1}{2}}$$

$$\times \left| I + U_1^{\frac{1}{2}} X_1 U_1^{\frac{1}{2}} + \ldots + U_n^{\frac{1}{2}} X_n U_n^{\frac{1}{2}} \right|^{-a} dU_1 \ldots dU_n. \tag{5.5.2}$$

Proof   Consider the transformation

$$Y_j = U_j^{\frac{1}{2}} X_j U_j^{\frac{1}{2}} \Rightarrow dY_j = |U_j|^{\frac{p+1}{2}} dX_j,$$

for fixed U_j and

$$|X_j| = |Y_j| \, |U_j|^{-1}.$$

Take the M-transform of the right side with respect to $X_1, \ldots, X_n$. Integrate out the X_j's by using a type-2 Dirichlet integral of Section 5.1.8. Then denoting the X_j-integrals by $I_{X_1 \ldots X_n}$ we have the following:

$$I_{X_1 \ldots X_n} = \int_{X_1 > 0} \ldots \int_{X_n > 0} |X_1|^{\rho_1 - \frac{p+1}{2}} \ldots |X_n|^{\rho_n - \frac{p+1}{2}}$$

$$\times \left| I + U_1^{\frac{1}{2}} X_1 U_1^{\frac{1}{2}} + \ldots + U_n^{\frac{1}{2}} X_n U_n^{\frac{1}{2}} \right|^{-a} dX_1 \ldots dX_n$$

$$= |U_1|^{-\rho_1} \ldots |U_n|^{-\rho_n}$$

$$\times \; \frac{\Gamma_p(\rho_1) \ldots \Gamma_p(\rho_n)\Gamma_p(a - \rho_1 - \ldots - \rho_n)}{\Gamma_p(a)} \tag{5.5.3}$$

for $\operatorname{Re}(\rho_j, \; j = 1, \ldots, n, \; a - \rho_1 - \ldots - \rho_n) > \frac{p-1}{2}$. Integrating out the U_j's by using type-1 beta integrals of Section 5.1.8 and denoting it by $I_{U_1 \ldots U_n}$

we have

$$
I_{U_1 \ldots U_n} = \int_0^I \ldots \int_0^I |U_1|^{b_1 - \rho_1 - \frac{p+1}{2}} \ldots |U_n|^{b_n - \rho_n - \frac{p+1}{2}}
$$

$$
\times \, |I - U_1|^{c_1 - b_1 - \frac{p+1}{2}} \ldots |I - U_n|^{c_n - b_n - \frac{p+1}{2}} \, \mathrm{d}U_1 \ldots \mathrm{d}U_n
$$

$$
= \frac{\left\{ \prod_{j=1}^n \Gamma_p(b_j - \rho_j) \right\} \left\{ \prod_{j=1}^n \Gamma_p(c_j - b_j) \right\}}{\left\{ \prod_{j=1}^n \Gamma(c_j - \rho_j) \right\}} \tag{5.5.4}
$$

for $\mathrm{Re}(b_j - \rho_j, \, c_j - b_j) > \frac{p-1}{2}$, $j = 1, \ldots, n$. From (5.5.3) and (5.5.4) the M-transform of the right side of (5.5.1) reduces to that of a f_A and hence the result.

Theorem 5.42

$$
f_A = \frac{1}{\Gamma_p(a)} \int_{T>0} e^{-\mathrm{tr}(T)} |T|^{a - \frac{p+1}{2}} {}_1F_1^*(b_1; c_1; -T^{\frac{1}{2}} X_1 T^{\frac{1}{2}})
$$

$$
\times \, \ldots {}_1F_1^* \left(b_n; c_n; -T^{\frac{1}{2}} X_n T^{\frac{1}{2}} \right) \mathrm{d}T \tag{5.5.5}
$$

where ${}_1F_1^*$ is a confluent hypergeometric function defined in Section 5.2.4.

Proof By using the M-transform of a hypergeometric function the X_j-integrals in the M-transform of the right side can be evaluated as follows after transforming $U_j = T^{\frac{1}{2}} X_j T^{\frac{1}{2}}$ for fixed T:

$$
I_{X_j} = \int_{X_j > 0} |X_j|^{\rho_j - \frac{p+1}{2}} {}_1F_1^* \left(b_j; c_j; -T^{\frac{1}{2}} X_j T^{\frac{1}{2}} \right) \mathrm{d}X_j
$$

$$
= |T|^{-\rho_j} \frac{\Gamma_p(c_j)}{\Gamma_p(b_j)} \frac{\Gamma_p(b_j - \rho_j) \Gamma_p(\rho_j)}{\Gamma_p(c_j - \rho_j)}
$$

for $\mathrm{Re}(b_j - \rho_j, \, c_j - \rho_j, \, \rho_j) > \frac{p-1}{2}$. Now integrate out T by using the gamma integral to see the result.

For proving the next result we need a Ψ_2-function of many matrix variables. This will be defined in terms of its M-transform.

Definition 5.34 **The Ψ_2-function of matrix arguments**

$$
\Psi_2 = \Psi_2(a; c_1, \ldots, c_n; -X_1, \ldots, -X_n)
$$

is defined as that class of functions for which the M-transform is the following.

$$
\begin{aligned}
M(\Psi_2) &= M_{\rho_1,...,\rho_n}[\Psi_2(a; c_1, ..., c_n; -X_1, ..., -X_n)] \\
&= \frac{\left\{\prod_{j=1}^n \Gamma_p(c_j)\right\}}{\Gamma_p(a)} \\
&\times \frac{\Gamma_p(a - \rho_1 - ... - \rho_n)\left\{\prod_{j=1}^n \Gamma_p(\rho_j)\right\}}{\left\{\prod_{j=1}^n \Gamma_p(c_j - \rho_j)\right\}}
\end{aligned}
\tag{5.5.6}
$$

for $\operatorname{Re}(\rho_j, \ c_j - \rho_j, \ a - \rho_1 - ... - \rho_n) > \frac{p-1}{2}$, $j = 1, ..., n$.

Theorem 5.43

$$
\begin{aligned}
f_A &= \left[\prod_{j=1}^n \Gamma_p(b_j)\right]^{-1} \int_{T_1>0} ... \int_{T_n>0} e^{-\operatorname{tr}(T_1 + ... + T_n)} \\
&\times |T_1|^{b_1 - \frac{p+1}{2}} ... |T_n|^{b_n - \frac{p+1}{2}} \\
&\times \Psi_2\left(a; c_1, ..., c_n; -T_1^{\frac{1}{2}} X_1 T_1^{\frac{1}{2}}, ..., -T_n^{\frac{1}{2}} X_n T_n^{\frac{1}{2}}\right) dT_1 ... dT_n.
\end{aligned}
$$

Proof   Take the M-transform of the right side with respect to $X_1, ..., X_n$ by taking $Y_j = T_j^{\frac{1}{2}} X_j T_j^{\frac{1}{2}}$ for fixed T and using Definition 5.34. Then evaluate the T_j-integrals by using the gamma integral to see the result.

The proofs in the next three results are similar to the one above and hence these theorems will be stated without proofs.

Theorem 5.44   *For* $\operatorname{Re}(a, \ d - a) > \frac{p-1}{2}$

$$
\begin{aligned}
f_A &= f_A(a, b_1, ..., b_n; c_1, ..., c_n; -X_1, ..., -X_n) \\
&= \frac{\Gamma_p(d)}{\Gamma_p(a)\Gamma_p(d-a)} \int_0^I |U|^{a - \frac{p+1}{2}} |I - U|^{d - a - \frac{p+1}{2}} \\
&\times f_A\left(d, b_1, ..., b_n; c_1, ..., c_n; -U^{\frac{1}{2}} X_1 U^{\frac{1}{2}}, ..., -U^{\frac{1}{2}} X_n U^{\frac{1}{2}}\right) dU.
\end{aligned}
$$

Theorem 5.45   For $\mathrm{Re}(b_j, \; d_j - b_j) > \frac{p-1}{2}, \; j = 1, ..., n$

$$f_A = f_A(a, b_1, ..., b_n; c_1, ..., c_n; -X_1, ..., -X_n)$$

$$= \frac{\left\{ \prod_{j=1}^n \Gamma_p(d_j) \right\}}{\left\{ \prod_{j=1}^n \Gamma_p(b_j) \right\} \left\{ \prod_{j=1}^n \Gamma_p(d_j - b_j) \right\}} \int_0^I \cdots \int_0^I |U_1|^{b_1 - \frac{p+1}{2}}$$

$$\times \; ...|U_n|^{b_n - \frac{p+1}{2}} |I - U_1|^{d_1 - b_1 - \frac{p+1}{2}} ...|I - U_n|^{d_n - b_n - \frac{p+1}{2}}$$

$$\times \; f_A \left(a, d_1, ..., d_n; c_1, ..., c_n; -U_1^{\frac{1}{2}} X_1 U_1^{\frac{1}{2}}, ..., -U_n^{\frac{1}{2}} X_n U_n^{\frac{1}{2}} \right) dU_1 ... dU_n.$$

Theorem 5.46   For $\mathrm{Re}(d_j, \; c_j - d_j) > \frac{p-1}{2}, \; j = 1, ..., n$

$$f_A = f_A(a, b_1, ..., b_n; c_1, ..., c_n; -X_1, ..., -X_n)$$

$$= \frac{\left\{ \prod_{j=1}^n \Gamma_p(c_j) \right\}}{\left\{ \prod_{j=1}^n \Gamma_p(d_j) \right\} \left\{ \prod_{j=1}^n \Gamma_p(c_j - d_j) \right\}} \int_0^I \cdots \int_0^I |U_1|^{d_1 - \frac{p+1}{2}}$$

$$\times \; ...|U_n|^{d_n - \frac{p+1}{2}} |I - U_1|^{c_1 - d_1 - \frac{p+1}{2}} ...|I - U_n|^{c_n - d_n - \frac{p+1}{2}}$$

$$\times \; f_A \left(a, b_1, ..., b_n; d_1, ..., d_n; -U_1^{\frac{1}{2}} X_1 U_1^{\frac{1}{2}}, ..., -U_n^{\frac{1}{2}} X_n U_n^{\frac{1}{2}} \right) dU_1 ... dU_n.$$

5.5.2   Lauricella function f_B of matrix arguments

Again a definition in terms of M-transform will be given and then we will establish a number of results on the integral representations of f_B.

Definition 5.35 f_B is denoted by

$$f_B = f_B(a_1, ..., a_n, b_1, ..., b_n; c; -X_1, ..., -X_n)$$

and is defined as that class of functions having the following M-transform.

$$M(f_B) = \frac{\Gamma_p(c)}{\left\{ \prod_{j=1}^n \Gamma_p(a_j)\Gamma_p(b_j) \right\}}$$

$$\times \; \frac{\left\{ \prod_{j=1}^n \Gamma_p(a_j - \rho_j)\Gamma_p(b_j - \rho_j)\Gamma_p(\rho_j) \right\}}{\Gamma_p(c - \rho_1 - ... - \rho_n)} \tag{5.5.7}$$

for $\text{Re}(\rho_j,\ a_j-\rho_j,\ b_j-\rho_j,\ c-\rho_1-...-\rho_n) > \frac{p-1}{2}$, $j=1,...,n$. A number of results for f_B will be established using this definition. These will be stated as theorems.

Theorem 5.47   *For* $\text{Re}(a_j,\ j=1,...,n,\ c-a_1-...-a_n) > \frac{p-1}{2}$

$$f_B = f_B(a_1,...,a_n,b_1,...,b_n;c;-X_1,...,-X_n)$$

$$= \frac{\Gamma_p(c)}{\Gamma_p(c-a_1-...-a_n)\left\{\prod_{j=1}^n \Gamma_p(a_j)\right\}} \int ... \int |U_1|^{a_1-\frac{p+1}{2}}$$

$$\times\ ...|U_n|^{a_n-\frac{p+1}{2}}|I-U_1-...-U_n|^{c-a_1-...-a_n-\frac{p+1}{2}}$$

$$\times\ \left|I+U_1^{\frac{1}{2}}X_1U_1^{\frac{1}{2}}\right|^{-b_1} ... \left|I+U_n^{\frac{1}{2}}X_nU_n^{\frac{1}{2}}\right|^{-b_n} dU_1...dU_n$$

where the integrals over U_j's *are such that* $0 < U_j < I$, $j=1,...,n$, $0 < U_1 + ... + U_n < I$.

Proof   Take the M-transform of the right side with respect to $X_1,...,X_n$ by using type-2 beta integrals. Then

$$\int_{X_j>0} |X_j|^{\rho_j-\frac{p+1}{2}} \left|I+U_j^{\frac{1}{2}}X_jU_j^{\frac{1}{2}}\right|^{-b_j} dX_j = |U_j|^{-\rho_j} \frac{\Gamma_p(\rho_j)\Gamma_p(b_j-\rho_j)}{\Gamma_p(b_j)}.$$

Now integrate out the U_j's by using a type-1 Dirichlet integral of Section 5.1.8 to see the result.

Theorem 5.48   *For* $\text{Re}(d_j,\ c-d_1-...-d_n) > \frac{p-1}{2}$, $j=1,...,n$

$$f_B = \frac{\Gamma_p(c)}{\Gamma_p(c-d_1-...-d_n)\left\{\prod_{j=1}^n \Gamma_p(d_j)\right\}}$$

$$\times \int ... \int |U_1|^{d_1-\frac{p+1}{2}}...|U_n|^{d_n-\frac{p+1}{2}}$$

$$\times |I-U_1-...-U_n|^{c-d_1-...-d_n-\frac{p+1}{2}}\ {}_2F_1^*\left(a_1,b_1;d_1;-U_1^{\frac{1}{2}}X_1U_1^{\frac{1}{2}}\right)$$

$$\times\ ...{}_2F_1^*\left(a_n,b_n;d_n;-U_n^{\frac{1}{2}}X_nU_n^{\frac{1}{2}}\right)dU_1...dU_n$$

where the integration is over $0 < U_j < I$, $j=1,...,n$, $U_1+...+U_n < I$.

Proof   Take the M-transform of the right side with the help of Section 5.2.4 and then evaluate by using the type-1 Dirichlet integral of Section 5.1.8 to see the result.

For proving the next result we need a Φ_2-function of matrix arguments which will be defined by using M-transform as follows.

Definition 5.36   **The Φ_2-function of matrix arguments**

$$\Phi_2 = \Phi_2(b_1, ..., b_n; c; -X_1, ..., -X_n)$$

is that class of functions for which

$$M(\Phi_2) = \frac{\Gamma_p(c)}{\left\{\prod_{j=1}^{n} \Gamma_p(b_j)\right\}} \frac{\left\{\prod_{j=1}^{n} \Gamma_p(b_j - \rho_j)\Gamma_p(\rho_j)\right\}}{\Gamma_p(c - \rho_1 - ... - \rho_n)}.$$

Theorem 5.49   *For* $\mathrm{Re}(a_j) > \frac{p-1}{2}$, $j = 1, ..., n$

$$f_B = f_B(a_1, ..., a_n, b_1, ..., b_n; c; -X_1, ..., -X_n)$$

$$= \left[\prod_{j=1}^{n} \Gamma_p(a_j)\right]^{-1}$$

$$\times \int_{T_1>0} ... \int_{T_n>0} e^{-\mathrm{tr}(T_1+...+T_n)} |T_1|^{a_1 - \frac{p+1}{2}} ... |T_n|^{a_n - \frac{p+1}{2}}$$

$$\times \Phi_2\left(b_1, ..., b_n; c; -T_1^{\frac{1}{2}} X_1 T_1^{\frac{1}{2}}, ..., -T_n^{\frac{1}{2}} X_n T_n^{\frac{1}{2}}\right) dT_1...dT_n.$$

Proof   Take the M-transform of the right side by using Definition 5.36 and integrate out the T_j's by using the gamma integrals to see the result.

A result on Φ_2 itself will be given as the next theorem without proof. The steps for proving this are similar to the ones in the previous theorems.

Theorem 5.50

$$\Phi_2 = \Phi_2(b_1, ..., b_n; c; -X_1, ..., X_n)$$

$$= \frac{\Gamma_p(c)}{\Gamma_p(c - d_1 - ... - d_n)\left\{\prod_{j=1}^{n} \Gamma_p(d_j)\right\}} \int ... \int |U_1|^{d_1 - \frac{p+1}{2}}$$

$$\times |U_n|^{d_n - \frac{p+1}{2}} |I - U_1 - ... - U_n|^{c - d_1 - ... - d_n - \frac{p+1}{2}}$$

$$\times {}_1F_1^*\left(b_1; d_1; -U_1^{\frac{1}{2}} X_1 U_1^{\frac{1}{2}}\right)...$$

$$\times {}_1F_1^*\left(b_n; d_n; -U_n^{\frac{1}{2}} X_n U_n^{\frac{1}{2}}\right) dU_1...dU_n$$

where the integration is over $0 < U_j < I$, $j = 1, ..., n$, $U_1 + ... + U_n < I$.

For establishing the next result we need some results on gamma functons and Bessel functions of matrix arguments. These will be stated as lemmas.

Lemma 5.4

$$\Gamma_2(2\alpha) = \pi^{-\frac{3}{2}} 2^{4\alpha - \frac{5}{2}} \Gamma_2\left(\alpha + \frac{1}{2}\right) \Gamma_2\left(\alpha + \frac{1}{4}\right).$$

This can be established from the definition of a $\Gamma_p(\alpha)$ and the duplication formula for gamma functions, namely,

$$\Gamma(2z) = \pi^{-\frac{1}{2}} 2^{2z-1} \Gamma(z) \Gamma\left(z + \frac{1}{2}\right).$$

Note that for a general p we do not have a formula giving $\Gamma_p(2\alpha)$ in terms of the product of two gammas of the type $\Gamma_p(\alpha \pm \delta)$ for some δ. As a consequence of this reduction formula for a $\Gamma_2(2\alpha)$ we have the following result which will be needed in two of the theorems later on.

$$\frac{\Gamma_2(a - 2b)}{\Gamma_2(a)} = 2^{-4b} \frac{\Gamma_2\left(\frac{a}{2} + \frac{1}{2} - b\right) \Gamma_2\left(\frac{a}{2} + \frac{1}{4} - b\right)}{\Gamma_2\left(\frac{a}{2} + \frac{1}{2}\right) \Gamma_2\left(\frac{a}{2} + \frac{1}{4}\right)}. \tag{5.5.8}$$

Lemma 5.5 *For $\text{Re}(\rho_j) > \frac{p-1}{2}$, $j = 1, ..., n$*

$$M(_0F_1) = \int_{X_1 > 0} \cdots \int_{X_n > 0} |X_1|^{\rho_1 - \frac{p+1}{2}} \cdots |X_n|^{\rho_n - \frac{p+1}{2}}$$

$$\times {}_0F_1\left(; c; -\frac{1}{4}(T_1 X_1 T_1 + ... + T_n X_n T_n)\right) dX_1 ... dX_n$$

$$= 4^{p(\rho_1 + ... + \rho_n)} |T_1|^{-2\rho_1} ... |T_n|^{-2\rho_n}$$

$$\times \frac{\Gamma_p(c)\Gamma_p(\rho_1) ... \Gamma_p(\rho_n)}{\Gamma_p(c - \rho_1 - ... - \rho_n)}. \tag{5.5.9}$$

Proof   Let $U_j = \frac{1}{4} T_j X_j T_j$, $U = U_1 + ... + U_n$. Then the factor

$$K_{T_1 ... T_n} = 4^{p(\rho_1 + ... + \rho_n)} |T_1|^{-2\rho_1} ... |T_n|^{-2\rho_n}$$

comes out. Now replace $_0F_1(\;;c;-U)$ by the inverse Laplace representation of a hypergeometric function in Section 5.2.1 to get the following:

$$M(_0F_1) = K_{T_1\ldots T_n} \int_{U_1>0} \cdots \int_{U_n>0} |U_1|^{\rho_1 - \frac{p+1}{2}} \ldots |U_n|^{\rho_n - \frac{p+1}{2}}$$

$$\times \left\{ \frac{\Gamma_p(c)}{(2\pi i)^{\frac{p(p+1)}{2}}} \int_{R(Z)=X>X_0} e^{\mathrm{tr}(Z)} e^{-\mathrm{tr}(UZ^{-1})} |Z|^{-c} dZ \right\}$$

$$\times \; dU_1 \ldots dU_n \tag{5.5.10}$$

where $Z = X + iY$, $i = \sqrt{-1}$, $X \in S_p^*$, $Y \in S_p^*$, $X > 0$, see also Notation 5.1. Take out $e^{-\mathrm{tr}(UZ^{-1})}$ from (5.5.10) and integrate out $U_1, \ldots, U_n$ to obtain $|Z|^{\rho_1+\cdots+\rho_n}\Gamma_p(\rho_1)\ldots\Gamma_p(\rho_n)$. Now integrate out Z by using the inverse Laplace representation of a gamma function to get $\Gamma_p(c)/\Gamma_p(c-\rho_1-\ldots-\rho_n)$. This establishes the lemma. This can also be done by using Theorem 5.11.

The following theorem holds only for $p = 2$. But it will be stated by using a general p to be consistent with the notations elsewhere.

Theorem 5.51 *For $p = 2$, $\mathrm{Re}(a_j) > \frac{p-1}{2}$, $j = 1, \ldots, n$*

$$f_{BB} = f_B \left(\frac{a_1}{2} + \frac{1}{2}, \ldots, \frac{a_n}{2} + \frac{1}{2}, \frac{a_1}{2} + \frac{1}{4}, \ldots, \frac{a_n}{2} + \frac{1}{4}; c; -X_1, \ldots, -X_n \right)$$

$$= \left[\prod_{j=1}^{n} \Gamma_p(a_j) \right]^{-1} \int_{T_1>0} \cdots \int_{T_n>0} |T_1|^{a_1 - \frac{p+1}{2}}$$

$$\times \; \ldots |T_n|^{a_n - \frac{p+1}{2}} e^{-\mathrm{tr}(T_1+\ldots+T_n)}$$

$$\times \; _0F_1^* \left(\;; c; -\frac{1}{4}(T_1 X_1 T_1 + \ldots + T_n X_n T_n) \right) dT_1 \ldots dT_n.$$

Proof Take the M-transform of the right side with respect to $X_1, \ldots, X_n$ and with the help of Lemma 5.5 to obtain

$$\left\{ \prod_{j=1}^{n} \frac{\Gamma_p(a_j - 2\rho_j)}{\Gamma_p(a_j)} \right\} \frac{\Gamma_p(c)}{\Gamma(c - \rho_1 - \ldots - \rho_n)} \Gamma_p(\rho_1)\ldots\Gamma_p(\rho_n) 4^{p(\rho_1+\ldots+\rho_n)}.$$

Use Lemma 5.4 to simplify $\Gamma_p(a_j - 2\rho_j)/\Gamma_p(a_j)$. Integrate out the T_j's by using gamma integrals to see the result.

Note that the result in Theorem 5.51 is different from the corresponding one in the scalar case. Also note that results on scalar variables involving higher powers and fractional powers of the variables need not have the corresponding matrix analogues because in some cases fractional powers of matrices are not uniquely defined and in some cases the Jacobians of transformations lead to completely different forms. As a simple example note that if $Y = X^2$ where $X = X' > 0$ is a real $p \times p$ matrix then $dY \neq 2|X|dX$. In fact the Jacobian is of the form $\prod_{j \leq k}(\lambda_j + \lambda_k)$ which does not have a nice representation in terms of $|X|$, where $\lambda_1, ..., \lambda_p$ are the eigenvalues of X.

5.5.3   Lauricella function f_C of matrix arguments

Definition 5.37

$$f_C = f_C(a, b; c_1, ..., c_n; -X_1, ..., -X_n)$$

is defined as that class of functions for which the M-transform is the following.

$$M(f_C) = \frac{\left\{\prod_{j=1}^{n} \Gamma_p(c_j)\right\}}{\Gamma_p(a)\Gamma_p(b)} \left\{\prod_{j=1}^{n} \Gamma_p(\rho_j)\right\}$$
$$\times \frac{\Gamma_p(a - \rho_1 - ... - \rho_n)\Gamma_p(b - \rho_1 - ... - \rho_n)}{\left\{\prod_{j=1}^{n} \Gamma_p(c_j - \rho_j)\right\}} \qquad (5.5.11)$$

for $\mathrm{Re}(\rho_j,\ c_j - \rho_j,\ a - \rho_1 - ... - \rho_n,\ b - \rho_1 - ... - \rho_n) > \frac{p-1}{2}$, $j = 1, ..., n$. By using this definition we will establish a few results on integral representations of f_C. These will be stated without proofs. Take the M-transform and use Lemmas 5.4 and 5.5 to see the results.

Theorem 5.52   For $\mathrm{Re}(a, b) > \frac{p-1}{2}$

$$f_C = f_C(a, b; c_1, ..., c_n; -X_1, ..., -X_n)$$
$$= [\Gamma_p(a)\Gamma_p(b)]^{-1} \int_{T_1 > 0} \int_{T_2 > 0} e^{-\mathrm{tr}(T_1 + T_2)}$$
$$\times |T_1|^{a - \frac{p+1}{2}} |T_2|^{b - \frac{p+1}{2}} {}_0F_1^* \left(; c_1; -T_1^{\frac{1}{2}} T_2^{\frac{1}{2}} X_1 T_2^{\frac{1}{2}} T_1^{\frac{1}{2}}\right)$$
$$\times ... {}_0F_1^* \left(; c_n; -T_1^{\frac{1}{2}} T_2^{\frac{1}{2}} X_n T_2^{\frac{1}{2}} T_1^{\frac{1}{2}}\right) dT_1 dT_2.$$

Theorem 5.53 *For* $p = 2$, $\operatorname{Re}(a) > \frac{p-1}{2}$

$$
f_C = f_C\left(\frac{a}{2} + \frac{1}{2}, \frac{a}{2} + \frac{1}{4}; c_1, ..., c_n; -X_1, ..., -X_n\right)
$$

$$
= [\Gamma_p(a)]^{-1} \int_{T>0} |T|^{a - \frac{p+1}{2}} e^{-\operatorname{tr}(T)}
$$

$$
\times {}_0F_1^*\left(;c_1; -\frac{1}{4}TX_1T\right)\cdots
$$

$$
\times {}_0F_1^*\left(;c_n; -\frac{1}{4}TX_nT\right)dT.
$$

5.5.4 Lauricella function f_D of matrix arguments

Definition 5.38

$$
f_D = f_D(a, b_1, ..., b_n; c; -X_1, ..., -X_n)
$$

is defined as that class of functions for which the M-transform is the following.

$$
M(f_D) = \frac{\Gamma_p(c)}{\Gamma_p(a)\left\{\prod_{j=1}^n \Gamma_p(b_j)\right\}}\left\{\prod_{j=1}^n \Gamma_p(\rho_j)\right\}
$$

$$
\times \frac{\Gamma_p(a - \rho_1 - ... - \rho_n)\left\{\prod_{j=1}^n \Gamma_p(b_j - \rho_j)\right\}}{\Gamma_p(c - \rho_1 - ... - \rho_n)} \tag{5.5.12}
$$

for $\operatorname{Re}(c,\ a,\ \rho_j,\ b_j - \rho_j,\ c - \rho_1 - ... - \rho_n,\ a - \rho_1 - ... - \rho_n) > \frac{p-1}{2}$, $j = 1, ..., n$.
A few results on integral representations of f_D will be established here.

Theorem 5.54 *For* $\operatorname{Re}(b_j,\ a - b_1 - ... - b_n) > \frac{p-1}{2}$, $j = 1, ..., n$

$$
f_D = f_D(a, b_1, ..., b_n; c; -X_1, ..., -X_n)
$$

$$
= \frac{\Gamma_p(c)}{\Gamma_p(c - b_1 - ... - b_n)\left\{\prod_{j=1}^n \Gamma_p(b_j)\right\}}\int \cdots \int |U_1|^{b_1 - \frac{p+1}{2}}
$$

$$
\times ... |U_n|^{b_n - \frac{p+1}{2}} |I - U_1 - ... - U_n|^{c - b_1 - ... - b_n - \frac{p+1}{2}}
$$

$$
\times \left|I + U_1^{\frac{1}{2}} X_1 U_1^{\frac{1}{2}} + ... + U_n^{\frac{1}{2}} X_n U_n^{\frac{1}{2}}\right|^{-a} dU_1 ... dU_n
$$

where the integration is over $0 < U_j < I, \ j = 1, ..., n, \ U_1 + ... + U_n < I$.

Proof   Take the M-transform of the right side by using a type-2 Dirichlet integral and then integrate out the U_j's by using a type-1 Dirichlet integral to establish the result.

Theorem 5.55   *For* $\mathrm{Re}(a, \ c - a) > \frac{p-1}{2}$

$$
f_D = \frac{\Gamma_p(c)}{\Gamma_p(a)\Gamma_p(c-a)} \int_0^I |U|^{a-\frac{p+1}{2}} |I - U|^{c-a-\frac{p+1}{2}}
$$
$$
\times \ \left|I + U^{\frac{1}{2}} X_1 U^{\frac{1}{2}}\right|^{-b_1} ... \left|I + U^{\frac{1}{2}} X_n U^{\frac{1}{2}}\right|^{-b_n} dU.
$$

Proof   Take the M-transform of the right side by using type-2 beta integrals and then integrate out U by using a type-1 beta integral to see the result.

The steps are similar and hence the next three theorems will be stated without proofs.

Theorem 5.56   *For* $\mathrm{Re}(a) > \frac{p-1}{2}$

$$
f_D = [\Gamma_p(a)]^{-1} \int_{T>0} e^{-\mathrm{tr}(T)} |T|^{a-\frac{p+1}{2}}
$$
$$
\times \ \Phi_2 \left(b_1, ..., b_n; c; -T^{\frac{1}{2}} X_1 T^{\frac{1}{2}}, ..., -T^{\frac{1}{2}} X_n T^{\frac{1}{2}}\right) dT
$$

where the Φ_2*-function is given in Definition 5.36.*

Theorem 5.57   *For* $\mathrm{Re}(d, \ c - d) > \frac{p-1}{2}$

$$
f_D = f_D(a, b_1, ..., b_n; c; -X_1, ..., -X_n)
$$
$$
= \frac{\Gamma_p(c)}{\Gamma_p(d)\Gamma_p(c-d)} \int_0^I |U|^{d-\frac{p+1}{2}} |I - U|^{c-d-\frac{p+1}{2}}
$$
$$
\times \ f_D \left(a, b_1, ..., b_n; d; -U^{\frac{1}{2}} X_1 U^{\frac{1}{2}}, ..., -U^{\frac{1}{2}} X_n U^{\frac{1}{2}}\right) dU.
$$

Theorem 5.58   For $\mathrm{Re}(b_j, d_j - b_j) > \frac{p-1}{2}$, $j = 1, ..., n$

$$f_D = f_D(a, b_1, ..., b_n; c; -X_1, ... - X_n)$$

$$= \frac{\left\{\prod_{j=1}^{n} \Gamma_p(d_j)\right\}}{\left\{\prod_{j=1}^{n} \Gamma_p(b_j)\Gamma_p(d_j - b_j)\right\}} \int_0^I \cdots \int_0^I |U_1|^{b_1 - \frac{p+1}{2}} \cdots$$

$$\times |U_n|^{b_n - \frac{p+1}{2}}|I - U_1|^{d_1 - b_1 - \frac{p+1}{2}} \cdots |I - U_n|^{d_n - b_n - \frac{p+1}{2}}$$

$$\times f_D\left(a, d_1, ..., d_n; c; -U_1^{\frac{1}{2}} X_1 U_1^{\frac{1}{2}}, ..., -U_n^{\frac{1}{2}} X_n U_n^{\frac{1}{2}}\right) dU_1 ... dU_n.$$

For proving the next theorem we need Theorem 5.11 of Section 5.3 which gives, as a particular case, the M-transform of a $_1F_1$ when the argument is a sum of many matrices.

Theorem 5.59   For $\mathrm{Re}(b_j) > \frac{p-1}{2}$, $j = 1, ..., n$

$$f_D = f_D(a, b_1, ...b_n; c; -X_1, ..., -X_n)$$

$$= [\Gamma_p(b_1)...\Gamma_p(b_n)]^{-1} \int_{T_1 > 0} \cdots \int_{T_n > 0}$$

$$\times e^{-\mathrm{tr}(T_1 + ... + T_n)}|T_1|^{b_1 - \frac{p+1}{2}} \cdots |T_n|^{b_n - \frac{p+1}{2}}$$

$$\times {}_1F_1^*\left(a; c; -T_1^{\frac{1}{2}} X_1 T_1^{\frac{1}{2}} - ... - T_n^{\frac{1}{2}} X_n T_n^{\frac{1}{2}}\right) dT_1 ... dT_n.$$

Proof   Evaluate the M-transform of the right side by using Theorem 5.11 and then integrate out the T_j's to see the result.

5.5.5   Some mixed results

By using the definitions for f_A, f_B, f_C, f_D one can obtain some mixed results which will be stated as the next theorems. For proving these take the M-transforms of the right sides and interpret the results.

Theorem 5.60   For $\mathrm{Re}(b_j, d - b_1 - ... - b_n) > \frac{p-1}{2}$, $j = 1, ..., n$

$$f_B = f_B(a_1, ..., a_n, b_1, ..., b_n; c; -X_1, ..., -X_n)$$

$$= \frac{\Gamma_p(d)}{\Gamma_p(d - b_1 - ... - b_n)\left\{\prod_{j=1}^{n} \Gamma_p(b_j)\right\}} \int \cdots \int |U_1|^{b_1 - \frac{p+1}{2}}$$

$$\times ... |U_n|^{b_n - \frac{p+1}{2}}|I - U_1 - ... - U_n|^{d - b_1 - ... - b_n - \frac{p+1}{2}}$$

$$\times f_D\left(d, a_1, ..., a_n; c; -U_1^{\frac{1}{2}} X_1 U_1^{\frac{1}{2}}, ..., -U_n^{\frac{1}{2}} X_n U_n^{\frac{1}{2}}\right) dU_1 ... dU_n$$

where the integration is over $0 < U_j < I$, $j = 1, ..., n$, $U_1 + ... + U_n < I$.

Theorem 5.61   *For* $\mathrm{Re}(d_j, \ c - d_1 - ... - d_n) > \frac{p-1}{2}$, $j = 1, ..., n$

$$f_D = f_D(a, b_1, ..., b_n; c; -X_1, ..., -X_n)$$

$$= \frac{\Gamma_p(c)}{\Gamma_p(c - d_1 - ... - d_n)\left\{\prod_{j=1}^n \Gamma_p(d_j)\right\}} \int ... \int |U_1|^{d_1 - \frac{p+1}{2}}$$

$$\times ... |U_n|^{d_n - \frac{p+1}{2}} |I - U_1 - ... - U_n|^{c - d_1 - ... - d_n - \frac{p+1}{2}}$$

$$\times f_A\left(a, b_1, ..., b_n; d_1, ..., d_n; -U_1^{\frac{1}{2}} X_1 U_1^{\frac{1}{2}}, ..., -U_n^{\frac{1}{2}} X_n U_n^{\frac{1}{2}}\right) dU_1 ... dU_n$$

where the integration is over $0 < U_j < I$, $j = 1, ..., n$, $U_1 + ... + U_n < I$.

Theorem 5.62   *For* $\mathrm{Re}(b_j, \ d - b_1 - ... - b_n) > \frac{p-1}{2}$, $j = 1, ..., n$

$$_1F_1^*(b_1; c_1; -X_1)..._1F_1^*(b_n; c_n; -X_n)$$

$$= \frac{\Gamma_p(d)}{\Gamma_p(d - b_1 - ... - b_n)\left\{\prod_{j=1}^n \Gamma_p(b_j)\right\}} \int ... \int |U_1|^{b_1 - \frac{p+1}{2}}$$

$$\times ... |U_n|^{b_n - \frac{p+1}{2}} |I - U_1 - ... - U_n|^{d - b_1 - ... - b_n - \frac{p+1}{2}}$$

$$\times \Psi_2\left(d; c_1, ..., c_n; -U_1^{\frac{1}{2}} X_1 U_1^{\frac{1}{2}}, ..., -U_n^{\frac{1}{2}} X_n U_n^{\frac{1}{2}}\right)$$

$$\times dU_1 ... dU_n$$

where the Ψ_2-*function is given in Definition 5.34 and the integration is over* $0 < U_j < I$, $j = 1, ..., n$, $U_1 + ... + U_n < I$.

Theorem 5.63   *For* $\mathrm{Re}(b_j, \ d - b_1 - ... - b_n) > \frac{p-1}{2}$, $j = 1, ..., n$

$$f_A = f_A(a, b_1, ..., b_n; c_1, ..., c_n; -X_1, ..., -X_n)$$

$$= \frac{\Gamma_p(d)}{\Gamma_p(d - b_1 - ... - b_n)\left\{\prod_{j=1}^n \Gamma_p(b_j)\right\}} \int ... \int |U_1|^{b_1 - \frac{p+1}{2}}$$

$$\times ... |U_n|^{b_n - \frac{p+1}{2}} |I - U_1 - ... - U_n|^{d - b_1 - ... - b_n - \frac{p+1}{2}}$$

$$\times f_C\left(a, d; c_1, ..., c_n; -U_1^{\frac{1}{2}} X_1 U_1^{\frac{1}{2}}, ..., -U_n^{\frac{1}{2}} X_n U_n^{\frac{1}{2}}\right) dU_1 ... dU_n$$

where the integration is over $0 < U_j < I$, $j = 1, ..., n$, $U_1 + ... + U_n < I$.

5.5.6   Some applications to statistical distributions

(a)   Distribution of a linear form

In many areas of statistical distribution theory functions of several matrix arguments, especially belonging to the Lauricella type, appear. Here we will consider a case where a Φ_2 function appears. A representation of the exact density of a linear function of independent real matrix-variate gamma variables, in terms of hypergeometric function of many matrix variables, will be considered here, see Sections 5.1 and 5.2 for a discussion of matrix-variate gamma and hypergeometric functions. The result will be obtained in terms of a Φ_2 function of many matrix variables. Such a linear function is also connected to generalized quadratic forms in Gaussian random variables.

One standard technique for deriving such distributions is to go through a generalized Laplace transform. Instead of inverting the Laplace transform, which in this case will be too complicated, two different representations of the Laplace transform are derived by using the properties of hypergeometric functions of many matrix variables and then by using the uniqueness property of the density, the exact density is written down from one of the representations.

In order to motivate the procedure we will consider a positive linear function of independent real scalar gamma variables first. Its density will be derived and represented in terms of a Φ_2 function of scalar variables. Let x_j, $j = 1, ..., k$ be independent real scalar gamma random variables with the densities

$$f_j(x_j) = \begin{cases} \frac{b_j^{-\alpha_j}}{\Gamma(\alpha_j)} x_j^{\alpha_j - 1} e^{-x_j/b_j}, & x_j > 0 \\ 0, & \text{elsewhere} \end{cases} \qquad (5.5.13)$$

for $b_j > 0$, $\alpha_j > 0$, $j = 1, ..., k$. Let

$$y = a_1 x_1 + ... + a_k x_k, \quad a_j > 0, \quad j = 1, ..., k \qquad (5.5.14)$$

where the a_j's are constants. For an account of the various applications of such a linear function see Mathai and Provost (1992). The density of y will be represented in terms of a multiple hypergeometric function of the Φ_2 type. For a discussion and the series form of Φ_2 in the scalar case see Exton(1976).

The Laplace transform of the density function of y, denoted by $L_y(t)$, is given by

$$L_y(t) = \prod_{j=1}^{k} (1 + b_j a_j t)^{-\alpha_j} \qquad (5.5.15)$$

for $1 + b_j a_j t > 0$, $j = 1, ..., k$. Let $\gamma > 0$ be an arbitrary scalar quantity. This γ can be used to accelerate the rate of convergence of some series later on. $L_y(t)$ can be rewritten as follows:

$$1 + b_j a_j t = 1 + \gamma b_j a_j - \gamma b_j a_j + b_j a_j t$$
$$= (1 - \gamma b_j a_j) + \gamma b_j a_j \left(1 + \gamma^{-1} t\right).$$

$$L_y(t) = \gamma^{-(\alpha_1 + ... + \alpha_k)} \left\{ \prod_{j=1}^{k} (b_j a_j)^{-\alpha_j} \right\}$$

$$\times \left(1 + \gamma^{-1} t\right)^{-(\alpha_1 + ... + \alpha_k)} \prod_{j=1}^{k} \left[1 - \delta_j \left(1 + \gamma^{-1} t\right)^{-1}\right]^{-\alpha_j}$$

where $(1 - \delta_j)^{-1} = \gamma a_j b_j$. One can adjust γ and t such that $\left(1 + \gamma^{-1} t\right) > 0$ and $\left| \delta_j \left(1 + \gamma^{-1} t\right)^{-1}\right| < 1$ for $j = 1, ..., k$ and hence one can expand by using a binomial expansion. That is,

$$\prod_{j=1}^{k} \left[1 - \delta_j \left(1 + \gamma^{-1} t\right)^{-1}\right]^{-\alpha_j} = \sum_{r_1=0}^{\infty} \cdots \sum_{r_k=0}^{\infty} \frac{(\alpha_1)_{r_1} ... (\alpha_k)_{r_k}}{r_1! ... r_k!}$$

$$\times \delta_1^{r_1} ... \delta_k^{r_k} \left(1 + \gamma^{-1} t\right)^{-(r_1 + ... + r_k)}.$$

Collecting the factors containing $(1 + \gamma^{-1} t)$ and inverting one gets the density of y. Note that the density corresponding to

$$\left(1 + \gamma^{-1} t\right)^{-(\alpha_1 + ... + \alpha_k + r_1 + ... + r_k)},$$

denoted by $h(y)$, is given by

$$h(y) = \frac{\gamma^{(\alpha_1 + ... + \alpha_k + r_1 + ... + r_k)}}{\Gamma\left(\alpha_1 + ... + \alpha_k + r_1 + ... + r_k\right)}$$
$$\times y^{\alpha_1 + ... + \alpha_k + r_1 + ... + r_k - 1} e^{-\gamma y}$$
$$= \frac{\gamma^{(\alpha_1 + ... + \alpha_k)}}{\Gamma\left(\alpha_1 + ... + \alpha_k\right)} y^{(\alpha_1 + ... + \alpha_k - 1)} e^{-\gamma y}$$
$$\times \frac{(\gamma y)^{r_1 + ... + r_k}}{(\alpha_1 + ... + \alpha_k)_{r_1 + ... + r_k}} \tag{5.5.16}$$

where, for example,

$$(a)_m = a(a+1)...(a+m-1), \quad (a)_0 = 1, \quad a \neq 0.$$

Note that the multiple sum can be written in terms of a generalized hypergeometric function of the Φ_2 type. That is,

$$\sum_{r_1=0}^{\infty} \cdots \sum_{r_k=0}^{\infty} \frac{(\alpha_1)_{r_1} \cdots (\alpha_k)_{r_k}}{(\alpha_1 + \ldots + \alpha_k)_{r_1 + \ldots + r_k}} \frac{(\gamma y \delta_1)^{r_1} \ldots (\gamma y \delta_k)^{r_k}}{r_1! \ldots r_k!}$$

$$= \Phi_2 \left(\alpha_1, \ldots, \alpha_k; \alpha_1 + \ldots + \alpha_k; \gamma y \delta_1, \ldots, \gamma y \delta_k \right). \qquad (5.5.17)$$

Hence, if the density of y is denoted by $g(y)$ then

$$g(y) = \left\{ \prod_{j=1}^{k} (b_j a_j)^{-\alpha_j} \right\} \frac{y^{\alpha_1 + \ldots + \alpha_k - 1} e^{-\gamma y}}{\Gamma(\alpha_1 + \ldots + \alpha_k)}$$

$$\times \Phi_2 \left(\alpha_1, \ldots, \alpha_k; \alpha_1 + \ldots + \alpha_k; \gamma y \delta_1, \ldots, \gamma y \delta_k \right), \qquad (5.5.18)$$

for $y > 0$ and $g(y) = 0$ elsewhere.

The representation in (5.5.18) is suitable for extending the results to linear functions of Wishart matrices or linear functions of matrix-variate real gamma variables.

Let the real symmetric positive definite $p \times p$ matrix $X_j = X_j' > 0$ be a matrix-variate real gamma random variable with the density

$$f_j(X_j) = \frac{|C_j|^{\alpha_j}}{\Gamma_p(\alpha_j)} |X_j|^{\alpha_j - \frac{p+1}{2}} e^{-\text{tr}(C_j X_j)} \qquad (5.5.19)$$

for $C_j = C_j' > 0$, $X_j = X_j' > 0$, $\text{Re}(\alpha_j) > \frac{p-1}{2}$ and $f_j(X_j) = 0$ elsewhere. Let $X_1, \ldots, X_k$ be independently distributed. Let

$$Y = A_1^{\frac{1}{2}} X_1 A_1^{\frac{1}{2}} + \ldots + A_k^{\frac{1}{2}} X_k A_k^{\frac{1}{2}} \qquad (5.5.20)$$

where $A_j = A_j' > 0$ and $A_j^{\frac{1}{2}}$ denotes the symmetric square root of A_j, $j = 1, \ldots, k$. Let $T = (\delta_{ij} t_{ij})$ be a $p \times p$ parameter matrix such that $t_{ij} = t_{ji}$, $\delta_{ij} = \frac{1}{2}$, $i \neq j$, $\delta_{jj} = 1$ for all $i, j = 1, \ldots, p$. Let the density of Y be denoted by $g(Y)$. Then the Laplace transform of the density of Y is given by

$$L_g(T) = E \left[e^{-\text{tr}(TY)} \right] \qquad (5.5.21)$$

where E denotes the expected value. This Laplace transform can be evaluated by integrating out over the joint density of $X_1, \ldots, X_k$. Then

$$L_g(T) = \left\{ \prod_{j=1}^{k} \frac{|C_j|^{\alpha_j}}{\Gamma_p(\alpha_j)} \right\} \int_{X_1 > 0} \cdots \int_{X_k > 0}$$

$$\times \left\{ \prod_{j=1}^{k} |X_j|^{\alpha_j - \frac{p+1}{2}} \right\} e^{-\operatorname{tr}(C_1 X_1 + \ldots + C_k X_k)}$$

$$\times e^{-\operatorname{tr}\left[T\left(A_1^{\frac{1}{2}} X_1 A_1^{\frac{1}{2}} + \ldots + A_k^{\frac{1}{2}} X_k A_k^{\frac{1}{2}} \right) \right]} dX_1 \ldots dX_k. \qquad (5.5.22)$$

Let

$$Y_j = C_j^{\frac{1}{2}} X_j C_j^{\frac{1}{2}} \Rightarrow dY_j = |C_j|^{\frac{p+1}{2}} dX_j,$$

and

$$X_j = C_j^{-\frac{1}{2}} Y_j C_j^{-\frac{1}{2}}.$$

Each Y_j can be integrated out by using a matrix-variate gamma integral to obtain

$$L_g(T) = \prod_{j=1}^{k} |I + B_j T|^{-\alpha_j}, \quad B_j = A_j^{\frac{1}{2}} C_j^{-1} A_j^{\frac{1}{2}} \qquad (5.5.23)$$

for $I + B_j T > 0$, $j = 1, \ldots, k$.

　　Evidently g is available by taking the inverse Laplace transform of the right side in (5.5.23). But this process is extremely difficult. One method will be to resort to an expansion in terms of generalized zonal polynomials. We will try to evaluate g by using some properties of functions of many matrix arguments. For this purpose we will rewrite the expression in (5.5.23) in a more convenient form. This form will then be identified with the Laplace transform of a known function of several matrix arguments thus identifying g since g is unique. Consider an arbitrary matrix $B = B' > 0$. Then

$$\begin{aligned}|I + B_j T|^{-\alpha_j} &= |I + B_j B - B_j B + B_j T|^{-\alpha_j} \\ &= |I - B_j B + B_j (B + T)|^{-\alpha_j} \\ &= |B_j|^{-\alpha_j} |B + T|^{-\alpha_j} \\ &\quad \times \left| I + \left(B_j^{-1} - B \right) (B + T)^{-1} \right|^{-\alpha_j} \qquad (5.5.24)\end{aligned}$$

for $I + \left(B_j^{-1} - B \right) (B + T)^{-1} > 0$ and $B + T > 0$. Hence

$$\begin{aligned}L_g(T) &= \prod_{j=1}^{k} |I + B_j T|^{-\alpha_j} \\ &= \left\{ \prod_{j=1}^{k} |B_j|^{-\alpha_j} \right\} |B + T|^{-(\alpha_1 + \ldots + \alpha_k)} \\ &\quad \times \prod_{j=1}^{k} \left| I + \left(B_j^{-1} - B \right) (B + T)^{-1} \right|^{-\alpha_j}. \qquad (5.5.25)\end{aligned}$$

In order to identify (5.5.25) with the Laplace transform of a known function, we will examine some properties of Lauricella functions and Φ_2 functions of many matrix arguments. Now recall Theorem 5.59 which gives an integral representation of f_D, namely,

$$f_D = \left[\prod_{j=1}^{n} \Gamma_p(b_j) \right]^{-1} \int_{T_1>0} \cdots \int_{T_n>0}$$
$$\times e^{-\text{tr}(T_1+\cdots+T_n)} |T_1|^{b_1 - \frac{p+1}{2}} \cdots |T_n|^{b_n - \frac{p+1}{2}}$$
$$\times {}_1F_1^* \left(a; c; -T_1^{\frac{1}{2}} U_1 T_1^{\frac{1}{2}} \cdots - T_n^{\frac{1}{2}} U_n T_n^{\frac{1}{2}} \right) dT_1 ... dT_n. \quad (5.5.26)$$

Note that when $a = c$,

$$ {}_1F_1^*(c; c; -X) = {}_0F_0^*(\ ; \ ; -X) = e^{-\text{tr}(X)}.$$

Thus when $a = c$ the result in (5.5.26) becomes

$$f_D (c, b_1, ..., b_n; c; -U_1, ..., -U_n) = \prod_{j=1}^{n} |I + U_j|^{-b_j} \quad (5.5.27)$$

by integrating out using a matrix-variate gamma after replacing the ${}_1F_1^*$ for $a = c$ by the exponential function.

Now rewriting Theorem 5.56 one has the following result. For $B = B' > 0$

$$f_D = f_D \left(a, b_1, ..., b_n; c; -(T+B)^{-1}U_1, ..., -(T+B)^{-1}U_n \right)$$
$$= \frac{|T+B|^a}{\Gamma_p(a)} \int_{U>0} e^{-\text{tr}((T+B)U)} |U|^{a - \frac{p+1}{2}}$$
$$\times \Phi_2 \left(b_1, ..., b_n; c; -U^{\frac{1}{2}}U_1U^{\frac{1}{2}}, ..., -U^{\frac{1}{2}}U_nU^{\frac{1}{2}} \right) dU. \quad (5.5.28)$$

For $a = c$ the left side of (5.5.28), in the light of (5.5.27), reduces to

$$\prod_{j=1}^{n} |I + (B+T)^{-1}U_j|^{-b_j}.$$

Since

$$e^{-\text{tr}((T+B)U)} = e^{-\text{tr}(TU)} e^{-\text{tr}(BU)}$$

one can look upon the right side of (5.5.28), excluding $|B + T|^a$, as the
Laplace transform of the function

$$\frac{1}{\Gamma_p(a)} e^{-\text{tr}(BU)} |U|^{a - \frac{p+1}{2}} \Phi_2 \left(b_1, ..., b_n; c; -U^{\frac{1}{2}} U_1 U^{\frac{1}{2}}, ..., -U^{\frac{1}{2}} U_n U^{\frac{1}{2}} \right).$$

Let $U_j = B_j^{-1} - B$, $a = \alpha_1 + ... + \alpha_k$, $n = k$. Compare (5.5.28) for
$a = c = \alpha_1 + ... + \alpha_k$ with the Laplace transform in (5.5.25) to see that
(5.5.25) is the Laplace transform of $g(Y)$ where

$$g(Y) = \frac{\left\{ \prod_{j=1}^k |B_j|^{-\alpha_j} \right\}}{\Gamma_p(\alpha_1 + ... + \alpha_k)} e^{-\text{tr}(BY)} |Y|^{\alpha_1 + ... + \alpha_k - \frac{p+1}{2}}$$

$$\times \Phi_2 \left(\alpha_1, ..., \alpha_k; \alpha_1 + ... + \alpha_k; -Y^{\frac{1}{2}} \left(B_1^{-1} - B \right) Y^{\frac{1}{2}}, \right.$$

$$\left. ..., -Y^{\frac{1}{2}} \left(B_k^{-1} - B \right) Y^{\frac{1}{2}} \right), Y = Y' > 0 \qquad (5.5.29)$$

and $g(Y) = 0$ elsewhere, where $B = B' > 0$ is an arbitrary matrix and the
B_j's are given in (5.5.23). Due to the uniqueness, the density of Y is the
$g(Y)$ given in (5.5.29). Note that $\int_{Y>0} g(Y) dY = 1$.

(b)   Distribution of a determinant

Let $\lambda = |Y|$ where Y is defined in (5.5.20). This λ is of interest in testing
statistical hypotheses. Let us examine the exact h-th moment of λ.

$$E(\lambda^h) = \int_{Y>0} |Y|^h g(Y) dY. \qquad (5.5.30)$$

Let

$$Z = B^{\frac{1}{2}} Y B^{\frac{1}{2}} \Rightarrow dZ = |B|^{\frac{p+1}{2}} dY.$$

Then

$$E(\lambda^h) = \frac{\left\{ \prod_{j=1}^k |B_j|^{-\alpha_j} \right\}}{\Gamma_p(\alpha_1 + ... + \alpha_k)} |B|^{-(\alpha_1 + ... + \alpha_k)}$$

$$\times \int_{Z>0} e^{-\text{tr}(Z)} |Z|^{\alpha_1 + ... + \alpha_k + h - \frac{p+1}{2}}$$

$$\times \Phi_2 \left(\alpha_1, ..., \alpha_k; \alpha_1 + ... + \alpha_k; -Z^{\frac{1}{2}} \left(B^{-\frac{1}{2}} B_1^{-1} B^{-\frac{1}{2}} - I \right) Z^{\frac{1}{2}}, \right.$$

$$\left. ... - Z^{\frac{1}{2}} \left(B^{-\frac{1}{2}} B_k^{-1} B^{-\frac{1}{2}} - I \right) Z^{\frac{1}{2}} \right) dZ$$

$$= \frac{\Gamma_p(\alpha_1 + ... + \alpha_k + h)}{\Gamma_p(\alpha_1 + ... + \alpha_k)} \left\{ \prod_{j=1}^{k} |B_j|^{-\alpha_j} \right\}$$

$$\times |B|^{-(\alpha_1 + ... + \alpha_k + h)} f_D(\alpha_1 + ... + \alpha_k + h, \alpha_1, ..., \alpha_k; \alpha_1 + ... + \alpha_k;$$
$$I - B^{-\frac{1}{2}} B_1 B^{-\frac{1}{2}}, ..., I - B^{-\frac{1}{2}} B_k B^{-\frac{1}{2}}) \tag{5.5.31}$$

for $\text{Re}(\alpha_1 + ... + \alpha_k + h) > \frac{p-1}{2}$. The result in (5.5.31) is obtained by using Theorem 5.56.

(c)   Generalized quadratic forms

Let X' be a $p \times n$, $n \geq p$, matrix with the columns X_j's of X being independently and identically distributed as real multivariate Gaussian, that is, $X_j \sim N_p(0, V)$, $V > 0$, $j = 1, ..., n$, see the Appendix for the notation and details. Consider the generalized quadratic form

$$Q = X'AX \tag{5.5.32}$$

where $A = A'$ is an $n \times n$ constant matrix. Let T be a parameter matrix as defined in (5.5.21). Then the Laplace transform of the density of Q, denoted by $L_Q(T)$, is available from Mathai and Provost (1992) as

$$L_Q(T) = \prod_{j=1}^{r} \left| I + 2\lambda_j V^{\frac{1}{2}} T V^{\frac{1}{2}} \right|^{-\frac{1}{2}} \tag{5.5.33}$$

where r is the rank of A and $\lambda_1, ..., \lambda_r$ are the nonzero eigenvalues of A. Let $\lambda_j > 0$, $j = 1, ..., r$. Comparing (5.5.33) and (5.5.23) one has

$$B_j = 2\lambda_j V, \quad \alpha_j = \frac{1}{2}, \quad j = 1, ..., r,$$

$$\prod_{j=1}^{r} |B_j|^{-\alpha_j} = 2^{-pr} (\lambda_1 ... \lambda_r)^{-\frac{p}{2}} |V|^{-\frac{r}{2}}.$$

Let the exact density of Q be denoted by $K(Q)$. Then from (5.5.29) one has

$$K(Q) = \left[2^{pr} (\lambda_1 ... \lambda_r)^{\frac{p}{2}} |V|^{\frac{r}{2}} \Gamma_p \left(\frac{r}{2} \right) \right]^{-1} e^{-\text{tr}(BQ)}$$

$$\times |Q|^{\frac{r}{2} - \frac{p+1}{2}} \Phi_2 \left(\frac{1}{2}, ..., \frac{1}{2}; \frac{r}{2}; -Q^{\frac{1}{2}} ((2\lambda_1 V)^{-1} - B) Q^{\frac{1}{2}}, \right.$$

$$\left. ..., -Q^{\frac{1}{2}} ((2\lambda_r V)^{-1} - B) Q^{\frac{1}{2}} \right)$$

for $Q = Q' > 0$, $r > p - 1$ and $K(Q) = 0$ elsewhere, where $B = B' > 0$ is an arbitrary constant matrix.

If some of the λ_j's are negative then Q can be treated as a difference of two positive generalized quadratic forms. Other representations for the exact and asymptotic distributions of Q in the central and noncentral cases as well as the distributions of generalized quadratic expressions are available in the literature, see for example, Mathai, Provost and Hayakawa (1995).

If one does not wish to use the M-transform to define Lauricella functions of matrix arguments then one can use some integral representations as definitions and then derive the other results from these definitions. For example the integral representation in Theorem 5.41 can be used to define f_A, Theorem 5.47 to define f_B and Theorem 5.55 to define f_D. Various properties of Lauricella functions of matrix arguments are derived from these definitions in Mathai (1993b). There is no convenient integral that can be used to define f_C but Theorem 5.52 is a good candidate. If computable representations for f_A, f_B, f_C and f_D are needed then one can define them in terms of invariant polynomials but the representations will be complicated to derive theoretical results and computations will also be quite difficult. Details on zonal polynomials, invariant polynomials, computability and functions of several matrix arguments defined in terms of these polynomials may be found in Mathai, Provost and Hayakawa (1995).

Exercises

5.5.1   Define $f_A = f_A(a, b_1, ..., b_n; c_1, ..., c_n; X_1, ..., X_n)$ by using the integral in Theorem 5.41 and then from this definition derive the results in Theorem 5.42 to Theorem 5.46.

5.5.2   Define $f_B = f_B(a_1, ..., a_n, b_1, ..., b_n; c; X_1, ..., X_n)$ by using the integral in Theorem 5.47 and then from this definition derive the results in Theorem 5.48 to Theorem 5.50.

5.5.3   Define $f_C = f_C(a, b; c_1, ..., c_n; X_1, ..., X_n)$ by using the integral in Theorem 5.52 and derive Theorem 5.53 from this definition.

5.5.4   Define $f_D = f_D(a, b_1, ..., b_n; c; X_1, ..., X_n)$ by using Theorem 5.55 and then establish Theorems 5.54, 5.57–5.59 from this definition.

5.5.5   Establish Theorems 5.60–5.63 from the definitions in Exercises 5.5.1 to 5.5.4 above.

CHAPTER 6

Functions of matrix argument
in the complex case

6.0 Introduction

Here we consider real valued scalar functions of a single matrix argument
of the type $\tilde{Z} = X + iY$ where X and Y are $p \times p$ matrices with real
elements and $i = \sqrt{-1}$, as well as scalar functions of many matrices $\tilde{Z}_j$, $j =$
$1, ..., k$ where each $\tilde{Z}_j$ is of the type $\tilde{Z}$ above. Most of the basic results in
Chapter 5 in the real case will be extended to the corresponding complex
case. In the real case we confined our discussion to the situation where the
argument matrix was real symmetric positive definite. This was done so
that fractional powers of matrices and functions of such matrices could be
uniquely defined. Corresponding properties are available if we restrict to
the class of hermitian positive definite matrices.

We will denote the conjugate of $\tilde{Z}$ by $\bar{\tilde{Z}}$ and the conjugate transpose
by $\tilde{Z}^* = \bar{\tilde{Z}}'$. If $\tilde{Z}$ is hermitian then $\tilde{Z} = \tilde{Z}^*$, that is,

$$\tilde{Z} = \tilde{Z}^* \Rightarrow X + iY = (X + iY)^* = X' - iY'$$
$$\Rightarrow X = X' \text{ and } Y = -Y'.$$

Thus X is symmetric and Y is skew symmetric. Further, if $\tilde{Z}$ is hermitian
positive definite then all the eigenvalues of $\tilde{Z}$ are real and positive. We will
use the notation $\tilde{Z} > 0$ to indicate that $\tilde{Z}$ is hermitian positive definite.
Since we are dealing with real matrices in Chapters 1, 2, 5 and complex
matrices in Chapters 3, 4, 6, in order to distinguish a complex matrix variable from a real one we will use a tilde for the complex variable. Constant
matrices will be written without a tilde whether the elements are real or
complex unless it has to be emphasized that the matrix involved has complex elements. Then in that case a constant matrix will also be written
with a tilde.

The basic tools for tackling real valued scalar functions of matrix argument in the complex case are already developed in Chapter 3 while considering Jacobians of matrix transformations in the complex case. Some of
these will be restated here for ready reference.

Definition 6.1   Matrix-variate gamma in the complex case

$$\tilde{\Gamma}_p(\alpha) = \pi^{\frac{p(p-1)}{2}}\Gamma(\alpha)\Gamma(\alpha-1)...\Gamma(\alpha-p+1) \tag{6.0.1}$$

$$= \int_{\tilde{Z}=\tilde{Z}^*>0} |\det(\tilde{Z})|^{\alpha-p}e^{-\mathrm{tr}(\tilde{Z})}\mathrm{d}\tilde{Z}, \ \mathrm{Re}(\alpha)>p-1.$$

See Example 3.5 of Chapter 3 for the integral representation given above.

Definition 6.2   Matrix-variate beta in the complex case

$$\tilde{B}_p(\alpha,\beta) = \tilde{\Gamma}_p(\alpha)\tilde{\Gamma}_p(\beta)/\tilde{\Gamma}_p(\alpha+\beta) \tag{6.0.2}$$

$$= \int_{0<\tilde{Z}=\tilde{Z}^*<I} |\det(\tilde{Z})|^{\alpha-p}|\det(I-\tilde{Z})|^{\beta-p}\mathrm{d}\tilde{Z}$$

$$= \int_{0<\tilde{T}<I} |\det(\tilde{T})|^{\beta-p}|\det(I-\tilde{T})|^{\alpha-p}\mathrm{d}\tilde{T}$$

$$= \int_{\tilde{U}>0} |\det(\tilde{U})|^{\alpha-p}|\det(I+\tilde{U})|^{-(\alpha+\beta)}\mathrm{d}\tilde{U}$$

$$= \int_{\tilde{V}>0} |\det(\tilde{V})|^{\beta-p}|\det(I+\tilde{V})|^{-(\alpha+\beta)}\mathrm{d}\tilde{V}$$

for $\mathrm{Re}(\alpha)>p-1$, $\mathrm{Re}(\beta)>p-1$. See also equation (3.2.4) of Chapter 3. One integral representation is available from this equation by using the fact that the density integrates out to unity because the total probability is one. This can also be established by steps parallel to the ones in Example 5.1 of Chapter 5.

6.1   Some elementary functions in the complex case

We will consider exponential function, gamma and beta functions, Laplace transform and M-transform for the complex case in this section and hypergeometric function in the next section. The exponential function is $e^{\pm\mathrm{tr}(\tilde{Z})}$ and in the notation of hypergeometric functions this will be written as

$$_0\tilde{F}_0(\ ;\ ;\pm\tilde{Z}) = e^{\pm\mathrm{tr}(\tilde{Z})}. \tag{6.1.1}$$

From (6.0.1) and (6.0.2) one can construct matrix-variate gamma and beta densities in the complex case as follows.

Definition 6.3   Matrix-variate gamma density in the complex case

Let

$$f(\tilde{Z}) = \frac{|\det(B)|^\alpha}{\tilde{\Gamma}_p(\alpha)} |\det(\tilde{Z})|^{\alpha-p} e^{-\mathrm{tr}(B\tilde{Z})}$$

for $B = B^* > 0$, $\tilde{Z} = \tilde{Z}^* > 0$, $\mathrm{Re}(\alpha) > p-1$ and $f(\tilde{Z}) = 0$ elsewhere. This $f(\tilde{Z})$ is known as the *matrix-variate gamma density in the complex case with parameters* (α, B), see also Example 3.6 of Chapter 3. When $\alpha = n$, a positive integer, $n \geq p$, and $B = V^{-1}$ with $V = V^* > 0$ then this $f(\tilde{Z})$ is known as the central Wishart density in the complex case with n degrees of freedom and parameter matrix V. This is the fundamental density in multivariate statistical analysis in the complex case.

Definition 6.4   Type-1 matrix-variate beta density in the complex case

$$f(\tilde{Z}) = \frac{\tilde{\Gamma}_p(\alpha + \beta)}{\tilde{\Gamma}_p(\alpha)\tilde{\Gamma}_p(\beta)} |\det(\tilde{Z})|^{\alpha-p} |\det(I - \tilde{Z})|^{\beta-p}$$

for $0 < \tilde{Z} < I$, $\mathrm{Re}(\alpha) > p - 1$, $\mathrm{Re}(\beta) > p - 1$ and $f(\tilde{Z}) = 0$ elsewhere is called *type-1 matrix-variate beta density in the complex case with the parameters* (α, β), see also equation (3.2.4) of Chapter 3.

Definition 6.5   Type-2 matrix-variate beta density in the complex case

$$f(\tilde{Z}) = \frac{\tilde{\Gamma}_p(\alpha + \beta)}{\tilde{\Gamma}_p(\alpha)\tilde{\Gamma}_p(\beta)} |\det(\tilde{Z})|^{\alpha-p} |\det(I + \tilde{Z})|^{-(\alpha+\beta)}$$

for $\tilde{Z} = \tilde{Z}^* > 0$, $\mathrm{Re}(\alpha) > p - 1$, $\mathrm{Re}(\beta) > p - 1$ and $f(\tilde{Z}) = 0$ elsewhere, is called a *type-2 matrix-variate beta density with the parameters* (α, β).

A derivation of type-1 beta density from gamma densities is given in Example 3.8 of Chapter 3. If $\tilde{Z} \sim$ type-1 beta observe that $I - \tilde{Z} \sim$ type-1 beta with α, β interchanged where $\sim$ indicates *distributed as*. Type-2 beta density is available from a type-1 beta of Definition 6.4 by using the transformation

$$\tilde{U} = (I - \tilde{Z})^{-\frac{1}{2}} \tilde{Z} (I - \tilde{Z})^{-\frac{1}{2}} = \left(\tilde{Z}^{-1} - I\right)^{-1} \Rightarrow$$

$$\tilde{Z}^{-1} = I + \tilde{U}^{-1} \Rightarrow$$
$$|\det(\tilde{Z}\tilde{Z}^*)|^{-p}d\tilde{Z} = |\det(\tilde{U}\tilde{U}^*)|^{-p}d\tilde{U} \Rightarrow$$
$$d\tilde{Z} = |\det((I + \tilde{U})(I + \tilde{U})^*)|^{-p}d\tilde{U}$$

by Theorem 3.8 of Chapter 3. Now substitute for $\tilde{Z}$ in terms of $\tilde{U}$ in $f(\tilde{Z})d\tilde{Z}$ to get $g(\tilde{U})d\tilde{U}$ where $g(\tilde{U})$ denotes the density of $\tilde{U}$. This gives a type-2 beta with the parameters (α, β). While substituting one has to make use of the property that

$$|\det(I + \tilde{U})|^{-\alpha+p} = |\det(I + \tilde{U})|^{-\alpha}|\det(I + \tilde{U})|^p$$

where

$$|\det(I + \tilde{U})|^p = |\det((I + \tilde{U})(I + \tilde{U})^*)|^{\frac{p}{2}}.$$

Type-2 beta density can also be obtained from a type-1 beta of Definition 6.4 by making use of the transformation

$$\tilde{V} = (I - \tilde{Z})^{-1} - I \Rightarrow$$
$$d\tilde{Z} = |\det((I + \tilde{V})(I + \tilde{V})^*)|^{-p}d\tilde{V}.$$

In this substitution we get a type-2 beta with the parameters (β, α). Similarly we can get a type-1 beta from the type-2 beta of Definition 6.5 by using the following transformation:

$$\tilde{W} = (I + \tilde{Z})^{-\frac{1}{2}}\tilde{Z}(I + \tilde{Z})^{-\frac{1}{2}} = \left(\tilde{Z}^{-1} + I\right)^{-1} \Rightarrow$$
$$\tilde{W}^{-1} = I + \tilde{Z}^{-1} \Rightarrow$$
$$|\det(\tilde{Z}\tilde{Z}^*)|^{-p}d\tilde{Z} = |\det(\tilde{W}\tilde{W}^*)|^{-p}d\tilde{W} \Rightarrow$$
$$d\tilde{Z} = |\det((I - \tilde{W})(I - \tilde{W})^*)|^{-p}d\tilde{W}.$$

Under this substitution $\tilde{W}$ is a type-1 beta with the parameters (α, β). If $\tilde{T} = (I + \tilde{Z})^{-1}$ where $\tilde{Z}$ is type-2 beta with the parameters (α, β) then it is easy to note that $\tilde{T}$ is type-1 beta with the parameters (β, α).

6.1.1   Laplace transform in the complex case

Before defining the Laplace and the Fourier transforms (French mathematician J.B.J. Fourier (1768–1830)) of a real valued scalar function of matrix argument in the complex case, let us examine some linear functions and trace of a product of two matrices. Let $\tilde{T}$ be a $p \times 1$ parameter vector of complex elements and $\tilde{X}$ a $p \times 1$ vector of complex variables. That is,

$$\tilde{T} = T_1 + iT_2 \text{ and } \tilde{X} = X_1 + iX_2$$

where T_1, T_2, X_1, X_2 are real $p \times 1$ vectors. Let $\tilde{T}^*$ be the conjugate transpose of $\tilde{T}$. Then

$$\tilde{T}^* \tilde{X} = (T_1 - iT_2)'(X_1 + iX_2)$$
$$= T_1'X_1 + T_2'X_2 + i(T_1'X_2 - T_2'X_1)$$

and

$$\mathrm{Re}(\tilde{T}^* \tilde{X}) = T_1'X_1 + T_2'X_2$$

where $\mathrm{Re}(\cdot)$ denotes the real part of $(\cdot)$. But observe that $T_1'X_1 + T_2'X_2$ gives a linear function of all the real variables in $\tilde{X}$. Thus if $f(\tilde{X})$ is a real valued scalar function of the complex p-vector $\tilde{X}$ then to be consistent with the definition of the multivariable Laplace transform of Definition 5.2 of Chapter 5, and the corresponding Fourier transform, one should be evaluating the integrals

$$\int_{\tilde{X}} e^{-\mathrm{Re}(\tilde{T}^* \tilde{X})} f(\tilde{X}) d\tilde{X} \text{ and } \int_{\tilde{X}} e^{-i\mathrm{Re}(\tilde{T}^* \tilde{X})} f(\tilde{X}) d\tilde{X}$$

to obtain the Laplace transform and Fourier transform of $f(\tilde{X})$ respectively. Thus if $f(\tilde{X})$ is the density of a $p \times 1$ vector random variable $\tilde{X}$ then its characteristic function, denoted by $\phi(\tilde{T})$, is given by

$$\phi(\tilde{T}) = \int_{\tilde{X}} e^{i\mathrm{Re}(\tilde{T}^* \tilde{X})} f(\tilde{X}) d\tilde{X}.$$

Definition 6.6   Fourier transform: vector case   Let $\tilde{T}$ be a $p \times 1$ complex parameter vector, $\tilde{X}$ a $p \times 1$ complex vector variable and $f(\tilde{X})$ a real valued scalar function of $\tilde{X}$. Then the Fourier transform of f, denoted by $F_f(\tilde{T})$, is given by

$$F_f(\tilde{T}) = \int_{\tilde{X}} e^{-i\mathrm{Re}(\tilde{T}^* \tilde{X})} f(\tilde{X}) d\tilde{X}.$$

If $f(\tilde{X})$ is a complex valued function of the complex variable $\tilde{X}$ then one can write

$$f(\tilde{X}) = f_1(\tilde{X}) + if_2(\tilde{X}), \ i = \sqrt{-1}$$

where $f_1(\tilde{X})$ and $f_2(\tilde{X})$ are real valued functions. Various transforms can be defined by using the above representations. But we will confine our discussions to real valued functions in this chapter.

Example 6.1   Evaluate the characteristic function of a p-variate nonsingular Gaussian complex random variable $\tilde{X}$ with the density

$$f(\tilde{X}) = \frac{e^{-(\tilde{X}-\mu)^*V^{-1}(\tilde{X}-\mu)}}{\pi^p|\det(V)|}, V = V^* > 0.$$

Solution   From the above discussion

$$\phi(\tilde{T}) = \int_{\tilde{X}} e^{i\operatorname{Re}(\tilde{T}^*\tilde{X})} f(\tilde{X})d\tilde{X}$$

$$= \frac{\int_{\tilde{X}} e^{i\operatorname{Re}(\tilde{T}^*\tilde{X})-(\tilde{X}-\mu)^*V^{-1}(\tilde{X}-\mu)}d\tilde{X}}{\pi^p|\det(V)|}$$

$$= \frac{e^{i\operatorname{Re}(\tilde{T}^*\mu)}}{\pi^p|\det(V)|} \int_{\tilde{X}} e^{i\operatorname{Re}(\tilde{T}^*(\tilde{X}-\mu))-(\tilde{X}-\mu)^*V^{-1}(\tilde{X}-\mu)}d\tilde{X}.$$

Since $V = V^* > 0$ we may write $V^{-1} = B^*B$ and consider the transformation

$$\tilde{Y} = B(\tilde{X}-\mu) \Rightarrow d\tilde{Y} = |\det(BB^*)|d\tilde{X} = |\det(V)|^{-1}d\tilde{X},$$

see also Theorem 3.1 of Chapter 3, and $\tilde{X} - \mu = B^{-1}\tilde{Y}$. Then

$$\phi(\tilde{T}) = \frac{e^{i\operatorname{Re}(\tilde{T}^*\mu)}}{\pi^p} \int_{\tilde{Y}} e^{i\operatorname{Re}(\tilde{T}^*B^{-1}\tilde{Y})-\tilde{Y}^*\tilde{Y}}d\tilde{Y}.$$

If $\tilde{T}^*B^{-1} = \alpha^* = (\alpha'_1 - i\alpha'_2)$ and $\tilde{Y} = Y_1 + iY_2$ where $\alpha_1, \alpha_2, Y_1, Y_2$ are real p-vectors then

$$i\operatorname{Re}(\tilde{T}^*B^{-1}\tilde{Y}) - \tilde{Y}^*\tilde{Y} = -[Y'_1Y_1 + Y'_2Y_2 - i\alpha'_1Y_1 - i\alpha'_2Y_2]$$

$$= -[(Y_1 - \frac{1}{2}i\alpha_1)'(Y_1 - \frac{1}{2}i\alpha_1)$$

$$+ (Y_2 - \frac{1}{2}i\alpha_2)'(Y_2 - \frac{1}{2}i\alpha_2)$$

$$- \frac{1}{4}(i\alpha_1)'(i\alpha_1) - \frac{1}{4}(i\alpha_2)'(i\alpha_2)].$$

Now integrating out over Y_1 and Y_2 one has π^p cancelling and the factor

$$e^{-\frac{1}{4}(\alpha'_1\alpha_1 + \alpha'_2\alpha_2)} = e^{-\frac{1}{4}\alpha^*\alpha} = e^{-\frac{1}{4}\tilde{T}^*V\tilde{T}}$$

coming out. Then

$$\phi(\tilde{T}) = e^{i\operatorname{Re}(\tilde{T}^*\mu)-\frac{1}{4}\tilde{T}^*V\tilde{T}}. \tag{6.1.2}$$

Replace $E\left(e^{i\text{Re}(\tilde{T}^*\tilde{X})}\right)$ by $E\left(e^{\text{Re}(\tilde{T}^*\tilde{X})}\right)$ and proceed to get the moment generating function $M_{\tilde{X}}(\tilde{T})$ of the Gaussian vector random variable $\tilde{X}$. Then

$$M_{\tilde{X}}(\tilde{T}) = e^{\text{Re}(\tilde{T}^*\mu) + \frac{1}{4}\tilde{T}^*V\tilde{T}}. \tag{6.1.3}$$

Now consider the case of matrix variables. Let $\tilde{X}$ be a $p \times p$ hermitian positive definite matrix of functionally independent variables. Let $\tilde{X} = X_1 + iX_2$ where X_1 and X_2 are real, $X_1 = X_1'$ and $X_2 = -X_2'$. Let $\tilde{T}$ be a $p \times p$ parameter matrix, $\tilde{T} = T_1 + iT_2$ with T_1, T_2 real. Consider

$$\begin{aligned}
\text{tr}(\tilde{T}^*\tilde{X}) &= \text{tr}((T_1' - iT_2')(X_1 + iX_2)) \\
&= \text{tr}(T_1'X_1 + T_2'X_2 + i(T_1'X_2 - T_2'X_1)).
\end{aligned}$$

Thus

$$\text{Re}[\text{tr}(\tilde{T}^*\tilde{X})] = \text{tr}(T_1'X_1 + T_2'X_2).$$

Obviously T_1 and T_2 are to be at least symmetric otherwise the number of real parameters in $\tilde{T}$ will be more than the number of real scalar variables in X_1 and X_2 combined. Let $T_1 = (t_{jk1})$, $T_2 = (t_{jk2})$, $X_1 = (x_{jk1})$, $X_2 = (x_{jk2})$ and let $T_1 = T_1'$, $T_2 = T_2'$. Then, for example,

$$\text{tr}(T_1'X_1) = \text{tr}(T_1X_1) = \sum_{j=1}^{p} t_{jj1}x_{jj1} + 2\sum_{j>k} t_{jk1}x_{jk1}$$

and

$$\text{tr}(T_2'X_2) = \text{tr}(T_2X_2) = -\sum_{j>k} t_{jk2}x_{jk2} + \sum_{j>k} t_{jk2}x_{jk2}$$

since $x_{jj2} = 0$ for all j, $x_{jk2} = -x_{kj2}$ for all $j \neq k$. Thus $\text{tr}(T_2X_2) = 0$ if $T_2 = T_2'$. But if $T_2 = -T_2'$ then

$$\text{tr}(T_2X_2) = 2\sum_{j>k} t_{jk2}x_{jk2} \text{ or } -2\sum_{j>k} t_{jk2}x_{jk2}.$$

But note that when $\tilde{T} = \tilde{T}^*$ and $\tilde{X} = \tilde{X}^*$ then T_2 and X_2 are skew symmetric and T_1 and X_1 are symmetric, and thus $\text{tr}(T_1X_2) = 0$ and $\text{tr}(T_2X_1) = 0$. Hence

$$\text{tr}(\tilde{T}^*\tilde{X}) = \text{tr}(T_1X_1) + \text{tr}(T_2X_2). \tag{6.1.4}$$

Thus when $\tilde{T}$ is hermitian, that is $\tilde{T} = \tilde{T}^*$, one can come up with a definition of Laplace transform, consistent with Definition 5.2 of Chapter 5, by modifying $\tilde{T}$ so that the factor 2 disappears from the nondiagonal elements.

Let $\tilde{T} = (\eta_{jk}\tilde{t}_{jk}) = (T_1 + iT_2)$ where $\eta_{jk} = 1$, $j = k$ and $\eta_{jk} = \frac{1}{2}$, $j \neq k$. Let $\tilde{T} = \tilde{T}^*$ and $\tilde{X} = \tilde{X}^*$. Then

$$\text{tr}(\tilde{T}^* \tilde{X}) = \sum_{j=1}^p t_{jj1}x_{jj1} + \sum_{j>k} t_{jk1}x_{jk1} + \sum_{j>k} t_{jk2}x_{jk2}. \qquad (6.1.5)$$

If the elements above the diagonal in T_2 appear with negative signs take the elements below the diagonal with negative signs in X_2 to have the structure in (6.1.5). Note also that there are p^2 real variables in $\tilde{X} = \tilde{X}^* > 0$.

Definition 6.7   Laplace transform: matrix case   Let $\tilde{T}$ and $\tilde{X}$ be $p \times p$ hermitian matrices where $\tilde{T} = T_1 + iT_2$ be a matrix of parameters as defined in (6.1.5). Let $\tilde{X} = \tilde{X}^* > 0$ be a matrix of functionally independent variables and $f(\tilde{X})$ a real valued scalar function of $\tilde{X}$. Then the Laplace transform of f is given by

$$F_f(\tilde{T}) = \int_{\tilde{X}=\tilde{X}^*>0} e^{-[\text{tr}(\tilde{T}^*\tilde{X})]} f(\tilde{X}) d\tilde{X}. \qquad (6.1.6)$$

If η_{jk} is removed from $\tilde{T}$ then $f(\tilde{X})$ in (6.1.6) is to be treated as a function of x_{jj1}, $j = 1, ..., p$ and $2x_{jk1}$, $2x_{jk2}$, $j > k$. Then the definition holds for a real valued function of the real variables x_{jj1}, $j = 1, ..., p$, $2x_{jk1}$, $2x_{jk2}$, $j > k$.

From the matrix-variate gamma of (6.0.1) or from the matrix-variate gamma density of Definition 6.3 we may notice the following identity. This identity will be useful in evaluating certain integrals.

$$|\det(B)|^{-\alpha} = \frac{1}{\tilde{\Gamma}_p(\alpha)} \int_{\tilde{A}=\tilde{A}^*>0} |\det(\tilde{A})|^{\alpha-p} e^{-\text{tr}(B\tilde{A})} d\tilde{A} \qquad (6.1.7)$$

for $B = B^* > 0$, $\text{Re}(\alpha) > p - 1$. A convolution theorem for the Laplace transform in the complex case, parallel to Theorem 5.2 of Chapter 5 in the real case, can be established as follows.

Theorem 6.1   *Let f_1 and f_2 be real valued scalar functions of the $p \times p$ hermitian positive definite matrix $\tilde{Z}$ with Laplace transforms $g_1(\tilde{T})$ and $g_2(\tilde{T})$ where $\tilde{T} = T_1 + iT_2$, $\tilde{T} = \tilde{T}^*$. Then the Laplace transform of f_3 is g_1g_2 where*

$$f_3(\tilde{Z}) = \int_{0<\tilde{S}<\tilde{Z}} f_1(\tilde{Z} - \tilde{S}) f_2(\tilde{S}) d\tilde{S}.$$

Proof From Definition 6.7 the Laplace transform of $f_3(\tilde{Z})$, denoted by $L_{f_3}(\tilde{T})$, is given by

$$L_{f_3}(\tilde{T}) = \int_{\tilde{Z}=\tilde{Z}^* > 0} e^{-\operatorname{tr}(\tilde{T}^* \tilde{Z})} f_3(\tilde{Z}) \mathrm{d}\tilde{Z}$$

$$= \int_{\tilde{Z}>0} e^{-\operatorname{tr}(\tilde{T}^* \tilde{Z})} [\int_{0<\tilde{S}<\tilde{Z}} f_1(\tilde{Z} - \tilde{S}) f_2(\tilde{S}) \mathrm{d}\tilde{S}] \mathrm{d}\tilde{Z}.$$

Note that $\tilde{Z} > 0$ and $0 < \tilde{S} < \tilde{Z}$ is equivalent to the statement $\tilde{Z} > \tilde{S}$ and $\tilde{S} > 0$. Change $\tilde{Z}$ to $\tilde{W}$ for a fixed $\tilde{S}$ by $\tilde{W} = \tilde{Z} - \tilde{S}$. Then $\mathrm{d}\tilde{W} = \mathrm{d}\tilde{Z}$ and $\tilde{W} > 0$. Thus

$$L_{f_3}(\tilde{T}) = \int_{\tilde{W}>0} \int_{\tilde{S}>0} e^{-\operatorname{tr}(\tilde{T}^*(\tilde{W}+\tilde{S}))} f_1(\tilde{W}) f_2(\tilde{S}) \mathrm{d}\tilde{S} \mathrm{d}\tilde{W}$$

$$= \int_{\tilde{W}>0} e^{-\operatorname{tr}(\tilde{T}^* \tilde{W})} f_1(\tilde{W}) \mathrm{d}\tilde{W} \int_{\tilde{S}>0} e^{-\operatorname{tr}(\tilde{T}^* \tilde{S})} f_2(\tilde{S}) \mathrm{d}\tilde{S}$$

$$= g_1 g_2.$$

Also the formula for the inverse Laplace transform can be written down as follows, going through the steps in Section 5.1.3 of Chapter 5. For $\tilde{Z} = X + iY$, $\tilde{Z} = \tilde{Z}^*$ and $\operatorname{Re}(\alpha) > p - 1$, let

$$h(\tilde{Z}) = \frac{|\det(\tilde{Z})|^\alpha}{\tilde{\Gamma}_p(\alpha)} \int_{\tilde{A}=\tilde{A}^* > 0} |\det(\tilde{A})|^{\alpha-p} e^{-\operatorname{tr}(\tilde{Z}^* \tilde{A})} f(\tilde{A}) \mathrm{d}\tilde{A}.$$

In $h(\tilde{Z})$ replace $\tilde{Z}$ by $\tilde{Z} = \tilde{Z}_1 + i\tilde{Z}_2$ where $\tilde{Z}_1$ is hermitian positive definite and $\tilde{Z}_2$ is skew hermitian. Note that $\tilde{Z}$ is still hermitian. Then

$$\frac{2^{p(p-1)}\tilde{\Gamma}_p(\alpha)}{(2\pi i)^{p^2}} \int_{\tilde{Z}=\tilde{Z}_{10}+i\tilde{Z}_2} |\det(\tilde{Z})|^{-\alpha} e^{\operatorname{tr}(\tilde{Z}\tilde{A})} h(\tilde{Z}) \mathrm{d}\tilde{Z}$$

$$= \begin{cases} f(\tilde{A}), & \tilde{A} = \tilde{A}^* > 0 \\ 0, & \text{elsewhere,} \end{cases} \qquad (6.1.8)$$

where the integral is over all $\tilde{Z}$ such that $\tilde{Z}_1$ is fixed at some hermitian positive definite matrix $\tilde{Z}_{10}$. Note that there are p^2 real variables in $\tilde{Z}$ when $\tilde{Z} = \tilde{Z}^*$ and a total of $p(p - 1)$ real variables in $\tilde{Z}$ with the factor 2. When

$$h(\tilde{Z}) = \tilde{\Gamma}_p(\alpha)|\det(\tilde{Z})|^{-\alpha}$$

we have

$$\frac{2^{p(p-1)}}{(2\pi i)^{p^2}} \int_{\operatorname{Re}(\tilde{Z})>X_0} |\det(\tilde{Z})|^{-\alpha} e^{\operatorname{tr}(\tilde{Z}^* \tilde{U})} \mathrm{d}\tilde{Z} = \frac{|\det(\tilde{U})|^{\alpha-p}}{\tilde{\Gamma}_p(\alpha)} \qquad (6.1.9)$$

for $\text{Re}(\alpha) > p - 1$, $\tilde{U} = \tilde{U}^* > 0$.

Remark 6.1   If $\tilde{Z} = (\eta_{jk}\tilde{z}_{jk})$, $\eta_{jk} = 1$, $j = k$ and $\eta_{jk} = \frac{1}{2}$, $j \neq k$ then the real variables in $\tilde{Z}$ are z_{jj1}, $j = 1, ..., p$ and z_{jk1}, z_{jk2}, $j > k$. In this case the factor $2^{p(p-1)}$ will be absent on the left sides of (6.1.8) and (6.1.9).

Example 6.2   Evaluate the Laplace transform of the gamma density given in Definition 6.3.

Solution   Let $\tilde{T} = \tilde{T}^* > 0$ be a parameter matrix and let the Laplace transform be denoted by $L_f(\tilde{T})$. Then from Definition 6.7

$$L_f(\tilde{T}) = \int_{\tilde{Z}=\tilde{Z}^*>0} e^{-\text{tr}(\tilde{T}^*\tilde{Z})}\frac{|\det(B)|^\alpha}{\tilde{\Gamma}_p(\alpha)}|\det(\tilde{Z})|^{\alpha-p}e^{-\text{tr}(B\tilde{Z})}d\tilde{Z}$$

$$= \frac{|\det(B)|^\alpha}{\tilde{\Gamma}_p(\alpha)}\int_{\tilde{Z}>0}|\det(\tilde{Z})|^{\alpha-p}e^{-\text{tr}[(B+\tilde{T}^*)\tilde{Z}]}d\tilde{Z}.$$

But note that since $B + \tilde{T}^*$ is hermitian positive definite, there exists a matrix C such that $(B + \tilde{T}^*) = CC^*$ and

$$\text{tr}(CC^*\tilde{Z}) = \text{tr}(C^*\tilde{Z}C)$$

with

$$\tilde{Y} = C^*\tilde{Z}C \Rightarrow d\tilde{Y} = |\det(CC^*)|^p d\tilde{Z}$$

see also Theorem 3.5 of Chapter 3. Also

$$|\det(\tilde{Z})| = |\det(CC^*)|^{-1}|\det(\tilde{Y})|$$

$$= |\det(B + \tilde{T}^*)|^{-1}|\det(\tilde{Y})|.$$

Hence

$$L_f(\tilde{T}) = \frac{|\det(B)|^\alpha|\det(B + \tilde{T}^*)|^{-\alpha}}{\tilde{\Gamma}_p(\alpha)}\int_{\tilde{Y}>0}|\det(\tilde{Y})|^{\alpha-p}e^{-\text{tr}(\tilde{Y})}d\tilde{Y}$$

$$= |\det(B)|^\alpha|\det(B + \tilde{T}^*)|^{-\alpha}$$

$$= |\det(I + B^{-1}\tilde{T}^*)|^{-\alpha}$$

by integrating out $\tilde{Y}$ with the help of (6.0.1).

From the discussion so far the reader may have noticed the similarities of the results in the complex case to those in the real case. $\Gamma_p(\alpha)$ in the real case corresponds to $\tilde{\Gamma}_p(\alpha)$ in the complex case. A determinant corresponds to the absolute value (norm) of a determinant. A symmetric positive definite matrix in the real case is replaced by a hermitian positive definite matrix in the complex case. In the integral representation for the matrix-variate gamma, $|X|^{\alpha - \frac{p+1}{2}}$ is replaced by $|\det(\tilde{X})|^{\alpha - p}$ in the complex case. Integrals over the full or semiorthogonal group is replaced by the ones over the full or semiunitary group. With these changes the results in Chapter 5 dealing with Laplace transform and its applications can be extended to the complex case. Extensions of zonal polynomials and M-transforms to the complex case will be discussed next.

6.1.2   Zonal polynomials in the complex case

The material in Section 5.1.5 of Chapter 5 can be given the following extension to the complex case. As before, let $K = (k_1, ..., k_p)$ denote a partition of the nonnegative integer k such that $k_1 \geq k_2 \geq ... \geq k_p \geq 0$ and $k_1 + ... + k_p = k$. Let us denote the general hypergeometric coefficient for the partition K by the following:

Definition 6.8   **General hypergeometric coefficient in complex case**

$$[a]_K = \prod_{j=1}^{p}(a - j + 1)_{k_j} = \frac{\tilde{\Gamma}_p(a, K)}{\tilde{\Gamma}_p(a)} \tag{6.1.10}$$

with $(a)_m = a(a+1)...(a + m - 1), (a)_0 = 1, a \neq 0$ and

$$\tilde{\Gamma}_p(a, K) = \pi^{\frac{p(p-1)}{2}} \prod_{j=1}^{p} \Gamma(a + k_j - j + 1)$$

$$= \tilde{\Gamma}_p(a) \prod_{j=1}^{p}(a - j + 1)_{k_j}. \tag{6.1.11}$$

Let A be a $p \times p$ hermitian matrix with eigenvalues $\lambda_1 > ... > \lambda_p$. Then the zonal polynomials in the complex case, denoted by $\tilde{C}_K(A)$, are defined by James (1964) as follows:

Definition 6.9   **Zonal polynomials: complex case**

$$\tilde{C}_K(A) = \chi_{[K]}(1)\chi_{\{K\}}(A) \tag{6.1.12}$$

where $\chi_{[K]}(1)$ is the dimension of the representation $[K]$ of the symmetric group on k symbols and $\chi_{\{K\}}(A)$ is the character of the representation $\{K\}$

of the linear group where

$$\chi_{[K]}(1) = \frac{k!\left\{\prod_{1=m<n}^{p}(k_m - k_n - m + n)\right\}}{\left\{\prod_{j=1}^{p}(k_j + p - j)!\right\}} \qquad (6.1.13)$$

and

$$\chi_{\{K\}}(A) = \left|(\lambda_m^{k_n+p-n})\right| / \left|(\lambda_m^{p-n})\right| \qquad (6.1.14)$$

where the explicit forms of the determinants are the following:

$$\left|(\lambda_m^{k_n+p-n})\right| = \begin{vmatrix} \lambda_1^{k_1+p-1} & \lambda_1^{k_2+p-2} & \cdots & \lambda_1^{k_p} \\ \vdots & \vdots & \ddots & \vdots \\ \lambda_p^{k_1+p-1} & \lambda_p^{k_2+p-2} & \cdots & \lambda_p^{k_p} \end{vmatrix}$$

and

$$\left|(\lambda_m^{p-n})\right| = \begin{vmatrix} \lambda_1^{p-1} & \lambda_1^{p-2} & \cdots & \lambda_1^0 \\ \vdots & \vdots & \ddots & \vdots \\ \lambda_p^{p-1} & \lambda_p^{p-2} & \cdots & \lambda_p^0 \end{vmatrix}$$

$$= \prod_{m<n}(\lambda_m - \lambda_n). \qquad (6.1.15)$$

When A is the $p \times p$ identity matrix I_p it can be shown, see also Khatri (1970), that

$$\chi_{\{K\}}(I_p) = \frac{\left\{\prod_{m<n}(k_m - k_n - m + n)\right\}}{\left\{\prod_{j=1}^{p}\Gamma(p - j + 1)\right\}} \qquad (6.1.16)$$

and

$$\tilde{C}_K(I_p) = \left\{\chi_{[K]}(1)\right\}^2 \frac{\tilde{\Gamma}_p(p, K)}{k!\,\tilde{\Gamma}_p(p)}, \qquad (6.1.17)$$

$$\tilde{\Gamma}_p(p, 0) = \tilde{\Gamma}_p(p).$$

In terms of $\tilde{C}_K(A)$ the exponential function and the binomial function have the following series expansions:

$$e^{\text{tr}(A)} = {}_0\tilde{F}_0(\ ;\ ; A) = \sum_{k=0}^{\infty}\sum_{K}\frac{\tilde{C}_K(A)}{k!} \qquad (6.1.18)$$

and

$$\left|\det(I - A)\right|^{-\alpha} = {}_1\tilde{F}_0(\alpha;\ ; A) = \sum_{k=0}^{\infty}\sum_{K}[\alpha]_K\frac{\tilde{C}_K(A)}{k!}. \qquad (6.1.19)$$

For evaluating integrals and establishing many results on functions of matrix argument in the complex case the following basic properties of zonal polynomials will be helpful. For $\tilde{Z}, \tilde{S}$ hermitian positive definite, $\text{Re}(\alpha) > p - 1$, $K = (k_1, ..., k_p)$, $k_1 \geq k_2 \geq ... \geq k_p \geq 0$, $k_1 + ... + k_p = k$,

$$\int_{\tilde{Z}=\tilde{Z}^* > 0} e^{-\text{tr}(\tilde{Z}\tilde{S})} |\det(\tilde{Z})|^{\alpha-p} \tilde{C}_K(\tilde{Z}\tilde{T}) d\tilde{Z}$$

$$= \tilde{\Gamma}_p(\alpha, K) |\det(\tilde{S})|^{-\alpha} \tilde{C}_K(\tilde{T}\tilde{S}^{-1}). \tag{6.1.20}$$

For $\tilde{Z} = \tilde{Z}^* > 0$, $0 < \tilde{Z} < I$, $\text{Re}(\alpha) > p - 1$, $\text{Re}(\beta) > p - 1$,

$$\int_{0 < \tilde{Z} < I} |\det(\tilde{Z})|^{\alpha-p} |\det(I - \tilde{Z})|^{\beta-p} \tilde{C}_K(\tilde{Z}\tilde{S}) d\tilde{Z}$$

$$= \frac{\tilde{\Gamma}_p(\alpha, K) \tilde{\Gamma}_p(\beta) \tilde{C}_K(\tilde{S})}{\tilde{\Gamma}_p(\alpha + \beta, K)}. \tag{6.1.21}$$

For $\tilde{X}$ an $n \times n$ arbitrary complex matrix and $\tilde{U}$ a unitary matrix on the unitary group $\tilde{O}_{(n)}$ of order n, that is, $\tilde{U} \in \tilde{O}_{(n)}$,

$$\int_{\tilde{O}_{(n)}} e^{\text{tr}(\tilde{X}\tilde{U} + \tilde{U}^* \tilde{X}^*)} d\tilde{U} = {}_0\tilde{F}_1(\ ; n; \tilde{X}\tilde{X}^*) \tag{6.1.22}$$

where $d\tilde{U}$ is the unitary invariant measure of the unitary group normalized to make the total volume unity.

6.1.3   The M-transform

Let $f(\tilde{Z})$ be a real valued scalar function defined on the class of $p \times p$ hermitian matrices which are either positive definite or negative definite or null and let f be symmetric in the sense $f(\tilde{Z}\tilde{W}) = f(\tilde{W}\tilde{Z})$ where $\tilde{W}$ also belongs to the same class. A consequence of symmetry is that

$$f(\tilde{Z}) = f(\tilde{U}\tilde{U}^* \tilde{Z}) = f(\tilde{U}^* \tilde{Z}\tilde{U}) = f(D)$$

where $\tilde{U}$ is a unitary matrix, $\tilde{U}\tilde{U}^* = I_p$, and $D = \text{diag}(\lambda_1, ..., \lambda_p)$ where $\lambda_1, ..., \lambda_p$ are the eigenvalues of $\tilde{Z}$. These λ_j's, $j = 1, ..., p$ are all either strictly positive or strictly negative real quantities or zero. The M-transform in the complex case will be defined as follows:

Definition 6.10   M-transform: complex case

$$M_\rho(f) = \int_{\tilde{Z}=\tilde{Z}^*>0} |\det(\tilde{Z})|^{\rho-p} f(\tilde{Z}) \mathrm{d}\tilde{Z}$$

whenever $M_\rho(f)$ exists. Here $M_\rho(f)$ is a function of the scalar variable ρ and for convenience it is written as $M_\rho(f)$.

A convolution theorem for the M-transform can be established as follows:

Theorem 6.2   *Let $f_1(\tilde{S})$ and $f_2(\tilde{S})$ be symmetric functions with the M-transforms $M_\rho(f_1) = g_1(\rho)$ and $M_\rho(f_2) = g_2(\rho)$ respectively. Let*

$$f_3(\tilde{S}) = \int_{\tilde{Z}=\tilde{Z}^*>0} |\det(\tilde{Z})|^\beta f_1(\tilde{Z}^{\frac{1}{2}} \tilde{S} \tilde{Z}^{\frac{1}{2}}) f_2(\tilde{Z}) \mathrm{d}\tilde{Z}$$

then the M-transform of f_3 is

$$M_\rho(f_3) = g_1(\rho) g_2(p + \beta - \rho)$$

where $\tilde{Z}^{\frac{1}{2}}$ denotes the hermitian positive definite square root of $\tilde{Z} = \tilde{Z}^* > 0$.

Exercises

6.1.1   Prove the results in (6.1.16) and (6.1.17) starting with (6.1.14) and then taking the limits $\lambda_j \to 1$, $j = 1, ..., p$.

6.1.2   Prove that

$$\tilde{B}_p(\alpha, \beta) = \int_{0<\tilde{Z}=\tilde{Z}^*<I} |\det(\tilde{Z})|^{\alpha-p} |\det(I - \tilde{Z})|^{\beta-p} \mathrm{d}\tilde{Z}$$
$$= \tilde{\Gamma}_p(\alpha) \tilde{\Gamma}_p(\beta) / \tilde{\Gamma}_p(\alpha + \beta).$$

6.1.3   Show that

$$\tilde{B}_p(\alpha, \beta) = \int_{\tilde{Z}=\tilde{Z}^*>0} |\det(\tilde{Z})|^{\alpha-p} |\det(I + \tilde{Z})|^{-(\alpha+\beta)} \mathrm{d}\tilde{Z}.$$

6.1.4   For a $p \times p$ hermitian positive definite matrix $\tilde{Z}$ such that $0 < \tilde{Z} < I$ show that

$$\int_0^I \mathrm{d}\tilde{Z} = \frac{[1!2!...(p-1)!]^2}{p!(p+1)!...(2p-1)!} \pi^{\frac{p(p-1)}{2}}$$

and verify it by direct integration for $p = 2$.

6.1.5   For a $p \times p$ hermitian positive definite matrix $\tilde{Z}$ show that

$$\int_{\tilde{Z}>0} e^{-\text{tr}(\tilde{Z})} d\tilde{Z} = (p-1)!(p-2)!...1! \; \pi^{\frac{p(p-1)}{2}}$$

and verify it by direct integration for $p = 2$.

6.2   Hypergeometric function of matrix argument: complex case

A general hypergeometric function in the complex case, with r upper parameters $a_1, ..., a_r$ and s lower parameters $b_1, ..., b_s$, will be denoted by

$$_r\tilde{F}_s = {_r\tilde{F}_s}(a_1, ..., a_r; b_1, ..., b_s; \tilde{A}). \tag{6.2.1}$$

As in the real case, the function in (6.2.1) can be defined by using zonal polynomials, through M-transform or through generalized Laplace transform. In all these definitions all the matrices appearing are assumed to be hermitian positive definite and the function itself is assumed to be symmetric in the sense described in Section 6.1.3.

Through zonal polynomials $_r\tilde{F}_s$ will be defined as the following multiple series.

$$_r\tilde{F}_s = \sum_{k=0}^{\infty} \sum_K \frac{[a_1]_K...[a_r]_K}{[b_1]_K...[b_s]_K} \frac{\tilde{C}_K(\tilde{A})}{k!} \tag{6.2.2}$$

where the symbols k, K and $[a]_K$ are defined in (6.1.10) and the zonal polynomials, in (6.1.12). The series converges for all $\tilde{A}$ when $s \geq r$ or for $\|\tilde{A}\| < 1$ when $r = s + 1$ where $\|(\cdot)\|$ denotes a norm of $(\cdot)$. The parameters are assumed to be such that the ratios make sense for all parameters, all k and all partitions K of k.

Through Laplace transform $_r\tilde{F}_s$ will be defined as that symmetric function which satisfies the following pair of integral equations:

$$_{r+1}\tilde{F}_s\left(a_1, ..., a_r, c; b_1, ..., b_s; -\tilde{Z}^{-1}\right) |\det(\tilde{Z})|^{-c} = \frac{1}{\tilde{\Gamma}_p(c)}$$

$$\times \int_{\tilde{\Lambda}=\tilde{\Lambda}^*>0} e^{-\text{tr}(\tilde{\Lambda}\tilde{Z})} {_r\tilde{F}_s}(a_1, ..., a_r; b_1, ..., b_s; -\tilde{\Lambda}) |\det(\tilde{\Lambda})|^{c-p} d\tilde{\Lambda} \tag{6.2.3}$$

and

$$
{}_r\tilde{F}_{s+1}(a_1, ..., a_r; b_1, ..., b_s, c; -\tilde{\Lambda})|\det(\tilde{\Lambda})|^{c-p} = \frac{\tilde{\Gamma}_p(c)2^{p(p-1)}}{(2\pi i)^{p^2}}
$$

$$
\times \int_{\tilde{Z}=\tilde{Z}_{10}+i\tilde{Z}_2} e^{\operatorname{tr}(\tilde{\Lambda}\tilde{Z})} {}_r\tilde{F}_s\left(a_1, ..., a_r; b_1, ..., b_s; -\tilde{Z}^{-1}\right) |\det(\tilde{Z})|^{-c} d\tilde{Z} \quad (6.2.4)
$$

where $\tilde{Z} = \tilde{Z}_1 + i\tilde{Z}_2$ with $\tilde{Z}_1$ hermitian positive definite and $\tilde{Z}_2$ hermitian. The integral is over all $\tilde{Z}$ such that $\tilde{Z}_1$ is fixed at some $\tilde{Z}_{10}$, see also (6.1.8). The nondiagonal elements in $\tilde{Z}$ are not multiplied by $\frac{1}{2}$ and hence the factor $2^{p(p-1)}$ is present on the right side of (6.2.4).

Through M-transform ${}_r\tilde{F}_s$ will be defined as that class of symmetric functions, denoted by

$$
{}_r\tilde{F}_s^*(-\tilde{Z}) = {}_r\tilde{F}_s^*(a_1, ..., a_r; b_1, ..., b_s; -\tilde{Z}),
$$

having the following M-transform.

$$
M_\rho({}_r\tilde{F}_s^*(-\tilde{Z})) = \frac{\left\{\prod_{j=1}^s \tilde{\Gamma}_p(b_j)\right\} \left\{\prod_{j=1}^r \tilde{\Gamma}_p(a_j - \rho)\right\} \tilde{\Gamma}_p(\rho)}{\left\{\prod_{j=1}^r \tilde{\Gamma}_p(a_j)\right\} \left\{\prod_{j=1}^s \tilde{\Gamma}_p(b_j - \rho)\right\}} \quad (6.2.5)
$$

whenever the gammas are defined.

Example 6.3   Establish the following result by using (1) M-transform, (2) zonal polynomial, (3) generalized Laplace transform.

$$
{}_1\tilde{F}_1(a; c; -\tilde{\Lambda}) = \frac{\tilde{\Gamma}_p(c)}{\tilde{\Gamma}_p(a)\tilde{\Gamma}_p(c-a)} \int_0^I |\det(\tilde{U})|^{a-p}
$$

$$
\times |\det(I - \tilde{U})|^{c-a-p} e^{-\operatorname{tr}(\tilde{\Lambda}\tilde{U})} d\tilde{U}.
$$

Solution   Take the M-transform of the right side taking ρ as the parameter. Integration over $\tilde{\Lambda}$ leads to the following:

$$
\int_{\tilde{\Lambda}>0} |\det(\tilde{\Lambda})|^{\rho-p} e^{-\operatorname{tr}(\tilde{\Lambda}\tilde{U})} d\tilde{\Lambda} = \tilde{\Gamma}_p(\rho)|\det(\tilde{U})|^{-\rho}
$$

by using a matrix-variate gamma integral of (6.0.1) after making a transformation $\tilde{U}^{\frac{1}{2}}\tilde{\Lambda}\tilde{U}^{\frac{1}{2}} = \tilde{Z}$. Now the integration of $\tilde{U}$ by using a type-1 beta integral of (6.0.2) leads to

$$
\frac{\tilde{\Gamma}_p(a-\rho)\tilde{\Gamma}_p(c-a)}{\tilde{\Gamma}_p(c-\rho)}.
$$

Comparing with the M-transform of the left side and identifying with the help of (6.2.5) the result follows.

For proving the result by using zonal polynomials expand

$$e^{-\text{tr}(\tilde{\Lambda}\tilde{U})} = \sum_{k=0}^{\infty} \sum_{K} \frac{\tilde{C}_K(-\tilde{\Lambda}\tilde{U})}{k!}$$

and integrate out $\tilde{U}$ by using (6.1.21) to get the right side, denoted by g, as follows:

$$g = \frac{\tilde{\Gamma}_p(c)}{\tilde{\Gamma}_p(a)} \sum_{k=0}^{\infty} \sum_{K} \frac{\tilde{\Gamma}_p(a, K)}{\tilde{\Gamma}_p(c, K)} \frac{\tilde{C}_K(-\tilde{\Lambda})}{k!}$$

$$= \sum_{k=0}^{\infty} \sum_{K} \frac{[a]_K}{[c]_K} \frac{\tilde{C}_K(-\tilde{\Lambda})}{k!}$$

$$= {}_1\tilde{F}_1(a; c; -\tilde{\Lambda})$$

by making use of the simplification in (6.1.10).

For establishing the result with the help of the Laplace transform, make the transformation

$$\tilde{\Lambda}^{\frac{1}{2}}\tilde{U}\tilde{\Lambda}^{\frac{1}{2}} = \tilde{V} \Rightarrow d\tilde{U} = |\det(\tilde{\Lambda}\tilde{\Lambda}^*)|^{-p}d\tilde{V}.$$

Under this substitution $|\det(\tilde{\Lambda})|^{-c+p}$ comes out from the right side. Rewrite the result to be established in the following form:

$$\frac{|\det(\tilde{\Lambda})|^{c-p} {}_1\tilde{F}_1(a; c; -\tilde{\Lambda})}{\tilde{\Gamma}_p(c)} = \frac{1}{\tilde{\Gamma}_p(a)\tilde{\Gamma}_p(c-a)}$$

$$\times \int_{0<\tilde{V}<\tilde{\Lambda}} |\det(\tilde{V})|^{a-p}|\det(\tilde{\Lambda} - \tilde{V})|^{c-a-p}$$

$$\times e^{-\text{tr}(\tilde{V})}d\tilde{V}. \tag{a}$$

Apply Theorem 6.1 to the right side of (a) by taking

$$f_1(\tilde{V}) = |\det(\tilde{V})|^{c-a-p}$$

and

$$f_2(\tilde{V}) = |\det(\tilde{V})|^{a-p}e^{-\text{tr}(\tilde{V})}$$

giving

$$g_1(\tilde{T}) = |\det(\tilde{T})|^{-(c-a)}\tilde{\Gamma}_p(c-a)$$

and

$$g_2(\tilde{T}) = |\det(I + \tilde{T})|^{-a}\tilde{\Gamma}_p(a).$$

Thus the Laplace transform of the right side of (a) is $|\det(\tilde{T})|^{a-c}|\det(I + \tilde{T})|^{-a}$. Now take the Laplace transform of the left side of (a) by using (6.2.3) to get

$$|\det(\tilde{T})|^{-c} \, _2\tilde{F}_1\left(a, c;, c; -\tilde{T}^{-1}\right).$$

But note that

$$_2\tilde{F}_1\left(a, c; c; -\tilde{T}^{-1}\right) = {}_1\tilde{F}_0\left(a; ; -\tilde{T}^{-1}\right)$$

$$= \left|\det\left(I + \tilde{T}^{-1}\right)\right|^{-a}$$

$$= |\det(\tilde{T})|^{a}|\det(I + \tilde{T})|^{-a}.$$

Observe that the Laplace transform of the left side of (a) is the same as that of the right side and hence the result.

Example 6.4   Evaluate the following result for the Gauss' hypergeometric function $_2\tilde{F}_1$:

$$_2\tilde{F}_1(a, b; c; -\tilde{Z}) = \frac{\tilde{\Gamma}_p(c)}{\tilde{\Gamma}_p(a)\tilde{\Gamma}_p(c - a)} \int_0^I |\det(\tilde{A})|^{a-p}$$

$$\times |\det(I - \tilde{A})|^{c-a-p}|\det(I + \tilde{Z}\tilde{A})|^{-b}d\tilde{A}$$

for $\mathrm{Re}(a) > p - 1$, $\mathrm{Re}(c - a) > p - 1$.

Solution   Take the M-transform of the right side. This can be taken inside since the integral over $\tilde{A}$ is of a type-1 beta type. Then

$$\int_{\tilde{Z}>0} |\det(\tilde{Z})|^{\rho-p}|\det(I + \tilde{Z}\tilde{A})|^{-b}d\tilde{Z}$$

$$= |\det(\tilde{A})|^{-\rho}\frac{\tilde{\Gamma}_p(\rho)\tilde{\Gamma}_p(b - \rho)}{\tilde{\Gamma}_p(b)}$$

by using the type-2 beta integral in (6.0.2). Now integrate out $\tilde{A}$ to get the final result of the M-transform of the right side as

$$\frac{\tilde{\Gamma}_p(c)}{\tilde{\Gamma}_p(a)\tilde{\Gamma}_p(b)}\;\frac{\tilde{\Gamma}_p(a - \rho)\tilde{\Gamma}_p(b - \rho)\tilde{\Gamma}_p(\rho)}{\tilde{\Gamma}_p(c - \rho)}$$

which from (6.2.5) is the M-transform of the left side and hence the result. The procedures remain the same as in Example 6.3 if zonal polynomials or Laplace transform are used to establish this result.

Exercises

6.2.1   Show that

$$\int_0^I |\det(\tilde{Z})|^{a-p}|\det(I - \tilde{Z})|^{b-p}{}_2\tilde{F}_1(\alpha, \beta; \gamma; -\tilde{T}\tilde{Z})d\tilde{Z}$$

$$= \frac{\tilde{\Gamma}_p(a)\tilde{\Gamma}_p(b)}{\tilde{\Gamma}_p(a+b)}{}_3\tilde{F}_2(\alpha, \beta, a; \gamma, a + b; -\tilde{T}).$$

6.2.2   Defining a Whittaker function $\tilde{W}_{.,.}(\tilde{A})$ for the complex case, where $\tilde{A}$ is hermitian positive definite, by the integral

$$\int_{\tilde{Z}>0} |\det(\tilde{Z})|^{\beta-\alpha-\frac{p}{2}}|\det(I + \tilde{Z})|^{\beta+\alpha-\frac{p}{2}} e^{-\text{tr}(\tilde{A}\tilde{Z})}d\tilde{Z}$$

$$= |\det(\tilde{A})|^{-\beta-\frac{p}{2}} \tilde{\Gamma}_p\left(\beta - \alpha + \frac{p}{2}\right) e^{\frac{1}{2}\text{tr}(\tilde{A})}\tilde{W}_{\alpha,\beta}(\tilde{A})$$

for $\text{Re}(\beta - \alpha) > \frac{p}{2} - 1$, establish the following results:

$$\int_{\tilde{Z}>0} |\det(I + \tilde{Z})|^{-\alpha}e^{-\text{tr}(\tilde{A}\tilde{Z})}d\tilde{Z}$$

$$= |\det(\tilde{A})|^{\frac{q}{2}-p} \tilde{\Gamma}_p(p)e^{\frac{1}{2}\text{tr}(\tilde{A})}\tilde{W}_{-\frac{q}{2},\frac{1}{2}(-\alpha+p)}(\tilde{A}); \qquad (a)$$

for $B = B^* > 0$, $U = U^* > 0$, $M = M^* > 0$, $\text{Re}(q) > \frac{p}{2}$,

$$\int_{\tilde{X}>\tilde{U}} |\det(\tilde{X} + B)|^{2\alpha-p}|\det(\tilde{X} - \tilde{U})|^{2q-p}e^{-\text{tr}(M\tilde{X})}d\tilde{X}$$

$$= \tilde{\Gamma}_p(2q)|\det(\tilde{U} + B)|^{\alpha+q-p}|\det(M)|^{-(\alpha+q)}$$

$$\times e^{\frac{1}{2}\text{tr}[(B-\tilde{U})M]}\tilde{W}_{(\alpha-q),(\alpha+q-\frac{p}{2})}(\tilde{T}), \qquad (b)$$

$$\tilde{T} = (\tilde{U} + B)^{\frac{1}{2}}M(\tilde{U} + B)^{\frac{1}{2}}.$$

6.2.3   Defining the Bessel function of a complex matrix, $\tilde{A}_\gamma(\tilde{S})$ for $\tilde{S} = \tilde{S}^* > 0$, as

$$\tilde{A}_\gamma(\tilde{S}) = \frac{1}{\tilde{\Gamma}_p(\gamma+p)}{}_0\tilde{F}_1(\; ; \gamma + p; -\tilde{S}), \quad \text{Re}(\gamma) > -1,$$

show that

$$\int_{\tilde{S}>0} e^{-\mathrm{tr}(\tilde{A}\tilde{S})} |\det(\tilde{S})|^{\delta-p} \tilde{A}_{\gamma}(\tilde{S}) d\tilde{S}$$
$$= \frac{\tilde{\Gamma}_p(\delta)}{\tilde{\Gamma}_p(\gamma+p)} |\det(\tilde{A})|^{-\delta}{}_1\tilde{F}_1\left(\delta; \gamma+p; -\tilde{A}^{-1}\right).$$

6.2.4   Defining the Laguerre function $\tilde{L}_\nu^\gamma(\tilde{Z})$ in the complex case, for a hermitian positive definite $p \times p$ matrix $\tilde{Z}$, as

$$\tilde{L}_\nu^\gamma(\tilde{Z}) = \frac{\tilde{\Gamma}_p(\gamma+\nu+p)}{\tilde{\Gamma}_p(\gamma+p)}{}_1\tilde{F}_1(-\nu; \gamma+p; \tilde{Z})$$

for $\mathrm{Re}(\gamma) > -1$, $\mathrm{Re}(\gamma+\nu) > -1$, $-\mathrm{Re}(\nu) > p-1$, show that

$$\tilde{L}_\nu^\gamma(\tilde{Z}) = \frac{e^{\mathrm{tr}(\tilde{Z})} |\det(\tilde{Z})|^{-\gamma}}{\tilde{\Gamma}_p(-\nu)}$$
$$\times \int_{0<\tilde{A}<\tilde{Z}} e^{-\mathrm{tr}(\tilde{A})} |\det(\tilde{A})|^{\gamma+\nu} |\det(\tilde{Z}-\tilde{A})|^{-\nu-p} d\tilde{A}.$$

6.2.5   If the Jacobi polynomial of matrix argument is defined by

$$\tilde{P}_\nu^{(\gamma,\delta)}(\tilde{Z}) = \frac{\tilde{\Gamma}_p(\gamma+\nu+p)}{\tilde{\Gamma}_p(\gamma+p)}{}_2\tilde{F}_1(-\nu, \gamma+\delta+\nu+p; \gamma+p; \tilde{Z})$$

for a $p \times p$ hermitian positive definite matrix $\tilde{Z}$, show that

$$\tilde{P}_\nu^{(\gamma,\delta)}(\tilde{Z}) = \frac{\tilde{\Gamma}_p(\gamma+\nu+p)}{\tilde{\Gamma}_p(\gamma+p)} |\det(I-\tilde{Z})|^{-\delta}$$
$$\times {}_2\tilde{F}_1(\gamma+\nu+p, -\delta-\nu; \gamma+p; \tilde{Z}).$$

6.3   G- and H-functions of matrix argument: complex case

As in the real case discussed in Chapter 5 we can extend the definitions of G- and H-functions to the complex case and the properties will be quite parallel. We will give the definitions and some simple examples for the sake of illustration.

Definition 6.11   H-function of matrix argument: complex case

The H-function of matrix argument in the complex case, denoted by

$$\tilde{H}^{m,n}_{r,s}(\tilde{Z}) = \tilde{H}^{m,n}_{r,s}\left(\tilde{Z}\Big|^{(a_1,\alpha_1),\ldots,(a_r,\alpha_r)}_{(b_1,\beta_1),\ldots,(b_s,\beta_s)}\right)$$

is defined as that class of functions having the following M-transform:

$$\int_{\tilde{Z}=\tilde{Z}^*>0}|\det(\tilde{Z})|^{\rho-p}\tilde{H}^{m,n}_{r,s}(\tilde{Z})d\tilde{Z} = \phi(\rho) \tag{6.3.1}$$

where

$$\phi(\rho) = \frac{\left\{\prod_{j=1}^{m}\tilde{\Gamma}_p(b_j+\beta_j\rho)\right\}\left\{\prod_{j=1}^{n}\tilde{\Gamma}_p(p-a_j-\alpha_j\rho)\right\}}{\left\{\prod_{j=m+1}^{s}\tilde{\Gamma}_p(p-b_j-\beta_j\rho)\right\}\left\{\prod_{j=n+1}^{r}\tilde{\Gamma}_p(a_j+\alpha_j\rho)\right\}}$$

whenever the gammas on the right exist where the α_j's and β_j's are real and positive. When $\alpha_1 = \ldots = \alpha_r = 1 = \beta_1 = \ldots = \beta_s$ in (6.3.1) one has the G-function, denoted by

$$\tilde{G}^{m,n}_{r,s}(\tilde{Z}) = \tilde{G}^{m,n}_{r,s}\left(\tilde{Z}\Big|^{a_1,\ldots,a_r}_{b_1,\ldots,b_s}\right).$$

Definition 6.12   G-function of matrix argument: complex case

It is that class of functions having the following M-transform:

$$\int_{\tilde{Z}=\tilde{Z}^*>0}|\det(\tilde{Z})|^{\rho-p}\tilde{G}^{m,n}_{r,s}(\tilde{Z})d\tilde{Z} = \phi_1(\rho) \tag{6.3.2}$$

where

$$\phi_1(\rho) = \frac{\left\{\prod_{j=1}^{m}\tilde{\Gamma}_p(b_j+\rho)\right\}\left\{\prod_{j=1}^{n}\tilde{\Gamma}_p(p-a_j-\rho)\right\}}{\left\{\prod_{j=m+1}^{s}\tilde{\Gamma}_p(p-b_j-\rho)\right\}\left\{\prod_{j=n+1}^{r}\tilde{\Gamma}_p(a_j+\rho)\right\}}$$

whenever the gammas on the right exist. Note that

$$|\det(\tilde{Z})|^{(\rho+\alpha)-p} = |\det(\tilde{Z})|^{\rho-p}|\det(\tilde{Z})|^{\alpha}. \qquad (6.3.3)$$

If ρ on the right side of (6.3.2) is replaced by $\rho + \alpha$ then it is equivalent to multiplying the G-function by $|\det(\tilde{Z})|^{\alpha}$. In the light of this observation one can write up a hypergeometric function in terms of a G-function defined in (6.3.2). Let

$$g(\tilde{Z}) = |\det(\tilde{Z})|^{-b_1} \tilde{G}_{r,s}^{1,r}\left(\tilde{Z}\Big|_{b_1,...,b_s}^{a_1,...,a_r}\right).$$

Then from (6.3.2), the M-transform of $g(\tilde{Z})$ is given by the following:

$$
\begin{aligned}
M_\rho(g) &= \int_{\tilde{Z}=\tilde{Z}^*>0} |\det(\tilde{Z})|^{\rho-p} g(\tilde{Z}) d\tilde{Z} \\
&= \frac{\tilde{\Gamma}_p(\rho)\left\{\prod_{j=1}^r \tilde{\Gamma}_p(b_1 + p - a_j - \rho)\right\}}{\left\{\prod_{j=2}^s \tilde{\Gamma}_p(b_1 + p - b_j - \rho)\right\}}.
\end{aligned}
\qquad (6.3.4)
$$

Compare the right side of (6.3.4) with the M-transform in (6.2.5) to see that

$$g(\tilde{Z}) = \frac{\left\{\prod_{j=1}^r \tilde{\Gamma}_p(a_j^*)\right\}}{\left\{\prod_{j=2}^s \tilde{\Gamma}_p(b_j^*)\right\}} {}_r\tilde{F}_{s-1}^*(a_1^*, ..., a_r^*; b_2^*, ..., b_s^*; -\tilde{Z}) \qquad (6.3.5)$$

where

$$a_j^* = b_1 + p - a_j, \; j = 1, ..., r \text{ and } b_j^* = b_1 + p - b_j, \; j = 2, ..., s. \qquad (6.3.6)$$

Thus we have the following result.

Theorem 6.3 *For ${}_r\tilde{F}_s^*$ and $\tilde{G}_{r,s}^{m,n}$ defined in (6.2.5) and (6.3.2) respectively*

$$
\begin{aligned}
\tilde{G}_{r,s}^{1,r}(\tilde{Z}) &= \tilde{G}_{r,s}^{1,r}\left(\tilde{Z}\Big|_{b_1,...,b_s}^{a_1,...,a_r}\right) \\
&= |\det(\tilde{Z})|^{b_1} \frac{\left\{\prod_{j=1}^r \tilde{\Gamma}_p(b_1 + p - a_j)\right\}}{\left\{\prod_{j=2}^s \tilde{\Gamma}_p(b_1 + p - b_j)\right\}} \\
&\quad \times {}_r\tilde{F}_{s-1}^*(a_1^*, ..., a_r^*; b_2^*, ..., b_s^*; -\tilde{Z})
\end{aligned}
\qquad (6.3.7)
$$

where a_j^*, b_j^* are defined in (6.3.6).

Example 6.5 Show that for $p \times p$ hermitian positive definite matrix $\tilde{X}$ such that $0 < \tilde{X} < I$,

$$|\det(\tilde{X})|^{\alpha}|\det(I - \tilde{X})|^{\beta} = \tilde{\Gamma}_p(\beta + p)\tilde{G}_{1,1}^{1,0}\left(\tilde{X}\Big|_{\alpha}^{\alpha+\beta+p}\right), \|\tilde{X}\| < 1$$

in the sense that both sides have the same M-transform.

Solution Take the M-transform of the left side with respect to the parameter ρ and evaluate it by using a type-1 beta integral of (6.0.2) to get

$$M_{\rho}(\text{left side}) = \tilde{\Gamma}_p(\beta + p)\frac{\tilde{\Gamma}_p(\alpha + \rho)}{\tilde{\Gamma}_p(\alpha + \beta + \rho + p)}.$$

Compare with (6.3.2) to see that this is the M-transform of the $\tilde{G}$ on the right side and hence the result.

Example 6.6 For the Whittaker function defined in Exercise 6.2.2 of Section 6.2 for a hermitian positive definite $p \times p$ matrix $\tilde{Z}$, show that

$$\tilde{W}_{\mu,\nu}(\tilde{Z}) = \frac{|\det(\tilde{Z})|^{-\alpha}e^{-\frac{1}{2}\text{tr}(\tilde{Z})}}{\tilde{\Gamma}_p\left(\frac{p}{2} - \mu + \nu\right)\tilde{\Gamma}_p\left(\frac{p}{2} - \mu - \nu\right)}$$

$$\times \tilde{G}_{1,2}^{2,1}\left(\tilde{Z}\Big|_{\frac{p}{2}+\alpha+\nu,\frac{p}{2}+\alpha-\nu}^{p+\alpha+\mu}\right)$$

whenever all the gammas are defined.

Solution From (6.3.2) and (6.3.3) note that $|\det(\tilde{Z})|^{-\alpha}$ can be taken inside the G-function. The net effect is the deletion of α from the parameters of the G-function. Thus the result to be proved is the following:

$$e^{\frac{1}{2}\text{tr}(\tilde{Z})}\tilde{W}_{\mu,\nu}(\tilde{Z}) = \frac{\tilde{G}_{1,2}^{2,1}\left(\tilde{Z}\Big|_{\frac{p}{2}+\nu,\frac{p}{2}-\nu}^{p+\mu}\right)}{\tilde{\Gamma}_p\left(\frac{p}{2} + \nu - \mu\right)\tilde{\Gamma}_p\left(\frac{p}{2} - \nu - \mu\right)}. \tag{a}$$

We will establish (a) by showing that both sides of (a) have the same M-transform thereby showing that the Whittaker function defined by the integral in Exercise 6.2.2 of Section 6.2 can have a representation in terms of a G-function of Definition 6.12. From Exercise 6.2.2 of Section 6.2 one has

$$e^{\frac{1}{2}\text{tr}(\tilde{A})}\tilde{W}_{\mu,\nu}(\tilde{A}) = \frac{|\det(\tilde{A})|^{\frac{p}{2}+\nu}}{\tilde{\Gamma}_p\left(\nu - \mu + \frac{p}{2}\right)}$$

$$\times \int_{\tilde{Z}>0}|\det(\tilde{Z})|^{\nu-\mu-\frac{p}{2}}|\det(I + \tilde{Z})|^{\mu+\nu-\frac{p}{2}}$$

$$\times e^{-\text{tr}(\tilde{A}\tilde{Z})}d\tilde{Z}. \tag{b}$$

Take the M-transform of the right side of (b) with parameter ρ, that is,

$$M_\rho(\text{right side}) = \int_{\tilde{A}=\tilde{A}^\bullet>0} |\det(\tilde{A})|^{\rho-p}(\text{right side})d\tilde{A}.$$

Integrating out $\tilde{A}$ one has

$$\int_{\tilde{A}>0} |\det(\tilde{A})|^{\frac{p}{2}+\nu+\rho-p}e^{-\text{tr}(\tilde{A}\tilde{Z})}d\tilde{A}$$

$$= \tilde{\Gamma}_p\left(\frac{p}{2}+\nu+\rho\right)|\det(\tilde{Z})|^{-(\frac{p}{2}+\nu+\rho)} \qquad (c)$$

for $\text{Re}(\nu+\rho) > \frac{p}{2}-1$. Now integrate out $\tilde{Z}$ by using a type-2 beta integral of (6.0.2) to get the following:

$$\int_{\tilde{Z}>0} |\det(\tilde{Z})|^{-\mu-\rho-p}|\det(I+\tilde{Z})|^{-(\frac{p}{2}-\mu-\nu)}d\tilde{Z}$$

$$= \frac{\tilde{\Gamma}_p(-\mu-\rho)\tilde{\Gamma}_p\left(\frac{p}{2}-\nu+\rho\right)}{\tilde{\Gamma}_p\left(\frac{p}{2}-\mu-\nu\right)} \qquad (d)$$

for $\text{Re}(\mu+\rho) < -p+1$, $\text{Re}(p+\nu) < -\frac{p}{2}+1$, $\text{Re}(-\nu+\rho) > \frac{p}{2}-1$. Note that parameters μ, ν, ρ exist so that all the conditions in (c) and (d) are satisfied. From (c) and (d) the M-transform of the right side of (b) is given by

$$M_\rho(\text{right side}) = \frac{\tilde{\Gamma}_p\left(\frac{p}{2}+\nu+\rho\right)\tilde{\Gamma}_p(-\mu-\rho)\tilde{\Gamma}_p\left(\frac{p}{2}-\nu+\rho\right)}{\tilde{\Gamma}_p\left(\nu-\mu+\frac{p}{2}\right)\tilde{\Gamma}_p\left(-\nu-\mu+\frac{p}{2}\right)}.$$

But

$$\tilde{\Gamma}_p(-\mu-\rho)\tilde{\Gamma}_p\left(\frac{p}{2}-\nu+\rho\right)\tilde{\Gamma}_p\left(\frac{p}{2}+\nu+\rho\right)$$

$$= M_\rho\left(\tilde{G}^{2,1}_{1,2}\left(\tilde{A}\Big|^{p+\mu}_{\frac{p}{2}+\nu,\frac{p}{2}-\nu}\right)\right)$$

and thus the result in (a) is established. Note that the corresponding result is available in the scalar case, that is for $p=1$, see, for example, Mathai (1993, page 129).

Example 6.7 For a $p \times p$ hermitian positive definite matrix $\tilde{Z}$ and for a Whittaker function defined in Exercise 6.2.2 of Section 6.2 show that

$$\tilde{W}_{\mu,\nu}(\tilde{Z}) = |\det(\tilde{Z})|^{-\alpha+\frac{p}{2}}e^{\frac{1}{2}\text{tr}(\tilde{Z})}$$

$$\times \tilde{G}^{2,0}_{1,2}\left(\tilde{Z}\Big|^{\frac{p}{2}+\alpha-\mu}_{\alpha+\nu,\alpha-\nu}\right).$$

Solution   Since $|\det(\tilde{Z})|^{-\alpha+\frac{p}{2}}$ can be taken inside the G-function what is to be shown is that

$$e^{-\frac{1}{2}\mathrm{tr}(\tilde{Z})}\tilde{W}_{\mu,\nu}(\tilde{Z}) = \tilde{G}^{2,0}_{1,2}\left(\tilde{Z}\Big|^{p-\mu}_{\frac{p}{2}+\nu,\frac{p}{2}-\nu}\right). \tag{a}$$

From the integral representation of the Whittaker function in Exercise 6.2.2 of Section 6.2 we have

$$e^{-\frac{1}{2}\mathrm{tr}(\tilde{A})}\tilde{W}_{\mu,\nu}(\tilde{A}) = \frac{|\det(\tilde{A})|^{\frac{p}{2}+\nu}e^{-\mathrm{tr}(\tilde{A})}}{\tilde{\Gamma}_p\left(\frac{p}{2}+\nu-\mu\right)}$$

$$\times \int_{\tilde{Z}>0}|\det(\tilde{Z})|^{\nu-\mu-\frac{p}{2}}|\det(I+\tilde{Z})|^{\mu+\nu-\frac{p}{2}}$$

$$\times e^{-\mathrm{tr}(\tilde{A}\tilde{Z})}d\tilde{Z}. \tag{b}$$

Take the M-transform of the right side of (b) with parameter ρ. Integration of $\tilde{A}$ yields

$$\int_{\tilde{A}>0}|\det(\tilde{A})|^{\frac{p}{2}+\nu+\rho-p}e^{-\mathrm{tr}[(I+\tilde{Z})\tilde{A}]}d\tilde{A}$$

$$= \tilde{\Gamma}_p\left(\frac{p}{2}+\nu+\rho\right)|\det(I+\tilde{Z})|^{-\left(\frac{p}{2}+\nu+\rho\right)} \tag{c}$$

for $\mathrm{Re}(\nu+\rho) > \frac{p}{2}-1$. Integrate out $\tilde{Z}$ to get

$$\int_{\tilde{Z}>0}|\det(\tilde{Z})|^{\frac{p}{2}+\nu-\mu-p}|\det(I+\tilde{Z})|^{-(\rho+p-\mu)}d\tilde{Z}$$

$$= \frac{\tilde{\Gamma}_p\left(\frac{p}{2}+\nu-\mu\right)\tilde{\Gamma}_p\left(\frac{p}{2}-\nu+\rho\right)}{\tilde{\Gamma}_p(p+\rho-\mu)} \tag{d}$$

for $\mathrm{Re}(\nu-\mu) > \frac{p}{2}-1$, $\mathrm{Re}(\rho-\nu) > \frac{p}{2}-1$. From (b), (c) and (d) the M-transform of the right side of (b) is

$$\frac{\tilde{\Gamma}_p\left(\frac{p}{2}+\nu+\rho\right)\tilde{\Gamma}_p\left(\frac{p}{2}-\nu+\rho\right)}{\tilde{\Gamma}_p(p-\mu+\rho)}$$

which is the M-transform of the G-function on the right side of (a). Hence the result.

From Examples 6.6 and 6.7 we can have the following result on a G-function by equating the expressions for a Whittaker function. For a $p \times p$ hermitian positive definite matrix $\tilde{Z}$ and for $\mathrm{Re}(\mu \pm \nu) < 1 - \frac{p}{2}$,

$$e^{-\mathrm{tr}(\tilde{Z})}\tilde{G}^{2,0}_{1,2}\left(\tilde{Z}\Big|^{p-\mu}_{\frac{p}{2}+\nu,\frac{p}{2}-\nu}\right)$$

$$= \frac{1}{\tilde{\Gamma}_p\left(\frac{p}{2}-\mu+\nu\right)\tilde{\Gamma}_p\left(\frac{p}{2}-\mu-\nu\right)}\tilde{G}^{2,1}_{1,2}\left(\tilde{Z}\Big|^{p+\mu}_{\frac{p}{2}+\nu,\frac{p}{2}-\nu}\right). \tag{6.3.8}$$

Exercises

6.3.1   Prove the result in (6.3.8) starting from the definition of a G-function.

6.3.2   For $\tilde{X} = \tilde{X}^* > 0$, $\text{Re}(\beta - \alpha) > p - 1$, $\text{Re}(\beta - \gamma) > p - 1$ show that

$$\tilde{G}_{1,2}^{1,1}\left(\tilde{X}\Big|_{\beta,\gamma}^{\alpha}\right) = \frac{\tilde{\Gamma}_p(p + \beta - \alpha)}{\tilde{\Gamma}_p(p + \beta - \gamma)}|\det(\tilde{X})|^{\beta}$$
$$\times {}_1\tilde{F}_1^*(p + \beta - \alpha; p + \beta - \gamma; -\tilde{X}).$$

6.3.3   For $\tilde{X} = \tilde{X}^* > 0$, $\text{Re}(a+c_1) > p-1$, $\text{Re}(a+c_2) > p-1$, $\text{Re}(a+b) > p - 1$, show that

$$\tilde{G}_{2,2}^{1,2}\left(\tilde{X}\Big|_{a-p,-b}^{-c_1,-c_2}\right) = \frac{\tilde{\Gamma}_p(a + c_1)\tilde{\Gamma}_p(a + c_2)}{\tilde{\Gamma}_p(a + b)}|\det(\tilde{X})|^{a-p}$$
$$\times {}_2\tilde{F}_1^*(a + c_1, a + c_2; a + b; -\tilde{X}).$$

6.3.4   For finite p, show that

$$\lim_{a \to \infty} \frac{a^{p\rho}\tilde{\Gamma}_p(a - \rho)}{\tilde{\Gamma}_p(a)} = 1 \text{ for finite } \rho; \qquad (i)$$

$$\lim_{a \to \infty} \frac{1}{\tilde{\Gamma}_p(a)}\tilde{G}_{1,2}^{1,1}\left(\frac{1}{a}\tilde{X}\Big|_{\beta,p+\beta-b}^{p+\beta-a}\right) = \frac{{}_0\tilde{F}_1^*(\ ; b; -\tilde{X})}{\tilde{\Gamma}_p(b)}. \qquad (ii)$$

6.3.5   For finite p, show that

$$\lim_{a \to \infty} \frac{1}{\tilde{\Gamma}_p(p + a)}\tilde{G}_{2,2}^{1,2}\left(\frac{1}{p+a}\tilde{X}\Big|_{0,-b}^{-a,-c}\right)$$

$$= \frac{\tilde{\Gamma}_p(p + c)}{\tilde{\Gamma}_p(p + b)}{}_1\tilde{F}_1^*(p + c; p + b; -\tilde{X}).$$

6.4   Functions of several matrix arguments: complex case

Here we consider functions of several hermitian positive definite matrices. The theory is parallel to that in Chapter 5 for the real case. Hence we will only define the functions of Dirichlet, Appell, Kampé de Fériet, and Lauricella types. A few illustrative examples will be given to show the parallel, and some properties are listed in the exercises at the end of this section.

6.4.1   Dirichlet functions: complex case

Let Ω denote the set of hermitian positive definite matrices $\tilde{X}_1, ..., \tilde{X}_k$ such that $0 < \tilde{X}_j < I$, $j = 1, ..., k$ and $0 < \tilde{X}_1 + ... + \tilde{X}_k < I$. Then the *type-1 Dirichlet integral* is given by

$$\int_\Omega \left\{ \prod_{j=1}^k |\det(\tilde{X}_j)|^{\alpha_j - p} \right\} |\det(I - \tilde{X}_1 - ... - \tilde{X}_k)|^{\alpha_{k+1} - p} d\tilde{X}_1 ... d\tilde{X}_k$$

$$= C_k \qquad (6.4.1)$$

for $\text{Re}(\alpha_j) > p - 1$, $j = 1, ..., k + 1$ where

$$C_k = \frac{\prod_{j=1}^{k+1} \tilde{\Gamma}_p(\alpha_j)}{\tilde{\Gamma}_p(\alpha_1 + ... + \alpha_{k+1})}. \qquad (6.4.2)$$

A *type-1 Dirichlet density* in the complex case is defined by

$$f(\tilde{X}_1, ..., \tilde{X}_k) = C_k^{-1} \left\{ \prod_{j=1}^k |\det(\tilde{X}_j)|^{\alpha_j - p} \right\}$$

$$\times |\det(I - \tilde{X}_1 - ... - \tilde{X}_k)|^{\alpha_{k+1} - p} \qquad (6.4.3)$$

for $\tilde{X}_1, ..., \tilde{X}_k$ defined in Ω and $f(\tilde{X}_1, ..., \tilde{X}_k) = 0$ elsewhere, where the C_k is given in (6.4.2). Note that for $k = 1$ one has the type-1 beta density in the complex case.

For $\tilde{X}_j = \tilde{X}_j^* > 0$, $j = 1, ..., k$ a *type-2 Dirichlet integral* is defined as follows:

$$\int_{\tilde{X}_1 > 0} ... \int_{\tilde{X}_k > 0} \left\{ \prod_{j=1}^k |\det(\tilde{X}_j)|^{\alpha_j - p} \right\} |\det(I + \tilde{X}_1 + ... + \tilde{X}_k)|^{-(\alpha_1 + ... + \alpha_{k+1})}$$

$$\times d\tilde{X}_1 ... d\tilde{X}_k = C_k \qquad (6.4.4)$$

for $\text{Re}(\alpha_j) > p - 1$, $j = 1, ..., k$ where C_k is given in (6.4.2). A *type-2 Dirichlet density* is given by

$$f(\tilde{X}_1, ..., \tilde{X}_k) = C_k^{-1} \left\{ \prod_{j=1}^{k} |\det(\tilde{X}_j)|^{\alpha_j - p} \right\}$$
$$\times |\det(I + \tilde{X}_1 + ... + \tilde{X}_k)|^{-(\alpha_1 + ... + \alpha_{k+1})} \qquad (6.4.5)$$

for $\tilde{X}_j = \tilde{X}_j^* > 0$, $\text{Re}(\alpha_j) > p - 1$, $j = 1, ..., k$ and $f(\tilde{X}_1, ..., \tilde{X}_k) = 0$ elsewhere.

For $k = 1$ one has the type-2 beta density in the complex case.

Example 6.8   If $\tilde{X}_1, ..., \tilde{X}_k$ have a type-1 Dirichlet density as in (6.4.3) show that $\tilde{U} = \tilde{X}_1 + ... + \tilde{X}_k$ and $\tilde{V} = I - \tilde{U}$ are type-1 beta distributed.

Solution   Make the transformation $\tilde{Y}_1 = \tilde{U} = \tilde{X}_1 + ... + \tilde{X}_k$, $\tilde{Y}_j = \tilde{X}_j$, $j = 2, ..., k$. The Jacobian is unity. The joint density of $\tilde{Y}_j$, $j = 1, ..., k$, denoted by $g(\tilde{Y}_1, ..., \tilde{Y}_k)$, is given by

$$g(\tilde{Y}_1, ..., \tilde{Y}_k) = C_k^{-1} |\det(I - \tilde{U})|^{\alpha_{k+1} - p} |\det(\tilde{U} - \tilde{Y}_2 - ... - \tilde{Y}_k)|^{\alpha_1 - p}$$
$$\times |\det(\tilde{Y}_2)|^{\alpha_2 - p} ... |\det(\tilde{Y}_k)|^{\alpha_k - p}. \qquad (a)$$

But

$$|\det(\tilde{U} - \tilde{Y}_2 - ... - \tilde{Y}_k)|^{\alpha_1 - p} = |\det(\tilde{U})|^{\alpha_1 - p}$$
$$\times \left| \det \left(I - \tilde{U}^{-\frac{1}{2}} \tilde{Y}_2 \tilde{U}^{-\frac{1}{2}} - ... - \tilde{U}^{-\frac{1}{2}} \tilde{Y}_k \tilde{U}^{-\frac{1}{2}} \right) \right|^{\alpha_1 - p}.$$

Let

$$\tilde{U}_j = \tilde{U}^{-\frac{1}{2}} \tilde{Y}_j \tilde{U}^{-\frac{1}{2}} \Rightarrow d\tilde{Y}_j = |\det(\tilde{U} \tilde{U}^*)|^{\frac{p}{2}} d\tilde{U}_j, \quad j = 2, ..., k.$$

Substituting in (a) one has

$$g(\tilde{U}, \tilde{Y}_2, ..., \tilde{Y}_k) d\tilde{U} d\tilde{Y}_2 ... d\tilde{Y}_k = C_k^{-1} |\det(I - \tilde{U})|^{\alpha_{k+1} - p}$$
$$\times |\det(\tilde{U})|^{\alpha_1 + ... + \alpha_k - p} |\det(\tilde{U}_2)|^{\alpha_2 - p} ...$$
$$\times |\det(\tilde{U}_k)|^{\alpha_k - p} |\det(I - \tilde{U}_2 - ... - \tilde{U}_k)|^{\alpha_1 - p}$$
$$\times d\tilde{U} d\tilde{U}_2 ... d\tilde{U}_k.$$

The marginal density of $\tilde{U}$ is available by integrating out $\tilde{U}_2, ..., \tilde{U}_k$ and this integral gives a constant. The explicit form is available from (6.4.1). But we do not need the explicit form here. Thus the density of $\tilde{U}$, denoted by $g_1(\tilde{U})$ is given by

$$g_1(\tilde{U}) = c \,|\det(\tilde{U})|^{\alpha_1 + \dots + \alpha_k - p}|\det(I - \tilde{U})|^{\alpha_{k+1} - p}$$

for $0 < \tilde{U} < I$ and hence the normalizing constant

$$c = \frac{\tilde{\Gamma}_p(\alpha_1 + \dots + \alpha_{k+1})}{\tilde{\Gamma}_p(\alpha_1 + \dots + \alpha_k)\tilde{\Gamma}_p(\alpha_{k+1})}.$$

This establishes the result for $\tilde{U}$. It is trivial to note that $\tilde{V} = I - \tilde{U}$ is also a type-1 beta with the parameters interchanged.

Example 6.9   Establish the result in (6.4.4) by direct integration.

Solution   Integration over $\tilde{X}_1$, denoted by $I_{\tilde{X}_1}$, yields

$$I_{\tilde{X}_1} = \int_{\tilde{X}_1} |\det(\tilde{X}_1)|^{\alpha_1 - p}|\det(I + \tilde{X}_1 + \dots + \tilde{X}_k)|^{-(\alpha_1 + \dots + \alpha_{k+1})} d\tilde{X}_1$$

$$= |\det(\tilde{U})|^{-(\alpha_1 + \dots + \alpha_{k+1})} \int_{\tilde{X}_1} |\det(\tilde{X}_1)|^{\alpha_1 - p}$$

$$\times \left|\det\left(I + \tilde{U}^{-\frac{1}{2}}\tilde{X}_1\tilde{U}^{-\frac{1}{2}}\right)\right|^{-(\alpha_1 + \dots + \alpha_{k+1})} d\tilde{X}_1,$$

where $\tilde{U} = I + \tilde{X}_2 + \dots + \tilde{X}_k$. Put

$$\tilde{Y}_1 = \tilde{U}^{-\frac{1}{2}}\tilde{X}_1\tilde{U}^{-\frac{1}{2}} \Rightarrow d\tilde{Y}_1 = |\det(\tilde{U}\tilde{U}^*)|^{-\frac{p}{2}}d\tilde{X}_1.$$

Then

$$I_{\tilde{X}_1} = |\det(\tilde{U})|^{-(\alpha_2 + \dots + \alpha_{k+1})} \int_{\tilde{U}} |\det(\tilde{U})|^{\alpha_1 - p}|\det(I + \tilde{U})|^{-(\alpha_1 + \dots + \alpha_{k+1})} d\tilde{U}$$

$$= |\det(\tilde{U})|^{-(\alpha_2 + \dots + \alpha_{k+1})} \frac{\tilde{\Gamma}_p(\alpha_1)\tilde{\Gamma}_p(\alpha_2 + \dots + \alpha_{k+1})}{\tilde{\Gamma}_p(\alpha_1 + \dots + \alpha_{k+1})},$$

for $\mathrm{Re}(\alpha_1) > p - 1$, $\mathrm{Re}(\alpha_2 + \dots + \alpha_{k+1}) > p - 1$ by evaluating the integral using a type-2 beta integral. Now the integration of $\tilde{X}_2$ gives the gamma product

$$\frac{\tilde{\Gamma}_p(\alpha_2)\tilde{\Gamma}_p(\alpha_3 + \dots + \alpha_{k+1})}{\tilde{\Gamma}_p(\alpha_2 + \dots + \alpha_{k+1})}, \quad \mathrm{Re}(\alpha_2) > p - 1, \ \mathrm{Re}(\alpha_3 + \dots + \alpha_{k+1}) > p - 1.$$

Integrating out the variables like this the result follows and the conditions reduce to $\mathrm{Re}(\alpha_j) > p - 1$, $j = 1, ..., k + 1$.

A result which will be useful in handling functions of several matrix arguments is the M-transform for a hypergeometric function with the argument being a sum of hermitian positive definite matrices. This will be stated as a theorem.

Theorem 6.4   *Let* $\tilde{S} = \tilde{X}_1 + ... + \tilde{X}_k$ *where* $\tilde{X}_j$, $j = 1, ..., k$ *be* $p \times p$ *hermitian positive definite matrices. Let the M-transform of a function* f *of* k *matrices be denoted by* $M(f) = M_{\rho_1 ... \rho_k}(f)$. *Then writing*

$$_r\tilde{F}_s^*(-\tilde{S}) = {}_r\tilde{F}_s^*(a_1, ..., a_r; b_1, ..., b_s; -\tilde{S}),$$

$$M(_r\tilde{F}_s^*(-\tilde{S})) = \int_{\tilde{X}_1 > 0} \cdots \int_{\tilde{X}_k > 0} |\det(\tilde{X}_1)|^{\rho_1 - p} ... |\det(\tilde{X}_k)|^{\rho_k - p}$$

$$\times {}_r\tilde{F}_s^*(a_1, ..., a_r; b_1, ..., b_s; -\tilde{S}) \mathrm{d}\tilde{X}_1 ... \mathrm{d}\tilde{X}_k$$

$$= \frac{\left\{ \prod_{j=1}^{s} \tilde{\Gamma}_p(b_j) \right\}}{\left\{ \prod_{j=1}^{r} \tilde{\Gamma}_p(a_j) \right\}} \left\{ \prod_{j=1}^{k} \tilde{\Gamma}_p(\rho_j) \right\}$$

$$\times \frac{\left\{ \prod_{j=1}^{r} \tilde{\Gamma}_p(a_1 - \rho_1 - ... - \rho_k) \right\}}{\left\{ \prod_{j=1}^{s} \tilde{\Gamma}_p(b_j - \rho_1 - ... - \rho_k) \right\}}$$

for $\mathrm{Re}(\rho_j) > p - 1$, $\mathrm{Re}(a_j - \rho_1 - ... - \rho_k) > p - 1$, $j = 1, ..., r$, $\mathrm{Re}(b_j - \rho_1 - ... - \rho_k) > p - 1$, $j = 1, ..., s$.

Proof

$$\tilde{X}_1 = \tilde{S} - \tilde{X}_2 - ... - \tilde{X}_k$$
$$= \tilde{S}^{\frac{1}{2}} \left[I - \tilde{S}^{-\frac{1}{2}} \tilde{X}_2 \tilde{S}^{-\frac{1}{2}} - ... - \tilde{S}^{-\frac{1}{2}} \tilde{X}_k \tilde{S}^{-\frac{1}{2}} \right] \tilde{S}^{\frac{1}{2}}.$$

Put $\tilde{Y}_j = \tilde{S}^{-\frac{1}{2}} \tilde{X}_j \tilde{S}^{-\frac{1}{2}}$ for fixed $\tilde{S}$. Then

$$\mathrm{d}\tilde{Y}_j = |\det(\tilde{S}\tilde{S}^*)|^{-\frac{p}{2}} \mathrm{d}\tilde{X}_j, \quad j = 2, ..., k$$

and

$$\tilde{X}_1 = \tilde{S}^{\frac{1}{2}} [I - \tilde{Y}_2 - ... - \tilde{Y}_k] \tilde{S}^{\frac{1}{2}}.$$

Note that $\mathrm{d}\tilde{X}_1 = \mathrm{d}\tilde{S}$ for fixed $\tilde{X}_2, ..., \tilde{X}_k$. Thus we have

$$M(_r\tilde{F}_s^*(-\tilde{S})) = \int_{\tilde{S} > 0} |\det(\tilde{S})|^{\rho_1 + ... + \rho_k - p} {}_r\tilde{F}_s^*(-\tilde{S}) \mathrm{d}\tilde{S}$$

$$\times \int ... \int |\det(\tilde{Y}_2)|^{\rho_2 - p} ... |\det(\tilde{Y}_k)|^{\rho_k - p}$$
$$\times |\det(I - \tilde{Y}_2 - ... - \tilde{Y}_k)|^{\rho_1 - p} d\tilde{Y}_2 ... d\tilde{Y}_k.$$

Evaluating the $\tilde{Y}_j$-integrals by using (6.4.1), one has

$$\left\{ \prod_{j=1}^{k} \tilde{\Gamma}_p(\rho_j) \right\} / \tilde{\Gamma}_p(\rho_1 + ... + \rho_k) \qquad (a)$$

for $\text{Re}(\rho_j) > p - 1$, $j = 1, ..., k$. Evaluating the $\tilde{S}$-integral by using the M-transform of a hypergeometric function given in (6.2.5) one has

$$\frac{\left\{ \prod_{j=1}^{s} \tilde{\Gamma}_p(b_j) \right\} \left\{ \prod_{j=1}^{r} \tilde{\Gamma}_p(a_j - \rho_1 - ... - \rho_k) \right\}}{\left\{ \prod_{j=1}^{r} \tilde{\Gamma}_p(a_j) \right\} \left\{ \prod_{j=1}^{s} \tilde{\Gamma}_p(b_j - \rho_1 - ... \rho_k) \right\}} \tilde{\Gamma}_p(\rho_1 + ... + \rho_k). \qquad (b)$$

From (a) and (b) the result follows.

6.4.2 Appell's functions: complex case

We will use the notations $\tilde{F}_1, \tilde{F}_2, \tilde{F}_3, \tilde{F}_4$ to distinguish them from the real case discussed in Section 5.3 of Chapter 5. Let

$$\tilde{F}_1 = \tilde{F}_1(a, b, b'; c; -\tilde{X}_1, -\tilde{X}_2)$$
$$\tilde{F}_2 = \tilde{F}_2(a, b, b'; c, c'; -\tilde{X}_1, -\tilde{X}_2)$$
$$\tilde{F}_3 = \tilde{F}_3(a, a', b, b'; c; -\tilde{X}_1, -\tilde{X}_2)$$
$$\tilde{F}_4 = \tilde{F}_4(a, b; c, c'; -\tilde{X}_1, -\tilde{X}_2).$$

These will be defined in terms of their M-transforms as those classes of symmetric functions having the following M-transforms. Here $M(\cdot) = M_{\rho_1 \rho_2}(\cdot)$.

$$M(\tilde{F}_1) = \frac{\tilde{\Gamma}_p(c)}{\tilde{\Gamma}_p(a)\tilde{\Gamma}_p(b)\tilde{\Gamma}_p(b')} \frac{\tilde{\Gamma}_p(\rho_1)\tilde{\Gamma}_p(\rho_2)}{\tilde{\Gamma}_p(c - \rho_1 - \rho_2)}$$
$$\times \tilde{\Gamma}_p(a - \rho_1 - \rho_2)\tilde{\Gamma}_p(b - \rho_1)\tilde{\Gamma}_p(b' - \rho_2) \qquad (6.4.6)$$

for $\text{Re}(\rho_1, \rho_2, a - \rho_1 - \rho_2, b - \rho_1, b' - \rho_2, c - \rho_1 - \rho_2) > p - 1$ which means the real part of each item in the brackets is greater than $p - 1$.

$$M(\tilde{F}_2) = \frac{\tilde{\Gamma}_p(c)\tilde{\Gamma}_p(c')}{\tilde{\Gamma}_p(a)\tilde{\Gamma}_p(b)\tilde{\Gamma}_p(b')} \frac{\tilde{\Gamma}_p(\rho_1)\tilde{\Gamma}_p(\rho_2)}{\tilde{\Gamma}_p(c - \rho_1)\tilde{\Gamma}_p(c' - \rho_2)}$$
$$\times \tilde{\Gamma}_p(b - \rho_1)\tilde{\Gamma}_p(b' - \rho_2)\tilde{\Gamma}_p(a - \rho_1 - \rho_2) \qquad (6.4.7)$$

for $\text{Re}(\rho_1, \rho_2, a - \rho_1 - \rho_2, b - \rho_1, b' - \rho_2, c - \rho_1, c' - \rho_2) > p - 1$.

$$M(\tilde{F}_3) = \frac{\tilde{\Gamma}_p(c)}{\tilde{\Gamma}_p(a)\tilde{\Gamma}_p(a')\tilde{\Gamma}_p(b)\tilde{\Gamma}_p(b')} \frac{\tilde{\Gamma}_p(\rho_1)\tilde{\Gamma}_p(\rho_2)}{\tilde{\Gamma}_p(c - \rho_1 - \rho_2)}$$
$$\times \tilde{\Gamma}_p(a - \rho_1)\tilde{\Gamma}_p(a' - \rho_2)\tilde{\Gamma}_p(b - \rho_1)\tilde{\Gamma}_p(b' - \rho_2) \qquad (6.4.8)$$

for $\text{Re}(\rho_1, \rho_2, a - \rho_1, a' - \rho_2, b - \rho_1, b' - \rho_2, c - \rho_1 - \rho_2) > p - 1$.

$$M(\tilde{F}_4) = \frac{\tilde{\Gamma}_p(c)\tilde{\Gamma}_p(c')}{\tilde{\Gamma}_p(a)\tilde{\Gamma}_p(b)} \frac{\tilde{\Gamma}_p(a - \rho_1 - \rho_2)}{\tilde{\Gamma}_p(c - \rho_1)\tilde{\Gamma}_p(c' - \rho_2)}$$
$$\times \tilde{\Gamma}_p(b - \rho_1 - \rho_2)\tilde{\Gamma}_p(\rho_1)\tilde{\Gamma}_p(\rho_2) \qquad (6.4.9)$$

for $\text{Re}(\rho_1, \rho_2, a - \rho_1 - \rho_2, b - \rho_1 - \rho_2, c - \rho_1, c' - \rho_2) > p - 1$.

Example 6.10   For $\text{Re}(b, b', c - b - b') > p - 1$, $\tilde{X}_j = \tilde{X}_j^* > 0$, $j = 1, 2$, show that

$$\tilde{F}_1 = \tilde{F}_1(a, b, b'; c; -\tilde{X}_1, -\tilde{X}_2)$$
$$= \frac{\tilde{\Gamma}_p(c)}{\tilde{\Gamma}_p(b)\tilde{\Gamma}_p(b')\tilde{\Gamma}_p(c - b - b')}$$
$$\times \int \int |\det(\tilde{U}_1)|^{b-p}|\det(\tilde{U}_2)|^{b'-p}|\det(I - \tilde{U}_1 - \tilde{U}_2)|^{c-b-b'-p}$$
$$\times \left|\det\left(I + \tilde{U}_1^{\frac{1}{2}}\tilde{X}_1\tilde{U}_1^{\frac{1}{2}} + \tilde{U}_2^{\frac{1}{2}}\tilde{X}_2\tilde{U}_2^{\frac{1}{2}}\right)\right|^{-a} d\tilde{U}_1 d\tilde{U}_2. \qquad (6.4.10)$$

Proof   Take the M-transform of the right side with respect to $\tilde{X}_1$ and $\tilde{X}_2$ and with parameters ρ_1 and ρ_2. Then

$$\int_{\tilde{X}_1 > 0} \int_{\tilde{X}_2 > 0} |\det(\tilde{X}_1)|^{\rho_1 - p}|\det(\tilde{X}_2)|^{\rho_2 - p}$$
$$\times \left|\det\left(I + \tilde{U}_1^{\frac{1}{2}}\tilde{X}_1\tilde{U}_1^{\frac{1}{2}} + \tilde{U}_2^{\frac{1}{2}}\tilde{X}_2\tilde{U}_2^{\frac{1}{2}}\right)\right|^{-a} d\tilde{X}_1 d\tilde{X}_2$$
$$= |\det(\tilde{U}_1)|^{-\rho_1}|\det(\tilde{U}_2)|^{-\rho_2} \frac{\tilde{\Gamma}_p(\rho_1)\tilde{\Gamma}_p(\rho_2)}{\tilde{\Gamma}_p(a)} \tilde{\Gamma}_p(a - \rho_1 - \rho_2)$$

for $\text{Re}(\rho_1, \rho_2, a - \rho_1 - \rho_2) > p - 1$ by evaluating it with the help of the type-2 Dirichlet integral (6.4.4) after making the transformation $\tilde{Y}_j = \tilde{U}_j^{\frac{1}{2}}\tilde{X}_j\tilde{U}_j^{\frac{1}{2}}$ for fixed $\tilde{U}_j$, $j = 1, 2$. Now integrate out $\tilde{U}_1$ and $\tilde{U}_2$ by using a type-1

Dirichlet integral of (6.4.1) to see that the M-transform agrees with $M(\tilde{F}_1)$ of (6.4.6).

Example 6.11   For $\mathrm{Re}(a, c-a) > p - 1$, $\tilde{X} = \tilde{X}^* > 0$, $j = 1, 2$ show that

$$\tilde{F}_1 = \tilde{F}_1(a, b, b'; c; -\tilde{X}_1, -\tilde{X}_2)$$

$$= \frac{\tilde{\Gamma}_p(c)}{\tilde{\Gamma}_p(a)\tilde{\Gamma}_p(c-a)} \int_0^I |\det(\tilde{U})|^{a-p}|\det(I - \tilde{U})|^{c-a-p}$$

$$\times \left|\det\left(I + \tilde{U}^{\frac{1}{2}}\tilde{X}_1\tilde{U}^{\frac{1}{2}}\right)\right|^{-b}$$

$$\times \left|\det\left(I + \tilde{U}^{\frac{1}{2}}\tilde{X}_2\tilde{U}^{\frac{1}{2}}\right)\right|^{-b'} d\tilde{U}. \qquad (6.4.11)$$

Solution   Take the M-transform of the right side with respect to $\tilde{X}_1$ and $\tilde{X}_2$ and with parameters ρ_1 and ρ_2. Then integrate out $\tilde{U}$ by using a type-1 beta integral to obtain the result.

One can also use some integral representations as definitions for these functions and then derive the M-transforms discussed above as well as other properties from those integrals as definitions. Observe that the functions defined through such integrals will be unique. To illustrate this point let us consider the following. Take the integral representation of $\tilde{F}_1$ from (6.4.11) as the definition of $\tilde{F}_1$. Then $\tilde{F}_1$ is a unique function. We will establish the following result by using the M-transform as well as by taking (6.4.11) as the definition of $\tilde{F}_1$.

Theorem 6.5   For $\mathrm{Re}(d, c - d) > p - 1$

$$\tilde{F}_1 = \tilde{F}_1(a, b, b'; c; -\tilde{X}_1, -\tilde{X}_2)$$

$$= \frac{\tilde{\Gamma}_p(c)}{\tilde{\Gamma}_p(d)\tilde{\Gamma}_p(c-d)} \int_0^I |\det(\tilde{U})|^{d-p}|\det(I - \tilde{U})|^{c-d-p}$$

$$\times \tilde{F}_1\left(a, b, b'; d; -\tilde{U}^{\frac{1}{2}}\tilde{X}_1\tilde{U}^{\frac{1}{2}}, -\tilde{U}^{\frac{1}{2}}\tilde{X}_2\tilde{U}^{\frac{1}{2}}\right) d\tilde{U}.$$

Proof   By using M-transform the result is easily established by taking the M-transform of the right side with respect to $\tilde{X}_1$ and $\tilde{X}_2$ and with parameters ρ_1 and ρ_2. In this case put

$$\tilde{U}^{\frac{1}{2}}\tilde{X}_j\tilde{U}^{\frac{1}{2}} = \tilde{Y}_j \Rightarrow d\tilde{X}_j = |\det(\tilde{U}\tilde{U}^*)|^{-\frac{p}{2}}d\tilde{X}_j, \; j = 1, 2$$

for fixed $\tilde{U}$. Then integrate out $\tilde{U}$ by using a type-1 beta integral to see the result.

For proving the result taking (6.4.11) as the definition of $\tilde{F}_1$ replace the $\tilde{F}_1$ inside the integral as follows:

$$\tilde{F}_1\left(a, b, b'; d; -\tilde{U}^{\frac{1}{2}}\tilde{X}_1\tilde{U}^{\frac{1}{2}}, -\tilde{U}^{\frac{1}{2}}\tilde{X}_2\tilde{U}^{\frac{1}{2}}\right)$$

$$= \frac{\tilde{\Gamma}_p(d)}{\tilde{\Gamma}_p(a)\tilde{\Gamma}_p(d-a)} \int_0^I |\det(\tilde{V})|^{a-p}|\det(I-\tilde{V})|^{d-a-p}$$

$$\times \left|\det\left(I + \tilde{V}^{\frac{1}{2}}\left(\tilde{U}^{\frac{1}{2}}\tilde{X}_1\tilde{U}^{\frac{1}{2}}\right)\tilde{V}^{\frac{1}{2}}\right)\right|^{-b}$$

$$\times \left|\det\left(I + \tilde{V}^{\frac{1}{2}}\left(\tilde{U}^{\frac{1}{2}}\tilde{X}_2\tilde{U}^{\frac{1}{2}}\right)\tilde{V}^{\frac{1}{2}}\right)\right|^{-b'} d\tilde{V}.$$

Note that

$$\left|\det\left(I + \tilde{V}^{\frac{1}{2}}\left(\tilde{U}^{\frac{1}{2}}\tilde{X}_j\tilde{U}^{\frac{1}{2}}\right)\tilde{V}^{\frac{1}{2}}\right)\right| = \left|\det\left(I + \tilde{U}^{\frac{1}{2}}\tilde{V}\tilde{U}^{\frac{1}{2}}\tilde{X}_j\right)\right|.$$

Let

$$\tilde{W} = \tilde{U}^{\frac{1}{2}}\tilde{V}\tilde{U}^{\frac{1}{2}} \Rightarrow d\tilde{W} = |\det(\tilde{U}\tilde{U}^*)|^{\frac{p}{2}}d\tilde{V}$$

for fixed $\tilde{U}$. Collect the factors containing $\tilde{U}$. Under this substitution the right side is the following:

$$\text{right side} = \frac{\tilde{\Gamma}_p(c)}{\tilde{\Gamma}_p(a)\tilde{\Gamma}_p(c-d)\tilde{\Gamma}_p(d-a)} \int_{\tilde{U}} \int_{\tilde{W}} |\det((I-\tilde{U})|^{c-d-p}$$

$$\times |\det(\tilde{W})|^{a-p}|\det(\tilde{U}-\tilde{W})|^{d-a-p}$$

$$\times |\det(I + \tilde{W}\tilde{X}_1)|^{-b}|\det(I + \tilde{W}\tilde{X}_2)|^{-b'}d\tilde{U}d\tilde{W}.$$

Note that $0 < \tilde{W} < \tilde{U} < I$. Change $\tilde{U}$ to $\tilde{Z}$ for fixed $\tilde{W}$ by $\tilde{U} - \tilde{W} = \tilde{Z}$. The factors containing $\tilde{Z}$ are given by

$$|\det(I - \tilde{Z} - \tilde{W})|^{c-d-p}|\det(\tilde{Z})|^{d-a-p}$$

$$= |\det(I - \tilde{W})|^{c-a-p}|\det(\tilde{T})|^{d-a-p}|\det(I-\tilde{T})|^{c-d-p}$$

for

$$\tilde{T} = (I - \tilde{W})^{-\frac{1}{2}}\tilde{Z}(I - \tilde{W})^{-\frac{1}{2}}$$

with fixed $\tilde{W}$. Now integrate out $\tilde{T}$ to get

$$\frac{\tilde{\Gamma}_p(d-a)\tilde{\Gamma}_p(c-d)}{\tilde{\Gamma}_p(c-a)}.$$

Now the right side reduces to an integral which is $\tilde{F}_1(a, b, b'; c; -\tilde{X}_1, -\tilde{X}_2)$ from (6.4.11) and hence the result.

Thus the unique function $\tilde{F}_1$ defined through the integral (6.4.11) is also in the class of functions defined through the M-transform (6.4.6). In practical situations it is easier to handle through M-transform rather than through integral representations.

6.4.3 Series forms: complex case

If series representations are needed for $\tilde{F}_1, \tilde{F}_2, \tilde{F}_3, \tilde{F}_4$ then these are not yet worked out from their M-transforms. For getting series forms start with convenient integral representations. For example, start with (6.4.11) as a definition of $\tilde{F}_1$. Thus $\tilde{F}_1(a, b, b'; c; \tilde{X}_1, \tilde{X}_2)$ has the factors

$$\left|\det\left(I - \tilde{U}^{\frac{1}{2}}\tilde{X}_j\tilde{U}^{\frac{1}{2}}\right)\right| = |\det(I - \tilde{U}\tilde{X}_j)|$$

for $j = 1, 2$. Expand by using the zonal polynomial expansion in (6.1.19). That is,

$$|\det(I - \tilde{U}\tilde{X}_1)|^{-b} = \sum_{k=0}^{\infty} \sum_{K} \frac{[b]_K}{k!} \tilde{C}_K(\tilde{U}\tilde{X}_1), \quad \|\tilde{X}_1\| < 1$$

and

$$|\det(I - \tilde{U}\tilde{X}_2)|^{-b'} = \sum_{m=0}^{\infty} \sum_{M} \frac{[b']_M}{m!} \tilde{C}_M(\tilde{U}\tilde{X}_2), \quad \|\tilde{X}_2\| < 1$$

where K and M are partitions of the nonnegative integers k and m respectively,

$$K = (k_1, ..., k_p), \quad k_1 \geq ... \geq k_p \geq 0, \quad k_1 + ... + k_p = k;$$
$$M = (m_1, ..., m_p), \quad m_1 \geq ... \geq m_p \geq 0, \quad m_1 + ... + m_p = m$$

and the generalized hypergeometric coefficients $[\cdot]_{(\cdot)}$ and zonal polynomials $\tilde{C}_{(\cdot)}(\cdot)$ are defined in (6.1.10) and (6.1.12) respectively. Now $\tilde{F}_1$ can be written as the following multiple series:

$$\tilde{F}_1 = \tilde{F}_1(a, b, b'; c'; \tilde{X}_1, \tilde{X}_2)$$
$$= \frac{\tilde{\Gamma}_p(c)}{\tilde{\Gamma}_p(a)\tilde{\Gamma}_p(c-a)} \sum_{k=0}^{\infty} \sum_{m=0}^{\infty} \sum_{K} \sum_{M} \frac{[b]_K}{k!} \frac{[b']_M}{m!}$$
$$\times \int_{0 < \tilde{U} < I} |\det(\tilde{U})|^{a-p} |\det(I - \tilde{U})|^{c-a-p}$$
$$\times \tilde{C}_K(\tilde{U}\tilde{X}_1)\tilde{C}_M(\tilde{U}\tilde{X}_2)d\tilde{U}. \tag{6.4.12}$$

This integral over $\tilde{U}$ can be evaluated in terms of invariant polynomials involving the eigenvalues of $\tilde{X}_1$ and $\tilde{X}_2$. Even though one can express $\tilde{F}_1$ as multiple series involving invariant polynomials numerical computations will be difficult. Hence we will not attempt to obtain series forms for Appell's functions as well as for the other functions to be defined in the next sections. For invariant polynomials involving several matrices see Davis (1981). Applications of these in econometric theory may be seen from Chikuse and Davis (1986). A large number of papers are available on the applications of zonal polynomials and invariant polynomials in statistical theory; some of these are listed in Mathai, Provost and Hayakawa (1995).

6.4.4   Humbert's functions: complex case

We will define these in terms of their M-transforms as the classes of symmetric functions having the following M-transforms. M-transform will be denoted by $M(\cdot) = M_{\rho_1\rho_2}(\cdot)$ The following notations, parallel to the standard notations in the real case, will be used.

$$\tilde{\Phi}_1 = \tilde{\Phi}_1(a, b; c; -\tilde{X}_1, -\tilde{X}_2),$$
$$\tilde{\Phi}_2 = \tilde{\Phi}_2(b, b'; c; -\tilde{X}_1, -\tilde{X}_2),$$
$$\tilde{\Phi}_3 = \tilde{\Phi}_3(b; c; -\tilde{X}_1, -\tilde{X}_2),$$
$$\tilde{\Psi}_1 = \tilde{\Psi}_1(a, b; c, c'; -\tilde{X}_1, -\tilde{X}_2),$$
$$\tilde{\Psi}_2 = \tilde{\Psi}_2(a; c, c'; -\tilde{X}_1, -\tilde{X}_2),$$
$$\tilde{\Xi}_1 = \tilde{\Xi}_1(a, a', b; c; -\tilde{X}_1, -\tilde{X}_2).$$
$$\tilde{\Xi}_2 = \tilde{\Xi}_2(a, b; c; -\tilde{X}_1, \tilde{X}_2),$$

with the following M-transforms:

$$M(\tilde{\Phi}_1) = \frac{\tilde{\Gamma}_p(c)}{\tilde{\Gamma}_p(a)\tilde{\Gamma}_p(b)} \frac{\tilde{\Gamma}_p(a - \rho_1 - \rho_2)}{\tilde{\Gamma}_p(c - \rho_1 - \rho_2)}$$
$$\times \tilde{\Gamma}_p(b - \rho_1)\tilde{\Gamma}_p(\rho_1)\tilde{\Gamma}_p(\rho_2)$$

for $\mathrm{Re}(\rho_1, \rho_2, b - \rho_1, a - \rho_1 - \rho_2, c - \rho_1 - \rho_2) > p - 1$;

$$M(\tilde{\Phi}_2) = \frac{\tilde{\Gamma}_p(c)}{\tilde{\Gamma}_p(b)\tilde{\Gamma}_p(b')} \frac{\tilde{\Gamma}_p(b - \rho_1)}{\tilde{\Gamma}_p(c - \rho_1 - \rho_2)}$$
$$\times \tilde{\Gamma}_p(b' - \rho_2)\tilde{\Gamma}_p(\rho_1)\tilde{\Gamma}_p(\rho_2)$$

for $\mathrm{Re}(\rho_1, \rho_2, b - \rho_1, b' - \rho_2, c - \rho_1 - \rho_2) > p - 1$;

$$M(\tilde{\Phi}_3) = \frac{\tilde{\Gamma}_p(c)}{\tilde{\Gamma}_p(b)} \frac{\tilde{\Gamma}_p(b - \rho_1)\tilde{\Gamma}_p(\rho_1)\tilde{\Gamma}_p(\rho_2)}{\tilde{\Gamma}_p(c - \rho_1 - \rho_2)}$$

for $\mathrm{Re}(\rho_1, \rho_2, b - \rho_1, c - \rho_1 - \rho_2) > p - 1$;

$$M(\tilde{\Psi}_1) = \frac{\tilde{\Gamma}_p(c)\tilde{\Gamma}_p(c')}{\tilde{\Gamma}_p(a)\tilde{\Gamma}_p(b)} \frac{\tilde{\Gamma}_p(a - \rho_1 - \rho_2)}{\tilde{\Gamma}_p(c - \rho_1)\tilde{\Gamma}_p(c' - \rho_2)}$$
$$\times \tilde{\Gamma}_p(b - \rho_1)\tilde{\Gamma}_p(\rho_1)\tilde{\Gamma}_p(\rho_2)$$

for $\mathrm{Re}(\rho_1, \rho_2, b - \rho_1, c - \rho_1, c' - \rho_2, a - \rho_1 - \rho_2) > p - 1$;

$$M(\tilde{\Psi}_2) = \frac{\tilde{\Gamma}_p(c)\tilde{\Gamma}_p(c')}{\tilde{\Gamma}_p(a)} \frac{\tilde{\Gamma}_p(a - \rho_1 - \rho_2)\tilde{\Gamma}_p(\rho_1)\tilde{\Gamma}_p(\rho_2)}{\tilde{\Gamma}_p(c - \rho_1)\tilde{\Gamma}_p(c' - \rho_2)}$$

for $\mathrm{Re}(\rho_1, \rho_2, c - \rho_1, c' - \rho_2, a - \rho_1 - \rho_2) > p - 1$;

$$M(\tilde{\Xi}_1) = \frac{\tilde{\Gamma}_p(c)}{\tilde{\Gamma}_p(a)\tilde{\Gamma}_p(a')\tilde{\Gamma}_p(b)} \frac{\tilde{\Gamma}_p(a - \rho_1)\tilde{\Gamma}_p(a' - \rho_2)}{\tilde{\Gamma}_p(c - \rho_1 - \rho_2)}$$
$$\times \tilde{\Gamma}_p(b - \rho_1)\tilde{\Gamma}_p(\rho_1)\tilde{\Gamma}_p(\rho_2)$$

for $\mathrm{Re}(\rho_1, \rho_2, a - \rho_1, a' - \rho_2, c - \rho_1 - \rho_2) > p - 1$;

$$M(\tilde{\Xi}_2) = \frac{\tilde{\Gamma}_p(c)}{\tilde{\Gamma}_p(a)\tilde{\Gamma}_p(b)} \frac{\tilde{\Gamma}_p(a - \rho_1)}{\tilde{\Gamma}_p(c - \rho_1 - \rho_2)}$$
$$\times \tilde{\Gamma}_p(b - \rho_1)\tilde{\Gamma}_p(\rho_1)\tilde{\Gamma}_p(\rho_2)$$

for $\mathrm{Re}(\rho_1, \rho_2, a - \rho_1, b - \rho_1, c - \rho_1 - \rho_2) > p - 1$.

Example 6.12 Show that for $\mathrm{Re}(b, b', c - b - b') > p - 1$,

$$\tilde{\Phi}_2 = \tilde{\Phi}_2(b, b'; c; -\tilde{X}_1, -\tilde{X}_2)$$
$$= \frac{\tilde{\Gamma}_p(c)}{\tilde{\Gamma}_p(b)\tilde{\Gamma}_p(b')\tilde{\Gamma}_p(c - b - b')} \int \int |\det(\tilde{U}_1)|^{b-p}$$
$$\times |\det(\tilde{U}_2)|^{b'-p}|\det(I - \tilde{U}_1 - \tilde{U}_2)|^{c-b-b'-p}$$
$$\times \exp\left\{-\mathrm{tr}\left(\tilde{U}_1^{\frac{1}{2}}\tilde{X}_1\tilde{U}_1^{\frac{1}{2}} + \tilde{U}_2^{\frac{1}{2}}\tilde{X}_2\tilde{U}_2^{\frac{1}{2}}\right)\right\} d\tilde{U}_1 d\tilde{U}_2. \quad (6.4.13)$$

Solution Take the M-transform of the right side with respect to $\tilde{X}_1$ and $\tilde{X}_2$ and with parameters ρ_1 and ρ_2 to get

$$\int_{\tilde{X}_1>0} \int_{\tilde{X}_2>0} |\det(\tilde{X}_1)|^{\rho_1-p}|\det(\tilde{X}_2)|^{\rho_2-p}$$
$$\times \exp\left\{-\mathrm{tr}\left(\tilde{U}_1^{\frac{1}{2}}\tilde{X}_1\tilde{U}_1^{\frac{1}{2}} + \tilde{U}_2^{\frac{1}{2}}\tilde{X}_2\tilde{U}_2^{\frac{1}{2}}\right)\right\} d\tilde{X}_1 d\tilde{X}_2$$
$$= |\det(\tilde{U}_1)|^{-\rho_1}|\det(\tilde{U}_2)|^{-\rho_2}\tilde{\Gamma}_p(\rho_1)\tilde{\Gamma}_p(\rho_2), \quad (a)$$

for $\text{Re}(\rho_1) > p - 1$, $\text{Re}(\rho_2) > p - 1$. Now integrate out $\tilde{U}_1$ and $\tilde{U}_2$ by using a type-1 Dirichlet integral of (6.4.1) to get

$$\frac{\tilde{\Gamma}_p(c)}{\tilde{\Gamma}_p(b)\tilde{\Gamma}_p(b')\tilde{\Gamma}_p(c-b-b')} \frac{\tilde{\Gamma}_p(b-\rho_1)\tilde{\Gamma}_p(b'-\rho_2)\tilde{\Gamma}_p(c-b-b')}{\tilde{\Gamma}_p(c-\rho_1-\rho_2)}. \qquad (b)$$

From (a) and (b) the M-transform of the right side is the M-transform of $\tilde{\Phi}_1$ and hence the result.

Convenient integral representations such as the one in Example 6.12 could be used as definitions for the various Humbert's functions. Then these functions will be defined uniquely through such integral representations. The M-transform as well as other properties of Humbert's functions could be studied from such integral representations. Definitions through integral representations will be slightly difficult to handle unless one is prepared to go through series expansions such as the one in (6.4.12) at intermediate stages of the derivations of results on these functions. Such a technique is usually used in the scalar cases but when the arguments are matrices the series forms are complicated as seen from (6.4.12). Thus this approach of expanding certain functions and then carrying out integration will lead to various types of difficulties especially convergence problems and the difficulties of handling integrals involving invariant polynomials. The technique of M-transform is very simple but, in this case, one is establishing properties enjoyed by a class of functions and this class also contains the unique function defined through an integral representation.

Note that results parallel to the ones in Theorems 5.17–5.31 of Chapter 5 can be established in the complex case also. Before considering Kampé de Fériet's function in the complex case we will illustrate the point of using a convenient integral form as a definition and then deriving other results from such a definition. It is easy to see by using M-transform that the following two integrals can be established for $\tilde{\Phi}_2$:

$$\begin{aligned}
\tilde{\Phi}_2 &= \tilde{\Phi}_2(b, b'; c; -\tilde{X}_1, -\tilde{X}_2) \\
&= \frac{\tilde{\Gamma}_p(c)}{\tilde{\Gamma}_p(d)\tilde{\Gamma}_p(d')\tilde{\Gamma}_p(c-d-d')} \int \int |\det(\tilde{U}_1)|^{d-p} |\det(\tilde{U}_2)|^{d'-p} \\
&\quad \times |\det(I - \tilde{U}_1 - \tilde{U}_2)|^{c-d-d'-p} \\
&\quad \times {}_1\tilde{F}_1^* \left(b; d; -\tilde{U}_1^{\frac{1}{2}} \tilde{X}_1 \tilde{U}_1^{\frac{1}{2}} \right) \\
&\quad \times {}_1\tilde{F}_1^* \left(b'; d'; -\tilde{U}_2^{\frac{1}{2}} \tilde{X}_2 \tilde{U}_2^{\frac{1}{2}} \right) d\tilde{U}_1 d\tilde{U}_2; \qquad (6.4.14) \\
\tilde{\Phi}_2 &= \tilde{\Phi}_2(b, b'; c; -\tilde{X}_1, -\tilde{X}_2)
\end{aligned}$$

$$= \frac{\tilde{\Gamma}_p(c)}{\tilde{\Gamma}_p(b)\tilde{\Gamma}_p(b')\tilde{\Gamma}_p(c-b-b')} \int \int |\det(\tilde{U}_1)|^{b-p} |\det(\tilde{U}_2)|^{b'-p}$$

$$\times |\det(I - \tilde{U}_1 - \tilde{U}_2)|^{c-b-b'-p}$$

$$\times \exp\left\{ -\text{tr}\left(\tilde{U}_1^{\frac{1}{2}} \tilde{X}_1 \tilde{U}_1^{\frac{1}{2}} + \tilde{U}_2^{\frac{1}{2}} \tilde{X}_2 \tilde{U}_2^{\frac{1}{2}} \right) \right\} d\tilde{U}_1 d\tilde{U}_2. \tag{6.4.15}$$

Example 6.13 By taking the integral representation in (6.4.14) as a definition for $\tilde{\Phi}_2$ establish (6.4.15).

Solution We start with (6.4.14). Consider an integral representation for $_1\tilde{F}_1^*$. This is available from Example 6.3 or from equation (6.2.3). That is, for $\tilde{V}_1 = \tilde{V}_1^* > 0$,

$$_1\tilde{F}_1^* \left(b; d; -\tilde{U}_1^{\frac{1}{2}} \tilde{X}_1 \tilde{U}_1^{\frac{1}{2}} \right)$$

$$= \frac{\tilde{\Gamma}_p(d)}{\tilde{\Gamma}_p(b)\tilde{\Gamma}_p(d-b)} \int_0^I |\det(\tilde{V}_1)|^{b-p} |\det(I - \tilde{V}_1)|^{d-b-p}$$

$$\times e^{-\text{tr}\left(\tilde{V}_1^{\frac{1}{2}} \left(\tilde{U}_1^{\frac{1}{2}} \tilde{X}_1 \tilde{U}_1^{\frac{1}{2}} \right) \tilde{V}_1^{\frac{1}{2}} \right)} d\tilde{V}_1.$$

Put

$$\tilde{W}_1 = \tilde{U}_1^{\frac{1}{2}} \tilde{V}_1 \tilde{U}_1^{\frac{1}{2}} \Rightarrow d\tilde{W}_1 = |\det(\tilde{U}_1)|^p d\tilde{V}_1$$

for fixed $\tilde{U}_1$. Under this substitution factors containing $\tilde{V}_1$ and $\tilde{U}_1$ change to

$$|\det(\tilde{W}_1)|^{b-p} |\det(\tilde{U}_1 - \tilde{W}_1)|^{d-b-p} e^{-\text{tr}(\tilde{X}_1 \tilde{W}_1)}$$

$$\times |\det(I - \tilde{U}_1 - \tilde{U}_2)|^{c-d-d'-p}, \quad 0 < \tilde{W}_1 < \tilde{U}_1 < I.$$

Change $\tilde{U}_1$ to $\tilde{Z}_1$ by $\tilde{U}_1 - \tilde{W}_1 = \tilde{Z}_1$ for fixed $\tilde{W}_1$. Then $0 < \tilde{Z}_1 < I - \tilde{W}_1$ and the above factors become

$$|\det(\tilde{W}_1)|^{b-p} |\det(\tilde{Z}_1)|^{d-b-p} e^{-\text{tr}(\tilde{X}_1 \tilde{W}_1)}$$

$$\times |\det(I - \tilde{Z}_1 - \tilde{W}_1 - \tilde{U}_2)|^{c-d-d'-p}.$$

Note that

$$|\det(I - \tilde{Z}_1 - \tilde{W}_1 - \tilde{U}_2)|^{c-d-d'-p}$$

$$= |\det(I - \tilde{W}_1 - \tilde{U}_2)|^{c-d-d'-p}$$

$$\times \left| \det\left(I - \tilde{Z}_1 \left(I - \tilde{W}_1 - \tilde{U}_2 \right)^{-1} \right) \right|^{c-d-d'-p}$$

for $0 < \tilde{Z}_1 < I - \tilde{W}_1 - \tilde{U}_2$. Put

$$\tilde{T}_1 = (I - \tilde{W}_1 - \tilde{U}_2)^{-\frac{1}{2}} \tilde{Z}_1 (I - \tilde{W}_1 - \tilde{U}_2)^{-\frac{1}{2}} \text{ for fixed } \tilde{W}_1, \tilde{U}_2$$
$$\Rightarrow d\tilde{T}_1 = |\det(I - \tilde{W}_1 - \tilde{U}_2)|^{-p} d\tilde{Z}_1.$$

Observe that $0 < \tilde{T}_1 < I$. Integrate out $\tilde{T}_1$ to get

$$|\det(I - \tilde{W}_1 - \tilde{U}_2)|^{c-b-d'-p} \frac{\tilde{\Gamma}_p(d-b)\tilde{\Gamma}_p(c-d-d')}{\tilde{\Gamma}_p(c-b-d')}$$

for $\mathrm{Re}(d - b) > p - 1$, $\mathrm{Re}(c - d - d') > p - 1$. Note that the constant part reduces to

$$\frac{\tilde{\Gamma}_p(c)}{\tilde{\Gamma}_p(b)\tilde{\Gamma}_p(d')\tilde{\Gamma}_p(c-b-d')}$$

and the factors containing $\tilde{U}_2$ reduce to

$$|\det(I - \tilde{W}_1 - \tilde{U}_2)|^{c-b-d'-p} |\det(\tilde{U}_2)|^{d'-p}$$
$$\times {}_1\tilde{F}_1^* \left(b'; d'; -\tilde{U}_2^{\frac{1}{2}} \tilde{X}_2 \tilde{U}_2^{\frac{1}{2}} \right).$$

Go through similar transformations

$$\tilde{U}_2 \to \tilde{W}_2, \ \tilde{U}_2 \to \tilde{Z}_2, \ \tilde{Z}_2 \to \tilde{T}_2$$

and integrate out $\tilde{T}_2$ to see that the constant part reduces to

$$\frac{\tilde{\Gamma}_p(c)}{\tilde{\Gamma}_p(b)\tilde{\Gamma}_p(b')\tilde{\Gamma}_p(c-b-b')}$$

and the integral part over $\tilde{W}_1$ and $\tilde{W}_2$ agrees with that in (6.4.15) and hence the result.

6.4.5   Kampé de Fériet's functions: complex case

These will be denoted by

$$\tilde{F}_{s:m:n}^{r:q:k} = \tilde{F}_{s:m:n}^{r:q:k} \left[-\begin{pmatrix} \tilde{X}_1 \\ \tilde{X}_2 \end{pmatrix} \middle| \begin{matrix} (a_r) : (b_q) : (c_k) \\ (\alpha_s) : (\beta_m) : (\gamma_n) \end{matrix} \right] \quad (6.4.16)$$

and will be defined in terms of its M-transform as that class of symmetric functions having the following M-transform:

$$M(\tilde{F}) = \int_{\tilde{X}_1 = \tilde{X}_1^* > 0} \int_{\tilde{X}_2 = \tilde{X}_2^* > 0} |\det(\tilde{X}_1)|^{\rho_1 - p} |\det(\tilde{X}_2)|^{\rho_2 - p}$$

$$\times \tilde{F}_{s:m:n}^{r:q:k} \left[-\begin{pmatrix} \tilde{X}_1 \\ \tilde{X}_2 \end{pmatrix} \middle| \begin{matrix} (a_r) : (b_q) : (c_k) \\ (\alpha_s) : (\beta_m) : (\gamma_n) \end{matrix} \right] d\tilde{X}_1 d\tilde{X}_2$$

$$= C \frac{\left\{ \prod_{j=1}^{r} \tilde{\Gamma}_p(a_j - \rho_1 - \rho_2) \right\} \left\{ \prod_{j=1}^{q} \tilde{\Gamma}_p(b_j - \rho_1) \right\}}{\left\{ \prod_{j=1}^{s} \tilde{\Gamma}_p(\alpha_j - \rho_1 - \rho_2) \right\} \left\{ \prod_{j=1}^{m} \tilde{\Gamma}_p(\beta_j - \rho_1) \right\}}$$

$$\times \frac{\left\{ \prod_{j=1}^{k} \tilde{\Gamma}_p(c_j - \rho_2) \right\}}{\left\{ \prod_{j=1}^{n} \tilde{\Gamma}_p(\gamma_j - \rho_2) \right\}} \tilde{\Gamma}_p(\rho_1) \tilde{\Gamma}_p(\rho_2) \tag{6.4.17}$$

for $\mathrm{Re}(\rho_1, \rho_2, a_j - \rho_1 - \rho_2, j = 1, ..., r, \alpha_j - \rho_1 - \rho_2, j = 1, ..., s, b_j - \rho_1, j = 1, ..., q, \beta_j - \rho_1, j = 1, ..., m, c_j - \rho_2, j = 1, ..., k, \gamma_j - \rho_2, j = 1, ..., n) > p - 1$
where

$$C = \frac{\left\{ \prod_{j=1}^{s} \tilde{\Gamma}_p(\alpha_j) \right\} \left\{ \prod_{j=1}^{m} \tilde{\Gamma}_p(\beta_j) \right\} \left\{ \prod_{j=1}^{n} \tilde{\Gamma}_p(\gamma_j) \right\}}{\left\{ \prod_{j=1}^{r} \tilde{\Gamma}_p(a_j) \right\} \left\{ \prod_{j=1}^{q} \tilde{\Gamma}_p(b_j) \right\} \left\{ \prod_{j=1}^{k} \tilde{\Gamma}_p(c_j) \right\}}.$$

Standard notations are used in writing up this function. For example, the symbol (a_r) stands for the sequence of parameters $a_1, ..., a_r$. Results parallel to the ones in (5.4.2)–(5.4.4) and Theorems 5.32–5.40 of Chapter 5 can also be obtained for the complex case.

Example 6.14   For $\mathrm{Re}(a_j, j = 1, ..., r, \alpha_j \neq -\lambda, \lambda = 0, 1, ..., j = 1, ..., s, s \geq r, b_j, \beta_j - b_j, j = 1, ..., q, c_j, \gamma_j - c_j, j = 1, ..., k) > p - 1$ show that

$$\tilde{F} = \tilde{F}_{s:q:k}^{r:q:k} \left[-\begin{pmatrix} \tilde{X}_1 \\ \tilde{X}_2 \end{pmatrix} \middle| \begin{matrix} (a_r) : (b_q) : (c_k) \\ (\alpha_s) : (\beta_q) : (\gamma_k) \end{matrix} \right]$$

$$= C_1 \int_0^I \cdots \int_0^I \left\{ \prod_{j=1}^{q} |\det(\tilde{U}_j)|^{b_j - p} |\det(I - \tilde{U}_j)|^{\beta_j - b_j - p} \right\}$$

$$\times \left\{ \prod_{j=1}^{k} |\det(\tilde{V}_j)|^{c_j - p} |\det(I - \tilde{V}_j)|^{\gamma_j - c_j - p} \right\}$$

$$\times {}_r\tilde{F}_s^*(a_1, ..., a_r; \alpha_1, ..., \alpha_s; -\tilde{Z}) d\tilde{U}_1 ... d\tilde{U}_q d\tilde{V}_1 ... d\tilde{V}_k$$

where

$$\tilde{Z} = \tilde{U}_1^{\frac{1}{2}}...\tilde{U}_q^{\frac{1}{2}}\tilde{X}_1\tilde{U}_q^{\frac{1}{2}}...\tilde{U}_1^{\frac{1}{2}} + \tilde{V}_1^{\frac{1}{2}}...\tilde{V}_k^{\frac{1}{2}}\tilde{X}_2\tilde{V}_k^{\frac{1}{2}}...\tilde{V}_1^{\frac{1}{2}},$$

$$C_1 = \left\{ \prod_{j=1}^{q} \frac{\tilde{\Gamma}_p(\beta_j)}{\tilde{\Gamma}_p(b_j)\tilde{\Gamma}_p(\beta_j - b_j)} \right\} \left\{ \prod_{j=1}^{k} \frac{\tilde{\Gamma}_p(\gamma_j)}{\tilde{\Gamma}_p(c_j)\tilde{\Gamma}_p(\gamma_j - c_j)} \right\}.$$

Solution   Take the M-transform of the right side with respect to $\tilde{X}_1$ and $\tilde{X}_2$ and with parameters ρ_1 and ρ_2 after making the transformation

$$\tilde{Y}_1 = \tilde{U}_1^{\frac{1}{2}}...\tilde{U}_q^{\frac{1}{2}}\tilde{X}_1\tilde{U}_q^{\frac{1}{2}}...\tilde{U}_1^{\frac{1}{2}}$$

and

$$\tilde{Y}_2 = \tilde{V}_1^{\frac{1}{2}}...\tilde{V}_k^{\frac{1}{2}}\tilde{X}_2\tilde{V}_k^{\frac{1}{2}}...\tilde{V}_1^{\frac{1}{2}}$$

for fixed $\tilde{U}_1, ..., \tilde{U}_q, \tilde{V}_1, ..., \tilde{V}_k$. Then integrate out the $\tilde{U}_j$'s and $\tilde{V}_j$'s by using type-1 beta integrals to see the result.

All the results in Section 5.4 of Chapter 5 allow direct extensions to the complex case.

6.4.6   Lauricella functions: complex case

These will be denoted by

$$\tilde{f}_A = \tilde{f}_A(a, b_1, ..., b_n; c_1, ..., c_n; -\tilde{X}_1, ..., -\tilde{X}_n)$$
$$\tilde{f}_B = \tilde{f}_B(a_1, ..., a_n, b_1, ..., b_n; c; -\tilde{X}_1, ..., -\tilde{X}_n)$$
$$\tilde{f}_C = \tilde{f}_C(a, b; c_1, ..., c_n; -\tilde{X}_1, ..., -\tilde{X}_n)$$
$$\tilde{f}_D = \tilde{f}_D(a, b_1, ..., b_n; c; -\tilde{X}_1, ..., -\tilde{X}_n).$$

These can be defined in terms of their M-transforms or by using some integral representations. Multiple series forms in terms of invariant polynomials, parallel to the one in (6.4.12), can also be obtained. Since the extensions of the results in Section 5.5 of Chapter 5 on Lauricella's functions in the real case to the complex case are parallel, we will define these by using M-transforms and illustrate a few properties through a few examples.

We define these as classes of symmetric functions having the following M-transforms with respect to $\tilde{X}_1, ..., \tilde{X}_n$ and with parameters $\rho_1, ..., \rho_n$:

$$M(\tilde{f}_A) = \int_{\tilde{X}_1 > 0} ... \int_{\tilde{X}_n > 0} |\det(\tilde{X}_1)|^{\rho_1 - p}...|\det(\tilde{X}_n)|^{\rho_n - p}$$
$$\times \tilde{f}_A(a, b_1, ..., b_n; c_1, ..., c_n; -\tilde{X}_1, ..., -\tilde{X}_n)d\tilde{X}_1...d\tilde{X}_n$$

$$= \frac{\left\{\prod_{j=1}^{n} \tilde{\Gamma}_p(c_j)\right\} \left\{\prod_{j=1}^{n} \tilde{\Gamma}_p(b_j - \rho_j)\right\}}{\tilde{\Gamma}_p(a) \left\{\prod_{j=1}^{n} \tilde{\Gamma}_p(b_j)\right\}}$$

$$\times \frac{\tilde{\Gamma}_p(a - \rho_1 - \ldots - \rho_n) \left\{\prod_{j=1}^{n} \tilde{\Gamma}_p(\rho_j)\right\}}{\left\{\prod_{j=1}^{n} \tilde{\Gamma}_p(c_j - \rho_j)\right\}} \tag{6.4.18}$$

for $\mathrm{Re}(b_j - \rho_j, c_j - \rho_j, \rho_j, a - \rho_1 - \ldots - \rho_n) > p - 1$, $j = 1, \ldots, n$;

$$M(\tilde{f}_B) = \frac{\tilde{\Gamma}_p(c)}{\left\{\prod_{j=1}^{n} \tilde{\Gamma}_p(a_j)\tilde{\Gamma}_p(b_j)\right\}}$$

$$\times \frac{\left\{\prod_{j=1}^{n} \tilde{\Gamma}_p(a_j - \rho_j)\tilde{\Gamma}_p(b_j - \rho_j)\tilde{\Gamma}_p(\rho_j)\right\}}{\tilde{\Gamma}_p(c - \rho_1 - \ldots - \rho_n)} \tag{6.4.19}$$

for $\mathrm{Re}(\rho_j, a_j - \rho_j, b_j - \rho_j, c - \rho_1 - \ldots - \rho_n) > p - 1$, $j = 1, \ldots, n$;

$$M(\tilde{f}_C) = \frac{\left\{\prod_{j=1}^{n} \tilde{\Gamma}_p(c_j)\right\}}{\tilde{\Gamma}_p(a)\tilde{\Gamma}_p(b)} \left\{\prod_{j=1}^{n} \tilde{\Gamma}_p(\rho_j)\right\}$$

$$\times \frac{\tilde{\Gamma}_p(a - \rho_1 - \ldots - \rho_n)\tilde{\Gamma}_p(b - \rho_1 - \ldots - \rho_n)}{\left\{\prod_{j=1}^{n} \tilde{\Gamma}_p(c_j - \rho_j)\right\}} \tag{6.4.20}$$

for $\mathrm{Re}(\rho_j, c_j - \rho_j, a - \rho_1 - \ldots - \rho_n, b - \rho_1 - \ldots - \rho_n) > p - 1$, $j = 1, \ldots, n$;

$$M(\tilde{f}_D) = \frac{\tilde{\Gamma}_p(c)}{\tilde{\Gamma}_p(a) \left\{\prod_{j=1}^{n} \tilde{\Gamma}_p(b_j)\right\}} \left\{\prod_{j=1}^{n} \tilde{\Gamma}_p(\rho_j)\right\}$$

$$\times \frac{\tilde{\Gamma}_p(a - \rho_1 - \ldots - \rho_n) \left\{\prod_{j=1}^{n} \tilde{\Gamma}_p(b_j - \rho_j)\right\}}{\tilde{\Gamma}_p(c - \rho_1 - \ldots - \rho_n)} \tag{6.4.21}$$

for $\mathrm{Re}(c, a, \rho_j, b_j - \rho_j, c - \rho_1 - \ldots - \rho_n, a - \rho_1 - \ldots - \rho_n) > p - 1$, $j = 1, \ldots, n$.

Example 6.15   For $\mathrm{Re}(b_j, a - b_1 - \ldots - b_n) > p - 1$, $j = 1, \ldots, n$ show that

$$\tilde{f}_D = \tilde{f}_D(a, b_1, \ldots, b_n; c; -\tilde{X}_1, \ldots, -\tilde{X}_n)$$

$$= \frac{\tilde{\Gamma}_p(c)}{\tilde{\Gamma}_p(c - b_1 - \ldots - b_n) \left\{\prod_{j=1}^{n} \tilde{\Gamma}_p(b_j)\right\}} \int \ldots \int |\det(\tilde{U}_1)|^{b_1 - p}$$

$$\times \ldots |\det(\tilde{U}_n)|^{b_n - p} |\det(I - \tilde{U}_1 - \ldots - \tilde{U}_n)|^{c - b_1 - \ldots - b_n - p}$$

$$\times \left|\det\left(I + \tilde{U}_1^{\frac{1}{2}} \tilde{X}_1 \tilde{U}_1^{\frac{1}{2}} + \ldots + \tilde{U}_n^{\frac{1}{2}} \tilde{X}_n \tilde{U}_n^{\frac{1}{2}}\right)\right|^{-a} d\tilde{U}_1 \ldots d\tilde{U}_n.$$

Proof   Take the M-transform of the right side by using a type-2 Dirichlet integral and then integrate out the $\tilde{U}_j$'s by using a type-1 Dirichlet integral to establish the result.

Example 6.16   For $\mathrm{Re}(a_j) > p - 1$, $j = 1, ..., n$ show that

$$
\tilde{f}_B = \tilde{f}_B(a_1, ..., a_n, b_1, ..., b_n; c; -\tilde{X}_1, ..., -\tilde{X}_n)
$$

$$
= \left[\prod_{j=1}^{n} \tilde{\Gamma}_p(a_j) \right]^{-1}
$$

$$
\times \int_{\tilde{T}_1 > 0} \cdots \int_{\tilde{T}_n > 0} e^{-\mathrm{tr}(\tilde{T}_1 + ... + \tilde{T}_n)} |\det(\tilde{T}_1)|^{a_1 - p} ... |\det(\tilde{T}_n)|^{a_n - p}
$$

$$
\times \tilde{\phi}_2\left(b_1, ..., b_n; c; -\tilde{T}_1^{\frac{1}{2}} \tilde{X}_1 \tilde{T}_1^{\frac{1}{2}}, ..., -\tilde{T}_n^{\frac{1}{2}} \tilde{X}_n \tilde{T}_n^{\frac{1}{2}} \right) d\tilde{T}_1 ... d\tilde{T}_n
$$

where the $\tilde{\phi}_2$-function

$$
\tilde{\phi}_2 = \tilde{\phi}_2(b_1, ..., b_n; c; -\tilde{X}_1, ..., -\tilde{X}_n)
$$

is defined by the following M-transform:

$$
M(\tilde{\phi}_2) = \frac{\tilde{\Gamma}_p(c)}{\left\{ \prod_{j=1}^{n} \tilde{\Gamma}_p(b_j) \right\}} \frac{\left\{ \prod_{j=1}^{n} \tilde{\Gamma}_p(b_j - \rho_j) \tilde{\Gamma}_p(\rho_j) \right\}}{\tilde{\Gamma}_p(c - \rho_1 - ... - \rho_n)}. \tag{6.4.22}
$$

Proof   Take the M-transform of the right side by using the above definition of the $\tilde{\phi}_2$-function and integrate out the $\tilde{T}_j$'s by using the gamma integrals to see the result.

Example 6.17   Show that

$$
\tilde{f}_A = \left[\prod_{j=1}^{n} \tilde{\Gamma}_p(b_j) \right]^{-1} \int_{\tilde{T}_1 > 0} \cdots \int_{\tilde{T}_n > 0} e^{-\mathrm{tr}(\tilde{T}_1 + ... + \tilde{T}_n)}
$$

$$
\times |\det(\tilde{T}_1)|^{b_1 - p} ... |\det(\tilde{T}_n)|^{b_n - p}
$$

$$
\times \tilde{\psi}_2\left(a; c_1, ..., c_n; -\tilde{T}_1^{\frac{1}{2}} \tilde{X}_1 \tilde{T}_1^{\frac{1}{2}}, ..., -\tilde{T}_n^{\frac{1}{2}} \tilde{X}_n \tilde{T}_n^{\frac{1}{2}} \right) d\tilde{T}_1 ... d\tilde{T}_n
$$

where the $\tilde{\psi}_2$-function of matrix arguments

$$
\tilde{\psi}_2 = \tilde{\psi}_2(a; c_1, ..., c_n; -\tilde{X}_1, ..., -\tilde{X}_n)
$$

is defined as that class of symmetric functions for which the M-transform is the following:

$$M(\tilde{\psi}_2) = M_{\rho_1,\ldots,\rho_n}(\tilde{\psi}_2(a; c_1, \ldots, c_n; -\tilde{X}_1, \ldots, -\tilde{X}_n))$$

$$= \frac{\left\{\prod_{j=1}^n \tilde{\Gamma}_p(c_j)\right\}}{\tilde{\Gamma}_p(a)}$$

$$\times \frac{\tilde{\Gamma}_p(a - \rho_1 - \ldots - \rho_n)\left\{\prod_{j=1}^n \tilde{\Gamma}_p(\rho_j)\right\}}{\left\{\prod_{j=1}^n \tilde{\Gamma}_p(c_j - \rho_j)\right\}} \qquad (6.4.23)$$

for $\text{Re}(\rho_j, c_j - \rho_j, a - \rho_1 - \ldots - \rho_n) > p - 1$, $j = 1, \ldots, n$.

Proof Take the M-transform of the right side with respect to $\tilde{X}_1, \ldots, \tilde{X}_n$ by letting $\tilde{Y}_j = \tilde{T}_j^{\frac{1}{2}} \tilde{X}_j \tilde{T}_j^{\frac{1}{2}}$ for fixed $\tilde{T}_j$, and by using the above definition for a $\tilde{\psi}_2$-function. Then evaluate the $\tilde{T}_j$-integrals by using the gamma integral to see the result.

Note that in the complex case the analogue of Lemma 5.4 of Chapter 5 will have a different structure and hence the corresponding results where this lemma is used will have structurally different analogues.

Some applications of real valued scalar functions of matrices in the complex case to various areas are pointed out at the end of Section 3.2 of Chapter 3 . Some more properties of hypergeometric functions, Hermite and Laguerre polynomials of matrix argument in the complex case may be seen from Chikuse (1976, 1992, 1992a). Some more sample references on the distributional aspects of random eigenvalues in the complex case are Al-Ani (1972) and Wigner (1965).

Exercises

6.4.1 For $\text{Re}(b, b', c - b, c' - b') > p - 1$, show that

$$\tilde{F}_2 = \tilde{F}_2(a, b, b'; c, c'; -\tilde{X}_1, -\tilde{X}_2)$$

$$= \frac{\tilde{\Gamma}_p(c)\tilde{\Gamma}_p(c')}{\tilde{\Gamma}_p(b)\tilde{\Gamma}_p(b')\tilde{\Gamma}_p(c - b)\tilde{\Gamma}_p(c' - b')} \int_0^I \int_0^I |\det(\tilde{U}_1)|^{b-p}$$

$$\times |\det(\tilde{U}_2)|^{b'-p}|\det(I - \tilde{U}_1)|^{c-b-p}|\det(I - \tilde{U}_2)|^{c'-b'-p}$$

$$\times \left|\det\left(I + \tilde{U}_1^{\frac{1}{2}}\tilde{X}_1\tilde{U}_1^{\frac{1}{2}} + \tilde{U}_2^{\frac{1}{2}}\tilde{X}_2\tilde{U}_2^{\frac{1}{2}}\right)\right|^{-a} d\tilde{U}_1 d\tilde{U}_2.$$

6.4.2   For $\text{Re}(b.b', c - b - b') > p - 1$, show that

$$\tilde{F}_3 = \tilde{F}_3(a, a', b, b'; c; -\tilde{X}_1, -\tilde{X}_2)$$

$$= \frac{\tilde{\Gamma}_p(c)}{\tilde{\Gamma}_p(b)\tilde{\Gamma}_p(b')\tilde{\Gamma}_p(c - b - b')} \int \int |\det(\tilde{U}_1)|^{b-p}$$

$$\times |\det(\tilde{U}_2)|^{b'-p} |\det(I - \tilde{U}_1 - \tilde{U}_2)|^{c-b-b'-p}$$

$$\times \left| \det \left(I + \tilde{U}_1^{\frac{1}{2}} \tilde{X}_1 \tilde{U}_1^{\frac{1}{2}} \right) \right|^{-a} \left| \det \left(I + \tilde{U}_2^{\frac{1}{2}} \tilde{X}_2 \tilde{U}_2^{\frac{1}{2}} \right) \right|^{-a'} d\tilde{U}_1 d\tilde{U}_2.$$

6.4.3   For $\text{Re}(a, b) > p - 1$, show that

$$\tilde{F}_4 = \tilde{F}_4(a, b; c, c'; -\tilde{X}_1, -\tilde{X}_2)$$

$$= \frac{1}{\tilde{\Gamma}_p(a)\tilde{\Gamma}_p(b)} \int_{\tilde{T}_1 > 0} \int_{\tilde{T}_2 > 0} e^{-\text{tr}(\tilde{T}_1 + \tilde{T}_2)} |\det(\tilde{T}_1)|^{a-p}$$

$$\times |\det(\tilde{T}_2)|^{b-p} {}_0\tilde{F}_1 \left(; c; -\tilde{T}_1^{\frac{1}{2}} \tilde{T}_2^{\frac{1}{2}} \tilde{X}_1 \tilde{T}_2^{\frac{1}{2}} \tilde{T}_1^{\frac{1}{2}} \right)$$

$$\times {}_0\tilde{F}_1 \left(; c'; -\tilde{T}_1^{\frac{1}{2}} \tilde{T}_2^{\frac{1}{2}} \tilde{X}_2 \tilde{T}_2^{\frac{1}{2}} \tilde{T}_1^{\frac{1}{2}} \right) d\tilde{T}_1 d\tilde{T}_2.$$

6.4.4   For $\text{Re}(a, c - a) > p - 1$, show that

$$\tilde{\Phi}_1 = \tilde{\Phi}_1(a, b; c; -\tilde{X}_1, -\tilde{X}_2)$$

$$= \frac{\tilde{\Gamma}_p(c)}{\tilde{\Gamma}_p(a)\tilde{\Gamma}_p(c - a)} \int_0^I |\det(\tilde{U})|^{a-p} |\det(I - \tilde{U})|^{c-a-p}$$

$$\times \left| \det \left(I + \tilde{U}^{\frac{1}{2}} \tilde{X}_1 \tilde{U}^{\frac{1}{2}} \right) \right|^{-b} e^{-\text{tr}(\tilde{U}\tilde{X}_2)} d\tilde{U}.$$

6.4.5   For $\text{Re}(b, b', c - b - b') > p - 1$, show that

$$\tilde{\Xi}_1 = \tilde{\Xi}_1(a, b', b; c; -\tilde{X}_1, -\tilde{X}_2)$$

$$= \frac{\tilde{\Gamma}_p(c)}{\tilde{\Gamma}_p(b)\tilde{\Gamma}_p(b')\tilde{\Gamma}_p(c - b - b')} \int \int |\det(\tilde{U}_1)|^{b-p} |\det(\tilde{U}_2)|^{b'-p}$$

$$\times |\det(I - \tilde{U}_1 - \tilde{U}_2)|^{c-b-b'-p} \left| \det \left(I + \tilde{U}_1^{\frac{1}{2}} \tilde{X}_1 \tilde{U}_1^{\frac{1}{2}} \right) \right|^{-a}$$

$$\times e^{-\text{tr}(\tilde{U}_2 \tilde{X}_2)} d\tilde{U}_1 d\tilde{U}_2.$$

6.4.6   Show that

$$\lim_{\epsilon \to 0} \tilde{F}_1 \left(\frac{1}{\epsilon}, b, b'; c' - \epsilon \tilde{X}_1, -\epsilon \tilde{X}_2 \right) = \tilde{\Phi}_2(b, b'; c' - \tilde{X}_1, -\tilde{X}_2).$$

Appendix

Appendix A
Some concepts from statistical distribution theory

Most of the illustrative examples in the book are taken from statistical distribution theory. Some of the basic statistical concepts and properties will be given here.

Definition A1   A density function: continuous case   Let $f(X) = f(x_1, ..., x_p)$ be a real scalar function of the p real variables $x_1, ..., x_p$ such that $f(X) \geq 0$ for all x_j's and $\int_X f(X)dX = 1$ where $\int_X$ means the integral over the elements x_j, $j = 1, ..., p$ of $X' = (x_1, ..., x_p)$ and $dX = dx_1...dx_p$. Then $f(X)$ is called the joint density of the p real continuous scalar random variables $x_1, ..., x_p$ or the density of X.

Definition A2   Expected value of a scalar function   Let $f(X)$ be the density of the $p \times 1$ real random vector X. Let $\phi(X)$ be a scalar function of X. Then the expected value of $\phi(X)$, denoted by $E[\phi(X)]$ or $E\phi(X)$, is defined as

$$E[\phi(X)] = \int_X \phi(X)f(X)dX$$

whenever the integral exists.

Hereafter E will denote the expected value operator when dealing with random variables. Some commonly used expected values will be defined here.

Definition A3   Mean value, covariance matrix and moment generating function   Let X be a $p \times 1$ real random vector, $X' = (x_1, ..., x_p)$. Then $E(X)$ or EX is called the *mean value of* X, that is,

$$E(X) = \begin{pmatrix} Ex_1 \\ Ex_2 \\ \vdots \\ Ex_p \end{pmatrix}.$$

Let $V = E[(X - EX)(X - EX)']$. This $p \times p$ matrix $V = (v_{ij})$ is called the *covariance matrix of* X or the *variance-covariance matrix of* X and it

401

is written as

$$V = \text{Cov}(X) = E[(X - EX)(X - EX)'].$$

That is, $v_{ij} = \text{Cov}(x_i, x_j) = $ covariance between the elements x_i and x_j, and $v_{ii} = \text{Cov}(x_i, x_i) = \text{Var}(x_i) = $ variance of x_i. Let

$$M_X(T) = E\left(e^{T'X}\right)$$

where T is a $p \times 1$ vector of parameters, that is, T is free of X. This $M_X(T)$ is called the *moment generating function (m.g.f.)* of X or the m.g.f. of $f(X)$ where $f(X)$ is the density of X. Note that if $f(X) = 0$ for $x_j \leq 0$, $j = 1, ..., p$ then $M_X(-T)$ represents the *Laplace transform of* $f(X)$ and it is written as

$$L_f(T) = E\left(e^{-T'X}\right).$$

Definition A4   Real Gaussian or normal density   Let $f(X)$ be the density of the real $p \times 1$ random vector X. Then X is said to have a real nonsingular normal density with parameters μ and V and written as $X \sim N_p(\mu, V)$ if

$$f(X) = \frac{e^{-\frac{1}{2}(X-\mu)'V^{-1}(X-\mu)}}{(2\pi)^{\frac{p}{2}}|V|^{\frac{1}{2}}}$$

for $-\infty < X < \infty$, $-\infty < \mu < \infty$, $V = V' > 0$ where $V = V' > 0$ means that V is symmetric positive definite and, for example, $-\infty < X < \infty$ means that $-\infty < x_j < \infty$, $j = 1, ..., p$, $X' = (x_1, ..., x_p)$.

If V is singular then X is said to have a singular normal distribution and it is written as $X \sim N_p(\mu, V), |V| = 0$ but in this case X does not have a density as given above. However properties of X can be studied by using the properties of the real nonsingular $r \times 1$ normal vector Y by writing $X = \mu + BY$ where $V = BB', B$ is $p \times r$ with r being the rank of V, and $Y \sim N_r(0, I)$ where I is the identity matrix of order r. It can be shown that the m.g.f. of X, whether X has a nonsingular or singular normal distribution, is the following:

$$M_X(T) = e^{T'\mu + \frac{1}{2}T'VT} \text{ for } X \sim N_p(\mu, V).$$

It is easy to note that $E(X) = \mu$ and $\text{Cov}(X) = V$ when $X \sim N_p(\mu, V)$.

Definition A5   Matrix-variate nonsingular real normal density
Let X be a $p \times q$ real matrix. Then X is said to have a real nonsingular
normal density with the parameters M, W, V if the joint density function
of the q columns of X is given by

$$f(X) = \frac{e^{-\frac{1}{2}\text{tr}[V^{-1}(X-M)W^{-1}(X-M)']}}{(2\pi)^{\frac{pq}{2}}|V|^{\frac{q}{2}}|W|^{\frac{p}{2}}}$$

with $V = V' > 0$, $W = W' > 0$, $M = (m_{jk})$, $X = (x_{jk})$, $-\infty < x_{jk} <$
∞, $-\infty < m_{jk} < \infty$ for all j and k.

Definition A6   Real matrix-variate gamma   Let $X = X' > 0$ be
a real $p \times p$ matrix. Then $\Gamma_p(\alpha)$ is defined as the *matrix-variate gamma*
where

$$\Gamma_p(\alpha) = \int_{X>0} |X|^{\alpha - \frac{p+1}{2}} e^{-\text{tr}(X)} dX \text{ for } \text{Re}(\alpha) > \frac{p-1}{2}$$

where $\text{Re}(\cdot)$ denotes the real part of $(\cdot)$. It can be shown that, see also
Example 1.24,

$$\Gamma_p(\alpha) = \pi^{\frac{p(p-1)}{4}} \Gamma(\alpha)\Gamma\left(\alpha - \frac{1}{2}\right)...\Gamma\left(\alpha - \frac{p-1}{2}\right), \text{ } \text{Re}(\alpha) > \frac{p-1}{2}.$$

Definition A7   Real matrix-variate gamma density   Let

$$f(X) = \begin{cases} \frac{|B|^{\alpha}|X|^{\alpha - \frac{p+1}{2}} e^{-\text{tr}(BX)}}{\Gamma_p(\alpha)}, \text{ for } X = X' > 0, \\ \\ B = B' > 0, \text{Re}(\alpha) > \frac{p-1}{2} \\ 0, \text{ elsewhere}, \end{cases}$$

where B is free of X. Then X is said to have a *real matrix-variate gamma
density* $f(X)$ *with the parameters* (α, B).

Definition A8   Real matrix-variate beta   Let $X = X' > 0$, $I -$
$X > 0$ be a $p \times p$ real matrix. Let

$$B_p(\alpha, \beta) = \int_0^I |X|^{\alpha - \frac{p+1}{2}} |I - X|^{\beta - \frac{p+1}{2}} dX, \text{ } \text{Re}(\alpha) > \frac{p-1}{2}, \text{ } \text{Re}(\beta) > \frac{p-1}{2}.$$

Then $B_p(\alpha, \beta)$ is called the *matrix-variate beta*. It can be shown that

$$B_p(\alpha, \beta) = \frac{\Gamma_p(\alpha)\Gamma_p(\beta)}{\Gamma_p(\alpha + \beta)}$$

$$= \int_{Y>0} |Y|^{\alpha - \frac{p+1}{2}} |I + Y|^{-(\alpha+\beta)} dY, \text{ } \text{Re}(\alpha) > \frac{p-1}{2}, \text{ } \text{Re}(\beta) > \frac{p-1}{2}.$$

Definition A9   Real matrix-variate beta density   Let

$$f(X) = \begin{cases} \dfrac{|X|^{\alpha-\frac{p+1}{2}}|I-X|^{\beta-\frac{p+1}{2}}}{B_p(\alpha,\beta)}, \text{ for } 0 < X = X' < I, \\ \qquad\qquad \text{Re}(\alpha) > \frac{p-1}{2},\ \text{Re}(\beta) > \frac{p-1}{2} \\ 0, \text{ elsewhere .} \end{cases}$$

Then $f(X)$ is called a *real type-1 matrix-variate beta density with the parameters* (α, β). Let

$$g(Y) = \begin{cases} \dfrac{|Y|^{\alpha-\frac{p+1}{2}}|I+Y|^{-(\alpha+\beta)}}{B_p(\alpha,\beta)}, Y = Y' > 0,\ \text{Re}(\alpha) > \frac{p-1}{2},\ \text{Re}(\beta) > \frac{p-1}{2} \\ 0, \text{ elsewhere.} \end{cases}$$

Then $g(Y)$ is called the *real type-2 matrix-variate beta density with the parameters* (α, β).

Appendix B

Some multiple integrals and multivariate densities

Definition B1   Real type-1 Dirichlet integral

$$\delta_1 = \int_\Omega x_1^{\alpha_1-1}...x_k^{\alpha_k-1}(1 - x_1 - ... - x_k)^{\alpha_{k+1}-1}dx_1...dx_k$$
$$= \frac{\Gamma(\alpha_1)...\Gamma(\alpha_{k+1})}{\Gamma(\alpha_1 + ... + \alpha_{k+1})}$$

for $\text{Re}(\alpha_j) > 0$, $j = 1, ..., k + 1$ is called the real type-1 Dirichlet integral where $\Omega = \{(x_1, ..., x_k)|0 < x_j < 1,\ j = 1, ..., k,\ x_1 + ... + x_k < 1\}$.

Definition B2   Real type-1 Dirichlet density

$$f(x_1, ..., x_k) = \frac{\Gamma(\alpha_1 + ... + \alpha_{k+1})}{\Gamma(\alpha_1)...\Gamma(\alpha_{k+1})}$$
$$\times x_1^{\alpha_1-1}...x_k^{\alpha_k-1}(1 - x_1 - ... - x_k)^{\alpha_{k+1}-1},$$

for $0 < x_j < 1$, $j = 1, ..., k$, $x_1 + ... + x_k < 1$, $\text{Re}(\alpha_j) > 0$, $j = 1, ..., k + 1$ and $f(x_1, ..., x_k) = 0$ elsewhere is called the *real type-1 Dirichlet density*.

Definition B3   Real type-2 Dirichlet integral

$$\delta_2 = \int_0^\infty \cdots \int_0^\infty x_1^{\alpha_1-1}...x_k^{\alpha_k-1}$$
$$\times (1+x_1+...+x_k)^{-(\alpha_1+...+\alpha_{k+1})}\, dx_1...dx_k$$
$$= \frac{\Gamma(\alpha_1)...\Gamma(\alpha_{k+1})}{\Gamma(\alpha_1+...+\alpha_{k+1})}$$

for $\mathrm{Re}(\alpha_j) > 0$, $j = 1,...,k+1$ is called the *real type-2 Dirichlet integral.*

Definition B4   Real type-2 Dirichlet density

$$f(x_1,...,x_k) = \frac{\Gamma(\alpha_1+...+\alpha_{k+1})}{\Gamma(\alpha_1)...\Gamma(\alpha_{k+1})}$$
$$\times x_1^{\alpha_1-1}...x_k^{\alpha_k-1}(1+x_1+...+x_k)^{-(\alpha_1+...+\alpha_{k+1})}$$

for $0 < x_j < \infty$, $j = 1,...,k$, $\mathrm{Re}(\alpha_j) > 0, j = 1,...,k+1$ and $f(x_1,...,x_k) = 0$ elsewhere is called the *real type-2 Dirichlet density.*

Appendix C

Some results on random variables in the complex case

Some basic concepts will be given here and then some basic distributions will also be listed. The multivariate normal and associated distributions will be given for the complex cases. When many terms are involved in an expression a bar to denote the complex conjugate will not look elegant and hence we will also use the following notation.

Notation C1 $A^* = \bar{A}'$: conjugate transpose of A.

Definition C1   Variances and covariances of complex variables
Let $\tilde{z} = x + iy$ be a scalar complex random variable where x and y are real random variables and $i = \sqrt{-1}$. Let $\tilde{z}_1 = x_1 + iy_1$ and $\tilde{z}_2 = x_2 + iy_2$ be two complex random variables. Then the variance and covariance are defined as follows:

$$\mathrm{Var}(\tilde{z}) = E\left\{[(\tilde{z}-E\tilde{z})(\tilde{z}-E\tilde{z})^*]\right\}$$
$$= E\left\{[(x-Ex)+i(y-Ey)]\left[(x-Ex)-i(y-Ey)\right]\right\}$$
$$= E\left\{[(x-Ex)^2+(y-Ey)^2]\right\} = \mathrm{Var}(x)+\mathrm{Var}(y)$$
$$= E\left\{|\tilde{z}-E\tilde{z}|^2\right\}$$

where E denotes the expected value and $|\tilde{z} - E\tilde{z}|$ denotes the absolute value or norm of $\tilde{z} - E\tilde{z}$.

$$\begin{aligned}
\text{Cov}(\tilde{z}_1, \tilde{z}_2) &= E\left[(\tilde{z}_1 - E\tilde{z}_1)(\tilde{z}_2 - E\tilde{z}_2)^*\right] \\
&= \left[\text{Cov}(x_1, x_2) + \text{Cov}(y_1, y_2)\right] + i\left[\text{Cov}(x_2, y_1) - \text{Cov}(x_1, y_2)\right].
\end{aligned}$$

Note that $\text{Cov}(\tilde{z}_1, \tilde{z}_2) \neq \text{Cov}(\tilde{z}_2, \tilde{z}_1)$. Extending these ideas one can define the covariance matrix or the variance-covariance matrix of a $p \times 1$ complex random vector as follows:

Definition C2   The covariance matrix   Let $\tilde{Z} = X + iY$ where X and Y are $p \times 1$ real vectors and $i = \sqrt{-1}$. Then the covariance matrix of $\tilde{Z}$, denoted by $\text{Cov}(\tilde{Z})$, is defined as

$$\begin{aligned}
\text{Cov}(\tilde{Z}) &= E\left\{(\tilde{Z} - E\tilde{Z})(\tilde{Z} - E\tilde{Z})^*\right\} \\
&= E\left\{[(X - EX) + i(Y - EY)]\,[(X - EX) - i(Y - EY)]'\right\} \\
&= E\{(X - EX)(X - EX)' + E(Y - EY)(Y - EY)' \\
&\quad - iE(X - EX)(Y - EY)' + iE(Y - EY)(X - EX)'\} \\
&= V_{11} + V_{22} + i\left(-V_{12} + V_{21}\right)
\end{aligned}$$

where $V_{12} = \text{Cov}(X, Y)$.

If $\tilde{Z}_1 = X_1 + iY_1$ and $\tilde{Z}_2 = X_2 + iY_2$ are a $p \times 1$ and a $q \times 1$ complex vectors respectively then $\text{Cov}(\tilde{Z}_1, \tilde{Z}_2)$ is defined as

$$\begin{aligned}
\text{Cov}(\tilde{Z}_1, \tilde{Z}_2) &= E\left[(\tilde{Z}_1 - E\tilde{Z}_1)(\tilde{Z}_2 - E\tilde{Z}_2)^*\right] \\
&= \text{Cov}(X_1, X_2) + \text{Cov}(Y_1, Y_2) + i\left[\text{Cov}(Y_1, X_2) - \text{Cov}(X_1, Y_2)\right].
\end{aligned}$$

Note that $\text{Cov}(\tilde{Z}_1, \tilde{Z}_2) \neq \text{Cov}(\tilde{Z}_2, \tilde{Z}_1)$ even if $p = q$.

Definition C3   Multivariate normal density in the complex case
A $p \times 1$ complex random vector $\tilde{Z}$ is said to have a p-variate complex normal density with parameters $\tilde{\mu}$ and $\tilde{\Sigma}$, denoted by $\tilde{Z} \sim \tilde{N}_p(\tilde{\mu}, \tilde{\Sigma})$, if the density function of $\tilde{Z}$ is of the following form:

$$f(\tilde{Z}) = \frac{e^{-(\tilde{Z} - \tilde{\mu})^* \tilde{\Sigma}^{-1}(\tilde{Z} - \tilde{\mu})}}{\pi^p |\det(\tilde{\Sigma})|}$$

where $\tilde{\Sigma}$ is a hermitian positive definite matrix of the form $\tilde{\Sigma} = \Sigma_1 + i\Sigma_2$ with Σ_1 and Σ_2 being real matrices such that $\Sigma_1 = \Sigma_1' > 0$, $\Sigma_2 = -\Sigma_2'$, that is, Σ_1 is symmetric positive definite and Σ_2 is skew symmetric and $|\det(\tilde{\Sigma})|$ denotes the absolute value of the determinant of $\tilde{\Sigma}$.

Notation C2 $\tilde{N}_p(\tilde{\mu}, \tilde{\Sigma})$: p-variate complex normal with parameters $\tilde{\mu}$ and $\tilde{\Sigma}$.

We can show that if $\tilde{Z} = X + iY$ and if the real $p \times 1$ vectors X and Y have a certain joint normal distribution then this joint density is the same as the density $f(\tilde{Z})$ given above. Let the $(2p) \times 1$ vector $U = \begin{pmatrix} X \\ Y \end{pmatrix}$ have a $2p$-variate nonsingular real normal density of the following form

$$f(U) = \frac{e^{-\frac{1}{2}(U-\mu)'V^{-1}(U-\mu)}}{(2\pi)^p |V|^{\frac{1}{2}}}$$

where

$$\mu = \begin{pmatrix} \mu_1 \\ \mu_2 \end{pmatrix}, \quad \mu_1 = E(X), \quad \mu_2 = E(Y),$$

$$V = \begin{bmatrix} \frac{1}{2}V_{11} & \frac{1}{2}V_{12} \\ \frac{1}{2}V_{21} & \frac{1}{2}V_{11} \end{bmatrix} = \frac{1}{2}\begin{bmatrix} V_{11} & V_{12} \\ V_{21} & V_{11} \end{bmatrix}$$

with V_{12} skew symmetric. That is, the covariance matrices of the $p \times 1$ vectors X and Y are each equal to $\frac{1}{2}V_{11}$ and the covariance between X and Y, $V_{12} = V_{21}'$, is skew symmetric which implies that $V_{12} = -V_{12}' = -V_{21}$. Let

$$\begin{bmatrix} V_{11} & -V_{12} \\ V_{12} & V_{11} \end{bmatrix}^{-1} = \begin{bmatrix} V^{11} & V^{12} \\ -V^{12} & V^{11} \end{bmatrix}.$$

Then

$$V^{-1} = 2\begin{bmatrix} V^{11} & V^{12} \\ -V^{12} & V^{11} \end{bmatrix}.$$

Let

$$W = U - EU = \begin{pmatrix} X - EX \\ Y - EY \end{pmatrix} = \begin{pmatrix} W_1 \\ W_2 \end{pmatrix}.$$

The exponent, denoted by δ, is

$$\delta = -\frac{1}{2}(U - \mu)'V^{-1}(U - \mu)$$

$$= -\frac{1}{2}(W_1', W_2')2\begin{bmatrix} V^{11} & V^{12} \\ -V^{12} & V^{11} \end{bmatrix}\begin{pmatrix} W_1 \\ W_2 \end{pmatrix}$$

$$= -\left[W_1'V^{11}W_1 + W_2'V^{11}W_2 - W_2'V^{12}W_1 + W_1'V^{12}W_2\right].$$

It is easy to note that the exponent is the same as $-(\tilde{Z} - E\tilde{Z})^{*}\tilde{\Sigma}^{-1}(\tilde{Z} - E\tilde{Z})$ with $\tilde{\Sigma} = V_{11} + iV_{12}$ where V_{11} is symmetric positive definite and V_{12} is skew symmetric. This can be seen from the following. Let $(V_{11} + iV_{12})^{-1} = U_{11} - iU_{12}$ where U_{11} is symmetric and U_{12} is skew symmetric. Then

$$(V_{11} + iV_{12})(U_{11} - iU_{12}) = I \Rightarrow$$
$$V_{11}U_{11} + V_{12}U_{12} = I \text{ and } V_{12}U_{11} - V_{11}U_{12} = 0.$$

Solving these we have

$$U_{11} = \left(V_{11} + V_{12}V_{11}^{-1}V_{12}\right)^{-1} = V^{11}$$
$$U_{12} = V_{11}^{-1}V_{12}V^{11} = V^{12}.$$

Note that

$$(\tilde{Z} - E(\tilde{Z}))^{*} = W_1' - iW_2'$$

and thus

$$(\tilde{Z} - E(\tilde{Z}))^{*}\tilde{\Sigma}^{-1}(\tilde{Z} - E(\tilde{Z})) = (W_1' - iW_2')(U_{11} - iU_{12})(W_1 + iW_2)$$

$$= W_1'U_{11}W_1 + W_2'U_{11}W_2 - W_2'U_{12}W_1 + W_1'U_{12}W_2.$$

Note that $W_1'U_{12}W_1$ and $W_2'U_{12}W_2$ are zeros due to the skew symmetry of U_{12}. Thus the exponent agrees with the exponent in the $2p$-variate real normal.

The constant part reduces to the following:

$$(2\pi)^{p}|V|^{\frac{1}{2}} = (2\pi)^{p}\begin{vmatrix} \frac{1}{2}V_{11} & -\frac{1}{2}V_{12} \\ \frac{1}{2}V_{12} & \frac{1}{2}V_{11} \end{vmatrix}^{\frac{1}{2}}$$

$$= \pi^{p}\begin{vmatrix} V_{11} & -V_{12} \\ V_{12} & V_{11} \end{vmatrix}^{\frac{1}{2}}.$$

But

$$\text{Cov}(\tilde{Z}) = E(\tilde{Z} - E\tilde{Z})(\tilde{Z} - E\tilde{Z})^{*}$$
$$= \text{Cov}(X) + \text{Cov}(Y) + i\left[\text{Cov}(Y, X) - \text{Cov}(X, Y)\right]$$
$$= \frac{1}{2}V_{11} + \frac{1}{2}V_{11} + i\left[\frac{1}{2}V_{21} - \frac{1}{2}V_{12}\right] \text{ with } V_{21} = -V_{12}$$
$$= V_{11} - iV_{12}.$$

Then from Lemma 3.1

$$|\text{Cov}(\tilde{Z})| = \begin{vmatrix} V_{11} & -V_{12} \\ V_{12} & V_{11} \end{vmatrix}^{\frac{1}{2}}.$$

Hence the joint density of X and Y, denoted by $f(U)$, and the density of $\tilde{Z}$, denoted by $f(\tilde{Z})$ are one and the same.

Example C1   Evaluate the univariate nonsingular complex normal density directly, and also as a bivariate real normal case.

Solution   Let $\tilde{z} = x + iy$ where x and y are real with $E(x) = \mu_x$, $E(y) = \mu_y$, $\text{var}(x) = \sigma_1^2$, $\text{var}(y) = \sigma_2^2$, $\sigma_1^2 = \sigma_2^2 = \frac{1}{2}\sigma^2$ for some σ^2, $\text{cov}(x,y) = \sigma_{12}$. Then

$$\text{var}(\tilde{z}) = E[(\tilde{z} - E(\tilde{z}))(\tilde{z} - E(\tilde{z}))^*] = \sigma_1^2 + \sigma_2^2 = \sigma^2,$$

the exponent is

$$(\tilde{z} - E(\tilde{z}))^*(\text{var}(\tilde{z}))^{-1}(\tilde{z} - E(\tilde{z})) = \frac{1}{\sigma^2}\left[(x - \mu_x)^2 + (y - \mu_y)^2\right]$$

and the density is given by

$$f(\tilde{z}) = \frac{e^{-\frac{1}{\sigma^2}[(x-\mu_x)^2 + (y-\mu_y)^2]}}{\pi\sigma^2}, \quad -\infty < x, y < \infty$$

$$= \frac{e^{-\frac{1}{\sigma^2}|\tilde{z} - E(\tilde{z})|^2}}{\pi\sigma^2}.$$

If considered as a bivariate real normal then its covariance matrix is given by

$$V = \begin{bmatrix} \frac{1}{2}\sigma^2 & 0 \\ 0 & \frac{1}{2}\sigma^2 \end{bmatrix}$$

since $\sigma_{12} = -\sigma_{21} \Rightarrow \sigma_{12} = 0$. Then

$$\frac{1}{2}[(x - \mu_x), (y - \mu_y)] \begin{bmatrix} \frac{1}{2}\sigma^2 & 0 \\ 0 & \frac{1}{2}\sigma^2 \end{bmatrix}^{-1} \begin{bmatrix} x - \mu_x \\ y - \mu_y \end{bmatrix}$$

$$= \frac{1}{\sigma^2}\left[(x - \mu_x)^2 + (y - \mu_y)^2\right].$$

Note that $(2\pi)|V|^{\frac{1}{2}} = \pi\sigma^2$. Hence the result is verified.

Bibliography

Al-Ani, S. (1972). On the i-th latent root of a complex matrix. *Canadian Mathematical Bulletin* 15(3), 323–327.

Amur, K. (1986). Parallel maps that preserve geometric objects of hypersurfaces. *Rocky Mountain Journal of Mathematics* 16(1), 103–109.

Angermüller, G. (1983). On some conditions for a polynomial map with constant Jacobian to be invertible. *Archiv der Mathematik* 40(5), 415–420.

Anderson, T.W. (1971). *An Introduction to Multivariate Statistical Analysis*. Wiley, New York.

Anderson, T.W. and Girshick, M.A. (1944).Some extensions of the Wishart distribution. *Annals of Mathematical Statistics* 15, 345–357.

Anderson, T.W. and Takemura, A. (1986). Why do noninvertible estimated moving averages occur? *Journal of Time Series Analysis* 7(4), 235–254.

Atiyah, M.F. and Todd, J.A. (1960). On complex Stiefel manifolds. *Proceedings of the Cambridge Philosophical Society* 56, 342–353.

Bellman, R. (1956). A generalization of some integral identities due to Ingham and Siegel, *Duke Mathematical Journal* 24, 571–578.

Bellman, R. (1960). *Introduction to Matrix Analysis*. McGraw-Hill, New York.

Benhabib, J. and Nishimura, K. (1979). On the uniqueness of steady states in an economy with heterogeneous capital goods. *International Economic Review* 20(1), 59–82.

Biyari, K.H. and Lindsey, W.C. (1991). Statistical distributions of hermitian quadratic form in complex Gaussian variables. *IEEE Transactions on Information Theory* 39(3), 1076–1082.

Biyari, K.H. and Lindsey, W.C. (1995). Diversity reception through complex non-Gaussian, noisy channels. *IEEE Transactions on Communications* 43(2,3,4), 318–328.

Bochner, S. (1944). Group invariance of Cauchy's formula in several variables. *Annals of Mathematics* 45, 686–722.

Bochner, S. (1951). Some properties of modular relations. *Annals of Mathematics* 53(2), 332–363.

Bronk, B.V. (1965). Exponential ensemble for random matrices. *Journal Mathematical Physics* 6, 228–237.

Buschman, R.G. (1965). Integrals of hypergeometric funtions. *Mathematische Zeitschrift* 89, 74–76.

Campbell, L.A. (1993). Decomposing Samuelson maps. *Linear Algebra and Its Applications* 187, 227–238.

Campbell, L.A. (1994). Rational Samuelson maps are univalent. *Journal of Pure and Applied Algebra* **92(3)**, 227–240.

Carter, E.M., Khatri, C.G. and Srivastava, M.S. (1976). Nonnull distribution of likelihood ratio criterion for reality of covariance matrix. *Journal of Multivariate Analysis* **6**, 176–184.

Chikuse, Y. (1976). Partial differential equations for hypergeometric functions of complex argument matrices and their applications. *Annals of the Institute of Statistical Mathematics* **28**, 189–199.

Chikuse, Y. (1992). Properties of Hermite and Laguerre polynomials in matrix argument and their applications. *Linear Algebra and Its Applications* **176**, 237–260.

Chikuse, Y. (1992a). Generalized Hermite and Laguerre polynomials in multiple symmetric matrix arguments and their applications. *Linear Algebra and Its Applications* **176**, 261–287.

Chikuse, Y. and Davis, A.W. (1986). A survey on the invariant polynomials with matrix arguments in relation to econometric distribution theory. *Econometric Theory* **2**, 232–248.

Chung, K.L. (1983) (edited). *Pao-Lu Hsu Collected Papers*. Springer-Verlag, New York.

Coleman, T.F. and More, J.J. (1983). Estimation of sparse Jacobian matrices and graph coloring problems. *SIAM Journal of Numerical Analysis* **20(1)**, 187–209.

Conradie, W.J. and Gupta, A.K. (1987). Quadratic forms in complex normal variables: basic results. *Statistica* **XLVII(1)**, 74–83.

Conradie, W.J. and Troskie, C.G. (1984). The exact non-central distribution of a multivariate complex quadratic form of complex normal variates. *South African Statistical Journal* **18**, 123–134.

Constantine, A.G. (1963). Some noncentral distribution problems in multivariate analysis. *Annals of Mathematical Statistics* **34**, 1270–1285.

Davis, A.W. (1979). Invariant polynomials with two matrix arguments extending the zonal polynomials: application to multivariate distribution theory. *Annals of the Institute of Statistical Mathematics* **31(A)**, 465–485.

Davis, A.W. (1981). On the construction of a class of invariant polynomials in several matrices, extending the zonal polynomials. *Annals of the Institute of Statistical Mathematics* **33(A)**, 297–313.

Dean, P. (1956). The spectral distribution of a Jacobian matrix. *Proceedings of the Cambridge Philosophical Society* **52**, 752–755.

Deemer, W.L. and Olkin, I. (1951). The Jacobians of certain matrix transformations useful in multivariate analysis based on lectures by P.L. Hsu. *Biometrika* **38**, 345–367.

Divsalar, D., Simon, M.K. and Shahshahani, M. (1990). The performance of trellis-coded MDPSK with multiple symbol detection. *IEEE Transactions on Communications* 38(9), 1391–1403.

Dwyer, P.S. (1967). Some applications of matrix derivatives in multivariate analysis. *Journal of the American Statistical Association* 62, 607–625.(correction: 62, p.1518).

Dwyer, P.S. and Macphail, M.S. (1948). Symbolic matrix derivatives. *Annals of Mathematical Statistics* 19, 517–534.

Exton, H. (1976). *Multiple Hypergeometric Functions and Applications*. Ellis Horwood, Chichester, U.K.

Fang, K.-T. and Anderson, T.W. (1990) (edited). *Statistical Inference in Elliptically Contoured and Related Distributions*. Allerton Press, New York.

Fang, K.-T., Kotz, S., and Ng, K.W. (1990). *Symmetric Multivariate and Related Distributions. Monograph on Statistics and Applied Probability*. Chapman and Hall, New York.

Fang, K.-T. and Zhang, Y.-T. (1990). *Generalized Multivariate Analysis*. Springer-Verlag, Berlin. (Sciences Press, Beijing).

Fleiss, M. (1993). Invertibility of causal discrete time dynamical systems. *Journal of Pure and Applied Algebra* 86(2), 173–179.

Floreanini, R., Lapointe, L. and Vinet, L. (1994). A quantum algebra approach to basic multivariable special functions. *Journal of Physics A: Mathematics General* 27, 6781–6797.

Fujikoshi, Y. (1971). Asymptotic expansions of the non-null distributions of two criteria for the linear hypothesis concerning complex multivariate normal populations. *Annals of the Institute of Statistical Mathematics* 23, 477–490.

Giri, N.C. (1965). On the complex analogues of T^2 and R^2-tests. *Annals of Mathematical Statistics* 36, 664–670.

Giri, N.C. (1972). On testing problems concerning mean of multivariate complex Gaussian distribution. *Annals of the Institute of Statistical Mathematics* 24, 245–250.

Girko, V.L. (1990). *Theory of Random Determinants*. Kluwer Academic, Boston.

Good, I.J. (1981). Generalized determinants and generalized Jacobians. *Journal of Statistical Computation and Simulation* 13, 60–62.

Goodman, N.R. (1957). *On the Joint Estimation of the Spectra, Cospectrum and Quadrature Spectrum of a Two-Dimensional Stationary Gaussian Process*. Ph.D. Thesis, Princeton University.

Goodman, N.R. (1963). Statistical analysis based on a certain multivariate complex Gaussian distribution (an introduction). *Annals of Mathemat-

ical Statistics **34**, 152–177.

Gradshteyn, I.S. and Ryzhik, I.M. (1980). *Table of Integrals, Series, and Products*. Academic Press, New York.

Gupta, A.K. (1971). Distribution of Wilks' likelihood ratio criterion in the complex case. *Annals of the Institute of Statistical Mathematics* **23**, 77–87.

Gupta, A.K. (1973). On a test for reality of the covariance matrix in a complex Gaussian distribution. *Journal of Statistical Computation and Simulation* **2**, 333–342.

Gupta, A.K. (1976). Nonnull distribution of Wilks' statistic for MANOVA in the complex case. *Communications in Statistics - Simulation and Computation* **B5(4)**, 177–188.

Gupta, R.D. and Richards, D. St. P. (1987). Multivariate Liouville distributions. *Journal of Multivariate Analysis* **23**, 233–256.

Hannan, E.J. (1970). *Multiple Time Series*. Wiley, New York.

Haubold, H.J. and Mathai, A.M. (1994). The determination of the internal structure of the Sun by the density distribution. *American Institute of Physics, Conference Proceedings* **320**, 89–101.

Hayakawa, T. (1972). On the distribution of the latent roots of a complex Wishart matrix (non-central case). *Annals of the Institute of Statistical Mathematics* **24**, 1–17.

Hayakawa, T. (1972a). The asymptotic distributions of the statistics based on the complex Gaussian distribution. *Annals of the Institute of Statistical Mathematics* **24**, 231–244.

Henderson, H.V. and Searle, S.R. (1979). Vec and vech operators for matrices, with some uses in Jacobians and multivariate statistics. *The Canadian Journal of Statistics* **7(1)**, 65–81.

Herz, C.S. (1955). Bessel functions of matrix argument. *Annals of Mathematics* **61(3)**, 474–523.

Hirakawa, F. (1975). Some distributions of the latent roots of a complex Wishart matrix variate. *Annals of the Institute of Statistical Mathematics* **27**, 357–363.

Huang, J. and Campbell, L.L.(1991). Trellis coded MDPSK in correlated and shadowed Rician fading channels. *IEEE Transactions on Vehecular Technology* **40(4)**, 786–797.

Jack, H. (1964-65). A generalization of Dirichlet's multiple integral. *Proceedings of the Edinburgh Mathematical Society* **14**, 233–237.

James, A.T. (1961). Zonal polynomials of the real positive definite symmetric matrices. *Annals of Mathematics* **74**, 456–469.

James, A.T. (1964). Distributions of matrix variates and latent roots derived from normal samples. *Annals of Mathematical Statistics* **35**, 475–

501.

James, A.T. (1968). Calculation of zonal polynomial coefficients by use of the Laplace-Beltrami operator. *Annals of Mathematical Statistics* **39**(5), 1711–1718.

James, A.T. (1976). Special functions of matrix and single argument in statistics. In *Theory and Applications of Special Functions*, (Askey ed), Academic Press, New York. pp. 497–520.

Jinadasa, K.G. and Tracy, D.S. (1986). Higher order moment of random vectors using matrix derivatives. *Stochastic Analysis and Applications* **4**, 399–407.

Kabe, D.G. (1966). Complex analogues of some classical noncentral multivariate distributions. *Australian Journal of Statistics* **8**, 99–103.

Kabe, D.G. (1966a). On the distribution of the complex analogue of Rao's U statistic. *Journal of the Indian Statistical Association* **4**, 189–194.

Kabe, D.G. (1968). Some aspects of analysis of variance and covariance theory for a certain multivariate complex Gaussian distribution. *Metrika* **13**, 86–97.

Khatri, C.G. (1964). Distribution of the largest or the smallest characteristic root under null hypothesis concerning complex multivariate normal population. *Annals of Mathematical Statistics* **35**, 1807–1810.

Khatri, C.G. (1965). A test for reality of a covariance matrix in certain complex Gaussian distribution. *Annals of Mathematical Statistics* **36**, 115–119.

Khatri, C.G. (1965a). Classical statistical analysis based on a certain multivariate complex Gaussian distribution. *Annals of Mathematical Statistics* **36**, 98–114.

Khatri, C.G. (1966). On certain distribution problems based on positive definite quadratic function in normal vectors. *Annals of Mathematical Statistics* **37**, 468–479.

Khatri, C.G. (1969). Noncentral distributions of i-th largest characteristic roots of three matrices concerning complex multivariate normal population. *Annals of the Institute of Statistical Mathematics* **21**, 23–32.

Khatri, C.G. (1970). On the moments of traces of two matrices in three situations for complex multivariate normal populations. *Sankhyā Series A* **32**, 65–80.

Krishnaiah, P.R. (1976). Some recent developments on complex multivariate distributions. *Journal of Multivariate Analysis* **6**, 1–30.

Lang, J. and Mandall, S. (1993). On Jacobian n-tuples in characterisitc p. *Rocky Mountain Journal of Mathematics* **23**(1), 271–279.

Lee, J.C., Krishnaiah, P.R. and Chang, T.C. (1977). Approximations to the distributions of the determinants of real and complex multivariate beta matrices. *South African Statistical Journal* **11**, 13–26.

Li, H.C., Pillai, K.C.S. and Chang, T.C. (1970). Asymptotic expansions for distributions of the roots of two matrices from classical and complex Gaussian populations. *Annals of Mathematical Statististics* **41(5)**, 1541–1556.

MacRae, E.C. (1974). Matrix derivatives with an application to adaptive linear decision problem. *The Annals of Statistics* **2(2)**, 337–346.

Magnus, J.R. and Neudecker, H. (1988). *Matrix Differential Calculus with Applications in Statistics and Econometrics*. Wiley, New York.

Mahalanobis, P.C., Bose, R.C. and Roy, S.N. (1937). Normalization of statistical variates and the use of rectangular coordinates in the theory of sampling distributions. *Sankhyā* **3**, 1–40.

Mathai, A.M. (1978). Some results on functions of matrix arguments. *Mathematische Nachrichten* **84**, 171–177.

Mathai, A.M. (1980). An alternate simpler method of evaluating the multivariate beta function and an inverse Laplace transform connected with Wishart distribution. In *Statistical Distributions in Scientific Work* **4**, Taillie, C., Patil, G.P. and Baldessari, B.A. (editors), Reidel, Boston, pp. 281–286.

Mathai, A.M. (1981). Distributions of the canonical correlation matrix. *Annals of the Institute of Statistical Mathematics* **33**, 35–43.

Mathai, A.M. (1989) (edited). *Distributions of Test Statistics: Exact & Asymptotic, Null & Non-Null, Methods/Comparisons/Research Frontiers*. American Sciences Press, Columbus, Ohio.

Mathai, A.M. (1993). *A Handbook of Generalized Special Functions for Statistical and Physical Sciences*. Oxford University Press, Oxford.

Mathai, A.M. (1993a). The residual effect of a growth-decay mechanism and the distribution of covariance structures. *The Canadian Journal of Statistics* **21(3)**, 277–283.

Mathai, A.M. (1993b). Lauricella functions of real symmetric positive definite matrices. *Indian Journal of Pure and Applied Mathematics* **24(9)**, 513–531.

Mathai, A.M. (1993c). Appell's and Humbert's functions of matrix arguments. *Linear Algebra and Its Applications* **183**, 201–221.

Mathai, A.M. (1996). Whittaker and G-functions of matrix argument in the complex case. *International Journal of Mathematical and Statistical Sciences*, (to appear)

Mathai, A.M. and Haubold, H.J. (1988). *Modern Problems in Nuclear and Neutrino Astrophysics*. Akademie-Verlag, Berlin.

Mathai, A.M. and Pederzoli, G. (1993). Kampé de Fériet's functions of matrix arguments. *Metron* **LI(3-4)**, 3–24.

Mathai, A.M. and Pederzoli, G. (1995). Hypergeometric functions of many matrix variables and distributions of generalized quadratic forms. *American Journal of Mathematical and Management Sciences*, **15**(3&4), 343–354.

Mathai, A.M. and Pederzoli, G. (1996). Some transformations for functions of matrix argument. *Indian Journal of Pure and Applied Mathematics* **27**(3), 277–284.

Mathai, A.M. and Pederzoli, G. (1997). Some properties of matrix-variate Laplace transform and matrix-variate Whittaker function. *Linear Algebra and Its Applications*, (to appear).

Mathai, A.M. and Pederzoli, G. (1997a). On Whittaker function of matrix argument. (preprint).

Mathai, A.M. and Provost, S.B. (1992). *Quadratic Forms in Random Variables: Theory and Applications*. Marcel Dekker, New York.

Mathai, A.M., Provost, S.B. and Hayakawa, T. (1995). *Bilinear Forms and Zonal Polynomials*. Springer-Verlag, Lecture Notes in Statistics, **102**, New York.

Mathai, A.M. and Rathie, P.N. (1980). The exact non-null distribution for testing equality of covariance matrices. *Sankhyā Series A* **42**, 78–87.

Mathai, A.M. and Saxena, R.K. (1973). *Generalized Hypergeometric Functions with Applications in Statistics and Physical Sciences*. Springer-Verlag, Lecture Notes, **348**, New York.

Mathai, A.M. and Saxena, R.K. (1978). *The H-function with Applications in Statistics and Other Disciplines*. Wiley Halsted, New York and Wiley Eastern, New Delhi.

Mathai, A.M. and Saxena, R.K. (1987). Various practical problems in probability and statistics where Lauricella's functions appear naturally. *Rajasthan Ganita Parishad* 1, 41–48.

McCulloch, C.E. (1982). Symmetric matrix derivatives with applications. *Journal of the American Statistical Association* **77**, 679-682.

Mehta, M.L. (1967). *Random Matrices and the Statistical Theory of Energy Levels*. Academic Press, New York.

Meisters, G.H. (1982). Jacobian problems in differential equations and algebraic geometry. *Rocky Mountain Journal of Mathematics* **12**(4), 679–705.

Mitra, S.K. (1970). Analogues of multivariate beta (Dirichlet) distributions. *Sankhyā Series A* **32**, 189–192.

Morris, R.A. and Wang, S.S.S. (1981). A Jacobian criterion for smoothness. *Journal of Algebra* **69**(2), 483–486.

Muirhead, R.J. (1982). *Aspects of Multivariate Statistical Theory*. Wiley, New York.

Nagarsenker, B.N. and Das, M.M. (1975). Exact distribution of sphericity criterion in the complex case and its percentage points. *Communications in Statistics* 4(4), 363–374.

Nel, D.G. (1980). On matrix differentiation in statistics. *South African Statistical Journal* 14(2), 137–193.

Neudecker, H. (1969). Some theorems on matrix differentiation with special reference to Kronecker matrix products. *Journal of the American Statistical Association* 64, 953–963.

Neudecker, H. and Wansbeek, T. (1983). Some results on commutation matrices with statistical applications. *The Canadian Journal of Statistics* 11(3), 221–231.

Nishimura, K. (1981). Kuhn's intensity hypothesis revisited. *The Economic Review* 48(2), 351–354.

Njoroge, M.M. (1988). *On Jacobians Connected with Marix-variate Random Variables*. M.Sc. Thesis, McGill University, Montreal, Canada.

Olkin, I. (1951). On distribution problems in multivariate analysis. *Institute of Statistics, Mimeograph Series* 43, 1–126.

Olkin, I. (1953). Note on the Jacobians of certain matrix transformations useful in multivariate analysis. *Biometrika* 40, 43–46.

Olkin, I. (1959). A class of integral identities with matrix argument. *Duke Mathematical Journal* 26, 207–213.

Olkin, I. and Roy, S.N. (1954). On multivariate distribution theory. *Annals of Mathematical Statistics* 25, 329–339.

Olkin, I. and Sampson, A.R. (1972). Jacobians of matrix transformations and induced functional equations. *Linear Algebra and Its Applications* 5, 257–276.

Parthasarathy, T. and Ravindran, G. (1986). The Jacobian matrix, global univalence and completely mixed games. *Mathematics of Operations Research* 11(4), 663–671.

Pederzoli, G. (1995). Some properties of Kampé de Fériet's function of matrix arguments and applications. *Metron* LIII(3-4), (to appear)

Pillai, K.C.S. and Hsu, Y.S. (1979). The distribution of the characteristic roots of $S_1 S_2^{-1}$ under violations in the complex case and power comparisons of four tests. *Annals of the Institute of Statistical Mathematics* 31, 445–463.

Pillai, K.C.S. and Jouris, G.M. (1971). Some distribution problems in the multivariate complex Gaussian case. *Annals of Mathematical Statistics* 42, 517–525.

Pillai, K.C.S. and Li, H.C. (1970). Monotonicity of the power functions of some tests of hypotheses concerning multivariate complex normal distributions. *Annals of the Institute of Statistical Mathematics* 22, 307–318.

Pillai, K.C.S. and Young, D.L. (1971). An approximation to the distribution of the largest root of a complex Wishart matrix. *Annals of the Institute Statistical Mathematics* 23, 89–96.

Polasek, W. (1985). A dual approach for matrix-derivatives. *Metrika* 32, 275–292.

Pollock, D.S.G. (1985). Tensor products and matrix differential calculus. *Linear Algebra and Its Applications* 67, 169–193.

Potter, H.S.A. (1951). The volume of a certain matric domain. *Duke Mathematical Journal* 18, 391–397.

Rao, C.R. (1973). *Linear Statistical Inference and Its Applications*. Wiley, New York.

Rasch, G. (1948). A functional equation for Wishart's distribution. *Annals of Mathematical Statistics* 19, 262–266.

Richards, D.St.P. (1984). Hyperspherical models, fractional derivatives and exponential distributions on matrix spaces. *Sankhyā Series A* 46, 155–165.

Rogers, G.S. (1980). *Matrix Derivatives*. Marcel Dekker, New York.

Roy, S.N. (1952). Some useful results on Jacobians. *Calcutta Statistical Association Bulletin* 4, 117–122.

Roy, S.N. (1957). *Some Aspects of Multivariate Analysis*. Wiley, New York.

Saw, J.G. (1973). Jacobians of singular transformations with applications to statistical distribution theory. *Communications in Statistics* 1(1), 81–91.

Saw, J.G. (1975). An improved analytic form for the Jacobian of a singular transformation. *Communications in Statistics* 4(3), 273–276.

Saxena, R.K. (1970). Integrals involving Kampé de Fériet function and Gauss' hypergeometric function. *Ricerca (Napoli)* 2, 21–27.

Shaman, P. (1980). The inverted complex Wishart distribution and its application to spectral estimation. *Journal of Multivariate Analysis* 10, 51–59.

Singh, Anita (1982). Exact distribution of Wilks' Lvc criterion and its percentage points in the complex case. *Communications in Statistics - Simulation and Computation* 11(2), 217–225.

Sitgreaves, R. (1952). On the distribution of two random matrices used in classification procedures. *Annals of Mathematical Statistics*, 23 , 263–270.

Slater, L.J. (1966). *Generalized Hypergeometric Functions*. Cambridge University Press, London.

Srivastava, M.S. (1965). On the complex Wishart distribution. *Annals of Mathematical Statistics* 36, 313–315.

Srivastava, M.S. and Khatri, C.G. (1979). *An Introduction to Multivariate Statistics.* North Holland, New York.

Srivastava, H.M. and Karlsson, P.W. (1985). *Multiple Gaussian Hypergeometric Series.* Ellis Horwood, Chichester, U.K.

Sugiyama, T. (1972). Distributions of the largest latent root of the multivariate complex Gaussian distribution. *Annals of the Institute of Statistical Mathematics* 24, 87–94.

Tracy, D.S. and Dwyer, P.S. (1969). Multivariate maxima and minima with matrix derivatives. *Journal of the American Statistical Association* 64, 1576–1594.

Tracy, D.S. and Jinadasa, K.G. (1988). Patterned matrix derivatives. *The Canadian Journal of Statistics* 16, 411–418.

Tracy, D.S. and Singh, R.P. (1971). Some modifications of matrix differentiation for evaluating Jacobians of symmetric matrix transformations. In *Symmetric Functions in Statistics. Proceedings of the Symposium in Honor of Paul S. Dwyer*, University of Windsor, Windsor, Ontario, Canada. pp. 203–224.

Tracy, D.S. and Sultan, S.A. (1993). Higher order moments of multivariate normal distribution using matrix derivatives. *Stochastic Analysis and Applications* 11(3), 337–348.

Waikar, V.B., Chang, T.C. and Krishnaiah, P.R. (1972). Exact distributions of a few arbitrary roots of some complex random matrices. *Australian Journal of Statistics* 14(1), 84–88.

Wang, S.S.S. (1980). A Jacobian criterion for separability. *Journal of Algebra* 65(2), 453–494.

Wigner, E.P. (1965). Distribution laws for the roots of a random hermitian matrix. In *Statistical Theories of Spectra: Fluctuations* (Porter, C.E. (edited)), pp. 446–461. Academic Press, New York.

Wooding, R.A. (1956). The multivariate distribution of complex normal variables. *Biometrika* 43, 212–215.

Wong, C.S. (1980). Matrix derivatives and its applications in statistics. *Journal of Mathematical Psychology* 22(1), 70–81.

Xavier, F. (1993). Invertibility of Bass-Connell-Wright polynomial maps. *Mathematische Annalen* 295(1), 163–166.

Ypma, T.J. (1987). Efficient estimation of sparse Jacobian matrices by differences. *Journal of Computational and Applied Mathematics* 18(1), 17–28.

Glossary of symbols

$\text{tr}(A) = \text{tr}A$	trace of the matrix A	section 1.0	1
$\lvert A \rvert$	determinant of A	section 1.0	1
$X = X' > 0$	X is symmetric positive definite	section 1.0	1
$0 < A < X < B$	$A > 0, B > 0, B - X > 0,$		
	and $X - A > 0$	section 1.0	1
$\frac{\partial X}{\partial x}$	derivative of a matrix X with		
	respect to the scalar x	section 1.1.1	2
$\frac{\partial}{\partial X}$	vector of partial derivatives	section 1.1.2	2
	or matrix of partial derivatives	section 1.1.3	6
$\text{diag}(X)$	diagonal matrix formed with the		
	diagonal elements of X	section 1.1.3	6
Δ, Δ_{ij}	positional derivatives	equation (1.1.8)	14
$J(Y : X)$	Jacobian	section 1.2.1	21
dX	wedge product of differentials	equation (1.2.8)	23
$\Lambda[\cdot]$	wedge products in $[\cdot]$	equation (1.2.8)	23
(dX)	matrix of differentials	equation (1.2.10)	24
$N_p(\cdot, \cdot)$	p-variate normal density	example 1.11	24
$\sim$	distributed as	example 1.11	24
$\lvert A \rvert_+$	magnitude of the determinant of A	example 1.11	24
$\text{Re}(\cdot)$	real part of $(\cdot)$	example 1.14	34
$X^{\frac{1}{2}}$	symmetric square root of		
	$X = X' > 0$	example 1.15	35
$A \otimes B$	Kronecker product of A and B	definition 1.1	48
$\text{vec}(X)$	vector formed with the columns		
	of X	definition 1.2	49
$\Gamma_p(\cdot)$	matrix-variate gamma, real case	example 1.24	56
$V_{p,q}$	Stiefel manifold	definition 2.1	115
$O_p = O_{(p)}$	orthogonal group	definition 2.2	115
$\in$	in	exercise 2.2.5	136
$\bar{A}$	complex conjugate of A	section 3.0	171
$\bar{A}^*$	conjugate transpose of $\bar{A}$	section 3.0	171
$\tilde{X}$	matrix with the elements as		
	complex mathematical variables		
	or random variables	section 3.0	171
$(d\tilde{X})$	matrix of differentials,		
	$\tilde{X}$ complex	section 3.0	171
$i = \sqrt{-1}$		section 3.0	171
$\lvert \det(A) \rvert$	absolute value (norm) of the		

421

Author index

Subject index

absolute value $|A|_+$ 24,26,62,63
absolute value (norm) 171,174,176,199,201,205
adjoint 204
analysis of variance 279
Appell's functions of matrix arguments, complex case 385
    real case 307,308,323
astrophysics 307,322,331

Bessel functions of matrix argument, complex case 373
    real case 284,291,305
beta integral, real case 230,267,270,273,315,325,329,331,344
beta integral for zonal polynomials 267
bilinear forms 279
binomial function of matrix argument, complex case 366
    real case 284

canonical correlation analysis 20
characteristic function 360
confluent hypergeometric function of matrix argument 285
conjugate 355,405
conjugate transpose 355,405
correlation matrix 58,217
covariance 401,405

density,
    elliptically contoured 280
    matrix-variate gamma, real case 34,35,87,94,98,112,149,157
    254,256,281
    matrix-variate gamma, complex case 188,189,194,198,212,356,357
    matrix-variate Gaussian, complex case 178
    matrix-variate Gaussian, real case 26,122,144,264
    matrix-variate Liouville 278,281
    matrix-variate type-1 beta, complex case 192,198,357
    matrix-variate type-1 beta, real case 80,118,150,259,276,280
    matrix-variate type-2 beta, complex case 198,357
    matrix-variate type-2 beta, real case 61,79,262,280
    normal (Gaussian), complex case 176,360,406
    normal (Gaussian), real case 24,39,353

429

www.ingramcontent.com/pod-product-compliance
Lightning Source LLC
Chambersburg PA
CBHW050633190326
41458CB00008B/2249